D0214801

Single- and Multi-Chip Microcontroller Interfacing

For the Motorola 68HC12

Academic Press Series in Engineering

Series Editor
J. David Irwin
Auburn University

Designed to bring together interdependent topics in electrical engineering, mechanical engineering, computer engineering, and manufacturing, the Academic Press Series in Engineering provides state-of-the-art handbooks, textbooks, and professional reference books for researchers, students, and engineers. This series provides readers with a comprehensive group of books essential for success in modern industry. A particular emphasis is given to the applications of cutting-edge research. Engineers, researchers, and students alike will find the Academic Press Series in Engineering to be an indispensable part of their design toolkit.

Published books in the series:
Industrial Controls and Manufacturing, 1999, E. Kamen
DSP Integrated Circuits, 1999, L. Wanhammar
Time Domain Electromagnetics, 1999, S. M. Rao
Single- and Multi-Chip Microcontroller Interfacing for the Motorola 68HC12, 1999, G. J. Lipovski
Control in Robotics and Automation, 1999, B. K. Ghosh, N. Xi, T. J. Tarn

Single- and Multi-Chip Microcontroller Interfacing

For the Motorola 68HC12

G. Jack Lipovski
Department of Electrical and Computer Engineering
University of Texas
Austin, Texas

ACADEMIC PRESS

San Diego London Boston
New York Sydney Tokyo Toronto

This book is printed on acid-free paper. ∞

Academic Press
525 B. St., Suite 1900, San Diego, California 92101-4495, USA
http://www.apnet.com

Academic Press
24–28 Oval Road, London NW1 7DX, UK
http://www.hbuk.co.uk/ap/

Library of Congress Catalog Card Number: 98-89451
ISBN: 0-12-451830-3

Printed in the United States of America

99 00 01 02 03 MV 9 8 7 6 5 4 3 2 1

Dedicated to my mother,
Mary Lipovski

About the Author

G. Jack Lipovski has taught electrical engineering and computer science at the University of Texas since 1976. He is a computer architect internationally recognized for his design of the pioneering database computer, CASSM, and the parallel computer, TRAC. His expertise in microcomputers is also internationally recognized by his being a past director of Euromicro and an editor of *IEEE Micro*. Dr. Lipovski has published more than 70 papers, largely in the proceedings of the annual symposium on computer architecture, the IEEE transactions on computers and the national computer conference. He holds eight patents, generally in the design of logic-in-memory integrated circuits for database and graphics geometry processing. He has authored seven books and edited three. He has served as chairman of the IEEE Computer Society Technical Committee on Computer Architecture, member of the Computer Society Governing Board, and chairman of the Special Interest Group on Computer Architecture of the Association for Computer Machinery. He has been elected Fellow of the IEEE and a Golden Core Member of the IEEE Computer Society. He received his Ph.D. degree from the University of Illinois, 1969, and has taught at the University of Florida, and at the Naval Postgraduate School, where he held the Grace Hopper chair in Computer Science. He has consulted for Harris Semiconductor, designing a microcomputer, and for the Microelectronics and Computer Corporation, studying parallel computers. He founded the company Linden Technology Ltd., and is the chairman of its board. His current interests include parallel computing, database computer architectures, artificial intelligence computer architectures, and microcomputers.

Contents

Preface

By 1980, the microcomputer had changed so many aspects of engineering that the cliche "microcomputer revolution" echoed from almost every other magazine article and learned paper in the field. It is a great tool. This book's ancestor, *Microcomputer Interfacing: Principles and Practices*, was written at that time to establish some design theory for this dynamic field. A successor book, which is this book's predecessor, *Single- and Multiple-Chip Microcomputer Interfacing*, was motivated by two significant changes: the evolution of powerful single-chip microcomputers and the IEEE Computer Society Curriculum Committee recommendation for a course on microcomputer interfacing and communication. The development of powerful single-chip microcomputers introduces a new design choice: to use either a microprocessor in a personal computer with some 16M bytes of memory and an operating system or a less costly single-chip microcomputer with much less overhead. This decision is largely based on the designer's understanding of the capabilities and limitations of the single-chip microcomputer. The development of a standard curriculum for a course lent stability to this field. The book aimed to teach the principles and practices of microcomputer systems design in general, and interfacing in particular, and to foster an understanding of single-chip microcomputers within the guidelines of the IEEE Computer Society Curriculum Committee's recommendations. This book was motivated by the development of the Motorola 6812, and its need for more sophisticated software. Since the 6812 featured so many on-chip I/O devices, which were already connected to the 6812's address and data buses, but which each had to be programmed, and managing a number of I/O devices often necessitated the use of time sharing, this book features programming in C and C++. However, the designer must be wary of high-level language statements that do not do what he or she intended. High-level languages are designed for algorithms, not for I/O interfacing, and optimizing high-level language compilers can "optimize you right out of business." The designer is shown how each high-level language statement is implemented in assembler language.

This book's predecessor evolved from a set of notes for a senior level course in microcomputer design. The course – which is still taught – focuses on the combined hardware/software design of microcomputer systems. It emphasizes principles of design because theory is as necessary for a solid foundation in design as theory is in any engineering discipline. However, it also emphasizes the practices – the details of how to get a system to work – because microcomputer system design requires hands-on experience. There is a remarkable difference between a student who merely reads about microcomputers and a student who has worked with one – clear evidence that theory has to be taught with practice. Practical experience is desirable in almost any engineering course. This is not always possible. But microcomputer systems are inexpensive enough that the school or the student can afford this hands-on opportunity; and the joy of seeing the principles work is so attractive that the student often can't get enough of the material to be satiated. The development of very powerful, inexpensive single-chip microcomputers furthers this opportunity. So the course, this book's predecessor, and this book, all emphasize both the principles and practices of microcomputer design.

The principles and practices of microcomputer design have to cover both hardware and software. A purely hardware approach might appeal to a seasoned digital system designer or an electrical engineering student, but it leads to poor choices that either do not take full advantage of software's tremendous power or force unnecessary constraints and therefore higher costs on its development. However, a purely software approach misses the opportunity to understand how and why things happen, and how to take advantage of the hardware. The IEEE Computer Society Curriculum Committee recommends a combined hardware/software approach.

A combined hardware/software approach does require more background. The course this book is based on is the second of a two-course sequence. The first course teaches how C statements are implemented in assembler-language programming. The second course builds on that background and also presumes some background in logic design, as would be obtained in a standard introductory course on that topic. This book, however, has three chapters that survey these topics. These chapters can be skimmed as a review of the required background material or carefully studied as a condensed tutorial if the reader has not had the earlier courses or their equivalents. Because they are intended as review or intensive tutorial material, to prepare the readers for the book's main subject, these three chapters are comparatively compressed and terse.

We make the practices discussed in this book concrete through detailed discussion of the 6812 microcontroller. However, these products are used as a means to the end of teaching principles and practices in general, rather than to promote the use of specific Motorola products. Applications notes and catalogues are available from Motorola to that end. Specific, detailed discussion encourages and aids the reader in learning through hands-on experience and vitally contributes to his or her enthusiasm for and understanding of the principles. The 6812 is used primarily because the MC68HC812A4 (abbreviated 'A4) and the MC68HC912B32 (abbreviated 'B32) are believed to be the most easily taught single-chip microcomputer. The 6812 instruction set is as complete as that of any other comparable machine, supporting C and C++ well enough to teach the intelligent use of data structures, and it is symmetrical and comparatively free of quirks and warts that detract from the subject under discussion. Moreover, the 'A4's 4 K bytes of EEPROM, which can be written 10,000 times, is well suited to university-level lab and project use. The less costly 'B32, with its 1 K bytes of SRAM, is also suited to simple lab experiments. Nevertheless, we stress that detailed comparisons between the 'A4 and other well-designed microcomputers clearly show that others may be better than the 'A4 or 'B32 for different applications. However, a comparative study of different microcomputers and applications is beyond the scope of this book.

As mentioned, the first three chapters quickly survey the background needed for the remainder of the book. Chapter 1 is a survey that covers computer architecture and the instruction set of the 6812. Chapter 2 covers some software techniques, including subroutine parameter and local variable conventions, that are very useful in microcomputers. Object-oriented programming is introduced for readers with advanced programming skills. Chapter 3 covers basic computer organization, but confines itself to those aspects of the topic that are particularly germane to microcomputers. For example, basic computer organization traditionally covers floating point arithmetic, but this chapter doesn't; and this chapter dwells on address decoders, a topic not often covered in computer organization.

The rest of the book covers three basic themes: input-output (I/O) hardware and software, analog signals, and communications. Parallel I/O ports are covered in Chapter 4, interrupts and alternatives in Chapter 5, analog components in Chapters 7 and 8, communications devices in Chapter 9, and disk storage and CRT display devices in Chapter 10. The simple parallel I/O port and the synchronous serial I/O port – especially attractive in interfacing slow devices because it economizes on pins – are displayed in Chapter 4. Hardware and software aspects are studied side by side. The reader need no longer be intimidated by the basic interface. Chapter 5 discusses interrupts and their alternatives. Hardware/software tradeoffs are analyzed, and different techniques are exemplified. Chapter 6 describes system configuration ports for control over memory and time. Chapter 7 surveys the traditional (voltage) analog components that are commonly used in microcomputer I/O systems. Sufficient information is provided that the reader can understand the uses, limitations, and advantages of analog components, and can springboard from this chapter to other texts, magazine articles, or discussions with colleagues to a fuller understanding of analog design. Chapter 8 introduces the counter/timer as an interface to frequency-analog signals. Techniques to generate signals with a frequency or to measure a signal with a frequency that is analog to some quantity are discussed. Moreover, the hardware/software alternatives to using this most interesting integrated circuit are discussed and evaluated. Chapter 9 describes communications devices. The universal asynchronous receiver transmitter and its cousins are thoroughly studied, and other communications techniques are described. Finally, Chapter 10 introduces the magnetic storage device and the CRT display device.

This book emphasizes the software aspect of interfacing because software design costs dominate the cost of designing microcontroller systems. Software design cost is reduced by using abstraction, which is facilitated by using high-level languages. Programming is in C and C++, and the latter utilizes object-oriented techniques for I/O software. Generally, techniques are first introduced in C, with explanations of what is done, using assembler language. Then, in separately designated and optional sections, the techniques are extended using C++. The optional sections are clearly designated by the use of "object-oriented" in their title. These optional sections can be skipped without loosing any necessary background in later non-optional sections and chapters. However, if a given optional section is to be understood, previous optional sections need to be read for the necessary background for it. We hope that this organization will permit a reader who does not have extensive skills in programming to understand the use, in I/O interfacing, of any high-level language in general and C in particular, while the reader who does have extensive skills in programming to understand the use of objects in C++.

Object-oriented programming will be shown generally as an alternative to using operating systems in a microcontroller, although it can also be used in conjunction with operating systems. Objects obviously provide encapsulation, protection, and factoring of common code, but they can also provide the entire infrastructure that is also provided by an operating system, such as memory and time management, including time-sharing. Frankly, an operating system is often overkill, while object-oriented programming can provide just the right amount of infrastructure for an application in a microcontroller such as the 6812. An operating system is desirable if the infrastructure that it provides, such as managing disks and networks, is useful. However, if the application merely needs some I/O, such as an A-to-D converter, using an operating system device driver is unnecessary, and the overhead it incurs is often objectionable. In this book, we show you how to write what amounts to a tailor-made real-time operating system for your microcontroller system. We show you how to write classes that are, in effect, operating system device drivers, and we provide an efficient time-sharing kernel, which is adequate for the kind of well-defined threads that run in a typical 6812 microcontroller.

After hearing a presentation about the use of object-oriented programming in this book, someone remarked that it was nice to show students who have learned object-oriented programming how it can be used in microcontroller interfacing. There is much more to it. In this book we use object-oriented programming to elevate the programmer to a level of systems design. This object-oriented methodology promises not only to save software design cost, but also to integrate software and microcontroller organization to make more intelligent hardware/software trade-offs and to more effficiently implement changes that might be indicated after these trade-offs are evaluated.

Some remarks on this book's style are offered. On the one hand, terms are formally introduced and used as carefully as possible. This is really necessary to make some sense out of a subject as broad and rich as microcomputer system design. There is no room for muddy terminology or the use of undefined jargon. Even though the terminology used in trade journals and manufacturers' applications notes is inconsistent and often contradictory, the terminology used in a text must be clear and consistent. On the other hand, a book full of definitions is too hard to read. The first version of the course notes that lead to this book tended to be ponderous. Moreover, students are more tuned to television colloquialism today, and are turned off by "third person boring" that is often the style of many learned textbooks. So we condescend to "first person conversational", and we enjoy it. The "we" in this book stands not only for the author but also for his colleagues and his teachers, as well as his students – who have taught him a great deal and who have collectively inspired and developed the principles and practices discussed in the book. But we admit to occasionally exploring *Webster's Collegiate* for just the right word because we enjoy the challenge and even allowing a pun or two where it does not interfere with the presentation of the material. We can't deny it: microcomputer design is fun, and we have fun talking about it. Please forgive us if it shows in this book's style.

G. J. Lipovski
Austin, Texas
February 1999

List of Figures

List of Tables

Acknowledgments

The author would like to express his deepest gratitude to everyone who contributed to the development of this book. The students of EE 345L at the University of Texas at Austin during Fall 1998 significantly helped correct this book; special thanks are due to Levent Og, Ed Limbaugh, and Greg McCasKill, who located most of the errors. This text was prepared and run off using a Macintosh and LaserWriter, running WriteNow. I am pleased to write this description of the Motorola 6812, which is an incredibly powerful component and a vehicle for teaching a considerable range of concepts.

G. J. L.

1

Microcomputer Architecture

Microcomputers, microprocessors, and microprocessing are at once quite familiar and a bit fuzzy to most engineers and computer scientists. When we ask the question: "What is a microcomputer?" we get a wide range of answers. This chapter aims to clear up these terms. Also, the designer needs to be sufficiently familiar with the microcomputer instruction set to be able to read the object code generated by a C compiler. Clearly, we have to understand these concepts to be able to discuss and design I/O interfaces. This chapter contains essential material on microcomputers and microprocessors needed as a basis for understanding the discussion of interfacing in the rest of the book.

We recognize that the designer must know a lot about basic computer architecture and organization. But the goal of this book is to impart enough knowledge so that the reader, on completing it, should be ready to design good hardware and software for microcomputer interfaces. We have to trade material devoted to basics for material needed to design interface systems. There is so much to cover and so little space that we will simply offer a summary of the main ideas. If you have had this material in other courses or absorbed it from your work or from reading those fine trade journals and hobby magazines devoted to microcomputers, this chapter should bring it all together. Some of you can pick up the material just by reading this condensed version. Others will get an idea of the amount of background needed to read the rest of the book.

For this chapter, we assume the reader is fairly familiar with some kind of assembly language on a large or small computer or is able to pick it up quickly. The chapter will present an overview of microcomputers in general and the MC68HC812A4, or MC68HC912B32, single-chip microcomputer in particular.

1.1 An Introduction to the Microcomputer*

Just what is a microcomputer and a microprocessor, and what is the meaning of "microprogramming" – which is often confused with "microcomputers"? This section will survey these concepts and other commonly misunderstood terms in digital systems

---

*Portions of §1.1 were adapted with permission from "Digital Computer Architecture" pp. 298-327 by G. J. Lipovski, and "Microcomputers," pp. 397-480 by G. J. Lipovski and T. K. Agerwala, in the Encyclopedia of Computer Science and Technology, 1978, Belzer et al., courtesy of Marcel Dekker, Inc.

design. It describes the architecture of digital computers and gives a definition of architecture. Note that all *italicized* words are in the index, and are listed at the end of each chapter; these serve as a glossary to help you find terms that you may need later.

Because the microcomputer is pretty much like other computers except it is smaller and less expensive, these concepts apply to large computers as well as microcomputers. The concept of the computer is presented first, and the idea of an instruction is scrutinized next. The special characteristics of microcomputers will be delineated last.

1.1.1 Computer Architecture

Actually, the first and perhaps the best paper on computer architecture, "Preliminary discussion of the logical design of an electronic computing instrument," by A. W. Burks, H. H. Goldstein, and J. von Neumann, was written 15 years before the term was coined. We find it fascinating to compare the design therein with all computers produced to date. It is a tribute to von Neumann's genius that this design, originally intended to solve nonlinear differential equations, has been successfully used in business data processing, information handling, and industrial control, as well as in numeric problems. His design is so well defined that most computers – from large computers to microcomputers – are based on it, and they are called *von Neumann computers.*

In the early 1960s a group of computer designers at IBM – including Fred Brooks – coined the term "architecture" to describe the "blueprint" of the IBM 360 family of computers, from which several computers with different costs and speeds (for example, the IBM 360/50) would be designed. The *architecture* of a computer is, strictly speaking, its instruction set and the input/output (I/O) connection capabilities. More generally, the architecture is the view of the hardware as seen by the programmer. Computers with the same architecture can execute the same programs and have the same I/O devices connected to them. Designing a collection of computers with the same blueprint or architecture has been done by several manufacturers. This definition of the term "computer architecture" applies to this fundamental level of design, as used in this book. However outside of this book, the term "computer architecture" has become very popular and is also rather loosely used to describe the computer system in general, including the implementation techniques and organization discussed next.

The *organization* of a digital system like a computer is usually shown by a block diagram that shows the registers, buses, and data operators in the computer. Two computers have the same organization if they have the same block diagram. For instance, Motorola manufactures several microcomputers having the same architecture but different organizations to suit different applications. Incidentally, the organization of a computer is also called its *implementation.* Finally, the *realization* of the computer is its actual hardware interconnection and construction. It is entirely reasonable for a company to change the realization of one of its computers by replacing the hardware in a section of its block diagram with a newer type of hardware, which might be faster or cheaper. In this case the implementation or organization remains the same while the realization is different. In this book, when we want to discuss an actual realization, we will name the component by its full part number, like MC68HC812A4PV8. But we are usually interested in the organization or the architecture only. In these cases, we will refer

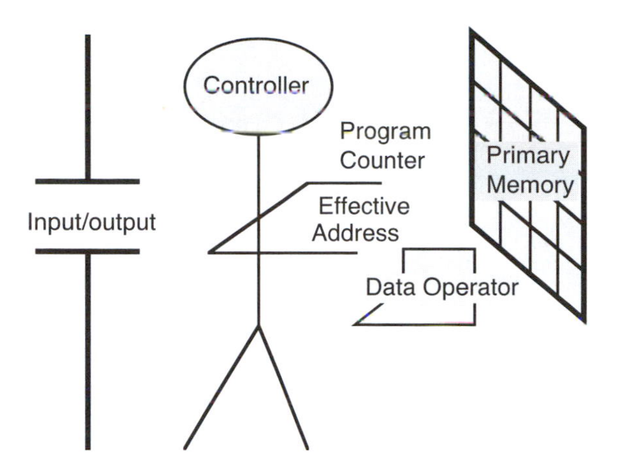

Figure 1.1. Analogy to the von Neumann Computer

to an organization as a partial name without the suffix, for example, the MC68HC812A4 (without PV8), which we abbreviate as 'A4, or the MC68HC912B32, abbreviated as 'B32 and refer to the architecture as a number 6812. This should clear up any ambiguity, while also being a natural, easy-to-read shorthand.

The architecture of von Neumann computers is disarmingly simple, and the following analogy shows just how simple. (For an illustration of the following terms, see Figure 1.1.) Imagine a person in front of a mailbox, with an adding machine and window to the outside world. The mailbox, with numbered boxes or slots, is analogous to the *primary memory;* the adding machine, to the *data operator* (arithmetic-logic unit); the person, to the *controller;* and the window, to *input/output* (I/O). The person's hands *access* the memory. Each slot in the mailbox has a paper that has a string of, say, eight 1s and 0s (*bits*) on it. A string of 8 bits is a *byte*. A string of bits – whether or not it is a byte – in a slot of the memory box is called a *word.*

The primary memory may be in part a *random access memory (RAM)* (so-called because the person is free to access words in any order at random, without having to wait any longer for a word because it is in a different location). RAM may be *static ram (SRAM)* if bits are stored in flip-flops, or *dynamic ram (DRAM)* if bits are stored as charges in capacitors. Memory that is normally written at the factory, never to be rewritten by the user, is called *read-only memory (ROM).* A *programmable read-only memory (PROM)* can be written once by a user, by blowing fuses to store bits in it. An *erasable programmable read-only memory (EPROM)* can be erased by ultraviolet light and then written electrically by a user. An *electrically erasable programmable read-only memory (EEPROM)* can be erased and then written by a user, but erasing and writing words in EEPROM takes several milliseconds. A variation of this memory, called *flash,* is less expensive but cannot be erased one word at a time.

With the left hand the person takes out a word from slot or box *n,* reads it as an instruction, and replaces it. Bringing a word from the mailbox (primary memory) to the person (controller) is called *fetching*. The hand that fetches a word from box *n* is analogous to the *program counter.* It is ready to take the word from the next box, box *n* + 1, when the next instruction is to be fetched.

An instruction in the 6812 is a *binary code* like 01000010. Consistent with the notation used by Motorola, binary codes are denoted in this book by a % sign, followed by 1s or 0s. (Decimal numbers, by comparison, will not use any special symbols.) Since all those 1s and 0s are hard to remember, a convenient format is often used, called *hexadecimal notation*. In this notation, a $ is written (to designate that the number is in hexadecimal notation, rather than decimal or binary), and the bits, in groups of 4, are represented as if they were "binary-coded" digits 0 to 9, or letters A, B, C, D, E, and F, to represent values 10, 11, 12, 13, 14, and 15, respectively. For example, %0100 is the binary code for 4, and %0010 is the binary code for 2, which in hexadecimal notation is represented as $2. The binary code 01000010, mentioned previously, is represented as $42 in hexadecimal notation. Whether the binary code or the simplified hexadecimal code is used, instructions written this way are called *machine-coded* instructions because that is the actual code fetched from the primary memory of the machine, or computer. However, this is too cumbersome. So a *mnemonic* (which means a memory aid) is used to represent the instruction. The instruction $42 in the 6812 actually increments (adds 1 to) accumulator `A`, so it is written as

```
INCA
```

(The 6812 accumulators and other registers are described in §1.2.1. – § means "section")

As better technology becomes available, and as experience with an architecture reveals its weaknesses, a new architecture may be crafted that includes most of the old instruction set and some new instructions. Programs written for the old computer should also run, with little or no change, on the new one, and more efficient programs can perhaps be written using the new features of the new architecture. Such new architecture is *upward compatible* from the old one if this property is preserved. If an architecture executes the same machine code the same way, it is fully upward compatible, but more generally, if it executes the same mnemonic instructions, even though they may be coded as different machine codes, then the architecture is *source code upward compatible.* In this book, we will focus on the 6812 architecture, which is source code upward compatible from the 6811 and its first ancestor, the 6800.

An *assembler* is a program that converts mnemonics into machine code so that the programmer can write in convenient mnemonics and the output machine code is ready to be put in primary memory to be fetched as an instruction. The mnemonics are therefore called *assembly-language* instructions. A *compiler* is a program that converts statements in a *high-level language* either to assembly language, to be input to an assembler, or to machine code, to be stored in memory and fetched by the controller.

While a lot of interface software is written in assembly language and many examples in this book are discussed using this language, most will be written in the high-level language C. However, quick fixes to programs are occasionally even written in machine code. Finally, an engineer should know exactly how an instruction is stored and how it is understood. Therefore, in this chapter we will show the assembly language and machine code for some assembly-language instructions that are important and that you might have some difficulty picking up on your own.

Many instructions in the 6812 are entirely described by one 8-bit word. However, some instructions require 16 bits, 24 bits, or more to fully specify them. They are stored in 8-bit words in consecutive primary memory locations (box numbers) so that when an instruction is fetched, each of the 8-bit words can be fetched one after another.

Now that we have some ideas about instructions, we resume the analogy to illustrate some things an instruction might do. For example, an instruction may direct the controller to take a word from a box *m* in the mailbox with the right hand, copy it into the adding machine (thus destroying the old word) and put the word back in the box. This is an example of an instruction called the *load* instruction. In the 6812 an instruction to load accumulator `A` with the word at location 256 in decimal, or $100 in hexadecimal, is fetched as three words

$B6
$01
$00

where the second word is the most significant byte, and the third is the least significant byte, of the address and is represented by mnemonics as

```
LDAA $100
```

in assembly language. The main operation – bringing a word from the mailbox (primary memory) to the adding machine (data operator) – is called *recalling* data. The right hand is used to get the word; it is analogous to the *effective address.*

As with instructions, assembly language uses a shorthand to represent locations in memory. A *symbolic address,* which is actually some address in memory, is a name that means something to the programmer. For example, location $100 might be called `ALPHA`. Then the assembly-language instruction above can be written as follows

```
LDAA ALPHA
```

We will be using the symbolic address `ALPHA` in most of our examples in this chapter, and it will represent location $100. Other symbolic addresses and other locations can be substituted, of course. It is important to remember that a symbolic address is just a representation of a number, which usually happens to be the numerical address of the word in primary memory to which the symbolic address refers. As a number, it can be added to other numbers, doubled, and so on. In particular, the instruction

```
LDAA ALPHA+1
```

will load the word at location $101 (`ALHPA + 1` is $100 + 1) into the accumulator.

Generally, after such an instruction has been executed, the left hand (program counter) is in position to fetch the next instruction in box $n + 1$. For example, the next instruction may give the controller directions to copy the number in the adding machine into a box in the mailbox, causing the word in that box to be destroyed. This is an example of a *store* instruction. In the 6812, the instruction to store accumulator `A` into location $100 can be written like this

```
STAA ALPHA
```

The main operation in this store instruction – putting a word from the adding machine (data operator) into a box in the mailbox (primary memory) – is called *memorizing* data. The right hand (effective address) is used to put the word into the box.

Before going on, we point out a feature of the von Neumann computer that is easy to overlook, but is at once von Neumann's greatest contribution to computer architecture and yet a major problem in computing. Because instructions and data both are stored in the primary memory, there is no way to distinguish one from the other except by which hand (program counter or effective address) is used to get the data. We can conveniently use memory not needed to store instructions – if few are to be stored – to store more data, and vice versa. It is possible to modify an instruction as if it were data, just before it is fetched, although a good computer scientist would shudder at the thought. However, through an error (*bug*) in the program, it is possible to start fetching data words as if they were instructions, which produces strange results fast.

A *program sequence* is a sequence of instructions fetched from consecutive locations one after another. To increment the word at location $100, we can load it into the accumulator using the `LDAA` instruction, increment it there using the `INCA` instruction, and then put it back using the `STAA` instruction. (A better way will be shown in the next section, but we do it in three instructions here to illustrate a point.) This program sequence is written in consecutive lines as follows

```
LDAA ALPHA
INCA
STAA ALPHA
```

Unless something is done to change the left hand (program counter), a sequence of words in contiguously numbered boxes will be fetched and executed as a program sequence. For example, a sequence of load and store instructions can be fetched and executed to copy a collection of words from one place in the mailbox to another place. However, when the controller reads the instruction, it may direct the left hand to move to a new location (load a new number in the program counter). Such an instruction is called a *jump,* which is an example of a *control instruction.* Such instructions will be discussed further in §1.2.3, where concrete examples using the 6812 instruction set are described. To facilitate the memory access functions, the effective address can be computed in a number of ways, called *addressing modes*. The 6812 addressing modes will be explained in §1.2.1.

1.1.2 The Instruction

In this section the concept of an instruction is described from different points of view. The instruction is discussed first with respect to the cycle of fetching, decoding, and sequencing of microinstructions. Then the instruction is discussed in relation to hardware-software trade-offs. Some concepts used in choosing the best instruction set are also discussed.

The controller fetches a word or a couple of words from primary memory and sends commands to all the modules to execute the instruction. An instruction, then, is essentially a complex command carried out under the direction of a single word or a couple of words fetched as an inseparable group from memory.

The bits in the instruction are broken into several fields. These fields may be the bit code for the instruction, for options in the instruction, for an address in primary memory, or for a data operator register address. For example, the complete instruction `LDAA ALPHA` may look like the bit pattern 101101100000000100000000. The leftmost bit and the fifth to eighth bits from the left – 1,0110 – tell the computer that this is a load instruction. Each instruction must have a different *opcode* word, like 1,0110, so the controller knows exactly which instruction to execute just by looking at the instruction word. The second bit from the left may identify the register that is to be loaded: 0 indicates that accumulator `A` is to be loaded. Bits 3 and 4 from the left, 11, indicate the address mode to access the word to be loaded. Finally, the last 16 bits may be a binary number address: 0000000100000000 indicates that the word to be loaded is to come from word number $100 (`ALPHA`). Generally, options, registers, addressing modes, and primary memory addresses differ for different instructions. The opcode code word – 1,0110, in this example – must be decoded before it can be known that the second bit from the left, 0, is a register address, the third and fourth bits are address mode designators, and so on.

The instruction can be executed by the controller as a sequence of small steps, called *microinstructions.* As opposed to instructions, which are stored in primary memory, microinstructions are usually stored in a small fast memory called *control memory.* A microinstruction is a collection of data transfer *orders* that are simultaneously executed; the *data transfers* that result from these orders are movements of, and operations on, words of data as these words are moved about the machine. While the control memory that stores the microinstructions is normally ROM, in some computers it can be rewritten by the user. The process of writing programs for the control memory is called *microprogramming.* It is the translation of an instruction's required behavior into the control of data transfers that carry out the instruction.

The entire execution of an instruction is called the *fetch-execute cycle* and is composed of a sequence of microinstructions. Access to primary memory is rather slow, so the microinstructions are grouped into *memory cycles,* which are fixed times when the memory fetches an instruction, memorizes or recalls a data word, or is idle. A *memory clock* beats out time signals, one clock pulse per memory cycle. The fetch-execute cycle is thus a sequence of memory cycles. The first cycle is the *fetch cycle,* when the instruction code is fetched. If the instruction is n bytes long, the first n memory cycles are usually fetch cycles. In some computers, the next memory cycle is a *decode cycle,* in which the instruction code is analyzed to determine what to do next. The 6812 doesn't need a separate cycle for this. The next cycle may be for *address calculations.* Data may be read from memory in one or more *recall cycles.* Then the instruction's main function is done in the *execute cycle.* Finally, the data may be memorized in the last cycle, the *memorize cycle.* This sequence is repeated indefinitely as each instruction is fetched and executed.

An instruction can be designed to execute a very complicated operation. Also, certain operations can be performed on execution of some address modes in an instruction that uses the address rather than additional instructions. It is also generally possible to fetch and execute a sequence of simple instructions to carry out the same net operation. The program sequence we discussed earlier can actually be done by a single instruction in the 6812

```
INC ALPHA
```

It recalls word $100, increments it, and memorizes the result in location $100 without changing the accumulator. If a useful operation is not performed in a single instruction like `INC ALPHA`, but in a sequence of simpler instructions like the program sequence already described, such a sequence is either a macroinstruction (macro) or a subroutine.

A sequence is a *macro* if the complete sequence of instructions is written every time in a program that the operation is required. It is a *subroutine* if the instruction sequence is written just once, and a jump to the beginning of this sequence is written each time the operation is required. In many ways macroinstructions and subroutines are similar techniques to get an operation done by executing a sequence of instructions. Perhaps one of the central issues in computer architecture design is this: What should be created as instructions or included as addressing modes, and what should be left out, to be carried out by macros or subroutines? On the one hand, it has been proven that a computer with just one instruction can do anything any existing computer can. It may take a long time to carry out an operation, and the program may be ridiculously long and complicated, but it can be done. On the other hand, programmers might find complex machine instructions that enable one to execute a high-level (for example, C) language statement desirable. Such complex instructions create undesirable side effects, however, such as long latency time for handling interrupts. However, the issue is overall efficiency. Instructions, which enable selected operations performed by a computer to be translated into programs, are chosen on the basis of which can be executed most quickly (speed) and which enable the programs to be stored in the smallest room possible (program density) without sacrificing low I/O latency (time to service an I/O request). (The related issue of storing data as efficiently as possible is discussed in §2.2.)

The choice of instructions is complicated by the range of requirements in two ways. Some applications need a computer to optimize speed, while others need their computer to optimize program density. For instance, if a computer is used like a desk calculator and the time to do an operation is only 0.1 second, there may be no advantage to doubling the speed because the user will not be able to take advantage of it, while there may be considerable advantage to doubling the program density because the cost of memory may be halved and the cost of the machine may drop substantially. But, for another example, if a computer is used in a computing center with plenty of memory, doubling the speed may permit twice as many jobs to be done, so that the computer center's income is doubled, while doubling the program density is not significant because there is plenty of memory available. Moreover, the different applications computers are put to require different proportions of speed and density.

No known computer is best suited to every application. Therefore, there is a wide variety of computers with different features, and there is a problem picking the computer that best suits the operations it will be used for. Generally, to choose the right computer from among many, a collection of simple, well-defined programs pertaining to the computer's expected use, called *benchmarks*, are available. Examples of benchmarks include multiplying two unsigned 16-bit numbers, moving some words from one location in memory to another, and searching for a word in a sequence of words. Programs are written for each computer to effect these benchmarks, and the speed and program density are recorded for each computer. A weighted sum of these values is used to derive a figure of merit for each machine. If storage density is studied, the weights are proportional to the number of times the benchmark (or programs similar to the benchmark) is expected to be stored in memory, and the figure of merit is called the *static efficiency*. If speed is studied, the weights are proportional to the number of times the benchmark (or similar routines) is expected to be executed, and the figure of merit is called the *dynamic efficiency*. These figures of merit, together with computer rental or purchase cost, available software, reputation for serviceability, and other factors, are used to select the machine.

The currently popular RISC (reduced instruction set computer) computer architecture philosophy exploits the concept of using many very simple instructions to execute a program most efficiently.

In this chapter and throughout the subject of software interface design, the issues of efficiency and I/O latency continually appear in the selection instructions for "good" programs. The 6812 has a very satisfactory instruction set, with several alternatives for many important operations. Readers are strongly encouraged to develop the skill of using the most efficient techniques. They should try to select instructions that execute the program the fastest, if dynamic efficiency is prized, or that can be stored in the least number of bytes, if static efficiency is desired.

1.1.3 Microcomputers

One can regard microcomputers as similar to the computers already discussed, but which are created with inexpensive technology. If the controller and data operator are on a single LSI integrated circuit or a small number of LSI integrated circuits, such a combination of data operator and controller is called a *microprocessor*. If memory and I/O module are added, the result is called a *microcomputer*. If the entire microcomputer (except the power supply and some of the hardware used for I/O) is in a single chip, we have a *single-chip microcomputer*. A *personal computer*, whether small or large, is any computer used by one person at a time. However, a computer using microprocessors, which are intended for industrial control rather than personal computing, is generally called a *microcontroller*. A microcontroller can be a single-chip or multiple-chip microcomputer. The 6812 is particularly useful for this book because it works as either, so it is suitable for illustrating the concepts of interfacing to both types of systems.

Ironically, this superstar of the 1970s through the 1990s, the microcomputer, was born of a broken marriage. At the dawn of this period, we were already putting pretty complicated calculators on LSI chips. So why not a computer? Fairchild and Intel made the PPS-25 and Intel 4004 respectively, which were almost computers but were not von Neumann architectures. Datapoint Corporation, a leading and innovative terminal manufacturer and one of the larger users of semiconductor memories, talked both Intel and Texas Instruments into building a microcomputer they had designed. Neither Intel nor Texas Instruments was excited about such an ambitious task, but Datapoint threatened to stop buying memories from them, so they proceeded. The resulting devices were disappointing – too expensive and much too slow. As a recession developed, Texas Instruments dropped the project, but did get the patent on the microcomputer. Datapoint decided it wouldn't buy the Intel 8008 after all, because it didn't meet specs. For some time, Datapoint was unwilling to use microcomputers. Once burned, twice cautious. It is ironic that two of the three parents of the microcomputer disowned the infant. Intel was a new company and could not afford to drop the project altogether. So Intel marketed the new machine as the 8008, and it sold. It is also ironic that Texas Instruments has the patent on the Intel 8008. The 8008 was incredibly clumsy to program and took so many additional supporting integrated circuits that it was about as large as a computer of the same power that didn't use microprocessors. Some claim it set back computer architecture at least ten years. But it was successfully manufactured and sold. It was in its way a triumph of integrated circuit technology because it proved a microcomputer was a viable product by creating a market where none had existed, and because the Intel Pentium, designed to be upward compatible to the 8008, is one of the most popular microcomputers in the world.

We will study the 6812 in this book. Larger microprocessors like the Motorola 68332 are used when more powerful software requiring more memory would be needed than would fit in a 6812. However, we chose to discuss 6812-based microcontrollers in this book because we encourage the reader to build and test some of the circuits we describe, and 6812-based microcontroller memories and processors are less expensive (especially if you connect power backward and pop the ICs) and the same concepts can be discussed as with larger microcomputers. We chose the 6812 because it has an instruction set that can efficiently execute programs written in C to illustrate good software practices. One implementation, the 'A4, has 4K bytes of EEPROM, and this novel component makes the 6812 architecture easy to experiment with. Another implementation, the 'B32, has 32K bytes of flash memory, and is suitable for systems requiring larger memory. A single-chip microcomputer can be used for a large variety of experiments. Nevertheless, other microcomputers have demonstrably better static and dynamic efficiency for certain benchmarks. Even if they have comparable (or even inferior) performance, they may be chosen because they cost less, have a better reputation for service and documentation, or are available, while the "best" chip does not meet these goals. The reader is also encouraged to be prepared to use other microcomputers if warranted by the application.

The microcomputer has unleashed a revolution in computer engineering. As the cost of microcomputers approaches ten dollars, computers become mere components. They are appearing as components in automobiles, kitchen appliances, toys, instruments, process controllers, communication systems, and computer systems. They replace larger computers in process controllers much as fractional horsepower motors replaced the large motor and belt shaft. They are "fractional horsepower" computers. This aspect of microcomputers will be our main concern through the rest of the book, since we will focus on how they can be interfaced to appliances and controllers. However, there is another aspect of microcomputers we will hardly have time to study, which will become equally important: their use in conventional computer systems. We are only beginning to appreciate their significance in computer systems. Microcomputers continue to spark startling innovations; however, the features of microcomputers, minicomputers, and large computers are generally very similar. In the following subsections the main features of the 6812, a von Neumann architecture, are examined in greater detail. Having learned basic principles on a 6812 microcontroller, you will be prepared to work with other similar microcontrollers.

1.2 The 6812 Instruction Set

A typical machine has six types of instructions and several addressing modes. Each type of instruction and addressing mode will be described in general terms. The types and modes indicate what an instruction set might look like. They also give concrete details about how the 6812 works, which help you understand the examples in this book.

This section describes the instruction set. It does not fully define the 6812's instruction set because you could get lost in details. The *CPU12 Reference Manual* available from Motorola (document CPU12RM/AD), should be used to fully specify the 6812 instruction set. We encourage you to experiment with the 6812. See the Appendix.

1.2.1 6812 Addressing Modes

The instructions discussed in §1.1 take a word from memory where the address of the word is given directly in the instruction. This mode of addressing, called *direct addressing,* is widely used on all computers. By the most commonly accepted definition, direct addressing must be able to effectively address any word in primary memory. The number of bits needed to directly address n words of memory is $\log_2 n$. For a standard-size 65,536-word memory, 16 bits are required to address each word. If the width of each word is 8 bits, an instruction may require 24 bits for the instruction code bit pattern and the address. This hurts static efficiency because a lot more bits are needed than with other modes introduced in this section. It also hurts dynamic efficiency because a lot of time is needed to pick up all those words. Then more efficient addressing modes are used to access most often needed words faster and to get to any word in memory by some means, without using as many bits as are needed to directly address each word of memory. This problem leads to the addressing modes that are especially important in small computers. In the remainder of this section, we discuss 6812 addressing modes. (See Table 1.1.)

Table 1.1. Addressing Modes for the 6812

Mode	Example	Use
Implied	`SWI`	Improve efficiency
Register	`INCA`	Improve efficiency
Immediate	`LDAA #12`	Initialize registers, provide constant operands
Page 0	`LDAA ALPHA`	Store global data (address 0 - $ff)
Direct	`LDAA ALPHA`	Access any word in memory
Index	`LDAA 5,X`	Address arrays
Double indexed	`LDAA D,Y`	Address arrays
Autoincrement	`LDAA 1,X+`	Access strings, queues
Autodecrement	`LDAA 1,-SP`	Access stacks
Indirect	`LDAA [D,X]`	Access data via variable addresses
Page relative	`BRA ALPHA` `LDAA 2,PCR`	Provide position independence

Motorola's addressing-mode notation differs from that used in other manufacturers' literature and in textbooks. In fact, Motorola's notation for its 6800-based microcontrollers differs from that of its 68000-based microcontrollers. Because we're using the 6812 to teach general principles, we'll use generally accepted terminology, rather than Motorola's 6800-based microcontroller terminology, throughout this book.

What everyone else calls direct addressing is referred to by Motorola as extended addressing, which you should know only if you want to read their literature. Motorola uses the term "direct addressing" for a short form of addressing that uses an 8-bit address, which we will call *page zero addressing.* Their terminology would be correct for a primary memory of only 256 words. Then direct addressing would just need to address a small memory, and addressing more memory would be called extended addressing. It seems the designers of the original 6800 assumed most systems would require only such a small (primary) memory. But as we now know, garbage accumulates to fill the container, so if we build a bigger container, it will soon be filled. Thus, we should also call it direct addressing when we use a 16-bit displacement as the effective address.

In the following discussion of addressing modes, the instruction bits used as an address, or added to get the effective address, are called the *displacement.* Also, in the following discussion an address is calculated the same way – in jump instructions – for the program counter as for the effective address, in such instructions as `LDAA` or `STAA.` Don't get confused about the addressing modes used in jump instructions; `JMP ALPHA` doesn't take a word from location `ALPHA` to put it into the program counter using direct addressing in the same way as in the instruction `LDAA ALPHA.` Rather, `JMP ALPHA` loads the address of `ALPHA` into the program counter. The simple analogy we used earlier makes it clear that the program counter is, like the effective address, a "hand" to address memory and is treated the same way by the addressing modes.

Some techniques improve addressing efficiency by avoiding the calculation of an address to memory. In *implied addressing,* the instruction always deals with the same memory word or register so that no instruction bits specify it. An example is a kind of jump to subroutine instruction called the software interrupt (`SWI`; this instruction will be further explained in §1.2.3). When the `SWI` is executed, the old value of the program counter is saved in a specific place (to be described later) and the new program counter value is gotten from two other specific places (memory location $FFF6, $FFF7). The instruction itself does not contain the usual bits indicating the address of the next instruction: the address is implied. Motorola and others also call this mode "inherent."

A similar mode uses registers as the source and destination of data for instructions. This is called *register addressing*. The 6812 has accumulators that can be so used, called *accumulator* `A`, *accumulator* `B`, and *accumulator* `D`. See Figure 1.2. Accumulator `D` is a 16-bit accumulator for 16-bit data operations and is actually the same as the two 8-bit accumulators `A` and `B` joined together. That is, if accumulator `A` has $3B and accumulator `B` has $A5, then accumulator `D` has $3BA5, and vice versa. In some instructions, such as `INC`, one can increment a memory byte using direct addressing, as in `INC ALPHA`; or one can increment a register, such as `INCA.` Thus, register addressing can be used instead of memory addressing to get data for instructions. This mode substantially improves both static and dynamic efficiency because fewer bits are needed to specify the register than a memory word and a register can be accessed without taking up a memory cycle.

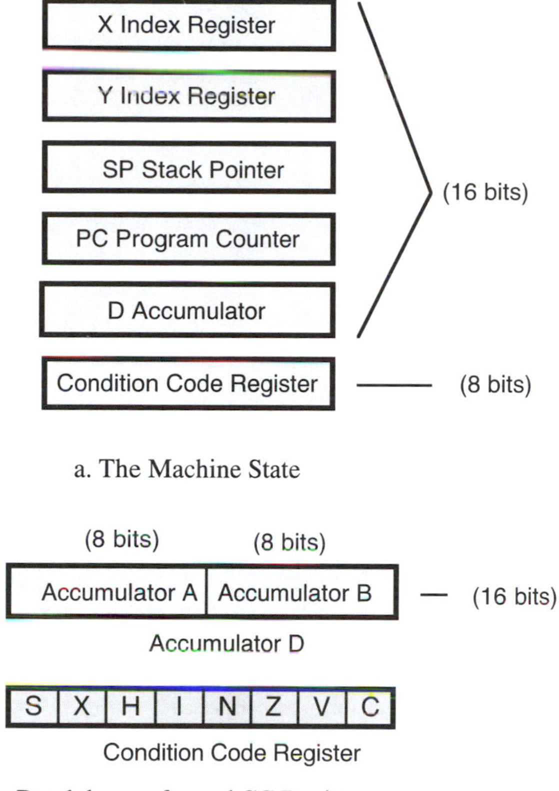

Figure 1.2. Registers in the 6812

Another nonaddressing technique is called *immediate addressing*. Herein, part of the instruction is the actual data, not the address of data. In a sense, the displacement is the data itself. For example, a type of load instruction,

```
LDAA #$10
```

puts the number \$10 into the accumulator. Using Motorola's notation, an immediate address is denoted by the # symbol. The LDAA instruction, with addressing mode bits for immediate addressing, is \$86, so this instruction is stored in machine code like this

\$86
\$10

The number \$10 is actually the second word (displacement) of the two-word instruction. This form of addressing has also been called literal addressing.

Page addressing is closely related to direct addressing. (Two variations of page addressing appear in the 6812, and will be discussed later.) If 8 bits can be used to give the address inside the instruction, then 2^8 or 256 contiguous words can be directly addressed with these bits. The 256 contiguous words in this example are called a *page.* In the 6812, data stored on page 0, where it can be accessed by page zero addressing (which Motorola calls direct addressing), can be accessed using an 8-bit displacement, which is the low byte, padded with 0s in the high byte to form a 16-bit address. Page zero addressing is used to get the global data more efficiently because a shorter (2-byte) instruction, rather than a longer direct addressed (3-byte) instruction, is used.

Although page 0 addressing permits one to access some data, one cannot access much of it. For example, if a page is 256 words, then a 20 X 20-word global array cannot be stored entirely on page 0. This is solved by *index addressing,* whereby fast registers in the controller module, called *index registers*, are used to obtain the address. The index registers in the 6812 are `X`, `Y`, and the *stack pointer* `SP` (see Figure 1.2). The effective address is the (signed) sum of a selected index register and part of the instruction called the *offset.* If `X` has $1000, then `LDAA -8,X` loads accumulator `A` with the contents of location $0ff8. As selected by some bits in the instruction, the offset can be 5 bits, 9 bits or 16 bits. Additionally, an accumulator, `A`, `B`, or `D`, can be added to an index register, `X`, `Y`, or `SP`, to get the effective address in *double-indexed addressing.* This mode is useful when, to read or write *V[i],* an index register such as `X` points to the lowest addressed word of vector *V* and the accumulator is the vector's index *i.* The contents of the index register or accumulator are unchanged by index addressing. A simple form of index addressing that uses an offset of 0, whose bit representation may be omitted from the instruction altogether, is called *pointer addressing,* and the index registers are called *pointer registers.* While pointer addressing can be used to recall any word in memory, it is more difficult to use than index addressing because the register must be reloaded each time to recall or memorize a word at a different location. However, *autoincrement addressing* uses `X`, `Y`, or `SP` as a pointer register, and increments the contents of the register after it is used in the address calculation. *Autodecrement addressing* with these registers decrements the contents of the register just before it is used in the address calculation. In the 6812, the instruction

```
STAA 1,X+
```

puts the word pointed to by `X` into accumulator `A`, then increments `X` by 1; the instruction

```
LDAA 1,-Y
```

decrements the index register `Y` by 1 and then loads accumulator `A` with the word pointed to by `Y` (after it is changed). In the 6812, an index register can be incremented or decremented by a number from 1 to 8 in place of the number 1 in the above examples.

A part of memory, called the *stack buffer,* is set aside for a *stack.* The stack pointer, `SP`, initially contains the lowest address of any word on the stack. *Pushing* a word that is in accumulator `A` onto this stack is accomplished by

```
STAA 1,-SP
```

and *pulling* (or popping) a word from this stack into accumulator `A` is done by

```
LDAA 1,SP+
```

The stack fills out, starting at high addresses and building toward lower addresses, in the stack buffer. If it builds into addresses lower than the stack buffer, a *stack overflow* error occurs, and if it is pulled too many times, a *stack underflow* error occurs. If no such errors occur, then the last word pushed onto the stack is the first word pulled from it, a property that sometimes labels a stack a LIFO (last-in, first-out) stack. Stack overflow or underflow often causes data stored outside of the stack buffer to be modified. This bug is hard to find.

The *jump to subroutine* instruction is a jump instruction that also pushes the 2-byte return address onto the stack, least significant byte first (which then appears at the higher address). The corresponding *return from subroutine* instruction pulls two words from the stack, putting them in the program counter. If nobody changes the stack pointer `SP`, or if the net number of pushes equals the net number of pulls (a *balanced* stack) between the jump to subroutine and the corresponding return from subroutine, then the last instruction causes the calling routine to resume exactly where it left off when it called the subroutine. This method of handling return addresses allows easy *nesting of subroutines*, whereby the main program jumps to a subroutine, say `Alpha`, and subroutine `Alpha` in turn jumps to subroutine `Beta`. When the main program jumps to subroutine `Alpha`, it pushes the main program's return address onto the stack. When `Alpha` jumps to subroutine `Beta`, it pushes `Alpha's` return address onto the stack, on top of (in lower memory words than) the other return address. When `Beta` is completed, the address it pulls from the stack is the return address to subroutine `Alpha`. And when `Alpha` is completed, what it pulls from the stack is the main program return address.

The stack in the 6812 is a good place to store *local data,* which is data used only by a subroutine and not by other routines. It can also be used to supply *arguments* (operands) for a subroutine and return results from it. To save local data, one can push it on the stack and pull it from the stack to balance it. A reasonable number of words can be stored this way. Note that the subroutine must pull as many words from the stack as it pushed before the return from subroutine instruction is executed, or some data will be pulled by that instruction into the program counter.

The stack pointer `SP` must be treated with respect. It should be initialized to point to the high address end of the stack buffer as soon as possible, right after power is turned on, and should not be changed except by incrementing or decrementing it to effectively push or pull words from it. Words above (at lower addresses relative to) the stack pointer must be considered garbage and may not be read after they are pulled.

An *indirect address* is where the instruction recalls a 16-bit word from memory, which it uses to read or write data in memory. Indirect addressing is denoted by square brackets [ ]. There are two indirect addresses in the 6812, using index addressing with accumulator `D` as in `LDAA [D,X]`, and index addressing with a 16-bit offset as in `LDAA [$1234,X]`. Indirect addressing is useful where the address of the data to be recalled or memorized is calculated and then stored in memory, at run time, but is not known when the program is assembled. Such cases are important when ROMs store the program.

Page relative addressing calculates the effective address by adding an 8-bit two's complement displacement to the program counter to get the address for a jump instruction (a *branch*) because one often jumps to a location that is fairly close to (on the same page as) where the jump is stored. The displacement is added to the program counter when it actually points to the beginning of the next instruction. This addressing mode only works if the jump address is within -128 to +127 locations of the next instruction's address after the branch instruction. Labels are followed by a semicolon (:). For example, if the place we want to branch to has a label `L` at $200

```
L: LDAA ALPHA
```

a branch to this location is denoted

```
BRA L
```

If `L` is at location $200 and the instruction is at location $1F0, then the program counter is at location $1F2 when the address is calculated, and the `BRA` instruction (whose instruction code is $20) will be assembled and stored in memory as

$20
$0E

Note that the assembly-language instruction uses the symbolic address L rather than the difference between the addresses, as in `BRA L-$F2`, and the assembly automatically determines the difference between the current program counter address and the effective address and puts this difference into the second word (displacement) of the instruction.

The program counter may be used in place of an index register in any instruction that uses index addressing, but like the preceding branch instruction, the symbolic address of the data is used rather than the difference between the current location and the address.

```
L: LDAA ALPHA,PCR
```

will load the data at location `ALPHA` into accumulator `A`, but the instruction will contain the difference between the address of alpha and the address of the next instruction, just like the `BRA` instruction bits contained the difference between the addresses.

1.2.2 6812 Data Operator Instructions

We now focus generally on the instruction set of a von Neumann computer, and in particular on the instructions in the 6812. There are a substantial number of special instructions in the 6212 architecture, such as those used for fuzzy logic. These include `ETBL`, `EMACS`, `MAXA`, `MAXD`, `MAXM`, `MEM`, `MINA`, `MINM`, `REV`, `REVW`, `TBL`, and `WAV`. We could attempt to cover all instructions, but these special instructions are not directly used in I/O interfacing, nor are they generated by a compiler. Therefore, in order to begin covering interfacing in earnest, we study only conventional 6812 instructions. The conventional instructions are grouped together in this section to see the available alternatives. The lowly but important move instruction is discussed first. The arithmetic and the logical instructions are covered next. Edit instructions such as shifts, and, finally, I/O instructions are covered in the remainder of this section. Control instructions such as jump will be discussed in §1.2.3.

Table 1.2. 6812 Move Instructions

LDAA	STAA	TFR	PSHA	CLRA
LDAB	STAB	SEX	PULA	CLRB
LDD	STD	TAB	PSHB	CLR
LDX	STX	TBA	PULB	TSTA
LDY	STY	TAP	PSHC	TSTB
LDS	STS	TPA	PULC	TST
MOVB	EXG	TSX	PSHD	
MOVW	XGDX	TXS	PULD	
LEAX	XGDY	TSY	PSHX	
LEAY		TYS	PULX	
LEAS			PSHY	
			PULY	

The simplest is a *move* instruction, such as load and store. This instruction moves a word to or from a register, in the controller or data operator, from or to memory. Typically, a third of program instructions are moves. If an architecture has good move instructions, it will have good efficiency for many benchmarks. (Table 1.2 lists the 6812's move instructions.)

We have discussed the `LDAA` and `STAA` instructions, which are in this class. New instructions, `MOVB`, `PSHA`, `PULA` (and so on), `XGDY`, and `TST` are 6812 move instructions, as is `CLR`, which is an alternative to one of these.

The load and store instructions can load or store any of the 8-bit registers `A` or `B` or the 16-bit registers `X`, `Y`, `SP`, or `D`. The instruction `LDX  ALPHA` puts the words at locations \$100 and \$101 into the `X` index register. The instruction `LDY  ALPHA` puts the words at locations \$100 and \$101 into the `Y` index register. Note that an index register can be used in an addressing mode calculation even when that register itself is being changed by the instruction, because in the fetch-execute cycle, the addresses are calculated before the data from memory is recalled and loaded into the register. In particular, the instruction

```
LDX 0,X
```

is both legal and very useful. If `X` has the value \$100 before the instruction is executed, the instruction gets 2 words from location \$100 and \$101, then puts them into the index register. Note that the richness of the addressing modes contributes greatly to the efficiency of the lowly but important move instructions.

The `MOVB` instruction moves an 8-bit byte and the `MOVW` instruction moves a 16-bit word. In effect it is a combination of a load and a store, and has addressing mode information for both (e.g., `MOVB  1,X+,1,Y+`). It does not affect the data in accumulators, which can be useful if the accumulators have valuable data in them. Using pointer addressing, the load effective address instructions, `LEAX`, `LEAY`, and `LEAS`, merely move the contents of `X`, `Y`, and `SP` into another of these registers.

Transfer and exchange instructions permit movement of data among registers of similar width. There is a general `TFR  R1,R2` instruction which transfers the data in R1 to

R2, where R1 and R2 can be any two registers. If the source is an 8-bit register and the destination is a 16-bit register, the instruction extends the sign of the 8-bit data as it loads the 16-bit register, so the instruction is called sign extend, or `SEX`. For example, `SEX B,D` will fill accumulator `A` with copies of the sign bit of accumulator `B`. To provide source code upward compatibility to the 6811, cases of this `TFR` instruction are given 6811-style mnemonics. `TAB` moves the contents of accumulator `A` to accumulator `B` without changing `A`. Similarly `TBA` moves `B` to `A`, `TAP` moves `A` to the condition code register, and `TPA` moves the condition code register to `A`. `TSX` moves `SP` to `X`, `TXS` moves `X` to `SP`, `TSY` moves `SP` to `Y`, and `TYS` moves `Y` to `SP`, with the following correction. There is a general `EXG R1,R2` instruction that exchanges the contents of the registers, where R1 and R2 can be any two registers. For source code upward compatibility with 6811 instructions, special cases of the `EXG` instructions are given 6811-style mnemonics: the `XGDX` instruction can exchange the 2 bytes in `X` with those in `D`, and the `XGDY` instruction can exchange the 2 bytes in `Y` with those in `D`.

The `PSHA` instruction pushes the byte in accumulator `A` onto the stack, and the `PULA` instruction pulls a byte from the stack into accumulator `A`; similarly, `PSHB` and `PULB` work on accumulator `B`, `PSHC` and `PULC` work on the condition code register, `PSHD` and `PULD` work on accumulator `D`, `PSHX` and `PULX` work on `X`, and `PSHY` and `PULY` work on `Y`. Note that `PSHD`, `PULD`, `PSHX`, `PULX`, `PSHY` and `PULY` push or pull two bytes.

The condition-code register bits control interrupts and save the results of operations. (See Figure 1.2b.) Condition codes are used to generalize the conditional jump capability of the computer and to control interrupts. When a result is obtained, the *zero bit* Z is usually set to 1 if the result was 0; to the *negative bit* N if the result was negative; to the *carry bit* C if an add operation produced a carry; and to an *overflow bit* V if an add operation produces a result considered invalid in the 2's complement number system because of overflow. These bits can be used in or tested by a later arithmetic or conditional jump instruction. A *half-carry bit* H is used in decimal arithmetic. Also, two *interrupt inhibit bits* (also called an *interrupt mask bit*) I and X are kept in the condition code; when they are set, interrupts are not permitted. Finally, a *stop disable bit* S is used to prevent execution of the `STOP` instruction operation, discussed later.

Load and store instructions change two condition codes used in conditional branch instructions. If you want to set the condition codes as in a load instruction, but the data is in an accumulator already, then the `TSTA` or `TSTB` instruction can be used. It is like half a load instruction because it sets the condition codes like a `LDAA` instruction but does not change `A` or `B`. Similarly, instruction `TST ALPHA` will set the condition codes like `LDAA ALPHA` but will not change `A`. Finally, because a load instruction with an immediate operand is used to initialize registers, since most initial values are 0, a separate instruction `CLR` is provided. Use `CLR` rather than `LDAA #0` to improve efficiency. However, `CLR` changes the C condition code bit, while `LDAA #0` doesn't change the C bit, so the longer `LDAA #0` instruction is often used to avoid altering this bit.

The *arithmetic* instructions add or subtract the value of the accumulator with the value of a word taken from memory, or multiply or divide values in the registers. The 6812 has arithmetic instructions to be used with 8-bit registers and some arithmetic instructions to be used with 16-bit registers. The 8-bit arithmetic instructions are discussed first, then the 16-bit instructions. Table 1.3 lists these arithmetic instructions.

Table 1.3. 6812 Arithmetic Instructions

ADDA, ADDB	INCA, INCB, INC	ADDD
ADCA, ADCB	DECA, DECB, DEC	SUBD
ABA	NEGA, NEGB, NEG	ABX, ABY
SUBA, SUBB	ASLA, ASLB, ASL	CPD, CPX
SBCA, SBCB	ASRA, ASRB, ASR	CPY, CPS
SBA	LSRA, LSRB, LSR	LEAX, LEAY, LEAS
CMPA, CMPB	LSLA, LSLB, LSL	INX, INY, INS
CBA		DEX, DEY, DES
		ASLD, LSRD, LSLD
(Special: MUL EMUL EMULS FDIV IDIV IDIVS EDIV EDIVS DAA)		

The basic 8-bit ADDA or ADDB instruction can add any word from memory into either accumulator A or accumulator B. The instruction is straightforward except for the setting of condition codes. The same instruction is used for unsigned adds as for 2's complement adds; only the testing of the codes differs. For example

```
ADDA ALPHA
```

will add the contents of accumulator A to word $100 of memory and put the result into accumulator A. Usually, the result is 1 bit wider than the operands, and the extra leftmost bit is put into the carry flip-flop. For unsigned numbers, the carry is often considered an overflow indicator; if the carry is 1, the result in the accumulator is incorrect because when the word is put back into the (8-bit-wide) memory, the ninth bit in the carry won't fit, and so the result in memory will also be incorrect. Also, the carry is used to implement multiple precision arithmetic, which is very important in a microcomputer, and will be discussed shortly. The N and Z condition codes are set, just as in the load and store instructions, to reflect that the result of the addition is negative or zero. A half-carry bit, used in decimal arithmetic (discussed shortly), is set in this instruction and the overflow bit V is set to 1 if the result is erroneous as a 2's complement number.

The instruction *add with carry* (ADCA or ADCB) is used to effect multiple precision arithmetic. It adds a number from memory into the accumulator and sets the condition codes, as in the ADDA or ADDB instruction, and also adds in the old carry flip-flop value in the least significant bit position. However, ADDD adds a 16-bit number to accumulator D. Addition of accumulator B to accumulator A is accomplished by ABA, and adding it (as an unsigned 8-bit number) to X or Y is accomplished by ABX and ABY. The LEAX, LEAY, and LEAS instructions, when used with double-indexed addressing modes, will add an accumulator value to the X, Y, or SP registers. However in these instructions, an 8-bit value will be sign-extended before adding to a 16-bit register.

The 6812, like most microcomputers, has a similar set of subtract instructions. The instruction

```
SUBA ALPHA
```

subtracts the word from location $100 from accumulator A and sets the condition codes as follows. N, Z, and V are set to indicate a negative result, 0 result, or 2's complement overflow, as in the ADDA or ADDB instruction. The carry flip-flop is actually the borrow indicator; it is set if subtraction requires a borrow from the next higher byte or if an unsigned underflow error exists because the result, a negative number, can't be represented as an unsigned number. SBA subtracts accumulator B from accumulator A. The instruction *subtract with carry* (SBCA or SBCB) behaves like ADCA or ADCB to implement multiple precision subtraction, and SUBD subtracts 16-bit values.

Subtraction is often used to compare two numbers, sometimes just to see if they are equal. The results are tested in conditional branch instructions. However, if we are comparing a given number against several numbers to avoid reloading the given number, it can be left in an accumulator, and a *compare* instruction, such as CMPA, can be used. CMPA is just like the subtract instruction, but it does not change the accumulator, so the number in it can be compared to others in later instructions. The condition codes are changed and can be tested by conditional branch instructions. There is a comparison instruction, CBA, to compare the contents of the two 8-bit accumulators. The 16-bit accumulator or index registers are often compared to 16-bit immediate operands or data; the CPD, CPX, CPY, or CPS instructions are used for these comparisons.

Because we often add or subtract just the constant 1, or negate a 2's complement number, special short instructions are provided to improve efficiency. The instructions INC, DEC, and NEG can increment, decrement, or negate either the accumulator or any word in memory. INC and DEC may increment or decrement a word in memory: Also, adding a number to itself and doubling it is an arithmetic left shift ASL, and dividing a 2's complement number by 2 and halving it is an arithmetic right shift ASR. Similarly, LSR divides an unsigned number by 2. The condition codes for ASL are set just as if you did add the number to itself, and the ASR and LSR instructions set N and Z as in the move instructions and shift the low-order bit that is shifted out of the word into the C bit. These edit instructions are also arithmetic instructions and can be applied to either accumulator or to any word in memory.

The LEAX, LEAY, and LEAS instructions, used with displacement index addressing, can add the displacement to index registers X, Y or SP. Therefore they can increment or decrement an index register. To maintain source code compatibility with the 6811, 6811-style instructions INX, INY, DEX, and DEY, which are fundamentally variations of these instructions, are included in the 6812; in fact they are further implemented as 1-byte instructions, whereas the LEAX, LEAY, and LEAS instructions are 2- or 3-byte instructions; but the latter can add any constant to or subtract any constant from their designated index register. Also, the latter instructions do not affect the V condition code.

Special arithmetic instructions act on the accumulators only. They enable us to multiply, divide, or add numbers using the binary-coded decimal number representation (BCD). To multiply two 8-bit unsigned numbers, put them in accumulators A and B, and execute the MUL instruction. The 16-bit result will be put in accumulator D. The EMUL instruction multiplies two unsigned 16-bit numbers to get a 32-bit product, and the EMULS instruction multiplies two signed 16-bit numbers to get a 32-bit product. The instruction FDIV divides accumulator D by index register X considered as a fraction, leaving the quotient in X and the remainder in D. The instruction IDIV divides accumulator D by index register X considered as an integer, leaving the quotient in X and

Table 1.4. 6812 Logic Instructions

EORA,	EORB	COMA,	COMB,	COM
ORAA,	ORAB	ORCC,	ANDCC	
ANDA,	ANDB	SEC,	SEI,	SEV
BITA,	BITB	CLC,	CLI,	CLV
		BSET,	BCLR	

the remainder in `D`. `EDIV` divides an unsigned 32-bit number by a 16-bit unsigned number, and `EDIVS` divides a signed 32-bit number by a 16-bit signed number. To execute arithmetic on binary-coded decimal numbers, with two BCD numbers in the left 4 bits and right 4 bits of a word, add them with the `ADDA` instruction or the `ADCA` instruction, followed immediately by the `DAA` (decimal adjust accumulator `A`) instruction. `DAA` uses the carry and half-carry to correct the number so that the sum is the BCD sum of the two numbers being added. Note, however, that accumulator `A` is an implied address for `DAA`, and the half-carry is only changed by the `ADDA` and `ADCA` instructions, so the `DAA` instruction only works after the `ADDA` and `ADCA` instructions.

Most *logic instructions* (see Table 1.4) are generally similar to arithmetic instructions except that they operate logically on corresponding bits of an accumulator (`A` or `B`) and an operand. The instruction

```
ANDA ALPHA
```

will logically "and," bit by bit, the word at $100 in memory to the accumulator `A`. We can "and" into either accumulator `A` or accumulator `B`. For example, if the word at location $100 were 01101010 and at accumulator `A` were 11110000, then after such an instruction is executed the result in accumulator `A` would be 01100000. A "bit test" instruction BIT "ands" an accumulator with a word recalled from memory but only sets the condition codes and does not change the accumulator. It may be used, like the `CMP` instructions, to compare a word – without destroying it – to many words recalled from memory to check if some of their bits are all 0. The complement instruction `COM` will complement each bit in the accumulator or any word in memory.

Bit-oriented instructions permit the setting of individual bits (`BSET`) and the clearing of those bits (`BCLR`). The instruction `BSET ALPHA,#4` will set bit 2 in word `ALPHA`, and `BCLR ALPHA,#4` will clear it. After a space, we put a field to indicate the pattern of bits to be ORed, or complemented and ANDed, into the word at the effective address. Note that more than 1 bit can be set or cleared in one instruction.

The condition codes are often set or cleared. The interrupt mask bit I may be set to prevent interrupts and cleared to allow them. The carry bit is sometimes used at the end of a subroutine to signal a special condition to the routine it returns to and is sometimes cleared before the instructions `ADC` or `SBC`. The carry bit is often set or cleared. The `ANDCC` and `ORCC` instructions clear or set any condition code bit or combination of condition code bits. But to preserve source code upward compatibility with the 6811, 6811-style instructions `CLC`, `CLI`, and `CLV` clear C, I, or V, respectively, and `SEC`, `SEI`, and `SEV` are provided to set C, I, or V.

Table 1.5. 6812 Edit Instructions

ASLA, ASLB, ASL	LSLA, LSLB, LSL	ROLA, ROLB, ROL
ASRA, ASRB, ASR	LSRA, LSRB, LSR	RORA, RORB, ROR
	LSLD,ASLD,LSRD	

The next class of instructions – the *edit* instructions – earrange the data bits without changing their meaning. The edit instructions in the 6812, shown in Table 1.5, can be used to shift or rotate either accumulator or a word in memory that is selected by any of the addressing modes. Most microcomputers have similar shift instructions. A right logical shift LSRA will shift the bits in the accumulator right one position, filling a 0 bit into the leftmost bit and putting the old rightmost bit into the C condition code register. Similarly, a logical left shift LSLA will shift the bits in the accumulator left one position, filling a 0 bit into the rightmost bit and putting the old leftmost bit into the C condition code register. A machine generally has several left and right shifts. The 6812 also has arithmetic shifts corresponding to doubling and halving a 2's complement number, as discussed with respect to arithmetic instructions. However, although there are different mnemonics for each, the LSLA and ASLA instructions do the same thing and have the same machine code. The rotate instructions ROLA and RORA circularly shift the 9 bits in accumulator A and the carry bit C 1 bit to the left or the right, respectively. They are very useful for multiple word shifts. For example, to shift the 16-bit word in accumulator D (accumulators A and B) 1 bit right, filling with a 0, we can execute this program sequence

```
LSRA
RORB
```

The RORB instruction rotates the bit shifted out of accumulator A, which is held in the C condition code bit, into accumulator B. Of course, this technique is more useful when more than 2 bytes must be shifted. Memory words can be shifted without putting them in the accumulator first. Since an 8-bit or a 16-bit word is often inadequate, multiple-precision shifting is common in microcomputers, and the RORA and ROLA instructions are very important. Also for this reason, microcomputers do not have the multiple shift instructions like LSRA 5, which would shift accumulator A right 5 bits in one instruction. Such an instruction would require saving 5 bits in the condition code registers to implement multiple precision shifts. That is generally too messy to use. Rather, a loop is set up, so inside the loop 1 bit is shifted throughout the multiple-precision word, and the loop is executed as many times as the number of bits to be shifted. However, the 6812 has instructions ASLD, LSLD, and LSRD to shift the 16-bit accumulator D the way ASL, LSL, and LSR shift 8-bit accumulators. Some computers have more complex edit instructions than the shifts discussed here, such as instructions to format strings of output characters for printing.

The next class of instructions is the I/O group for which a wide variety of approaches is used. In most computers, there are 8-bit and 16-bit registers in the I/O devices and control logic in the registers. In other computers there are instructions to transfer a byte or 16-bit word from the accumulator to the register in the I/O device; to transfer a byte or 16-bit word from the register to the accumulator; and to start, stop, and test the device's control logic. In the 6812 architecture, there are no special I/O instructions; rather, I/O registers appear as words in primary memory (*memory-mapped I/O*). The `LDAA` or `LDAB` instructions serve to input a byte from an input register, and `STAA` or `STAB` serves to output a byte; while `LDD`, `LDX`, and similar instructions serve to input a 16-bit word from an input register, and `STD`, or `STX`, and similar instructions serve to output a 16-bit word. Moreover, instructions like `INC ALPHA` will, if `ALPHA` is the location of a (readable) output register, modify that register in place so the word is not brought into the accumulator to modify it. `CLR`, `ASR` (which happens to be a test and set instruction needed to coordinate multiple processors), and `DEC` can operate directly on (readable) output registers in a memory-mapped I/O architecture. Indirect addressing can be used by programs in read-only memory so they can work with I/O registers even if they are at different locations. The indirect address can be in read-write memory and can be changed to the address of the I/O device. That way, a program in a read-only memory can be used in systems with different I/O configurations. Thus, the production of less expensive read-only memory software becomes feasible. This aspect of the architecture is of central importance to this book and will be dealt with extensively.

1.2.3 6812 Control Instructions

A final group contains the *control instructions* that affect the program counter. (See Table 1.6.) Next to move instructions, control instructions are most common, so their performance has a strong impact on a computer's performance. Also, microcomputers with an instruction set missing such operations as floating point arithmetic, multiple word shifts, and high-level language operations, such as switch statements, implement these "instructions" as subroutines rather than macros to save memory space. Unconditional jumps and no-operations are considered first, then the conditional branches and finally the subroutine calls are scrutinized.

Table 1.6. 6812 Control Instructions

Unconditional	Conditional simple	Conditional 2's complement	Conditional unsigned	Bit conditional	Subroutine and interrupt
JMP	BEQ, LBEQ	BGT, LBGT	BHI, LBHI	BRSET	JSR
BRA, LBRA	BNE, LBNE	BGE, LBGE	BHS, LBHS	BRCLR	BSR
BRN, LBRN	BMI, LBMI	BEQ, LBEQ	BEQ, LBEQ		CALL
NOP	BPL, LBPL	BLE, LBLE	BLS, LBLS		RTS
SKIP1	BCS, LBCS	BLT, LBLT	BLO, LBLO		RTI
SKIP2	BCC, LBCC				RTC
	BVS, LBVS	Count and loop			SWI
	BVC, LBVC				WAI
		DBEQ, DBNE, IBEQ,			STOP
		IBNE, TBEQ, TBNE			BGND

The left column of Table 1.6 shows unconditional jumps. As noted earlier, the JMP instruction can use direct (16-bit) or indexed-addressing mode, but the effective address is put in the program counter. The *no-operation* instructions do absolutely nothing. Why have them, you ask? Programs providing signals to the "outside world" – known as real time programs – may need time to execute a program segment for timing the length of a pulse. No-operation instructions provide delays for that purpose. `NOP` and branch never (`BRN`) delay one memory cycle, and long branch never (`LBRN`) delays three. Also, when we test a program, these instructions can be placed to save room for other instructions that we will later insert. An interesting instruction unconditionally skips over one or two words that might be executed as instructions if a jump is made directly to them. `CMPA` and `CPX` using immediate addressing do this, except they change the condition codes (usually no problem). Therefore, they can be called `SKIP1` and `SKIP2`.

If we move the program intact from one place in memory to another, their relative address remains unchanged. You may use page relative addressing of a `BRA` or a `LBRA` in place of the direct addressing used in a jump instruction. If a program does not use direct addressing in jump instructions but rather uses branch instructions, it has a characteristic called *position independence.* This means a program can be loaded anywhere in memory, and it will run without change, thus simplifying program and subroutine loading. This also means that a ROM can be loaded with the program and the same ROM will work wherever it is addressed. Position independence permits ROMs to be usable in a larger range of multiple-chip microcontrollers where the ROMs are addressed at different places to avoid conflicts with other ROMs, so they can be sold in larger quantities and will therefore cost less. Relative branch instructions simplify position independence.

When reading machine code, many programmers have difficulty with relative branch instructions that branch backward. We recommend using 16's complement arithmetic to determine the negative branch instruction displacement. The *sixteen's complement* is to hexadecimal numbers as the 2's complement is to binary numbers. To illustrate this technique consider a program that begins at location $200; the address of each word is shown on the left, with the value shown on the right in each line. All numbers are in hexadecimal.

```
200 86
201 80
202 E6
203 00
204 08
205 18
206 E7
207 00
208 18
209 08
20A 4A
20B 26
20C xx
```

The displacement used in the branch instruction, the last instruction in the program, is shown as xx. It can be determined as follows. When the branch is executed, the program counter has the value $20D, and we want to jump back to location $202. The difference, $20D - $202, is $0B, so the displacement should be -$0B. A safe way to calculate the displacement is to convert to binary, negate, and then convert to hexadecimal. $0B is 00001011, so the 2's complement negative is 11110101. In hexadecimal, this is $F5. That is not hard to see, but binary arithmetic gets rather tedious. A faster way takes the 16's complement of the hexadecimal number. Just subtract each digit from $F (15), digit by digit, then add 1 to the whole thing. -$0B is then ($F - 0),($F - B) + 1 or $F4 + 1, which is $F5. That's pretty easy, isn't it!

A branch instruction may direct the person in the analogy to branch, for instance, only if the number in the adder is positive. If that number is not positive, the next instruction is fetched and executed because the left hand is not moved. This is a *conditional branch*. The 6812 has only conditional branch instructions, rather than conditional jumps or conditional subroutine calls and conditional subroutine returns (as does the 8080). The conditional branch tests one or more condition codes, then branches to another location specified by the displacement if the condition is true, using relative addressing. For each conditional branch, which uses an 8-bit offset, there is also a conditional long branch, which uses a 16-bit offset. The instruction

```
BCC L
```

branches to location L if the carry bit is cleared, otherwise the instruction does nothing.

A set of simple branches test any one of the condition codes, branching if the bit is set or clear. For example, `BCC L` will branch to location `L` if the carry bit is clear, while `BCS L` will branch there if the carry bit is set. Other sets test combinations of condition codes (the Z, N, and V bits) that indicate 2's complement inequalities. The last set tests combinations of the Z and C bits that indicate unsigned number inequalities. Column 2 of Table 1.6 tests each condition code separately. The `BMI` and `BPL` instructions check the sign bit and should be used after `LDAA`, `STAA`, and `TST` (or equivalent) to check the sign of a 2's complement number that was moved. The `BCC` and `BCS` instructions test the carry bit, which indicates an overflow after adding unsigned numbers, or the bit shifted out after a shift instruction. The `BVS` and `BVC` instruction set tests the V condition code, set if an overflow occurs on adding 2's complement numbers. The Z bit is also tested easily, but since we often compare two numbers to set the Z bit if the two numbers are equal, the instruction is called `BEQ` and the complementary instruction is `BNE`. `BEQ` and `BNE` are also used in the 2's complement and unsigned number branches discussed next.

A 2's complement overflow will occur if the two numbers being added have the same sign and the result has a different sign. Have you ever added two positive numbers and gotten a negative number? That's an overflow. Or if you add two negative numbers and get a positive number, that too is an overflow. But if you add two numbers of different signs, an overflow cannot occur. In using these condition codes in branch instructions, we must be careful to test the carry bit, not the overflow bit, after an unsigned binary add, since the carry bit is set if an unsigned overflow occurs; and we must remember to test the overflow bit V after a 2's complement add, because it is set if the result is erroneous as a 2's complement number. The branches listed in the middle column of Table 1.6 are used after a compare (or subtract) instruction to sense the inequalities of the two numbers being compared as 2's complement numbers. After a CMPA ALPHA instruction, if the 2's complement number in accumulator A is greater than the number at ALPHA, then the branch is taken. If the 2's complement number in accumulator A is less than or equal to the number in location ALPHA, the branch instruction does nothing.

The fourth column shows an equivalent set of branches that senses inequalities between unsigned numbers. The program segment just presented could, by putting the instruction BHI in place of BGT, compare the unsigned numbers in accumulator A against the number at location ALPHA, putting $20 in accumulator B if the register was higher than the word and otherwise putting $10 in accumulator B. These instructions test combinations of the C and Z bits and should only be used after a compare or subtract instruction to sense the inequalities of unsigned numbers. To test 2's complement numbers after a compare, use the branches in the middle column of Table 1.6; and to test the sign of numbers after a load, store, or test, use the BPL or BMI instructions.

As a memory aid in using these conditional branch instructions, remember that signed numbers are greater than or less than and unsigned numbers are higher than or lower than (SGUH). Also, when comparing any register with a word from memory, a branch like BGT branches if the register is (greater than) the memory word.

The 6812 has some combined arithmetic-logic conditional branches, including bit test and count and loop instructions. Analogous to the logic instructions BCLR and BSET are bit test instructions BRCLR (branch if clear) and BRSET (branch if set). BRCLR ALPHA,#4,L branches to location L if bit 2 of location ALPHA is clear; BRSET ALPHA,#4,L branches to L if it is set. This instruction may test several bits at once. Count and loop branches are used in "DO-loops". A DO-loop repeats a given program segment a given number, say *n*, times. DBNE decrements an accumulator or index register and branches if the value (after decrementing) is nonzero. DBEQ decrements and branches if the value is zero, IBNE increments and branches if the value is nonzero, IBEQ increments and branches if the value is zero, TBNE branches if the value is nonzero, and TBEQ branches if the value is zero.

The instruction jump to subroutine (JSR) is used to execute a program segment called a subroutine, which is located in a different part of memory, and then return to the instruction right below this instruction. It not only changes the program counter like a jump, but also saves the old value of the program counter so that when the subroutine is finished, the old value of the program counter is restored (to return to the routine right after the jump to subroutine instruction). The last instruction of the subroutine – a return from subroutine instruction – causes the program counter to be restored.

Subroutines are called using JSR, or the relative-addressed BSR instructions. The 6812 pushes the program counter on the stack and jumps to the address provided by the instruction. These subroutines end in the RTS instruction, which pops the return address to the program counter to resume the calling program. The 6812 has a memory expansion mechanism so the memory can exceed the 64K bytes allowed by a 16-bit address. See §6.2.3. An *expansion program page* register *PPAGE* holds the high-order address bits. The CALL instruction saves *PPAGE* along with the program counter and loads both *PPAGE* and the program counter with new values. Return from call, RTC, pops the saved values back into the program counter and *PPAGE*.

Compare immediate can be used to skip over a 1-byte or 2-byte instruction. The machine code $8FC603 is `CPS #50691`, but if the program counter is set to the immediate operand, the instruction that is executed is `LDAB #3`. Thus, `CPS #` can skip a 2-byte instruction. Similarly `CMPA #` can skip a 1-byte instruction. These "instructions", `SKIP1` and `SKIP2`, are used by HIWARE's C compiler.

The hardware, or I/O, interrupt is an architectural feature that is very important to I/O interfacing. Basically, it is invoked when an I/O device needs service, either to move some more data into or out of the device, or to detect an error condition. *Handling* an interrupt stops the program that is running, causes another program to be executed to service the interrupt, and then resumes the main program exactly where it left off. The program that services the interrupt (called an *interrupt handler* or *device handler*) is very much like a subroutine, and an interrupt can be thought of as an I/O device for tricking the computer into executing a subroutine. An ordinary subroutine called from an interrupt handler is called an *interrupt service routine.* However, a handler or an interrupt service routine should not disturb the current program in any way. The interrupted program should get the same result no matter when the interrupt occurs.

I/O devices may request an interrupt in any memory cycle. However, the data operator usually has bits and pieces of information scattered around and is not prepared to stop the current instruction. Therefore, interrupts are always recognized at the end of the current instruction, when all the data are organized into accumulators and other registers (the machine state) that can be safely saved and restored. The time from when an I/O device requests an interrupt until the data that it wants moved is moved, or until the error condition is reported or fixed, is called the *latency time.* Fast I/O devices require low latency interrupt service. The lowest latency that can be guaranteed is limited to the duration of the longest instruction because the I/O device could request an interrupt at the beginning of such an instruction's execution.

The condition code register, accumulators, program counter, and other registers in the controller and data operator are collectively called the *machine state* and are saved and restored whenever an interrupt occurs. Hardware interrupts and the `SWI` instructions save the machine state by pushing them onto the stack. After completion of a handler entered by any `SWI` or hardware interrupt, the last instruction executed is *return from interrupt* (`RTI`). It pulls the top 9 words from the stack, replacing them in the registers the interrupt took them from.

The instruction `WAI` saves the registers on the stack as if an interrupt occurred, and then waits for an interrupt. It can significantly reduce I/O latency. The `STOP` instruction can push the registers like `WAI` and stop the computer until either a reset signal or interrupts are received. It can be used to shut down a microcontroller to save power.

The `SWI` instruction, which pushes all the registers (except `SP`) on the stack and puts the contents of locations $fff6 and $fff7 into the program counter, is very useful for testing programs as a *breakpoint*. A breakpoint is used to stop a program that is being tested so one can execute a *monitor* program that examines or changes the contents of registers or memory words. Because it is 1 byte long, it can be put in place of any instruction. Suppose we tried to use a `JSR` instruction to jump to the monitor so we could replace a single-length instruction like `INCA`, and we also jumped to the instruction just below it from somewhere else. Since the instruction just below must be replaced by the second word of the `JMP` instruction, and since it also was jumped to from somewhere else, it would jump into the middle of the `JMP` instruction and do some damage. This is a difficult problem to resolve, especially since a breakpoint often is used in a program that doesn't work right in the first place. So a `SWI` instruction can be used without fear that some jump instruction might jump into the second word, which might happen with a longer instruction. This instruction saves all the registers automatically, thus making it easy in the `SWI` handler to analyze them by examining the top nine words on the stack. However, this marvelous trick does not work if the program is in read-only memory, because an instruction in ROM can't be replaced by the `SWI` instruction. But EEPROM can have a breakpoint temporarily inserted into it.

An illegal instruction can also be useful as a convenient subroutine call to execute I/O operations and other complex "instructions," like floating point add. Its handler's address is put in locations $FFF8 and $FFF9. Some illegal instructions can be used for special subroutine calls to a monitor, to input or output data from or to a terminal or a personal computer that the designer uses to debug his or her design.

If a subroutine is currently being executed, and the same subroutine is called from within an interrupt handler or an interrupt service routine, data from the program that was interrupted could get mixed up with data used in the subroutine called from the handler or interrupt service routine, producing errors. If this is avoided, then the subroutine is said to be *reentrant* because it can be entered again, even when it is entered and not finished. Reentrancy is important in designing software for interfaces. Related to it is *recursion* - a property whereby a subroutine can call itself as many times as it wants. While recursion is a nice abstract property and useful in working with some data structures discussed in §2-2, it is not generally useful in interfacing; however, recursive subroutines are usually reentrant, and that is important. If the subroutine is reentered the local data for the subroutine's first execution are saved on the stack as new local data are pushed on top of them, and the new data are used by the subroutine's second execution. When the first execution is resumed, it uses the old data. Keeping all local data on the stack this way simplifies implementation of reentrancy.

1.3 Assembly-Language Directives

To read the output of a compiler, and to modify a compiler's output to improve static and dynamic efficiency, you often need to use some *assembler directives*. These appear just like instructions in an assembly-language program, but they tell the assembler to do something other than create the machine code for an instruction. They often control the placement of data. (See Table 1.7 for a list of directives.) Although these directives differ a little bit among various development systems, they are sufficiently similar that a brief discussion of one such set of directives will enable you to quickly learn another set of such directives. Most of the directives we need are used to allocate space for data storage.

Assembler directives allocate storage in one way or another. To *allocate* means to find room for a variable or program in memory. The place an assembler puts data is the value of a *location counter*. The *origin* statement sets the value of the location counter, thus telling the assembler where to put the next word it generates after `ORG`.

Table 1.7. Assembly-Language Directives for the 6812

Directive	Meaning
ORG N	Sets the origin to location N, so succeeding code is written starting at location N
L: DS.B N	Allocates N words and designates L as the label of the first word of the area being allocated
L: EQU N	Declares label L to have value N
L: DC.B N1,N2	Defines (generates) one byte per argument; assigns label L to the first byte
L: DC.B 'ABC'	Defines (generates) ASCII-coded characters for each character between quotes, assigns label L to the first character
L: DC.W N1,N2	Defines (generates) two bytes per argument, assigns label L to the first byte of the first argument

```
ORG     $100
LDAA    ALPHA
```

will put the instruction code word for `LDAA` at location \$100 (when the program is loaded in memory and each succeeding word in consecutive locations) by incrementing the location counter as each byte is generated. By using the `ORG` directive to insert words further down in memory than they should be without the `ORG` directive, an area of memory can be left aside to store data.

A second directive – *define storage (bytes)* `DS.B` – can be used to allocate an area of memory. As an example,

```
L: DS.B $100
```

allocates \$100 words for some data and lets you refer to it (actually the first byte in it) using the label `L`. Recall that labels are followed by a colon (:). The value to the right of the `DS.B` mnemonic (which may be an algebraic expression) is added to the location counter. The assembler will skip over the \$100 words to put its next word \$100 words further down (at higher addresses) than it would have. This can obviously be used to allocate storage. The words in this allocated area can be accessed by using the label for the `DS.B` directive with an offset. For instance, to load the first word from this area into accumulator `A`, use `LDAA L`; to load the second word, use `LDAA L+1`; and so on. Incidentally, the number of words can be 0 in an `RMB` directive; this can be used to put a label on a line without putting an instruction on it.

A third way to allocate words of memory for data is to use the *equate* directive `EQU`. A directive like

```
ALPHA: EQU   $100
```

can be put anywhere in the program. This will tell the assembler that wherever `ALPHA` appears in the program, the number \$100 is to be substituted. `EQU` directives are useful ways to tell the assembler where variables are located and are especially useful to label

I/O registers in memory and locations in other programs to jump or branch to. In an EQU's expression, the asterisk (*) is often used to indicate the current location counter.

The ORG, DS.B, and EQU directives determine where areas of data are to be put but do not fill those areas with initial values. The following directives not only provide room for variables but also *initialize* them with constants when the program is loaded.

The *define constant (byte)* directive DC.B will put a byte in memory for each operand of the directive. The value of an operand is put into memory when the location counter specifies the address and the location counter is incremented for each operand. DC.B 10 will put $0A in a word in memory. The directive

```
L: DC.B 1,2,3
```

will initialize 3 bytes in memory to be

01
02
03

and will tell the assembler that L is the symbolic address of the first word, whose initial value is $01. The location counter is incremented three times. ASCII characters can be inserted as DC.B arguments by putting them between matching quotes. *Define constant (word)* DC.W will initialize two consecutive 16-bit words for each argument. The value of each operand is put in two consecutive words and the location counter is incremented by two for each operand. For example, the directive

```
L: DC.W 1,2,3
```

will initialize six consecutive bytes in memory, as follows

00
01
00
02
00
03

and will tell the assembler that L is the address of the first word in this area, whose value is $00. The location counter is incremented six times. The DC.W directive is especially useful in putting addresses in memory so that they can be used in indirect addressing or picked up into an index register. If ALPHA is $100 because an EQU directive set it to that value or because it is a label of an instruction or directive like DC.B that begins at location $100, then the directive

```
DC.W ALPHA
```

will generate the following 2 bytes in memory:

01
00

1.4 Organization of 6812 Microcontrollers

In this section, we describe the block diagram of the 'A4's and 'B32's hardware – specific implementations of the 6812 architecture – considering its general implementation and its particular input/output hardware. Because we will present hardware descriptions in a style similar to the block diagrams programmers commonly see, we will also call our descriptions *block diagrams.* After then discussing the memory and I/O organization, we introduce the memory map, which explains the location of memory and I/O devices so programmers can access them.

1.4.1 Notation for Block Diagrams

A *block diagram* is used to describe hardware organization from the programmer's point of view (see §1.1.1). It is especially useful for showing how ICs work so that a programmer can focus on the main ideas without being distracted by details unrelated to the software. In this memory-mapped I/O architecture, a register is a location in memory that can be read or written as if it were a word in memory. A block diagram shows modules and registers as rectangles, with the most important inputs and outputs shown around the perimeter. Also, the effects of software instructions can be illustrated nicely on a block diagram; for instance, if the `LDX $4000` instruction reads a 16-bit word from a certain register or module, this can be shown, as in Figure 1.3. The `LDX` instruction could be replaced by `LDY`, `LDD`, or any instruction that can read 16 bits of data in two consecutive bytes. The instruction and the arrow away from the module show it can be read (a readable register). If an instruction like `STX $4000` appears there and an arrow is shown into the module, it can be written (a writable register). And if both are shown (`LDX/STX` and a double arrow), the register is read-write. If an instruction like `LDAA` appears by the line to a register, an 8-bit word can be read, and if `STAA` appears there, then an 8-bit word can be written in the register. Finally, a range of addresses can be shown as in `LDAA/STAA` $8000/$8003; this means the module can read-write for addresses $8000, $8001, $8002, $8003.

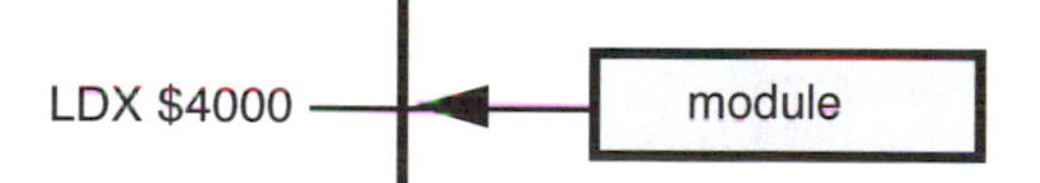

Figure 1.3. Block Diagram Showing the Effect of an Instruction

1.4.2 6812 Microcontroller I/O and Memory Organization

As §3.2 will clarify, most microcomputers, as von Neumann computers, are organized around memory and a bus between memory and the controller and data operator. This can be explained by a block diagram like Figure 1.4, without showing instructions and addresses as in Figure 1.3. The controller sends an address to memory on an address bus

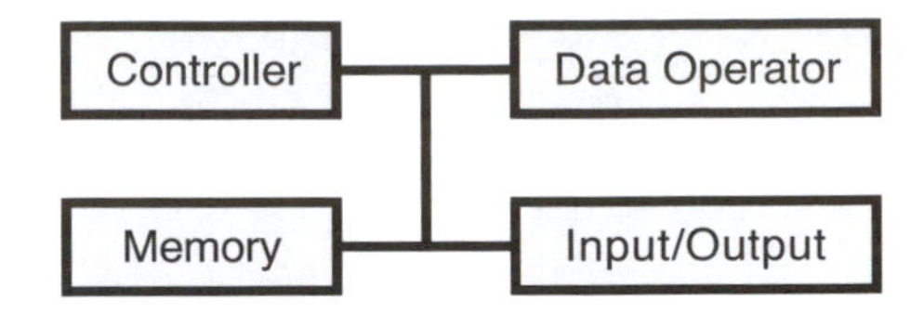

Figure 1.4. Organization of a von Neumann Computer

and a command to read or write. If the command is to write, the data to be written is sent on the data bus. If the command is to read as when the processor fetches an op code, the controller waits for memory to supply a word on the data bus and then uses it; and if the command is to read as when the processor recalls a data word, the data operator puts the word from the bus into some register in it. Memory-mapped I/O uses a "trick": it looks like memory so the processor writes in it or reads from it just as if it reads or writes memory words.

The 'A4 can operate in the single-chip mode or the expanded bus mode. In the *single-chip mode*, the 'A4 can be the only chip in a system, for it is self-sufficient. The processor, memory, controller, and I/O are all in the chip. (See Figure 1.5.) The controller and data operator execute the 6812 instruction set discussed earlier. The memory consists of 1K words of RAM and 4K words of EEPROM. The I/O devices include a dozen parallel I/O registers (described in §4.2.1) a serial peripheral interface (SPI) (described in §4.4.4), a serial communication interface (SCI), described in §9.3.5, a timer (described in §8.1), and an A/D converter (described in §7.5.3).

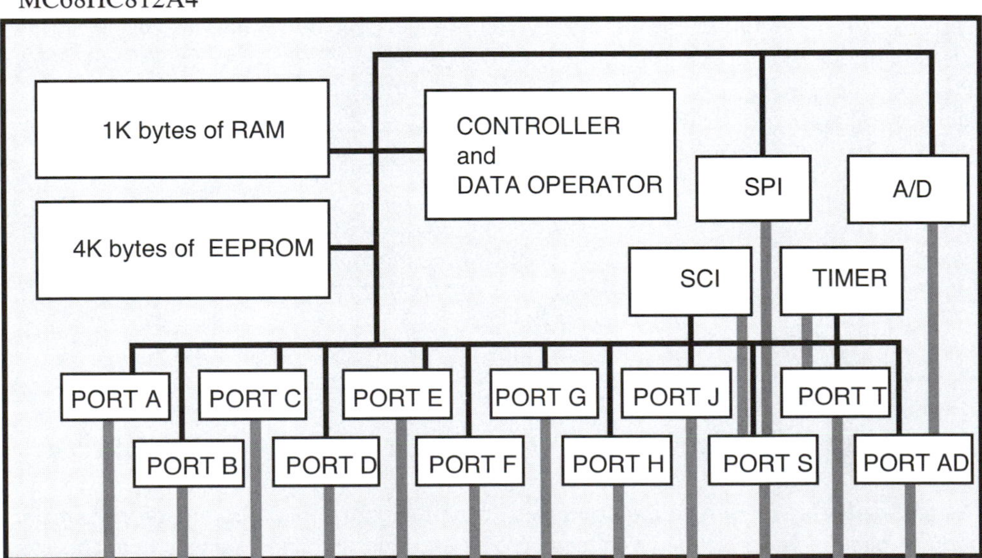

Figure 1.5. Single-Chip Mode of the MC68HC812A4

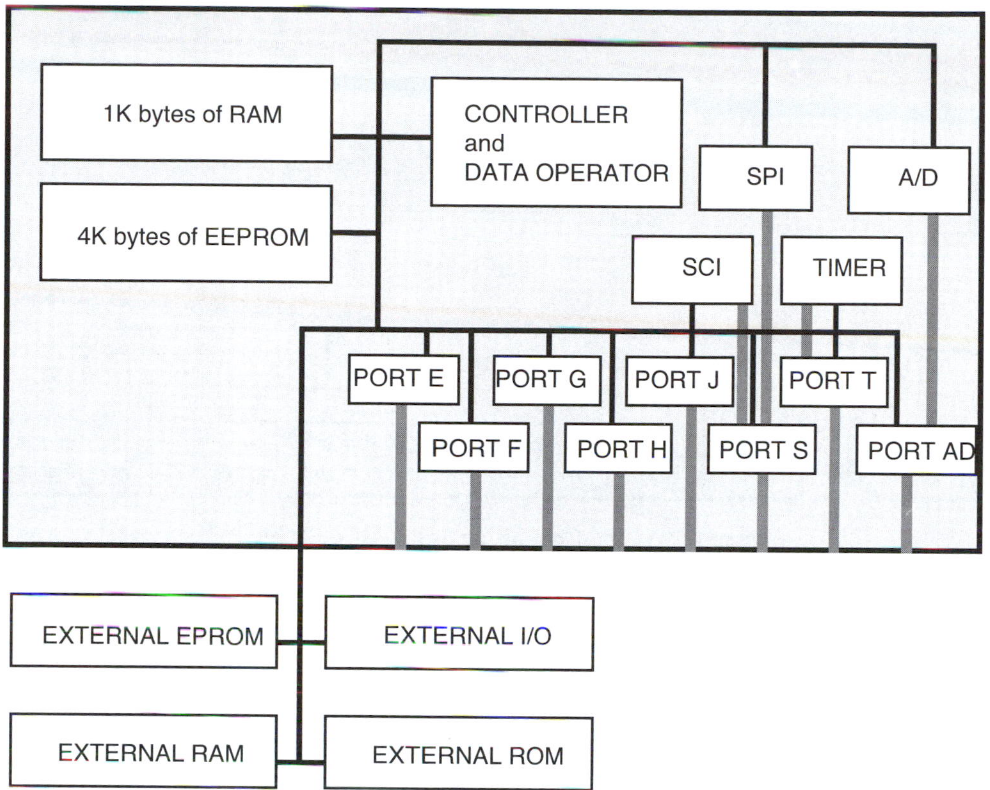

Figure 1.6. Wide Expanded Bus Mode of the MC68HC812A4

The *expanded bus mode* of the 'A4 removes three or four of the parallel ports, using their pins to send the address and data buses to other chips. RAM, ROM, and PROM can be added to this expanded bus. In a narrow expanded mode, ports A and B are removed for address lines, and port C is an 8-bit data bus. Port D is available for parallel I/O. In a wide expanded mode (see Figure 1.6), ports A and B are removed and their pins are used for address lines, and ports C and D are a 16-bit data bus. Port D is unavailable for parallel I/O. In both modes, ports E, F, and G can be used for bus control, chip selects, and memory expansion signals, or else for parallel I/O.

The 'B32 also operates in single-chip mode or expanded bus mode, but in the latter mode, address and data are time multiplexed on the same pins. In the single-chip mode, the 'B32 can be the only chip in a system. The processor, memory, controller, and I/O are all in the chip. (See Figure 1.7.) The controller and data operator execute the 6812 instruction set discussed earlier. The memory consists of 1K words of RAM, 768 bytes of EEPROM, and 32K words of *flash memory,* which is like EEPROM. The I/O devices include eight parallel I/O registers, a serial peripheral interface (SPI), a serial communication interface (SCI), a timer, a *pulse-width modulator* (PWM), a *byte data link communication* module (BDLC), and an A/D converter.

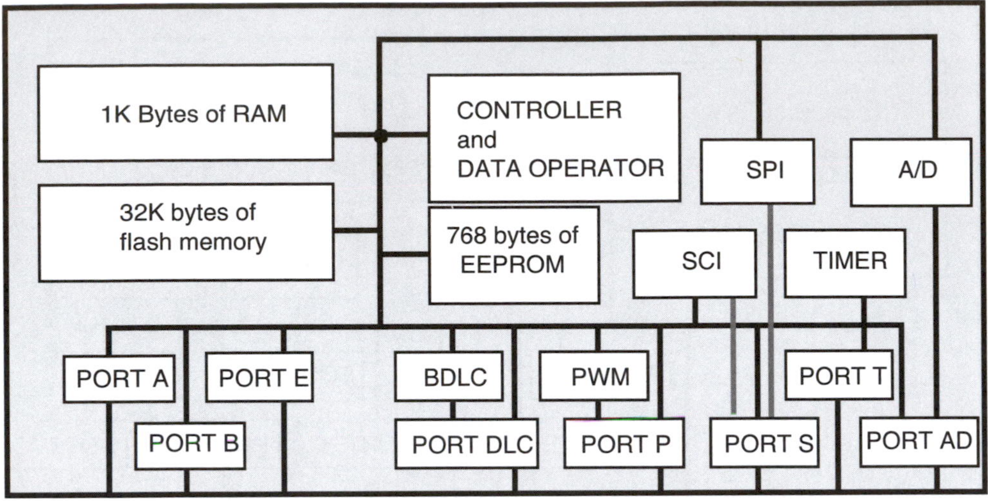

Figure 1.7. Single-Chip Mode of the MC68HC912B32

MC68HC912B32

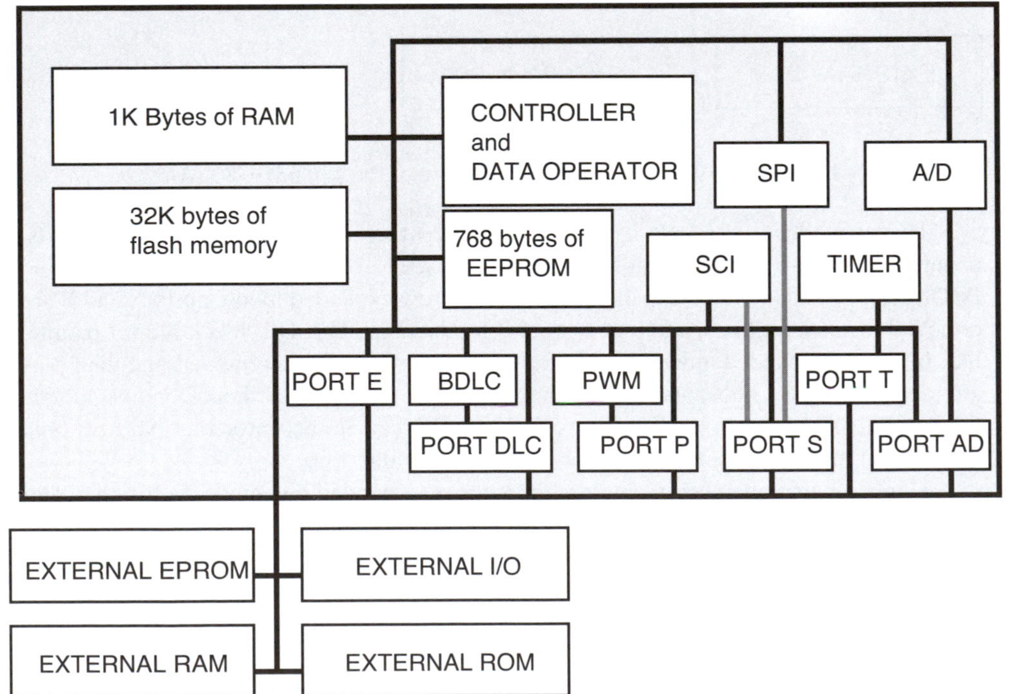

Figure 1.8. Wide Expanded Multiplexed Bus Mode of the MC68HC912B32

The expanded bus mode of the 'B32 removes two of the parallel ports, ports A and B, using their pins to send the time-multiplexed address and data buses to other chips. The address and data buses are time-multiplexed; in the first part of each memory cycle, the 16-bit address is output on the pins, and in the second part, data is output or input on the indicated pins. In a narrow expanded mode, port A is used for an 8-bit data bus. In a wide expanded mode (see Figure 1.8), ports A and B pins are used for a 16-bit data bus. In both modes, port E can be used for bus control, or else for parallel I/O. RAM, ROM, and PROM can be added to the expanded bus.

A significant advantage of the 'A4 and 'B32 is that each can be used in the single-chip or in either the narrow or wide expanded multiplexed bus mode. The former mode is obviously useful when the resources within the microcontroller are enough for the application – that is, when there is enough memory and I/O devices for the application. The latter mode is required when more memory is needed, when a program is in an EPROM and has to be used with the 'A4 or 'B32, or when more or different I/O devices are needed than are in the 'A4 or 'B32. We are excited about using it to teach interfacing, because it can show single-chip computer interfacing concepts as well as those of conventional multiple-chip system interfacing.

1.4.3 The MC68HC812A4 and MC68HC912B32 Memory Maps

A *memory map* is a description of the memory showing what range of addresses is used to access each part of memory or each I/O device. Figure 1.9 presents memory maps for the 'A4 and 'B32.

Actually, EEPROM, RAM, flash memory, and I/O may be put anywhere in memory (on a 2K or 4K boundary), but we will use them in the locations shown in Figure 1.9 throughout this text. I/O is at the lowest address to take advantage of page 0 addressing, and RAM is at $800 to $bff. In the 'A4, the EEPROM is at $F000; and in the 'B32, the flash memory at $8000 may have a monitor. Usually your data are put in

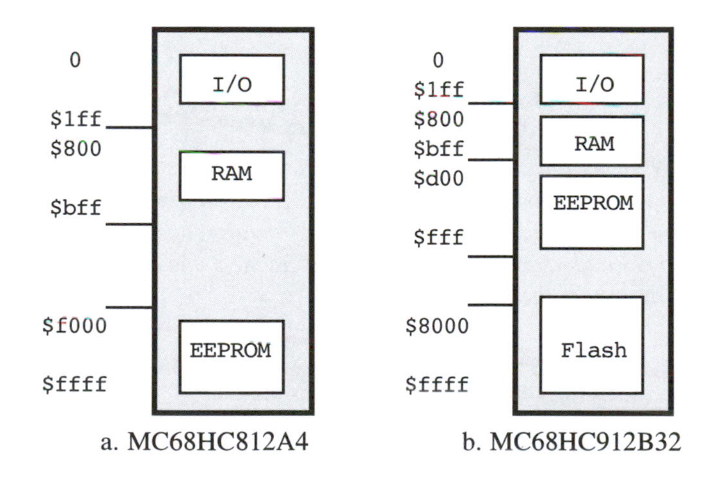

Figure 1.9. Memory Maps of 6812 Microcontrollers

RAM and your program may be put in RAM, EEPROM, or flash memory. If you are using a chip with a monitor, it uses some bytes of RAM for its data, and some bytes of EEPROM or flash memory for its program and interrupt vectors, so you can use the remaining low bytes of RAM for your program or data and low bytes of EEPROM or flash memory for your program.

Figure 1.9a presents a memory map for the 'A4. I/O is at the lowest address to take advantage of page 0 addressing, and RAM is at $800 to $bff. A 4K EEPROM at $f000 to $ffff often has a monitor. Usually your data is put in RAM and your program may be put in RAM or in EEPROM memory. Figure 1.9b presents a memory map for the 'B32. I/O is at the lowest address to take advantage of page 0 addressing, and RAM is at $800 to $bff. A small EEPROM is at $d00 to $fff. Flash memory at $8000 to $ffff has a monitor. Usually your data is put in RAM and your program may be put in RAM, in the small EEPROM, or in flash memory.

1.5 Conclusions

In this chapter, we have surveyed the background in architecture needed for microcomputer interfacing. The first section covered bare essentials about von Neumann computers, instructions and what they do, and microcomputers. You will find this background helpful as you begin to learn precisely what happens in an interface.

The middle section covered addressing modes and instructions that you may expect in any microcomputer, discussing those in the 6812 in more detail. The general comments there should help if you want to learn about another machine. And the 6812 comments should help you read the examples and do some of the experiments suggested in this book. A short section provides additional information needed for reading and modifying assembly language programs that are generated by a compiler.

The final section described some of the hardware used in interfacing, as seen from the programmer's perspective. You need to know this to write interfacing programs, described in more detail later.

Do You Know These Terms?

Following is a list of all italicized words in this chapter. You should check these terms to be sure that you recognize their meaning before going on to the next chapter. These terms also appear in the index, with page numbers for reference, so you can look up those that you do not understand. Here, they appear in the same order, down each column, as they appear in the text.

von Neumann computer
architecture
organization
implementation
realization
compiler
high-level language
load
recall
symbolic address
store
index addressing
index register
stack pointer
offset
double-indexed addressing
logic
edit
memory-mapped I/O
no-operation
position independence

primary memory
data operator
controller
input/output
access
bit
byte
word
random access memory (RAM)
static ram (SRAM)
dynamic ram (DRAM)
read-only memory (ROM)
programmable read-only memory (PROM)
erasable programmable read-only memory (EPROM)
electrically erasable programmable read-only memory (EEPROM)
flash memory
fetch
program counter
binary code
hexadecimal notation
machine coded
mnemonic
upward compatible
source code upward compatible
assembler
assembly language
memorize
bug
program sequence
jump
control instruction
addressing modes
opcode
microinstruction
control memory
data transfer order
data transfer
microprogramming
fetch execute cycle
memory cycle
memory clock
fetch cycle
decode cycle
address calculation
recall cycle
execute cycle
memorize cycle
macro
benchmark
static efficiency
dynamic efficiency
microprocessor
microcomputer
single-chip microcomputer
personal computer
microcontroller
direct addressing
displacement
implied addressing
register addressing accumulator
immediate addressing
page zero addressing
page
pointer addressing
pointer registers
autoincrement addressing
autodecrement addressing
stack buffer
stack
push
pull
stack overflow
stack underflow
jump to subroutine
return from subroutine
balanced stack
nesting of subroutines
local data
argument
indirect address
page relative addressing
branch
move
zero bit
negative bit
carry bit
overflow bit
half carry bit
interrupt inhibit bit
interrupt mask bit
stop disable bit
arithmetic
add with carry
subtract with carry
compare
sixteen's complement
conditional branch
subroutine
expansion program page (PPAGE)
hardware interrupt
I/O interrupt
handling an interrupt
interrupt handler
device handler
interrupt service routine
latency time
machine state
return from interrupt
breakpoint
monitor
reentrant
recursion
assembler directives
allocate
location counter
origin
define constant
define storage
equate
initialize
block diagram
single-chip mode
expanded bus mode
flash memory
pulse-width modulator (PWM)
byte data link communication (BDLC)
memory map

Problems

Problems 1 through 3 in this chapter and many problems in later chapters are paragraph correction problems. We use the following guidelines for all these problems.

These paragraph correction problems have been proven useful in helping students understand concepts and definitions. The paragraph in each problem has some correct and some erroneous sentences. Your task is to rewrite the paragraph so the whole paragraph is correct, deleting any sentences that do not fit into the paragraph's theme. However, if a sentence is correct, you should not change it, and you cannot use the word "not" or its equivalent to correct the sentence. Consider the first sentence in problem 1: "The architecture is the block diagram of a computer." This is incorrect. It can be made correct by changing "architecture" to "organization," or by changing "block diagram" to either "programmer's view" or "instruction set and I/O connection capabilities." Any of these corrections would be acceptable. The second sentence is correct, however, and should not be rewritten. Try to complete the problems without referring to the chapter, then check your answers by looking up the definitions. If you get a couple of sentences wrong, you're doing fine. But if you have more trouble, you should reread the sections the problem covers.

1. * The architecture is the block diagram of a computer. Von Neumann invented the architecture used on microcomputers. In it, the controller is analogous to the adding machine. We recall words from primary memory into the controller using the program counter (left hand). Symbolic addresses are used in assembly languages to represent locations in this memory. A macro is a program in another part of memory that is called by a program, so that when the macro is done, the calling program resumes execution at an instruction below the jump to macro. An I/O interrupt is like a subroutine that is requested by an I/O device. The latency time is the time needed to completely execute an interrupt. To optimize the speed of execution, choose a computer with good static efficiency. A microcomputer is a controller and data operator on a single LSI chip, or on a few LSI chips.

2. * Addressing modes are especially important because they affect the efficiency of the most common class of instructions, the arithmetic class. Direct addressing has the operand data in a part of the instruction called the displacement, and the displacement would be 8 bits long for an instruction using it to load an 8-bit accumulator. Indirect addressing allows programs to be position independent. The 6812 has direct page addressing, which is a "quick-and-dirty" index-addressing mode. Index addressing is especially useful for jumping to nearby locations. If we want to move data around in memory during execution of a program, indirect addressing is the only mechanism that can efficiently access single words as well as arrays.

3. * The 6812 has 96 bits of register storage, where the D accumulator is really the same as the X index register. The X register serves as an additional stack pointer, and instructions to push or pull can use X. Add with carry is used in multiple precision arithmetic. It can add into accumulator D. The 6812 has an instruction to divide one unsigned number into another unsigned number. The DAA instruction can be used after an INCA instruction to increment a decimal number in accumulator A. BGT, BLT, BGE, and BLE can be used after comparing 2's complement numbers. The SWI instruction is particularly useful as a subroutine call to a fast, short subroutine because it is a fast instruction.

4. Identify which applications would be concerned about storage density and which about speed. Give reasons for your decisions.

a. Pinball machine game
b. Microwave oven control
c. Home security monitor
d. Fast Fourier transform (FFT) module for a large computer
e. Satellite communications controller

5. Write the (hexadecimal) code for a `BRA L` instruction, where the instruction code is at locations $12A and $12B and

a. L is location $12F.
b. L is location $190.
c. L is location $12A.
d. L is location $120.
e. L is location $103.

6. Write the op code for the following instructions, assuming the first byte of the op code is at $208A and L is at $2095, X is $1000, and Y is $8000; and explain in words what happens when it is executed (including the effects on the condition codes).

a. `BSET $52,#$12`
b. `BCLR $34,X,#5`
c. `BSET 3,Y,#$7F`
d. `BRCLR $EE,#6,L`
e. `BRSET 3,Y,#8,L`

7. Write the op code for the following instructions, assuming that A is 5, B is 0x80, X is $1000, and Y is $8000; write the effective address and write the final values of X and Y.

a. `LDAA 2,X`
b. `STD 8,-Y`
c. `INC A,X`
d. `JMP [D,X]` (assuming the 16-bit word at $1580 is $1234)
e. `ROL [$12,Y]` (assuming the 16-bit word at $8012 is $1234)

8. Suppose a memory is filled, except for the program that follows, like this: the word at address $WXYZ is $YZ (for example, location $2538 has value $38). Assuming that an address in X never points to the program, what will the value of X be after each instruction is executed in this program?

```
LDX #$1
LDX 0,X
LDX 4,X
```

9. Suppose the condition code register is clear and the ADDA APLHA instruction is executed. Give the value in the condition code register if

a. Accumulator A is $77, ALPHA is $77.
b. Accumulator A is $C8, ALPHA is $77.
c. Accumulator A is $8C, ALPHA is $C8.
d. Repeat part c for SUBA ALPHA.

10. Explain under what conditions the H, N, Z, V, and C bits in the condition code register are set. Also explain the difference between overflow and carry.

11. If the accumulator A contains the value $59 and ALPHA contains the value $6C, what will be the value of the condition code register after the following instructions? (Assume the condition code register is clear before each instruction.)

a. ADDA ALPHA
b. SUBA ALPHA
c. TSTA
d. COMA
e. BITA ALPHA
f. EORA ALPHA

12. Repeat problem 11, assuming that the accumulator A contains the value $C9, ALPHA contains the value $59, and the condition code register is set to the value $FF before each instruction.

13. Give the shortest 6812 instruction sequences that perform the same operation as the following nonexistent 16-bit 6812 instructions. State whether the condition codes are set properly or not.

a. ASRD (shift D right arithmetically)
b. INCD ALPHA (increment 16-bit ALPHA)
c. NEGD (negate accumulator D)
d. DECD ALPHA (decrement 16-bit ALPHA)
e. MULS (multiply signed 8-bit A times 8-bit B to get 16-bit result in D)

14. Give the shortest 6812 instruction sequences to implement 32-bit arithmetic operations for each case given below. In each case, the data arrive in register Y (high 16 bits) and register D (low 16 bits), and are returned in the same way.

a. Complement b. Increment c. Decrement d. Shift right logical
e. Shift right arithmetic f. Shift left (arithmetic or logical)

15. Explain, in terms of condition code bits, when the branch is taken for the following conditional branch instructions:

a. BEQ
b. BGT
c. BHI
d. BHS
e. BLE
f. BPL

16. How many times does the following loop get repeated when the instruction CND is

a. BNE?
b. BPL?
c. BLT?

```
        LDAA            #200
LOOP:   statement list
        DECA
        CND             LOOP
```

Show calculations or explain your answers.

17. What is the value of accumulator B after the following program ends, when the instruction COND is

a. BEQ?
b. BMI?
c. BGT?
d. BVS?

```
        LDAA    #200
        CLRB
LOOP:   DECA
        COND    EXIT
        INCB
        BRA     LOOP
EXIT:   SWI
```

Show calculations or explain answers.

18. Convert the following high-level programming language construct into the shortest 6812 assembly-language instructions, assuming that the variable A is already assigned to the accumulator A. As long as the expression in the while statement is true, the statements inside braces are repeated.

```
A = 10;
while (A > 3)
{ statements;
 A = A -1; }
```

19. Repeat problem 18 with the following high-level programming language construct.

```
A = 10;
do
{ statements;
 A = A - 1; }
while ( A > 3)
```

20. `BETA` initially contains the value $9C. What is the value of BETA after the following instructions?

a. `BCLR BETA $15`
b. `BSET BETA $38`

21. `BETA` initially contains the value $9C. Will a branch occur after these instructions?

a. `BRCLR BETA,#$64,HERE`
b. `BRSET BETA,#$64,HERE`
c. `BRCLR BETA,#$61,HERE`
d. `BRSET BETA,#$84,HERE`

22. What are the hexadecimal values of registers `D` and `X` after the `FDIV` instruction if they contain these values, respectively?

a. $4000, $8000 b. 0.5, 0.75

23. Repeat problem 22 with `IDIV` instruction and

a. `D` = $0064, `X` = $0002. b. `D` = $0064, `X` = $0003.

24. What are the values of registers `A` and `B` after the `MUL` instruction if they contain these values, respectively?

a. $80, $80 b. $8C, $45

25. Write a shortest assembly-language program that adds five 8-bit unsigned numbers stored in consecutive locations, the first of which is at location $802, and put the 8-bit sum in $810. If any unsigned number addition errors occur, branch to ERROR.

26. Repeat problem 25 where the numbers are signed.

27. Repeat problem 25 where the numbers are 16-bit unsigned numbers.

28. Repeat problem 25 where the numbers are 16-bit signed numbers.

29. Write a shortest assembly-language program that adds a 5-byte unsigned number stored in consecutive locations, the first of which is at location $802, to a five-byte unsigned number stored in consecutive locations, the first of which is at location $812. If any unsigned number addition errors occur, branch to ERROR.

30. Give the shortest 6812 subroutines to implement 32-bit arithmetic operations for each case given below. In each case, one operand arrives in register `Y` (high 16 bits) and register `D` (low 16 bits), and the result is returned in the same way; and the other operand is pushed on the stack, high-order byte at lowest address, just before the subroutine is entered, and is removed by the subroutine just before it returns. Compare returns a value `SNGD` (bits 7 to 4) and `USGD` (bits 3 to 0), where `SNGD` and `USGD` have value 0 if less, 1 if equal, and 2 if greater, for signed and for unsigned comparisons of two arguments.

a. Add b. Subtract (stack value from register) c. Multiply d. Compare

The MC68HC912B32 die.

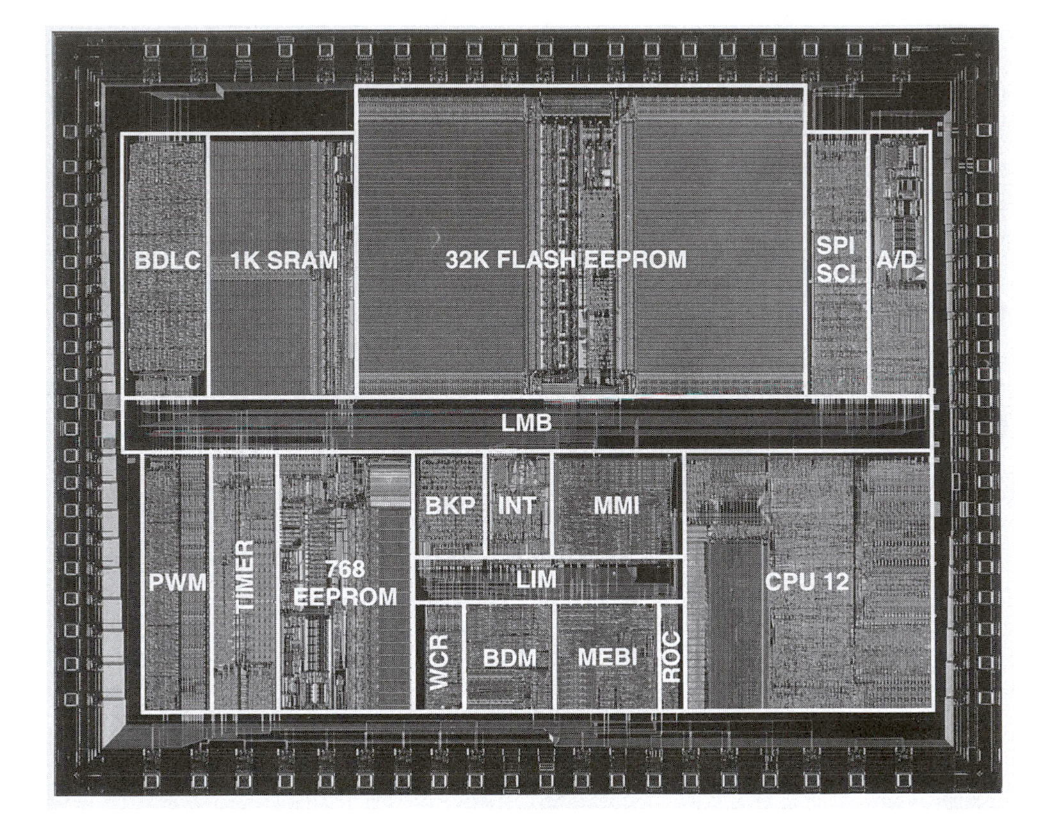

2

Programming Microcomputers

We now consider programming techniques used in I/O interfacing. The interface designer must know a lot about them. As the industry matures, the problems of matching voltage levels and timing requirements, discussed in §3.2, are being solved by better-designed chips, but the chips are getting more complex, requiring interface designers to write more software to control them.

The state-of-the-art 6812 clearly illustrates the need for programming I/O devices in a high-level language as well as for programming them in object-oriented languages. The dozen I/O ports and its plethora of SPI, SCI, A-to-D, and timer ports may be a challenge to many assembler language programmers. But the 4K-byte EEPROM memory is large enough to support high-level language programs. Also, object-oriented features like modularity, information hiding, and inheritance will further simplify the task of controlling 6812 systems.

This book develops C and C++ interfacing techniques. Chapter 1, describing the architecture of a microcomputer, has served well to introduce assembler language, although a bit more will be done in this chapter. We introduce C in this chapter. The simplest C programming constructs are introduced in the first section. The handling of data structures is briefly covered in the next section. Programming styles, including the writing of structured, modular, and object-oriented programming, will be introduced in the last section. Subroutines will be further studied as an introduction to programming style. The use of classes in C++ will be introduced at the end of this chapter. While this introduction is very elementary and rather incomplete, it is adequate for the discussion of interfacing in this text. Clearly, these concepts must be well understood before we discuss and design those interfaces.

For this chapter, the reader should have programmed in some high-level language. From it, he or she should learn general fundamentals of programming in C or C++ to become capable of writing and debugging tens of statements with little difficulty, and should learn practices specifically applicable to the 6812 microprocessor. If you have covered this material in other courses or absorbed it from experience this chapter should bring it all together. You may pick up the material just by reading this condensed version. Others should get an idea of the amount of background needed to read the rest of the book.

2.1 Introduction to C

I/O interfacing has long been done in assembly language. However, experience has shown that the average programmer can write something like ten lines of (debugged and documented) code per day, whether the language is assembler or higher level. But a line of high-level language code produces about 6 to 20 useful lines of assembly-language code, so if the program is written in a high-level language, we might become six to 20 times more efficient. We can write the program in a high-level language like C or C++. However, assembly-language code produced by a high-level language is significantly less statically and dynamically efficient, and somewhat less precise, than the best code produced by writing in assembly language, because it generates unnecessary code. Thus in smaller microcontrollers using the 6811, after a high-level language program is written it may be converted first to assembly language, where it is tidied up; then the assembly-language program is assembled into machine code. As a bonus, the original high-level language can be used to provide comments to the assembly-language program. In microcontrollers using processors designed for efficient high-level language programming, such as the 6812, C or C++ can control the device without being converted to assembly language. This has the advantage of being easier to maintain because changes in a C program do not have to be manually translated into and optimized in assembly language. Or, a small amount of assembly language, the part actually accessing the I/O device, can be embedded in a larger C program. This approach is generally easier to maintain because most of the program is implemented in C, and yet is efficient and precise in the small sections where the I/O device is accessed.

We will explain the basic form of a C *procedure,* the simple and the special numeric operators, conditional expression operators, and conditional and loop statements and functions. However, we do not intend to give all the rules of C that you need to write good programs. A C program consists of one or more procedures, of which the first to be executed is called `main`, and the others "subroutines" or "functions" if they return a value. All the procedures, including `main`, are written as follows:

```
declaration of global variable;
declaration of global variable;
      .
procedure_name(parameter_1, parameter_2,...)
      .
{
    declaration of local variable;
    declaration of local variable;
        ...

    statement;
    statement;
        ...

}
```

Table 2.1. Conventional C Operators Used in Expressions

=	make the left side equal to the expression on its right
+	add
-	subtract
*	multiply
/	divide
%	modulus (remainder after division)
&	logical bit-by-bit AND
|	logical bit-by-bit OR
~	logical bit-by-bit negation
<<	shift left
>>	shift right

Each *declaration of a parameter or a variable* and each statement ends in a semicolon (;), and more than one of these can be put on the same line. Carriage returns and spaces (except in names and numbers) are not significant in C and can be used to improve readability. The ellipsis points (. . .) in the example do not appear in C programs, but are meant here to denote that one or more declaration or statement may appear.

Parameters and variables used in the 6812 are usually 8-bit (`char`), 16-bit (`int`), or 32-bit (`long`) signed integer types. They can be declared unsigned by putting the word `unsigned` in front of `char`, `int`, or `long`. More than one variable can be put in a declaration; the variables are separated by commas (,). A vector having *n* elements is denoted by the name and square brackets around the number of elements *n*, and the elements are numbered 0 to *n* - 1. For example, `int a,b[10];` shows two variables, a scalar variable `a` and a vector `b` with ten elements. Variables declared outside the procedure – e.g., before the line with `procedure-name` – are global, and those declared within a procedure – e.g., between the braces "{" and "}" after `procedure-name` – are local. Parameters are discussed in §2.3.1. A *cast* redefines a value's type. A cast is put in parentheses before the value. If `i` is an `int`, `(char)i` is a `char`.

Statements may be algebraic expressions that generate assembly-language instructions to execute the procedure's activities. A *statement* may be replaced by a sequence of statements within a pair of curly braces "{" and "}". This will be useful in conditional and loop statements, which are discussed soon. Operators used in statements include addition, subtraction, multiplication, and division, as well as a number of very useful operators that convert efficiently to assembly-language instructions or program segments. Table 2.1 shows the conventional C operators that we will use in this book. Although they are not all necessary, we use a lot of parentheses so that we will not have to learn the precedence rules of C grammar. The following simple procedure `main` has (signed) 16-bit local variables `a` and `b`, and 8 bit global `c` and vector `d`, it puts 5 into `a`, 1 into `b`, and then the `(a+b)` th element of vector `d` into 8-bit unsigned global `c`, and returns nothing (`void`) as is shown to the left of the procedure name.

```
unsigned char c,d[10];
void main() {  int a, b;
    a =5; b=1; c = d[a+b];
}
```

We use the HIWARE C++ compiler to generate code for the 6812. The compiled and linked program is disassembled using the DECODE program. We will show disassembled machine code produced by the DECODE program from high-level procedures compiled by it. We hasten to note, however, that different code is produced when different optimization options are selected, or when a different version of the compiler is used. The assembly language, as shown below, is typical, however, of what the compiler produces.

```
    4: void main() { int a,b;
0000086A 1B9C          LEAS  -4,SP
    6:       a = 5; b = 1; c = d[a+b];
0000086C C60A          LDAB  #10
0000086E 87            CLRA
0000086F 6C82          STD   2,SP
00000871 C601          LDAB  #1
00000873 6C80          STD   0,SP
00000875 C605          LDAB  #5
00000877 E380          ADDD  0,SP
00000879 B745          TFR   D,X
0000087B E6E209E1      LDAB  2529,X
0000087F 7B09E0        STAB  $09E0
    7: }
00000882 1B84          LEAS  4,SP
00000884 3D            RTS
```

After the procedure `main` is called, the procedure's first instruction, `leas -4,sp`, makes room for *(allocates)* local variables on the stack; the last two instructions `leas 4,sp` and `rts`, remove room for *(deallocates)* local variables on the stack and return to the calling routine. Note that the first `leas` instruction subtracts 4 from the stack pointer SP. Parameters and local variables will be obtained by index addressing with SP. Global variables are obtained using direct addressing. This procedure first assigns or initializes 16-bit local variables `a` and `b` and then it just uses these to read an element from the vector `d` into global variable `c` in a manner discussed in the next section.

Some very powerful special operators are available in C. Table 2.2 shows the ones we use in this book. For each operator, an example is given together with its equivalent result, using the simple operators of Table 2.2. The assignment operator `=` assigns the value on its right to the variable named on its left and returns the value it assigns so that value can be used in an expression to the left of the assignment operation: the example shows that 0 is assigned to `c` and value (0) is assigned to `b`; then that value is assigned to `a`. The increment operator `++` can be used without an assignment operator (e.g., `a++` just increments `a`). It can also be used in an expression in which it increments its operand after the former operand value is returned to be used in the expression. For example, `b=a[i++]` will use the old value of `i` as an index to put `a[i]` into `b`, then it will increment `i`. Similarly, the decrement operator `--` can be used in expressions. If the `++` or `--` appear in front of the variable, then the value returned by the expression is the updated value; `a[++i]` will first increment `i`, then use the incremented value as an index into *a*. The next row shows the use of the + and = operators used together to represent adding to a variable. The following rows show -, |, and & appended in front of = to represent subtracting from, ORing to, or ANDing to a variable. Shift << and >> can be used in front of the = sign too. This form of a statement avoids the need to twice write the name of, and twice compute addresses for, the variable being added to or subtracted from. The last two rows of Table 2.2 show shift left and shift right operations and their equivalents in terms of multiplication or division by powers of 2. However, rather than allowing the use of a slower machine instruction, they force the use of the faster logical shift instructions.

A statement involving several operations saves intermediate values on the stack. The statement `i = (i << 3) + (i << 1) + c - '0';` where `i` and `c` are local variables declared as `unsigned char c; int i;` is compiled into the following assembly language:

```
00000867 EC80       LDD   0,SP
00000869 3B         PSHD
0000086A 59         ASLD
0000086B 59         ASLD
0000086C 59         ASLD
0000086D B745       TFR   D,X
0000086F 3A         PULD
00000870 59         ASLD
00000871 1AE6       LEAX  D,X
00000873 E682       LDAB  2,SP
00000875 87         CLRA
00000876 1AE6       LEAX  D,X
00000878 1AE1D0     LEAX  65488,X
0000087B 6EB2       STX   3,SP+
```

Temporary results are transferred to or exchanged with other registers, or pushed on the stack. The `LEAX` instruction is used in lieu of the `ADDD` instruction to add temporary results in an index register. The stack pointer offset, to access a local variable, changes as temporary results are saved on the stack.

Table 2.2. Special C Operators

Operator	Example	Equivalent to
=	*a=b=c=0;*	*a=0;b=0;c=0;*
++	*a ++;*	*a=a+1;*
- -	*a --;*	*a=a-1;*
+=	*a += 2;*	*a=a+2;*
- =	*a -= 2;*	*a=a-2;*
|=	*a |= 2;*	*a=a|2;*
&=	*a &= 2;*	*a=a&2;*
<<=	*a <<= 3*	*a=a<<3*
>>=	*a >>= 3*	*a=a>>3;*

Table 2.3.
Conditional Expression Operators

&&	AND
||	OR
!	NOT
>	Greater than
<	Less than
> =	Greater than or equal
< =	Less than or equal
= =	Equal to
! =	Not equal to

A statement can be conditional, or it can involve looping to execute a sequence of statements that are written within it many times. We will discuss these control flow statements by giving the flow charts for them. See Figure 2.1 for conditional statements, Figure 2.2 for case statements, and Figure 2.3 for loop statements. These simple standard forms appear throughout the book, and we will refer to them and their figures.

Simple conditional expressions of the form *if then* (shown in Figure 2.1a), full conditionals of the form *if then else* (shown in Figure 2.1b), and extended conditionals of the form *if then else if then else if then . . . else* (shown in Figure 2.1c), use conditional expression operators (shown in Table 2.3). In the last expression, the *else if* part can be repeated as many times as needed, and the last part can be an optional *else*. Variables are compared using *relational operators* (> and <), and these are combined using *logical operators* (`&&`). For example, `(a > 5) && (b < 7)` is true if `a > 5` and `b < 7`.

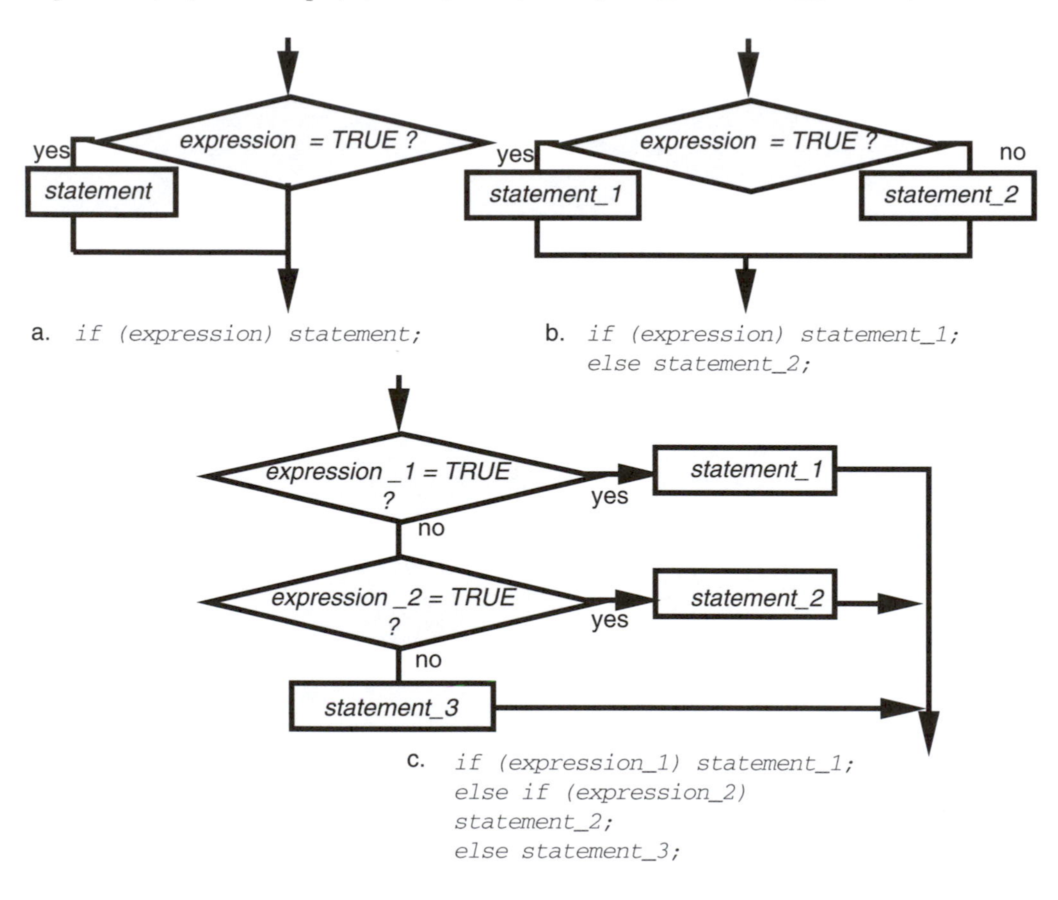

Figure 2.1. Conditional Statements

Consider a decision tree using conditional expressions, like `if(alpha!=0) beta=10; else if(gamma ==0) delta++; else if((epsilon!=0)&&(zeta==1)) beta=beta<<3;`, where each variable is local and of type `char`. This example contains many operators just discussed. This can be coded in assembly language as

```
        3:      if(alpha!=0) beta=10;
00000867 A685          LDAA  5,SP
00000869 2706          BEQ   *+8    ;abs = 0871
0000086B C60A          LDAB  #10
0000086D 6B80          STAB  0,SP
0000086F 2017          BRA   *+25   ;abs = 0888
        4:      else if(gamma ==0) delta++;
00000871 A684          LDAA  4,SP
00000873 2604          BNE   *+6    ;abs = 0879
00000875 6283          INC   3,SP
00000877 200F          BRA   *+17   ;abs = 0888
        5:      else if((epsilon!=0)&&(zeta==1)) beta=beta<<3;
00000879 A682          LDAA  2,SP
0000087B 270B          BEQ   *+13   ;abs = 0888
0000087D A681          LDAA  1,SP
0000087F 042006        DBNE  A,*+9   ;abs = 0888
00000882 6880          ASL   0,SP
00000884 6880          ASL   0,SP
00000886 6880          ASL   0,SP
```

The *case* statement is a useful alternative to the conditional statement. (See Figure 2.2.) A numerical expression is compared to each of several possible values; the match determines which statement will be executed next. The case statement (such as the simple one in Figure 2.2a) jumps into the statements where the variable matches the comparison value and executes all the statements below it. The *break* statement (shown in Figure 2.2b) exits the whole case statement, in lieu of executing its remaining statements.

An expression `switch(n){ case 1:i=1; break; case 3:i=2; break; case 6: i=3;break;}` is coded in assembly language by comparing the same number (the switch operand) against different case constants, conditionally branching to the case statement's code.

```
00000867 A681          LDAA  1,SP
00000869 8106          CMPA  #6
0000086B 2218          BHI   *+26   ;abs = 0885
0000086D 8101          CMPA  #1
0000086F 270A          BEQ   *+12   ;abs = 087B
00000871 8103          CMPA  #3
00000873 2709          BEQ   *+11   ;abs = 087E
00000875 8106          CMPA  #6
00000877 2708          BEQ   *+10   ;abs = 0881
00000879 200A          BRA   *+12   ;abs = 0885
        5:             case 1: i=1; break;
0000087B C601          LDAB  #1
0000087D 8FC602        CPS   #50690 ; is SKIP2 LDAB #2
        6:             case 3: i=2; break;
00000880 8FC603        CPS   #50691 ; is SKIP2 LDAB #3
00000883 6B80          STAB  0,SP
```

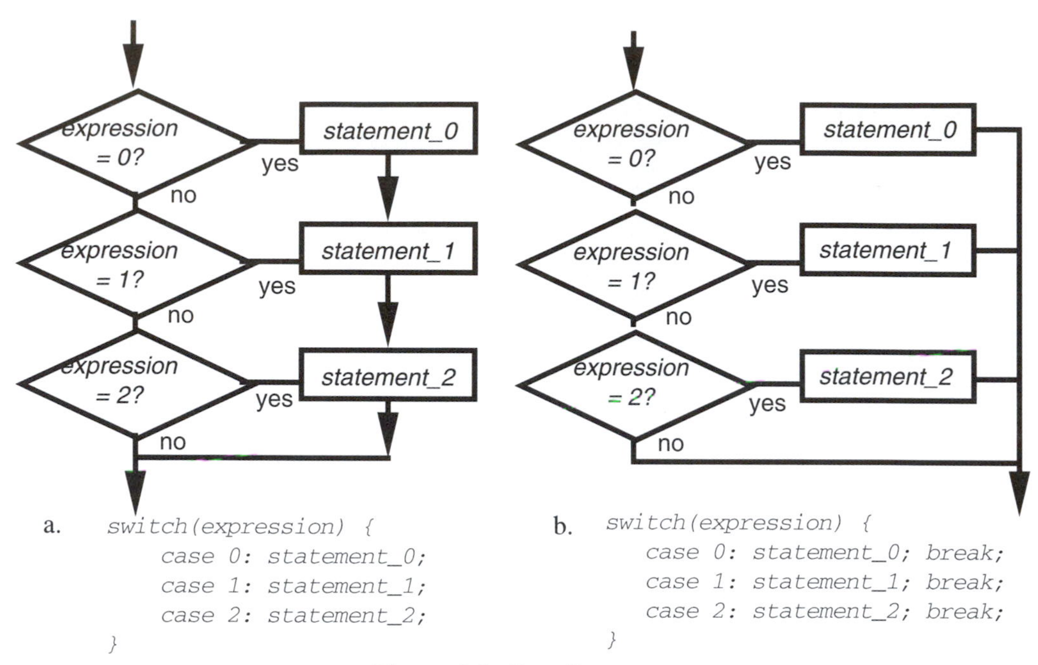

a.
```
switch(expression) {
    case 0: statement_0;
    case 1: statement_1;
    case 2: statement_2;
}
```

b.
```
switch(expression) {
    case 0: statement_0; break;
    case 1: statement_1; break;
    case 2: statement_2; break;
}
```

Figure 2.2. Case Statements

Alternative to the conditional branch statements shown above, *case* can be implemented by a subroutine followed by a list of case values and relative branch offsets. When a sequence of consecutively numbered cases from 0 to N - 1 are presented, HIWARE's C compiler uses one of its seven such subroutines, followed by a list of relative branch offsets. For instance, the expression `switch(n){ case 0:i=1; break; case 1:i=2; break; case 2: i=3;break;}` is coded by calling the subroutine `_CASE_DIRECT_BYTE`. The essential case statement's calling routine is implemented as

```
00000867 EC81            LDD   1,SP
00000869 072D            BSR   NEAR _CASE_DIRECT_BYTE
0000086B 03        L:    DC.B  L0-L
0000086C 06              DC.B  L1-L
0000086D 09              DC.B  L2-L
    5:               case 1: i=1; break;
0000086E C601     L0:    LDAB  #1
00000870 8FC602          CPS   #50690 ; is SKIP2 L1:LDAB #2
    6:               case 3: i=2; break;
00000873 8FC603          CPS   #50691 ; is SKIP2 L2:LDAB #3
00000876 6B80            STAB  0,SP
```

The subroutine `_CASE_DIRECT_BYTE` is

```
00000898 30              PULX
00000899 E6E6            LDAB D,X
0000089B 05E5            JMP B,X
```

Loop statements can be used to repeat a statement until a condition is met. A statement within the loop statement will be executed repeatedly. The expressions in both the following loop statements are exactly like the expressions of the conditional statements, using operators as shown in Table 2.3.

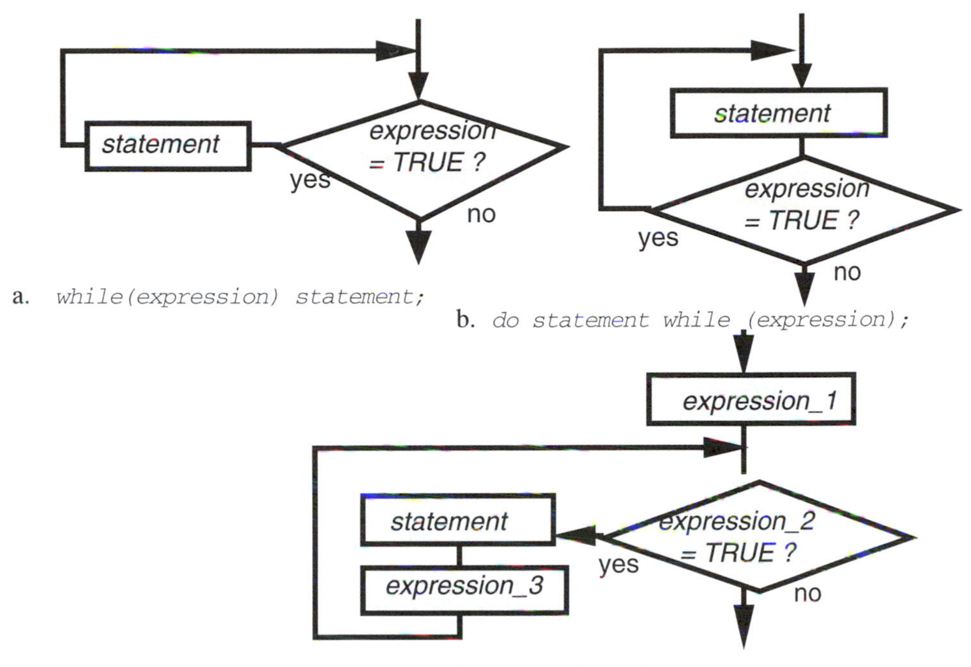

Figure 2.3. Loop Statements

The *while statement* of Figure 2.3a tests the condition before the loop is executed and is useful if, for example, a loop may have to be done 0 times. Assume `i` is initially cleared. Then the `while` statement can clear the array `alpha[10]`. The statement `while(i<10) alpha[i++]=0;` is compiled into assembly language as

```
          4:       while(i<10) alpha[i++]=0;
00000867 200E            BRA   *+16    ;abs = 0877
00000869 A680            LDAA  0,SP
0000086B 36              PSHA
0000086C 42              INCA
0000086D 6A81            STAA  1,SP
0000086F 33              PULB
00000870 87              CLRA
00000871 1A81            LEAX  1,SP
00000873 1AE6            LEAX  D,X
00000875 6A00            STAA  0,X
00000877 A680            LDAA  0,SP
00000879 810A            CMPA  #10
0000087B 25EC            BCS   *-18    ;abs = 0869
```

The *do while statement* (shown in Figure 2.3b) tests the condition after the loop is executed at least once, thus it tests results of the loop's activities. For instance, `do alpha[i++] = 0; while(i < 10);` clears `alpha`; it compiles into

```
      4:        do alpha[i++] = 0; while(i < 10);
00000867 A680          LDAA  0,SP
00000869 36            PSHA
0000086A 42            INCA
0000086B 6A81          STAA  1,SP
0000086D 33            PULB
0000086E 36            PSHA
0000086F 87            CLRA
00000870 1A82          LEAX  2,SP
00000872 1AE6          LEAX  D,X
00000874 6A00          STAA  0,X
00000876 33            PULB
00000877 C10A          CMPB  #10
00000879 25EC          BCS   *-18    ;abs = 0867
```

Generally the *do while() statement* is generally more efficient than the `while()` statement, because the latter has an extra branch instruction to jump to its end. But the HIWARE compiler optimizes `while` statements by removing this initial branch if the initial value of the condition is determined at compile time to be true.

The more general *for statement* (shown in Figure 2.3c) has three expressions separated by semicolons (;). The first expression initializes variables used in the loop; the second tests for completion in the same style as the while statement; and the third updates the variables each time after the loop is executed. Any of the expressions in the `for` statement may be omitted. For example, `for(i=0;i<10;i++) alpha[i]=0;` will clear the array `alpha` as the above loops did it. It is compiled as follows:

```
      4:        for(i=0;i<10;i++) alpha[i]=0;
0000086B 6980          CLR   0,SP
0000086D E680          LDAB  0,SP
0000086F 87            CLRA
00000870 1A81          LEAX  1,SP
00000872 1AE6          LEAX  D,X
00000874 6A00          STAA  0,X
00000876 6280          INC   0,SP
00000878 E680          LDAB  0,SP
0000087A C10A          CMPB  #10
0000087C 25EF          BCS   *-15    ;abs = 086D
```

This program segment is not particularly efficient. A more efficient assembly-language program equivalent to `for(i = 10; i ! = 0; i-- ) alpha[i - 1] = 0;` is

```
    ldx #10
L1: clr alpha-1,x
    dbne x,L1
```

However, the C compiler may not actually generate the assembly language shown above. If you need tight code, you will have to insert the assembly-language code into a C procedure, in a manner to be shown in §2.3.5. Note also that, as in this example if it were efficiently compiled, the clearest C program does not always lead to the most efficient assembly-language program. A less clear program may generate better code.

The *break* statement will cause the `for`, `while`, or `do while` loop to terminate just as in the *case* statement, and may be used in a conditional statement. For instance, `for(;;) {i++; if(i==30) break;}` executes the statement `{i++; if(i==30) break;}` indefinitely, but the loop is terminated when `i` is 30.

An important feature of C, extensively used to access I/O devices, is its ability to describe variables and addresses of variables. If *a* is a variable, then `&a` is the address of *a*. If *a* is a variable that contains an address of another variable `b`, then `*a` is the contents of the word pointed to by `a`, which is the contents of *b*. (Note that `a*b` is *a* times `b` but `*b` is the contents of the word pointed to by `b`.) Whenever you see `&`, read it as "address of"; and whenever you see *, read it as "contents of thing pointed to by." In a declaration statement, the statement `char *p;` means that the thing pointed to by `p` is a character, and `p` points to (contains the address of) a character. In an assignment statement, `*p = 1;` means that 1 is assigned to the value of the thing pointed to by `p`, whereas `p = 1;` means that the pointer `p` is given the value 1. Similarly, `a = *p;` means that *a* is given the value of the thing pointed to by `p`, while `a = p;` means `a` gets the value of the pointer `p`. Some C compilers will give an error message when you assign an integer to a pointer. If that occurs, use a cast. Write `p = (int *)0x4000;` to tell the compiler 0x4000 is really a pointer value to an integer and not an integer itself.

Finally a comment is anything enclosed by `/*` and `*/`. We strongly encourage you to supply comments to document your code.

2.2 Data Structures

Data structures are at least as important as programming techniques, for if the program is one-half of the software, the data and their structures are the other half. When we discuss storage density as an architecture characteristic, we discuss only the amount of memory needed to store the program. We are also concerned about data storage and its impact on static and dynamic efficiency, as well as the size of memory needed to store the data. Prudent selection of the data structures a program uses can shorten or speed up the program. These considerations about data structures are critical in microcontrollers.

A data structure is one among three views of data. The *information structure* is the view of data the end user sees. For instance, the end user may think of his or her data as a table, like Table 2.1 in this book. The programmer sees the same data as the *data structure:* strongly related to the way the data are accessed but independent of details such as size of words and position of bits. It is rather like a painter's template, which can be filled in with different colors. So the data structure may be an array of characters that spell out the words in Table 2.1. The *storage structure* is the way the information is actually stored in memory, right down to the bit positions. So the table may appear as an array of 8-bit words in the storage structure.

The data structure concept is a bit hard for some to accept. Its usefulness lies in its ability to provide a level of abstraction, allowing us to make some overall observations of how we store things, which can be applied to similar storage techniques. For instance, if we can develop a concept of how to access an array, we can use similar ideas to access arrays of 8-bit or 24-bit data, even though the programs could be quite different. But here we must stress that a data structure is simply a kind of template that tells us how data are stored, and it is also a menu of possible ways the data can be written or read. Two data structures are different if they have different templates that describe their general structure or if the menus of possible access techniques are different.

Constants are often used with data structures, for instance, to declare a size of a vector and to use that same number in `for` loops. They can be defined by *define statements* or *enum statements* and put before any declarations or statements to equate names to values. The `#define` statement begins with the characters `#define` and does not end with a semicolon.

```
#define ALPHA 100
```

Thenceforth, we can use the label `ALPHA` throughout the program, and 100 will effectively be put in place of `ALPHA` just before the program is actually compiled. This permits the program to be better documented, using meaningful labels, and easier to maintain, so that if the value of a label is changed it is changed everywhere it occurs.

A number of constants can be created using the `enum` statement. Unless reinitialized with an equal sign (=), each member has one greater than the value of the previous member. The first member has value 0. Hexadecimal values are prefixed with zero ex (0x):

```
enum { BETA, GAMMA, DELTA = 0x5};
```

defines `BETA` to have value 0, `GAMMA` to have value 1, and `DELTA` to have value 5.

The declaration of any scalar variable can be initialized by use of the = and a value. For instance, if scalar integers *i, j,* and *k* have initial values 1, 2, and 3, we write a global declaration:

```
int i = 1, j = 2, k = 3;
```

C procedures access global variables using direct addressing, and such global variables may be initialized in a procedure *ccmain* that is executed just before `main` is started. Initialized local variables of a procedure should generate machine code to initialize them just after they are allocated, each time the procedure is called. The procedure

```
void fun(){
     int i, j, k; /* allocate local variables */
     i=1; j=2; k=3; /* initialize local variables */
}
```

is equivalent to the procedure

```
void fun(){
     int i=1, j=2, k=3; /* allocate and initialize local variables */
}
```

Data structures divide into three main categories: indexable, sequential, and linked. Indexable and sequential, discussed here, are more important. Linked structures are very powerful, but are not as easy to discuss in abstract terms. They will be discussed later.

2.2.1 Indexable Data Structures

Indexable structures include vectors, lists, arrays, and tables. A *vector* is a sequence of elements, where each element is associated with an index *i* used to access it. To make address calculations easy, C associates the first element with the index 0, and each successive element with the next integer (*zero-origin indexing*). Also, the elements in a vector are considered numbers of the same *precision* (number of bits or bytes needed to store an element). We will normally consider one-word precision vectors, although we also show an example of how the ideas can be extended to *n*-word vectors. Finally, the *cardinality* of a vector is the number of elements in it. A vector is fully specified if its origin, precision, and cardinality are given. A zero-origin, 16-bit, three-element vector 31, 17, and 10 is generated by a declaration `int v[3]` and stored in memory as (hexadecimal)

001F
0011
000A

and we can refer to the first element as `v[0]`, which happens to be 31. However, the same sequence of values could be put in a zero-origin vector of three 8-bit elements, generated by a declaration `char u[3]` and stored in memory as

1F
11
0A

The declaration of a global vector variable can be initialized by the use of `=` and a list of values, in braces. For instance, the three-element global integer vector *v* can be allocated and initialized by

```
int v[3] = {31, 17, 10};
```

The vector *u* can be similarly allocated and initialized by the declaration

```
char u[3] = {31, 17, 10};
```

The procedure `main()` in §2.1 illustrated the accessing of elements of vectors in expressions. The expression `c = d[a+b];` accessed the `(a+b)`th element of the 8-bit, 10-element vector *d.* The term "vector" is a general term, and similar declarations and statements can be used for elements of 16 bit, 32 bit, or other precision, and for other cardinality vectors. The concept of "data structure" is to generalize the storage and access used in one instance to cover other instances of the same kind of data-handling technique. When reading the assembly code generated by C, be wary of the implicit multiplication of the vector's precision (in bytes) when calculating offset addresses of elements of the vector. And because C does not check that indexes are within the cardinality of a vector, your C program must be able to implicitly or explicitly ensure this to avoid nasty bugs - when a vector's data is inadvertently stored outside the memory allocated to a vector.

A *list* is like a vector, accessed by means of an index, but the elements of a list can be any combination of different precision words, code words, and so on. For example, the list can have three elements: the one-byte number 5, the two-byte number 7, and the one-byte number 9. This list is stored in machine code as follows:

```
05
0007
09
```

The powerful *structure* mechanism is used in C to implement lists. The mechanism is implemented by a declaration that begins with the word *struct* and has a definition of the structure within braces, and a list of variables of that structure type after the brackets, as in

```
struct { char l1; int l2; char l3;} list;
```

A globally defined list can be initialized as we did with vectors, as in

```
struct { char l1; int l2; char l3;} list={5,7,9};
```

The data in a list are identified by dot notation, where a dot (.) means "element." For instance, `list.l1` is the `l1` element of the list `list`. If *P* is a pointer to a `struct`, then arrow notation, such as `P->l1`, can access the element `l1` of the list. The `typedef` statement, though it can be used to create a new data type in terms of existing data types, is often used with `struct`s. If `typedef a struct { char l1; int l2; char l3;} list;` is written, then `list` is a data type, like *int* or *char*, and can be used in declarations such as `list b;` that declares `b` to be an instance of type `list`. The *typedef statement* is quite useful when a `struct` has to be declared many times and pointers to it need to be declared too. A structure can have *bit fields,* which are unsigned integer elements having less than 16 bits. Such a structure as

```
struct {unsigned a:1, b:2, c:3;}l;
```

has a one-bit field `l.a`, two-bit field `l.b`, and three-bit field `l.c`. A *linked list* structure, a list in which some elements are addresses of (the first word in) other lists, is flexible and powerful and is widely used in advanced software.

An *array* is a vector whose elements are vectors of the same length. We normally think of an array as a two-dimensional pattern, as in

1	2	3
4	5	6
7	8	9
10	11	12

An array is considered a vector whose elements are themselves vectors. The array is stored in *row major order:* in this arrangement a row is stored with its elements in consecutive memory locations. (In *column major order* a column is stored with its elements in consecutive memory locations.) For instance, the global declaration

```
int ar1[4][3]={{1,2,3},{4,5,6},{7,8,9},{10,11,12}};
```

allocates and initializes a row-major-ordered array `ar1`, and `a=ar1[i][j];` puts the row-`i` column-`j` element of `ar1` into `a`, as shown in the previous example of an array.

A *table* is to a list as an array is to a vector. It is a vector of identically structured lists (rows). Tables often store characters, where either a single character or a collection of *n* consecutive characters are considered elements of the lists in the table. Index addressing is useful for accessing elements in a row of a table, especially if the table is stored in row major order. If the address register points to the first word of any row, then the displacement can be used to access words in any desired column. Also, autoincrement addressing can be used to select consecutive words from a row of the table.

In C, a table `tbl` is considered a vector whose elements are structures. For instance, the declaration

```
struct {char l1;int l2;char l3;} tbl[3];
```

allocates a table whose rows are similar to the list `list` above. The dot notation with indexes can be used to access it, as in

```
a = tbl[2].l1;
```

In simple compilers, multidimensional arrays and *structs* are not implemented. They can be reasonably simulated using one-dimensional vectors. The user becomes responsible for generating vector index values to access row-column elements or *struct* elements.

2.2.2 Sequential Data Structures

Another important class of data structures is sequential structures, which are accessed by relative position. Rather than having an index *i* to get to any element of the structure, only the "next" element to the last one accessed may be accessed in a sequential structure. Strings, stacks, queues, and deques are sequential structures important in microcontrollers.

A *string* is a sequence of elements such that after the *i*th element has been accessed, only the (*i* + 1)th element or the (*i* - 1)th element can be accessed. In particular, a *character string,* storing characters using the *ASCII code* (Table 2.4), is used to store text. The ASCII code of a character is stored as a 7-bit code in a *char* variable. Character constants are enclosed by single quotes around the character, as `'A'` is the character A. Special characters are *null* `'\0'`, *line feed* `'\n'`, *form feed* `'\f'` (begin new page), *carriage return* `'\r'`, and `' '` space. Strings are allocated and used in C as if they were `char` vectors, are initialized by putting the characters in double quotes, and end in the null character `'\0'`. (Allow an extra byte for it.)

One can initialize a global character `c` to be the code for the letter `a` and a global string `s` to be `ABCD` with the global declaration

```
int c='a', s[5]="ABCD";
```

Table 2.4. ASCII Codes

<table>
<tr><th></th><th>00</th><th>10</th><th>20</th><th>30</th><th>40</th><th>50</th><th>60</th><th>70</th></tr>
<tr><td>0</td><td>'\0'</td><td></td><td>' '</td><td>0</td><td>@</td><td>P</td><td>`</td><td>p</td></tr>
<tr><td>1</td><td></td><td></td><td>!</td><td>1</td><td>A</td><td>Q</td><td>a</td><td>q</td></tr>
<tr><td>2</td><td></td><td></td><td>"</td><td>2</td><td>B</td><td>R</td><td>b</td><td>r</td></tr>
<tr><td>3</td><td></td><td></td><td>#</td><td>3</td><td>C</td><td>S</td><td>c</td><td>s</td></tr>
<tr><td>4</td><td></td><td></td><td>$</td><td>4</td><td>D</td><td>T</td><td>d</td><td>t</td></tr>
<tr><td>5</td><td></td><td></td><td>%</td><td>5</td><td>E</td><td>U</td><td>e</td><td>u</td></tr>
<tr><td>6</td><td></td><td></td><td>&</td><td>6</td><td>F</td><td>V</td><td>f</td><td>v</td></tr>
<tr><td>7</td><td></td><td></td><td>'</td><td>7</td><td>G</td><td>W</td><td>g</td><td>w</td></tr>
<tr><td>8</td><td></td><td></td><td>(</td><td>8</td><td>H</td><td>X</td><td>h</td><td>x</td></tr>
<tr><td>9</td><td></td><td></td><td>)</td><td>9</td><td>I</td><td>Y</td><td>i</td><td>y</td></tr>
<tr><td>A</td><td>'\n'</td><td></td><td>*</td><td>:</td><td>J</td><td>Z</td><td>j</td><td>z</td></tr>
<tr><td>B</td><td></td><td></td><td>+</td><td>;</td><td>K</td><td>[</td><td>k</td><td>{</td></tr>
<tr><td>C</td><td>'\f'</td><td></td><td>,</td><td><</td><td>L</td><td>\</td><td>l</td><td>|</td></tr>
<tr><td>D</td><td>'\r'</td><td></td><td>-</td><td>=</td><td>M</td><td>]</td><td>m</td><td>}</td></tr>
<tr><td>E</td><td></td><td></td><td>.</td><td>></td><td>N</td><td>^</td><td>n</td><td>~</td></tr>
<tr><td>F</td><td></td><td></td><td>/</td><td>?</td><td>O</td><td>_</td><td>o</td><td></td></tr>
</table>

Strings are also very useful for input and output when debugging C programs; we will discuss the use of strings when we describe the `printf` function later. However, a source-level debugger for a C compiler provides better debugging tools. Even so, some discussion of string-oriented input and output is generally desirable for human interfacing. This discussion of `IoStreams` will be done in §4.3.6 and §4.4.6.

Characters in strings can be accessed by indexing or by pointers. An index can be incremented to access each character, one after another, from first to last character. Or a pointer to a character such as `p` can be used; `*p` is the character that it points to and `*(p++)` returns the character pointed to and then increments the pointer `p` to point to the next character in the string. Alternatively, an index can be used.

The characters you type on a terminal are usually stored in memory in a character string. You can use a typed word as a command to execute a routine, with unique words for executing each routine. A C program comparing a string stored in memory, and pointed to by `p`, to a string stored in `start_word` is shown as `main()`.

```
char *p,start_word[6]="START"; /* assume p points to a string stored elsewhere */
void main() { int i,nomatch;
  for(i = nomatch = 0;i<5;i++){
        if(*(p++) != start_word[i]){ nomatch=1; break;}
  }
  if (nomatch==0) strt(); /* if the string is START then execute the strt proc */
}
```

Inside the loop, we compare a character at a time of the input string against the string `start_word`. If we detect any difference, we set local variable `nomatch` because the user did not type the string `start_word`. But if all five characters match up – the user did type the word `start_word` – the program calls `strt`, presumably to start something.

The assembly language for this C program is listed below.

```
main:
00000865 1B9C          LEAS -4,SP
    6:      for(i = nomatch = 0;
00000867 C7            CLRB
00000868 87            CLRA
00000869 6C82          STD   2,SP
0000086B 6C80          STD   0,SP
    7: if(*(p++) != start_word[i]){ nomatch=1; break;}}
0000086D FE0806        LDX   $0806
00000870 A630          LDAA  1,X+
00000872 7E0806        STX   $0806
00000875 EE80          LDX   0,SP
00000877 A1E20800      CMPA  start_word,X
0000087B 2707          BEQ   *+9    ;abs = 0884
0000087D C601          LDAB  #1
0000087F 87            CLRA
00000880 6C82          STD   2,SP
00000882 2008          BRA   *+10    ;abs = 088C
    6:      for( ... ;i<5;i++){
00000884 08            INX
00000885 6E80          STX   0,SP
00000887 8E0005        CPX   #5
0000088A 2DE1          BLT   *-29    ;abs = 086D
    8:      if (nomatch==0) strt();
0000088C EC82          LDD   2,SP
0000088E 2603          BNE   *+5    ;abs = 0893
00000890 160896        JSR   $0896
    9: }
00000893 1B84          LEAS  4,SP
00000895 3D            RTS
   10: void strt(){};
strt:
00000896 3D            RTS
```

Besides character strings, bit strings are important in microcontrollers. In particular, a very nice coding scheme called the *Huffman code* can pack characters into a bit stream and achieve about a 75% reduction in storage space when compared to storing the characters directly in an ASCII character string. It can be used to store characters more compactly and can also be used to transmit them through a communications link more efficiently. As a bonus, the encoded characters are very hard to decode without a code description, so you get a more secure communications link using a Huffman code. Furthermore, we need to handle data structures and bit shifting using the << and >> operators, and bit masking using the & operator, in many I/O procedures. Procedures for Huffman coding and decoding provide a rich set of examples of these techniques.

We recommend that you test your ability to read C by studying the procedures to follow. We also suggest that you compile these procedures and step through them using a high-level debugger. In this example, we are particularly interested in pointing out that strings may have elements other than characters (here they are bits). Further, the elements of strings can themselves be strings or other data structures, provided such data are decipherable.

The code is rather like Morse code in that frequently used characters are coded as short strings of bits, just as the often-used letter "e" is a single dot in Morse code. To ensure that code words are unique and to suggest a decoding strategy, the code is defined by a tree having two branches at each branching point (*binary tree*), as shown in Figure 2.4. The letters at each end (leaf) are represented by the pattern of 1s and 0s along the branches from the left end (root) to the leaf. Thus, the character string MISSISSIPPI can be represented by the bit string 111100010001011011010. Note that the ASCII string would take 88 bits of memory, while the Huffman string would take 21 bits. When you decode the bit string, start at the root and use each successive bit of the bit string to guide you up (if 0) or down (if 1) the next branch until you get to a leaf. Then copy the letter and start over at the root of the tree with the next bit of the bit string. The bit string has equal probabilities of 1s and 0s, so techniques used to decipher the code based on probabilities won't work. It is particularly hard to break a Huffman code.

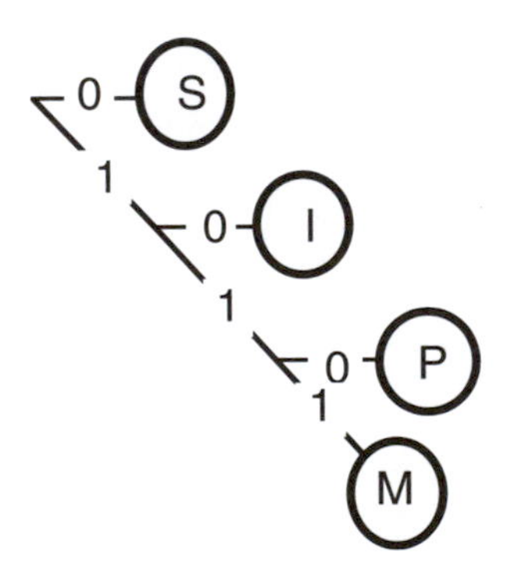

Figure 2.4. A Huffman Coding Tree

A C program for Huffman coding is shown below. The original ASCII character string is stored in the `char` vector `strng`. We will initialize it to the string *MISSISSIPPI* for convenience, although any string of M, I, S, and P letters could be used. The procedure converts this string into a Huffman-coded 48-bit bit string stored in the vector `code[3]`. It uses `shift()` to shift a bit into `code`. This procedure, shown at the end, is also used by the decoding procedure shown between these procedures.

```
extern int shift(void);

int code[3], bitlength; /* output code and its length */
char strng[12] = "MISSISSIPPI"; /* input code, terminated in a NULL character */
struct table{ char letter; char charcode[4]; } codetable[4]
     = { 'S',"XX0", 'I',"X10", 'P',"110",'M',"111" };
```

```
void main(){  int row, i; char *point, letter;
   for (point=strng; *point ; point++ ){
       for (row = 0; row < 4; row ++){
           if (((*point) & 0x7f) == codetable[row].letter){
               for (i = 0; i < 3; i++){
                   letter = codetable[row].charcode[i];
                   if (letter != 'X')
                       { shift(); code [2] |= (letter & 1); bitlength++;}
               }
           }
       }
   }
   i= bitlength; while((i++)<48)shift(); /* shift out unchanged bits */
}
```

Huffman decoding, using the same *shift()*, is done as follows:

```
int code[3] = {0xf116, 0xd000, 0}, bitlength = 21; /* input string */
   char tbl[3][2] = {{'S',1},{'I',2},{'P','M'}};
   char strng[20]; /* output string */
   void main(){  int row,entry; char *point;
       point=strng; row =0;
       while((bitlength--)>=0){
           if((entry = tbl[row][shift()]) < 0x20) row = entry;
           else {row =0; *(point++) = entry &0x7f; }
       }
       *point = '\0'; /* terminate C string with NULL character */
   }
   int shift() {
       int i;
       i=0; if( 0x8000 & code[0]) i=1; code[0] = code[0] <<1;
       if (code[1] & 0x8000) code[0] |=1; code[1] = code[1] <<1;
       if (code[2] & 0x8000) code[1] |=1; code[2] = code[2] <<1;
       return(i);
   }
```

Now that we have shown how nice the Huffman code is, we must admit to a few problems with it. To efficiently store some text, the text must be statistically analyzed to determine which letters are most frequent, to assign these the shortest codes. Note that S is most common, so we gave it a short code word. There is a procedure for generating the best Huffman code, that is presented in many information theory books, but you have to get the statistics of each letter's occurrences to get the code. Nevertheless, though less than perfect, one can use a fixed code that is based on other statistics if the statistics are reasonably similar. Finally, although the code is almost unbreakable without the decoding tree, if any bit in the bit string is erroneous, your decoding routine can get completely lost. This may be a risk you decide to avoid because the code has to be sent through a communications link that is as error free as possible.

A *deque* is a generalized data structure that includes two special cases: the stack and the queue. A deque (pronounced deck) is a sequence of elements with two ends we call the top and the bottom. You can only access the top or bottom elements on the deque. You can *push* an element on top by placing it on top of the top element, which makes it the new top element; or you can *push* an element on the bottom, making it the new bottom element; or you can *pull* (or *pop*) the top element, deleting the top element from the deque, making the next to top element the top element and putting the old top element somewhere else; or *pull* (or *pop*) the bottom element in like manner.

Deques are theoretically infinite, so you can push as many elements as you want on either the top or bottom. But practical deques have a maximum capacity. If this capacity is exceeded, we have an *overflow error.* Also, if you pull more elements than you push, an *underflow error* exists.

In the procedure below, note that you cannot really associate the *i*th word from either end of a deque with a particular location in memory. In fact, in a pure sense, you can only access the deque's top and bottom words and cannot read or write any other word in the deque. In practice, we sometimes access the *i*th element from the deque's top or bottom by using a displacement with the indexes that points to the top and bottom words – but this is not a pure deque. We call it an *indexable deque* to give it some name.

C declarations and programs for initializing, pushing, and pulling words are shown below. The *buffer* is an area of memory set aside for use as the deque expands, that cannot be used by any other data or program code. The programmer allocates as much room for the buffer as appears necessary for the worst-case (largest) expansion of the deque. Two indexes are used to read or write on the top or bottom, and a counter is used to detect overflow or underflow. The deque buffer is implemented as a ten-element global vector `deque`, and the indexes as global `unsigned char`s `top` and `bottom` initialized to the first element of the `deque`, as in the C declaration

```
unsigned char deque [10], size, error, top, bottom;
```

As words are pulled from top or bottom, more space is made available to push words on either the top or bottom. To take advantage of this, we think of the buffer as a ring or loop of words, so that the next word below the bottom of the buffer is the word on the top of the buffer. That way, as words are pulled from the top, the memory locations can become available to store words pushed on the bottom as well as words pushed on the top, and vice versa. Then to push or pop data into or from the top or bottom of it, we can execute the following procedures:

```
void pstop(int item_to_push) {
    if ((++size) > 10) error = 1;
    if (top ==10) top = 0; deque[top++] = item_to_push;
}

int pltop() {
    if ((-- size ) < 0) error = 1;
    if (top==0) top = 10;
    return(deque[- - top]);
}
```

```
void psbot(int item_to_push) {
    if ((++size) > 10) error = 1;
    if (bottom == 0) bottom = 10;
    deque[-- bottom] = item_to_push;
}

int plbot(){
    if ((-- size) < 0) error = 1;
    if (bottom == 10) bottom = 0;
    return( deque[bottom++] );
}
```

A *stack* is a deque in which you can push or pull on only one end. We have discussed the stack accessed by the stack pointer SP, which permits the machine to push or pull words from the top of the stack to save registers for procedure calls, as well as SWI and hardware interrupts. Now we consider the stack as a special case of a deque. (Actually, the 6812 stack can be made a special case of indexable deque using SP with index addressing.) It is an example of a stack that pushes or pulls elements from the top only. Another equally good stack can be created that pushes or pulls elements only from the bottom of the deque. In fact, if you want two different stacks in your memory, have one that pushes and pulls from the top and another that pushes and pulls from the bottom. Then both stacks can share the same buffer, as one starts at the top of this buffer (lowest address) and builds downward, while the other starts at the bottom (highest address) and builds upward. A stack overflow exists when the top pointer of the stack that builds upward is equal to the bottom pointer of the stack that builds downward. Note that if one stack is shorter, then the other stack can grow longer before an overflow exists, and vice versa. You only have to allocate enough words in the buffer for the maximum number of words that will be in both at the same time.

Programs to push or pull on the two stacks are simpler than the general program that operates on the deque, because pointers do not roll around the top or bottom of the buffer.

The final structure that is important in microcomputer systems is the *queue*. This is a deque in which we can push data on one end and pull data from the other end. In some senses, it is like a shift register, but it can expand if more words are pushed than are pulled (up to the size of the buffer). In fact, it has been called an elastic shift register. Or conversely, a shift register is a special case of a queue, a fixed-length queue. Queues are used to store data temporarily, such that the data can be used in the same order in which they were stored. We will find them very useful in interrupt handlers and in procedures that interact with them.

One of the rather satisfying results of the notion of data structures is that the stack and queue, actually quite different concepts, are found to be special cases of the deque. The two structures are, in fact, handled with similar programs. Other data structures – such as multidimensional arrays, trees, partially ordered sets, and graphs such as lattices and banyans – are important in general programming. You are invited to pursue the study of data structures to improve your programming skills. However, this section has covered the data structures we have found most useful in microcomputer interface software.

2.3 Writing Clear C Programs

6812 systems are almost always large enough for programmers to consider the advantages of writing clear assembly-language programs over the expense of writing short programs. There may be reason for concern about static efficiency in smaller microcontrollers. An implementation of the Motorola 6805 has so little memory – 2K words of ROM – and the programs are so short that static efficiency is paramount and readability is less important to a good programmer, who can comprehend even poorly written programs. Readability is significant for programs larger than 16K, even for a good programmer, because it may have to be written by several programmers who read one another's code and may have to be maintained long after the original programmers have gone.

A significant technique for writing clear programs is good documentation, such as using comments and flow charts. Of course, these do not take up memory in the machine code, so they can be used when static or dynamic efficiency must be optimized. Another technique is the use of consistent programming styles that constrain the programmer, thereby reducing the chance of errors and increasing the reader's ease of understanding. Also, a major idea in clear programming methodology is modular top-down design. We also need to develop the concept of object-oriented programming. In order to discuss these ideas, we first need to further refine our understanding of procedures and arguments.

2.3.1 C Procedures and Their Arguments

Conceptually, *arguments* or *parameters* are data passed from or to the calling routine to or from a C procedure, like the `x` `sin(x)`. Inside the procedure, the parameter is declared as a variable, such as `y`, as in `sin(int y){ ... }`. It is called the *formal parameter.* Each time the procedure is called the calling routine uses different variables for the parameter. The variable in the calling routine is called the *actual parameter.* For example, at one place in the program we put `sin(alpha)`, in another, `sin(beta)`, and in another, `sin(gamma)`. Variables `alpha`, `beta`, and `gamma` are actual parameters.

At the conceptual level, arguments are called by value, result, reference, or name. In *call by value,* the actual parameters themselves are passed into the procedure. In *call by result,* a formal parameter inside the procedure is usually left in a register (D). This value is usually then used in an expression or stored into its actual parameter after the procedure is finished.

In *call by reference,* or *call by name,* the data remain stored in the calling routine and are not actually moved to another location, but the address of the data is given to the procedure and the procedure uses this address to get the data whenever it needs to. The addresses of the actual parameters are input to the procedure. Inside the procedure, the address can be used to read the actual parameter data or to modify it. Large vectors, lists, arrays, and other data structures can be more effectively called by reference so they don't have to be copied into and out of the subroutine's local variables.

A procedure in C may be called by another procedure in C as a function. The arguments may be the data themselves, which are call by value, or the address of the data, which is call by reference or name.

HIWARE's C compiler passes a function's single input and output argument in a register. An 8-bit argument is passed in accumulator B, a 16-bit argument in D, a 24-bit in B (high byte) and index register X (low 16 bits), and a 32-bit argument in D (high 16 bits) and X (low 16 bits). Consider `int square(int i) { return i*i; }`.

```
00000887 B746        TFR   D,Y
00000889 13          EMUL
0000088A 3D          RTS
```

If an input operand needs to be saved within the procedure, it is generally pushed on the stack upon entry, like a hidden local variable. In the following example, `unsigned int swap(unsigned int i) { return (i >> 8) | (i << 8); }`, shows this.

```
00000887 3B          PSHD
00000888 B784        EXG   A,D
0000088A 3B          PSHD
0000088B EC82        LDD   2,SP
0000088D B710        TFR   B,A
0000088F C7          CLRB
00000890 EA81        ORAB  1,SP
00000892 AAB3        ORAA  4,SP+
00000894 3D          RTS
```

If more than one input argument is declared, all but the last argument are pushed on the stack before the subroutine is called and pulled after the subroutine returns. Consider `RaisePower( &k, &j, i)`, returning `j` to the power `i` in *k* where `i`, `j`, and `k` are integers; `i` is passed by value, while `j` and `k` are passed by name.

```
void RaisePower (int *k, int *j, int i) {
    for(*k = 1; i--; ) *k = *k * *j;
}
```

The called procedure is implemented as

```
RaisePower:
00000887 EE84        LDX   4,SP
00000889 CD0001      LDY   #1
0000088C 6D00        STY   0,X
0000088E B745        TFR   D,X
00000890 2011        BRA   *+19    ;abs = 08A3
   3:              for(*k = 1; i--; ) *k = *k * *j;
00000892 ED84        LDY   4,SP
00000894 EC40        LDD   0,Y
00000896 34          PSHX
00000897 EE84        LDX   4,SP
00000899 3B          PSHD
0000089A EC00        LDD   0,X
0000089C B765        TFR   Y,X
0000089E 31          PULY
```

```
0000089F 13          EMUL
000008A0 6C00        STD   0,X
000008A2 30          PULX
000008A3 191F        LEAY  -1,X
000008A5 8E0000      CPX   #0
000008A8 B765        TFR   Y,X
000008AA 26E6        BNE   *-24    ;abs = 0892
    4: }
000008AC 3D          RTS
    5: void main(){ int i, j = 2;
main:
000008AD C602        LDAB  #2
000008AF 87          CLRA
000008B0 6CAC        STD   4,-SP
    6:              RaisePower( &i, &j, 3);
000008B2 1A82        LEAX  2,SP
000008B4 34          PSHX
000008B5 1A82        LEAX  2,SP
000008B7 34          PSHX
000008B8 52          INCB
000008B9 07CC        BSR   *-50    ;abs = 0887
000008BB 1B88        LEAS  8,SP
    7: }
000008BD 3D          RTS
```

Call by value, as `i` is passed, does not allow data to be output from a procedure, but any number of call-by-value input parameters can be used in a procedure. Actual parameters passed by name in the calling procedure have an ampersand (&) prefixed to them to designate that the address is put in the parameter. In the called procedure, the formal parameters generally have an asterisk (*) prefixed to them to designate that the data at the address are accessed. Observe that call-by-name formal parameters *j* or *k* used inside the called procedure all have a prefix asterisk. A call-by-name parameter can pass data into or out of a procedure, or both. Data can be input to a procedure using call by name, because the address of the result is passed into the procedure and the procedure can read data at the given address. A result can be returned from a procedure using call by name, because the address of the result is passed into the procedure and the procedure can write new data at the given address to pass data out of the procedure. Any number of call-by-name input/output parameters can be used in a procedure.

A procedure may be used as a function which returns exactly one value and can be used in the middle of algebraic expressions. The value returned by the function is put in a *return statement.* For instance, the function *power* can be written

```
int power(int i, int j) { int k,n;
     for(n=1, k=0;k<j;k++) n=n*i; return n;
}
```

This function can be called within an algebraic expression by `a=power(b,2)`. The output of the function named in the *return* statement is passed by call by result.

In C, the address of a character string can be passed into a procedure, which uses a pointer inside it to read the characters. For example, the string *s* is passed to a procedure *puts* that outputs a string by outputting to the user's display screen one character at a time using a procedure `putchar`. The procedure puts is written

```
void puts(s) char *s; {
    while(*s!=0) putchar(*(s++));
}
```

It can be called in either of three ways, as shown side by side:

```
void main(){                  void main(){                 void main(){
     char s[6]="ALPHA";            char s[6]="ALPHA";           puts("ALPHA");
     puts(&s[0]);                  puts(s);                }
}                             }
```

The first calling sequence, though permissible, is clumsy. The second is often used to pass different strings to the procedure, while the third is preferred when the same constant string is passed to the procedure in the statement of the calling program. The third calling sequence is often used to write prompt messages out to the user and to pass a format string to a formatted input or output procedure like *printf* (described shortly).

A *prototype* for a procedure can be used to tell the compiler how arguments are passed to and from it. At the beginning of a program we write all prototypes, such as

```
extern void puts(char *);
```

The word *extern* indicates that the procedure `puts()` is not actually here but is elsewhere. The procedure itself can be later in the same file or in another file. The argument `char *` indicates that the procedure uses only one argument and it will be a pointer to a character (i.e., the argument is called by name). In front of the procedure name a type indicates the procedure's result. The type *void* indicates that the procedure does not return a result. After the prototype has been declared, any calls to the procedure will be checked to see if the types match. For instance, a call `puts('A')` will cause an error message because we have to send the address of a character (string), not a value of a character to this procedure. The prototype for `power()` is:

```
extern int power(int, int);
```

to indicate that it requires two arguments and returns one result, all of which are call-by-value-and-result 16-bit signed numbers. The compiler will use the prototype to convert arguments of other types if possible. For instance, if `x` and `y` are 8-bit signed numbers (of type `char`) then a call `power(x,y)` will automatically extend these 8-bit numbers to 16-bit signed numbers before passing them to the procedure. If a procedure has a `return` `n` statement that returns a result, then the type statement in front of the procedure name indicates the type of the result. If that type is declared to be *void,* as in the `puts()` procedure, there may not be a `return n` statement that returns a result.

At the beginning of each file, prototypes for all procedures in that file should be declared. While writing a procedure name and its arguments twice, once in a prototype and later in the procedure itself, may appear clumsy, it lets the compiler check for improper arguments and, where possible, instructs it to convert types used in the calling routine to the types expected in the called routine. We recommend the use of prototypes.

The *macro* is similar to a procedure, but is either evaluated at compile time or is inserted into the program wherever it is used, rather than being stored in one place and jumped to whenever it is called. The macro in C is implemented as a *#define* construct. As `#define`s were earlier used to define constants, macros are also expanded just before the program is compiled. The macro has a name and arguments, rather like a procedure, and the rest of the line is the body of the macro. For instance

```
#define f( a, b, c) a = b * 2 + c
```

is a macro with name `f` and arguments `a`, `b`, and `c`. Wherever the name appears in the program, the macro is expanded and its arguments are substituted. For instance if `f( x, y, 3)` appeared, then `x = y * 2 + 3` is inserted into the program. Macros with constant arguments are evaluated at compile time, generating a constant at run time.

The procedures `getchar` and `gets` input characters and character strings; `InDec` and `InHex` input decimal numbers and hexadecimal numbers, and `putchar`, `puts`, and `printf` output characters, character strings, and formatted character strings. These very powerful functions are actually executed in a host computer on whose keyboard the user is typing and on whose screen the user is reading the results, rather than in the target 6812 microcomputer, to avoid the loading of a lot of extra machine code for these I/O functions along with the program, which may be a serious problem if the target computer's memory is limited. They are not available when the 6812 microcontroller is used without a host computer in a real stand-alone system. The target 6812 computer being debugged is designed to carry out predetermined illegal instructions to actually execute these procedures in the host computer. When the monitor gets an illegal instruction "interrupt," the host computer reads the instruction from the target computer's memory. If it is one of the predetermined illegal instructions selected to call these input-output procedures, the host will execute the procedures, examining the target memory and writing data into it as needed, and then resume the program after the illegal instruction. The C procedures `strcpy` and `strcat` are also very useful but can be easily loaded into the target computer to manipulate strings being input or output.

The procedure `printf` requires a character string format as its first parameter and may have any number of additional parameters as required by the format string. The format string uses a percent sign (%) to designate the input of a parameter, and the characters following the % establish the format for the output of the parameter value. While there are a large number of formats, we generally use only a few. The string "%d" will output the value in decimal. For instance, if `i` has the value 123, then

```
printf("The number is %d", i);
```

will print on the terminal

The number is 123

Similarly, "%X" will output the value in hexadecimal. If *i* has the value 0x1A, then

```
printf("The number is 0x%X", i);
```

will print on the terminal

0x1A

If a number is put between the % and the d or X letters, that number gives the maximum number of characters that will be printed. If that number begins in a zero, it specifies the exact number of characters that are printed for the corresponding parameter.

Similarly, "%s" will output a string of characters passed as a parameter. For instance,

```
char st[6] ="ALPHA";
printf("%s", st);
```

will print on the terminal

ALPHA

Observe that the integers for decimal or hexadecimal output are passed by value, but the string is passed by name, as we discussed earlier in this section. Thus, for example,

```
printf("Hi There\nHow are you?");
```

will print on the terminal

Hi There
How are you?

Decimal numbers can be input using `inDec`, and hexadecimal numbers are input using `inHex`. They stop inputting when a nondigit character is typed in. Character strings input using `gets` can be analyzed and disassembled using indexes in or pointers to the strings. Character strings can be assembled for output using `puts` by the procedures `strcpy` and `strcat`. The procedure `strcpy(s1,s2);` will copy string `s2` (up to the null character at its end) into string `s1`. The procedure `strcat(s1,s2);` will concatenate string `s2` (up to the null character at its end) onto the end of string `s1`. These simple procedures are shown below.

```
strcpy(s1,s2) char *s1, s2; { while(*s2) *s1++ = *s2++; }

strcat(char *s1, char *s2) { while(*s1)s1++; while(*s2) *s1++=*s2++; }
```

We have examined techniques for calling subroutines and passing arguments. We have also learned to use some simple tools for input and output in C. We should now be prepared to write subroutines for interface software.

2.3.2 Programming Style

We conform our programming techniques to some style to make the program easier to read, debug, and maintain. The use of a consistent style is recommended, especially in longer programs where static efficiency is not paramount. For instance, we can rigidly enforce reentrancy and use some conventions to make this rather automatic. Another programming style, *structured programming,* uses only simple conditional and loop operations and avoids GOTO statements. After this we discuss top-down and bottom-up programming, which leads to an introduction to object-oriented programming.

An element of structured programming is the use of single entry point, single exit point program segments. This style makes the program much more readable because to get into a program segment, there are no circuitous routes that are hard to debug and test. The use of C for specification and documentation can force the use of this style. The conditional and loop statements described in §2.1 are single entry point, single exit point program segments, and they are sufficient for almost all programs. The *while* loop technique is especially attractive because it tests the termination condition before the loop is done even once, so programs can be written that accommodate all possibilities, including doing the loop no times. And the *for* loop is essentially a beefed-up *while* loop. You can use just these constructs. That means avoiding the use of GOTO statements. Several years ago, Professor Edsger Dijkstra made the then controversial remark, "GOTOs Considered Harmful!" Now, most good programmers agree with him. We heard a story (from Harold Stone) that Professor Goto in Japan has a sign on his desk that says "Dijkstras Considered Harmful!" (Professor Goto denies this.) A significant exception is the reporting of errors. We sometimes GOTO an error-reporting or error-correcting routine if an error is detected, such as an abnormal exit point for the program segment. Errors can alternatively be reported by a convention such as using the carry bit to indicate the error status: you exit the segment with carry clear if no errors are found and exit with carry set if an error is found. Thus, all segments can have single exit points.

Top-down design produces programs more quickly than ad hoc and haphazard writing. You write a main program that calls subroutines (or just program segments) without writing the subroutines (or segments). A procedure is used if a part of the program is called many times, and a program segment – not a procedure – is used if a part of the program is used only once. The abstract specification is translated into a main program, which is executed to check that the subroutines and segments are called up in the proper order under all conditions. Then the subroutines (or segments) are written in lower-level subroutines (or segments) and tested. This is continued until the lowest-level subroutines (or segments) are written and tested. Superior documentation is needed in this methodology to describe the procedure and program segments so they can be fully tested before being written. Also, subroutine inputs and outputs have to be carefully specified.

The inverse of top-down design is *bottom-up design,* in which the lowest-level subroutines or program segments are written first and then fully debugged. These are built up, bottom to top, to form the main program. To test the procedure, you write a short program to call the procedure, expecting to discard this program when the next higher-level program is written. Bottom-up design is especially useful in interface design. The lowest-level procedure that actually interfaces to the hardware is usually the trickiest to debug. This methodology lets you debug that part of the program with less interference from other parts. Bottom-up design is like solving an algebra problem with three separate equations, each equation in one unknown. Arbitrarily putting all the software and hardware together before testing any part of it is like simultaneously solving three equations in three unknowns. As the first algebraic problem is much easier, the use of bottom-up design is also a much easier way to debug interfacing software. In a senior level interfacing course at the University of Texas, students who tried to get everything working at once spent 30 hours a week in the lab, while those who used bottom-up design spent less than 10 hours a week on the same experiments.

Combinations of top-down and bottom-up design can be used. Top-down design works well with the parts of the program that do not involve interfacing to hardware, and bottom-up design works better with the parts that do involve interfacing.

2.3.3 Object-Oriented Programming

The concept of object-oriented programming was developed to program symbolic processes, database storage and retrieval systems, and user-friendly graphic interfaces. However, it is ideally suited to the design of I/O devices and systems that center on them. It provides a programming and design methodology that simplifies interface design.

Object-oriented programming began with the language SMALLTALK. Programmers using C wanted to use object-oriented techniques. Standard C cannot be used, but a derivative of C, called C++, has been developed to utilize objects with a syntax similar to that of C. A 6812 C++ compiler written by HIWARE was used to generate code for 6812-based microcontrollers to check out the ideas described below.

C++ has a few differences from C. C++ permits declarations inside expressions, as in `for (int i = 0; i < 10; i++)`. Parameters can be *called by reference* using a PASCAL-like convention; & in front of a formal parameter is like VAR. See the actual parameter *a* and corresponding formal parameter *b* below.

```
void main(){ char a;                void f(char &b) {
      f(a);                                  b = '1';
}                                   }
```

The advantage of call by reference over call by name, used in C or C++, is clarity. You do not need to put the `&` in front of the actual parameter in the calling procedure, nor do you need to put the `*` in front of the formal parameters where they appear inside the procedure.

An object's data are *data members* and its procedures are *function members;* data and function members are *encapsulated* together in an *object*. Combining them is a good idea because the programmer becomes aware of both together and logically separates them from other objects. As you get the data, you automatically get the function members used on it. In the class for a queue shown below, observe that data members `QSize`, `Qlen`, `error`, `QIn`, `QOut`, and `QEnd` are declared much as in a C `struct`, and function members `push`, `pull`, and `error` are declared pretty much like prototypes are declared in C. Protection terms, `protected`, `public`, and `virtual`, will be soon explained.

```
class Queue { protected: char Error; int *QIn, *QOut, *QEnd;
   public: char QSize, Qlen; Queue(char); virtual void push(int);
   virtual int pull(void); virtual char error(void);
};
```

A class's function members are written rather like C procedures, with the return type and class name in front of two colons and the function member name.

```
void Queue:: push (int i) {
    if((Qlen += 2) > QSize) Error = 1;
    if(QEnd == QIn) QIn -= QSize;
    *(QIn++)=i;
}

int Queue:: pull () {
    if((Qlen -= 2) < 0) Error = 1;
    if(QEnd == QOut) QOut -= QSize;
    return (QOut++);
}

char Queue::error()
    { char i; i=Error; Error=0; return i; }
```

Any data member, such as `QSize`, may be accessed inside any function member of class *queue*, such as `push()`. Inside a function member, when a name appears in an expression, the variable's name is first searched against local variables and function formal parameters. If the name matches, the variable is local or an argument. Then the variable is matched against the object data members, and finally against the global variables. In a sense, object data members are global among the function members because each of them can get to these same variables. However, it is possible that a data member and a local variable or argument have the same name, such as `size`. The data member can be identified as `this->size`, using the keyword `this` as a pointer to the object that called the function member, while the local variable or argument is just `size`.

C++ uses constructors, allocators, destructors, and deallocators. An *allocator* allocates data member storage; it has the same function name as the class name. A *constructor* initialize these variables, a *destructor* terminates the use of an object, and a *deallocator* recovers storage for data members for later allocation. We do not use a deallocator in our experiments; it is easier to reset the 6812 to deallocate storage. A destructor has the same function name as the class name but has a tilde (~) in front of the function member name. We will use destructors later. Here is `Queue`'s constructor:

```
Queue::Queue(short i)  {
    QEnd = (QIn = QOut = (int*)allocate(i) + (QSize = i));
    Qlen = Error = 0;
}
```

Throughout this text, a conventional C procedure *allocate* provides buffer storage for an object's data members, as its allocator, and for an object's additional storage such as its queues. The contents of global variable `free` are initialized to the address just above the last global variable; storage between `free` and the stack pointer is subdivided into buffers for each object by the `allocate` routine. Note that the stack used for return addresses and local variables builds from one end and the allocator builds from the other end of a common RAM buffer area. The procedure `allocate`'s return type `void *` indicates a pointer to anything.

```
char *free = 0xb80;
void *allocate(int i) {
    void *p = free;
    free += i;
    return p;
}
```

A global object of a class is declared and then used as shown below:

```
Queue Q(10);
void main() { int i;
    Q.push(1);
    i = Q.pull();
}
```

The object's data members, `QSize`, `Qlen`, `Error`, `QIn`, `QOut`, and `QEnd`, are stored in global memory just the way a global *struct* is stored. If a data member could be accessed in `main`, as in `i = Q.Error` or `i = Qptr->Error` (we see later that it can't be so accessed), the data member is accessed by using a predetermined offset from the base of the object exactly as a member of a C `struct` is accessed. Function members can be called using a notation like that used to access `struct` data; `Q.push(1)` calls `Q`'s *push* function member to push 1 onto `Q`'s queue. The "`Q.` " in front of the function member is rather like a first actual parameter, as in `push(Q,1)`, but can be used to select the function member to be run, as we will see later, so it appears before the function.

The class's constructor is executed before the main procedure is executed, to initialize the values of data members of the object. This declaration `Q(10)` passes actual parameter `10` to the constructor, which uses it, as formal parameter *i*, to allocate 10 bytes for the queue. The queue is stored in a buffer assigned by the *allocate* routine.

Similarly a local object of a class can be declared and then used as shown.

```
void main() { int i;
    Queue Q(10);
    Q.push(1);
    i = Q.pull();
}
```

The data members `QSize`, `Qlen`, `Error`, `QIn`, `QOut`, and `QEnd` are stored on the stack, and the constructor is called just after *main* is entered to initialize these data members; it then calls `allocate` to find room for the queue. The function members are called the same way, as in the first example when the object was declared globally.

Alternatively, a pointer `Qptr` to an object can be declared globally or locally; thus an object is set up and then used as shown.

```
void main() { int i;
   Queue * Qptr;
   Qptr = new Queue (20);
   Qptr ->push(1);
   i = Qptr ->pull();
}
```

In the first line, `Qptr,` a pointer to an object of class *queue,* is declared here as a local variable. (Alternatively it could have been declared as a global variable pointer.) The expression `Qptr = new Queue (20);` is put anywhere before the object is used. This is called *blessing* the object. The allocator and then the constructor are both called by the operator *new.* The `allocate` routine automatically provides room for the data members `QSize`, `Qlen`, `Error`, `QIn`, `QOut`, and `QEnd`. The constructor explicitly calls up the allocate procedure to obtain room for the queue itself and then initializes all the object's data members. After it is thus blessed, the object can be used in the program. An alternative way to use a pointer to an object is with a `#define` statement to insert the asterisk as follows:

```
#define Q (*Qptr)
void main() { int i;
   Queue *Qptr = new Queue(20);
   Q.push(1); i = Qptr.pull();
}
```

Wherever the symbolic name `Q` appears, the compiler substitutes `(*Qptr)` in its place. Note that `*ptr.member` is the same as `ptr->member`. So this makes the syntax of the use of pointers to objects match the syntax of the use of objects most of the time. However, the blessing of the object explicitly uses the pointer name.

A hierarchy of derived and base classes, inheritance, overriding, and factoring are all related ideas. A class can be a *derived class* (also called *subclass*) of another class, and a hierarchy of classes can be built up. We create derived classes to use some of the data or function members of the base class, but we can add members to, or replace some of the members of, the base class in the derived class. For instance, the aforementioned class *Queue* can have a derived class `CharQueue` for `char` variables; it declares a potentially modifiable constructor, and different function members `pull` and `push` for its queue. When defining the class `CharQueue`, the *base class* (also called *superclass*) of `CharQueue` is written after its name and a colon as `:Queue`. A class such as `Queue`, with no base class, is called a *root class;* its declaration has no colon and base class.

```
class CharQueue : public Queue {
   public:CharQueue(char); virtual void push(int);
   virtual int pull(void);
};
```

```
CharQueue::CharQueue(char i) : Queue(i) {}

void CharQueue:: push (int i)  {
   if((Qlen++)>(QSize))Error=1;
   if(QEnd==QIn)QIn-=QSize;*(((char*)QIn)++)=i;
}

int CharQueue:: pull () {
   if((Qlen--)==0)Error=1;
   if(QEnd==QOut)QOut-=QSize;
   return *(((char*)QOut)++);
}
```

The notion of *inheritance* is that an object will have data and function members defined in the base class(es) of its class as well as those defined in its own class. The derived class inherits the data members or function members of the parent that are not redefined in the derived class. If we execute `Qptr->error(10);` then the function member *Queue:: error* is executed because `CharQueue` does not declare a different `error` function member. If a function member cannot be found in the class which the object was declared or blessed for, then its base class is examined to find the function member to be executed. In a hierarchy of derived classes, if the search fails in the class's base class, the base class's base class is searched, and so on, up to the root class. *Overriding* is the opposite of inheritance. If we execute `Qptr->push(1);` the function member `CharQueue:: push` is executed rather than `Queue:: push` because the class defines an overriding function member. Although this derived class uses no additional variables, these rules of inheritance and overriding apply to data members as to function members.

Most programmers face the frustration of several times rewriting a procedure, such as one that outputs characters to a terminal, only to wish they had saved an earlier copy for use in later programs. Frequently reused procedures can be kept in a library. However when we collect such common routines, we will notice some universal parts in different routines. Common parts of these library procedures can be put in one place by *factoring*. Factoring is common to many disciplines – for instance, to algebra. If you have $ab + ac$ you can factor out the shared term a and write $a\,(b + c)$, which has fewer multiplies. Similarly, if a large number of classes use the same function member, instead of reproducing the function member in each, such a function member can be in one place in a base class where all derived classes inherit it. Also, if an error were discovered and corrected in a base class's function member, it is automatically corrected for use in all the derived classes that use the common function member. We will use these ideas of factoring and inheritance to develop a library of classes for 6812 I/O interfacing. The `CharQueue`'s constructor, using the notation *:Queue(i)* just after the constructor's name `CharQueue::CharQueue(char i)` and before the constructor's body in `{}`, calls the base class's constructor before its own constructor is executed. In fact, `CharQueue`'s constructor does nothing else, as is denoted by the empty procedure `{}`. All derived classes declare their constructor, even if that constructor does nothing but call its base class's constructor. Other function members can call their base's function members by the keyword *inherited* as in `inherited::push(i);` or by explicitly naming the class, in front of the call to the function, as in `Queue::push(i);`.

Consider the hypothetical situation where a program can declare classes `Queue` and `CharQueue`. Inside `main`, are a number of statements `Qptr->push(1);` and `i = Qptr->pull();`. At compile time, either of the objects can be declared for either *Queue* or `CharQueue`, using conditional compilation; for instance, see the program on the left,

```
void main(){ int i;
#ifdef mode
    Queue Q(10);
#else
    CharQueue Q(10);
#endif
    Q.push(1); i = Q.pull();
}
```

```
void main(){ int i; Queue *Qptr;
#ifdef mode
    Qptr = new Queue(10);
#else
    Qptr = new CharQueue (10);
#endif
    Qptr->push(1); i = Qptr->pull();
}
```

which declares *Q* a class `Queue` object if `mode` is `#declared;` otherwise it is a class `CharQueue` object. Then the remainder of the program is written unchanged. Alternatively, at compile time, a pointer to objects can be blessed for either the `Queue` or the `CharQueue` class. The program above right shows this technique.

Moreover, a pointer can be blessed to be objects of different classes at run time. At the very beginning of `main`, assume a variable called *range* denotes the actual maximum data size saved in the queue:

```
void main(){ int i, range; Queue *Qptr;
    if(range<128) Qptr = new CharQueue(10);
    else Qptr = new Queue(10);
    Qptr->push(1);
    i = Qptr->pull();
}
```

`Qptr->push(1);` and `i = Qptr->pull();` will use the queue of 8-bit members if the range is small enough to save space, otherwise it will use a queue that has enough room for each element to hold the larger data, as will be explained shortly.

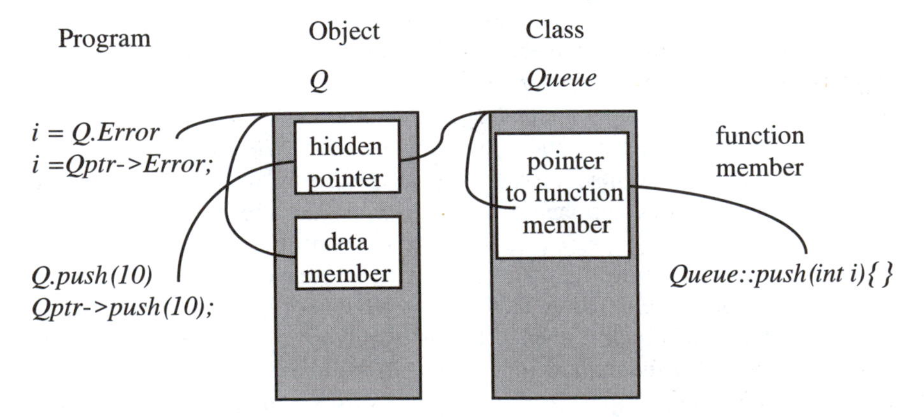

Figure 2.5. An Object and Its Pointers

Polymorphism means that any two classes – especially a class and one of its inherited classes – can declare the same function member name and argument. It means that simple intuitive names like `push` can be used for interchangeable function members of different classes. Polymorphism will be used later when we substitute one object for another object; the function member names and arguments do not have to be changed. You don't have to generate obscure names for functions to keep them separate from each other. Moreover, in C++, the number and types of operands, called the function's *signature*, are part of the name when determining if two functions have the same name. For instance, `push(char a)` is a different function than `push(int a)`.

When a pointer to an object is able to be blessed at runtime as a pointer to different classes, the *virtual* function becomes very useful. If we do not insert `virtual` in front of a function member in the class declaration, then the function is directly called by means of a `JSR`, `BSR`, or `LBSR` instruction, just like a normal C procedure. If a function member is declared virtual, then to call it, we look its address up in a table shown on the right side of Figure 2.5. This table is used because generally, a lot of objects of the same class might be declared or blessed, and they might have many virtual function members. For instance there could be queues for input and for output, and queues holding temporary results in the program. As Figure 2.5 shows, to avoid storing the pointers to virtual function members with every object that uses them, their pointers are collected together and put in a common table for the class. This is accomplished by having the *new* operator at runtime put the address of a different table into a hidden pointer (Figure 2.5) that points to a table of virtual function member addresses, depending on the run-time value of *range* in the last example. Then, data members are easily accessed by the pointer, and virtual function members are almost as easily accessed by means of a pointer to a pointer. If a function member is executed, as in `Q.Push(i)` or `Qptr->push(i)`, the object's hidden pointer has the address of a table of function members; the specific function member is jumped to by using a predetermined offset from the hidden pointer. Different objects of the same class point to the same table. Note that data members of different objects of a class are different data items, but function members of different objects of a class are common to all the objects of the same class via this table.

The user may want to enable or disable virtual functions throughout an application. We suggest writing `VIRTUAL` in place of `virtual` in the class definition. If we then write `#define VIRTUAL virtual`, the class member functions are declared virtual. But if we write `#define VIRTUAL /**/` then the member functions are not virtual.

When object pointers are blessed at run time and have virtual function members, if a virtual function member appears for a class and is overridden by function members with the same name in its derived classes, the sizes and types of all the arguments should be the same, due to the fact that the compiler does not know how an object will be blessed at run time. If they were not, the compiler would not know how to pass arguments to the function members. For this reason, we defined the arguments of `CharQueue's push` and `pull` function members to be `int` rather than `char`, so that the same function-member name can be used for the `int` version, or a `char` version, of the queue. This run-time selection of which class to assign to an object isn't needed with declarations of objects, but only with blessing of object pointers, since the run-time program can't select at compile time which of several declarations might be used. Also, the pointer to the object must be declared an object of a common base class if it is to be used for several classes.

Information hiding limits access to data or function members. A member can be declared *public,* making it available everywhere; *protected,* making it available only to function members of the same class or a derived class of it; or *private,* making it available only to the same class's function members and hiding it from other functions. These words appearing in a class declaration apply to all members listed after them until another such word appears; the default if no such words appear is *private.* The data member *Error* in the class *Queue* cannot be accessed by a pointer in `main` as in `i = Q.Error` or `i = Q->Error` because it is not `public`, but only through the public function member `error()`. This way, the procedure `main` can read (and automatically clear) the *Error* variable, but cannot accidentally or maliciously set `Error`, nor can it read it, forgetting to clear it. You should protect your data members to make your program much more bug-proof. Declare all data and function members as *private* if they are only to be used by the class's own function members, declare them *protected* if they might be used by derived classes, and declare them *public* if they are used outside the class and its derived classes.

Templates generalize object-oriented programming. A *template class* is a class that is defined for an arbitrary data type, which is selected when the object is blessed or declared. The class declaration and the function members have a prefix like `template <class T>` to allow the user to bless or declare the object for a specific class having a particular data type, as in `Q = new Queue<char>(10)`. The generalized class definition is given below; you can substitute the word `char` for the letter `T` everywhere in the declarations or the class function members.

The following class also exhibits another feature of C++ – especially a class and its inherited classes – the ability to write the function member inside the declaration of the class. The function is written in place of the prototype for the function. This is especially useful when templates are used with short function members because otherwise the notation `template <class T>` and the class name *Queue::* would have to be repeated before each function member.

```
template <class T> class Queue {

  private: T *QIn, *QOut, *QEnd, QSize, Qlen; char Error;

  public : Queue<T>::Queue(short i)
     { QEnd=(QIn=QOut=(T*)allocate(i) + (QSize=i)); Qlen=Error= 0; }

  VIRTUAL Queue<T>:: push (T i)
     { if((++Qlen)>=QSize)Error=1;if(QEnd==QIn)QIn-=QSize;*(QIn++)=i; }

  VIRTUAL T pull ()
     {if((--Qlen)<0)Error=1;if(QEnd==QOut)QOut-=QSize;return*(QOut ++);}

  VIRTUAL char error() { char i; i=Error; Error=0; return i; }

};
```

If you declare Queue<char> Q(10); or bless Qptr = new Queue<char>(10); then a queue is implemented that stores 8-bit data; but if you declare Queue<int> Q(10) or bless Qptr = new Queue<int>(10);, then a queue is implemented that stores 16-bit data. Clearly, templates permit us to define one generalized class that can be declared or blessed to handle 8-bit, 16-bit, or 32-bit signed or unsigned data when the program is compiled. This selection must be made at compile time, since it generates different calls.

Operator overloading means that the same operator symbol generates different effects, depending on the type of the data it operates on. The C compiler already effectively loads its operators. The + operator generates an ADDB instruction when adding data of type char and an ADDD instruction when adding data of type int. What C++ does but C cannot do is to overload operators to do different things when an operand is an object, which depends on the object's definition. In effect, the programmer provides a new part of the compiler that generates code for symbols, depending on the types of data used with the symbols. For instance, the << operator used for shift can be used for input or output if an operand is an I/O device. The expression Q << a can be defined to output the character *a* to the object Q, and Q >> a can be defined to input into a a character from the object Q. This type of operator overloading is used in I/O streams for inputting or outputting formatted character strings. Without this feature, we simply have to write our function calls as a=Q.Input() and Q.Output(a) rather than Q << a or Q >> a. However, with overloading we write a simpler program; for instance we can write an I/O stream Q << a << " is the value of " << b;. Overloading creates arithmetic-looking expressions that use function members to evaluate them. Besides operators like + and -, C++ considers cast and assignment (=) to be operators. In the following example, we overload the cast operator as shown by *operator T ();* and the assignment operator as shown by T operator = (T); T will be a cast, like char, so operator T (); will become operator char ();. Whenever the compiler has an explicit cast like *(char)i,* where *i* is an object, or an implicit cast where object i appears in an expression needing a char, the compiler calls the user-defined overloaded operator to perform the cast function. Similarly, wherever the compiler has calculated an expression that has a char value but the assignment statement has an object i on its left, the compiler calls up the overloaded = operator the user specifies with T operator = (T);.

```
template <class T> class Queue {

    private: T *QIn, *QOut, *QEnd, QSize, Qlen; char errors;

    public: Queue(short i)  {
        QEnd=(QIn=QOut=(T*)allocate(i)+(QSize=i)); Qlen=errors=0;
    }

    VIRTUAL void push(T i) {
        if((++Qlen)>QSize)errors=1;
        if(QEnd==QIn)QIn-=QSize;
        *(QIn++)=i;
```

```
    VIRTUAL T pull(void) {
        if((--Qlen)>0)errors=1;
        if(QEnd==QOut)QOut-=QSize;
        return *(QOut ++);
    }

    VIRTUAL char error(void)
        { char i; i = errors; errors=0; return i; }

    operator T () { return pull(); };                /* cast operator */

    T operator=(T data){push(data);return data;};/* assignment */
};
```

Now, whenever the compiler sees an object on the left side of an equal sign when it has evaluated a number for the expression on the right side and it would otherwise be unable to do anything correctly, the compiler looks at your declaration of the overloaded assignment operator to determine that the number will be pushed onto the queue. The expression `Q = 1;` will do the same thing as `Q.push(1);` and `*Qptr = 1` will do the same thing as `Qptr->push(1)`. Similarly, whenever the compiler sees an object anywhere on the right side of an equal sign when it is trying to get a number and it would otherwise be unable to do anything correctly, the compiler looks at your declaration of the overloaded cast operator, to determine that the number will be pulled from the queue. The expression `i = Q;` will do the same thing as `i = Q.pull()`, and `i = *Qptr` will do the same thing as `i = Qptr->pull()`. Now if a queue `Q` returns a temperature in degrees centigrade, you can write an expression like `degreeF = (Q * 9) / 5 + 32;` or `degreeF = (*Qptr * 9) / 5 + 32;`, and the compiler will pull an item from the queue each time it runs into the `Q` symbolic name. While overloading of operators isn't necessary, it simplifies expressions so they look like common algebraic formulas.

A derived class usually defines an overloaded assignment operator even if its base class has defined an overloaded assignment operator in exactly the same way, because the (Metrowerks) C++ compiler can get confused with the =. If `Q1` and `Q2` are objects of class ***Queue*****<*****char*****>**, then `Q1 = Q2;` won't pop an item from `Q2` and push it onto `Q1`, as we would wish when we use overloaded assignment and cast operators, but "clones" the object, copying `device2`'s contents into `device1` as if the object were a `struct`. That is, if ***Q1***'s class's base class overrides = but `Q1`'s class itself does not override =, `Q1 = Q2;` causes `Q2` to be copied into `Q1`. However, if = is overridden in `Q1`'s class definition, the compiler treats = as an overridden assignment operator, and `Q1 = Q2;` pops an item from `Q2` and pushes it onto `Q1`. The derived class has to override = to push data. The = operator, though quite useful, needs to be carefully handled. All our derived classes explicitly define `operator =` if "`=`" is to be overridden.

C++ object-oriented programming offers many useful features. Encapsulation associates variables with procedures that use them in classes; inheritance permits factoring out of procedures that are common to several classes; overriding permits the redefinition of procedures; polymorphism allows common names to be used for procedures; virtual functions permit different procedures to be selected at run time; information hiding protects data; template classes generalize classes to use different data types; and operator overloading permits a program to be written in the format of algebraic expressions. If the programmer doesn't have C++ but has a minimal C compiler, many of the features of object-oriented programming can be simulated by adhering to a set of conventions. For instance, in place of a C++ call `Queue.push()`, one can write instead `QueuePush()`. Information hiding can be enforced by only accessing variables like `QptrQSize` in procedures like `QueuePush()`. C++ gives us a good model for useful C conventions.

2.3.4 Optimizing C Programs Using Declarations

We will discuss some of C++'s techniques that can be used to improve your interface software, and some techniques that you can use to get around its limitations for this application. While the techniques discussed here are specific to a C++ compiler, if you are using another compiler or cross-compiler, similar ideas can be implemented.

C and C++ have some additional declaration keywords. If the word *register* is put in front of a local variable, that variable will be stored in a register. Data variables declared as in `register int i` or `register char c`, and pointer variables declared as in `register int *i` or `register char *c`, are stored in a register such as `Y`. Putting often-used local variable in registers instead of on the stack obviously speeds up procedures. It also puts them in known places that can be used in embedded assembly language, which is discussed in the next subsection. You can check your understanding of the use of these registers by writing a C procedure with embedded assembly language and then disassembling your program.

If the word *static* is put in front of a local variable, that variable will be initialized, stored, and accessed as is a global variable, but will only be "known" to the procedure as a local variable is. If the word *static* is put in front of a global variable or a procedure name, that variable or procedure will only be known within the file and not linked to other files. For instance if a C++ project is composed of several files of source code such as file1.c, file2.c, and so on, then if a procedure `fun()` in file1.c is declared static, it cannot be called from a procedure in file2.c. However both file1.c and file2.c can have procedures `fun()` in them without creating duplicate procedure names when they are linked together to run (or download) the procedures. If the word *static* is put in front of a class data member, that variable is common to all the objects of the class, just as function members are common to all objects of the class.

2.3.5 Optimizing C Programs with Assembly Language

Assembly language can be embedded in a C program. It is the only way to insert some instructions, like `CLI`, which is used to enable interrupts. It can be used to implement better procedures than are produced from C source code by the compiler. For instance, the `DBNE` instruction can be put in assembly language embedded in C to get a faster *do-while loop*. Finally, the `DC.B` or `DC.W` directives can be used to build the machine code of instructions that are in the 6812 instruction set, such as the use of illegal instructions as calls to debug routines, and are thus not generated by C.

Many C++ compilers restrict your ability to insert assembly language into its C++ procedures. Having implemented protection using *private, protected,* and *public* declarations, these compilers don't want the programmer to get around this protection using embedded assembly language. Generally, to embed assembly language in them, the procedure body is completely written in machine code in *inline procedures,* as in *inline f() = 0x1234;*, or in a list, as in *int inline f(int) = {0x1234, 0x5678, 0x9abcd};* (parameters are optional). However, HIWARE's C++ compiler permits embedded assembly language.

A single embedded assembly-language instruction is put after an expression *asm.* For instance, the line

```
asm SEI
```

will insert the SEI instruction in the C procedure. However, no other statements may appear after this construction. Also, several assembly-language instructions can be put on consecutive lines; the first line is preceded by *asm{* and the last is followed by a matching *}*. These first and last lines should have no assembly-language statements on them. Each intervening line will have a different assembly-language statement on it.

Parameters can be used in assembly-language instructions. For instance, a procedure clr to quickly clear a block of N bytes starting at location A can be written

```
void clr(char *A, int N) { asm {
                LDY A,SP
        l1:     CLR 1,Y+
                DBNE D,l1
        }
}
```

The *char* pointer variable *A is pushed on the stack before the subroutine is called; it happens to be at an offset of 2,SP when the subroutine is running. The symbolic name A is used in LDY A,SP, so you don't have to manually compute the stack offset to get the parameter values. The rightmost argument, label *N,* is passed in accumulator D.

Consult Chapter 1 and the CPU12 Reference Manual to understand the machine coding of 6812 instructions and the meaning of instructions. Using these resources, you should be able to insert assembly language and machine code into your C++ programs.

2.4 Conclusions

In this chapter, we have surveyed some software background needed for microcomputer interfacing. The first section introduced C. It was followed by the description and handling of data structures. C constructs were introduced in order to make the implementation of storage structures concrete. Indexed and sequential structures were surveyed. We then covered programming style and procedure-calling and argument-passing techniques. We then discussed structured, top-down, and bottom-up programming, introducing object-oriented C. Finally, we showed some techniques used to improve C procedures or insert necessary corrections to the code produced by the compiler.

If you found any section difficult, we can recommend additional readings. *The C Programming Language* by Kernighan and Richie, who were the original developers of C, remains the bible and fundamental reference for C. A fundamental reference for C++ is Tutorials on object-oriented programming are available from the IEEE Computer Society Press. Other fine books are available on these topics, and more are appearing daily. We might wish to contact a local college or university instructor who teaches architecture, microprocessors, or C programming for the most recent books on these topics.

Do You Know These Terms?

See page 36 for instructions.

procedure
main
declaration of a parameter or a variable
procedure-name
cast
statement
allocate
deallocate
relational operator
logical operator
case
while
do while
for
break
information structure
data structure
storage structure
define
enum
vector
zero-origin indexing
precision
cardinality
list
structure
struct
typedef
bit field
linked list
array
row major order
column major order
table
string
ASCII code
character string
Huffman code
binary tree
deque
overflow error
underflow error
indexable deque
buffer
public
stack
queue
arguments
parameters
formal parameter
actual parameter
call by value
call by result
call by reference
call by name
return statement
prototype
extern
void
macro
#define
getchar
gets
InDec
InHex
putchar
puts
printf
strcpy
strcat
structured programming
top-down design
bottom-up design
data members
function members
encapsulate
object
allocator
constructor
destructor
deallocator
blessing
new
derived class
subclass
base class
superclass
root class
inheritance
overriding
inherited
factoring
range
polymorphism
virtual
information hiding
protected
private
template class
operator overloading
register
static

Problems

Problem 4 is a paragraph correction problem (see the guidelines on page 38). Other problems in this chapter and many in later chapters are C and C++ language programming problems. We recommend the following guidelines for problems answered in C: In `main()` or "self-initializing procedures", each statement must be limited to C operators and statements described in this chapter, should include all initialization operations, and should have comments as noted at the end of §2.1. C subroutines should follow the C++ style for C procedures recommended in §2.3.1. Unless otherwise noted, you should write programs with the greatest static efficiency.

1. Write a shortest C or C++ procedure *void main()* that will find `x` and `y` if `ax + by = c` and `dx + ey = f`. Assume `a, b, c, d, e, f` are global integers that somehow are initialized with the correct parameters, and your answers, *x* and *y*, are stored in local variables in `main()`. (You might verify your program with a source-level debugger).

2. Write a shortest C or C++ procedure `void main()` that will sort five numbers in global integer vector `a[5]` using an algorithm that executes four passes, where each pass compares each `a[i]`, for all `i` running from 0 to 3, with `a[j]`, for `j = i + 1` to `3` and puts the smaller element in `a[i]` and larger in `a[j]`.

3. Write a C or C++ procedure *void main()* to generate the first five Fibonacci numbers **F**(i), (**F**(0) = **F**(1) = 1 and for i > 1, **F**(i) = **F**(i-1) + **F**(i-2)) in global `int a0, a1, a2, a3, s4` so that `ai` is **F**(i). Compute **F**(2), **F**(3), and **F**(4).

4.* The information structure is the way the programmer sees the data and is dependent on such details as the size of words and positions of bits. The data structure is the way the information is actually stored in memory, right down to the bit positions. A queue is a sequence of elements with two ends, in which an element can be pushed or pulled from either the top or bottom. A stack is a special case of queue, where an element can only be pushed from one end and pulled from the other. An important element in constricted programming is the use of single-entry single-exit point in a program segment. A calling routine passes the address of the arguments, called formal parameters, to the procedure. In call by value and call by result, the data are not actually moved to another location, but the address of the data is given to the procedure. Large vectors, lists, and arrays can be more effectively called by reference than by value.

5. A two-dimensional array can be simulated using one-dimensional vectors. Write a shortest C or C++ procedure *void main()* to multiply two 3 x 3 integer matrices, **A** and **B**, putting the result in **C**, all stored as one-dimensional vectors in row major order. Show the storage declarations/directives of the matrices, so that A and B are initialized as

$$\mathbf{A} = \begin{matrix} 1 & 2 & 3 \\ 4 & 5 & 6 \\ 7 & 8 & 9 \end{matrix} \qquad \mathbf{B} = \begin{matrix} 10 & 13 & 16 \\ 11 & 14 & 17 \\ 12 & 15 & 18 \end{matrix}$$

6. A `long` can be simulated using one-dimensional `char` vectors. Suppose *A* is a zero-origin 5 x 7 array of 32-bit numbers, with each number stored in consecutive bytes most significant byte first, and the matrix stored in row major order in a 140-byte `char` vector. Write a C or C++ procedure `int get(char *a, unsigned char i, unsigned char j, char *v)` where `a` is the storage array, `i` and `j` are row and column, and `v` is the vector result. If `0 ≤ i < 5` and `0 ≤ i < 7`, this procedure puts the `i`th row, `j`th column 32-bit value into locations `v, v+1, v+2,` *and* `v+3`, most significant byte first, and returns 1; otherwise it returns a 0 and does not write into `v`.

7. A *struct* can be simulated using one-dimensional arrays *char* vectors. The `struct{long v1; unsigned int v2:4, v3:8, v4:2,vv5:1};` has, tightly packed, a 32-bit element *v1*, a 4-bit element *v2*, an 8-bit element *v3*, a 2-bit element *v4*, a 1-bit element *v5*, and an unused bit to fill out a 16-bit *unsigned int*. Write shortest C or C++ procedures `void getV1(char *s, char *v), void getV2( char *s, char *v), void getV3(char *s, char *v), void getV4(char *s, char *v), void getV5(char *s, *v), void putV1(char *s, char *v), void putV2(char *s, char *v), void putV3(char *s, char *v), void putV4(char *s, char *v), void putV5(char *s, *v),` in which `get` will copy the element from the *struct* to the vector and *put* will copy the vector into the `struct`. For example, `getV2(s, v)` copies element `V2` into `v`, and `putV5(s, v)` copies `v` into element `V5`.

8. Write a shortest C or C++ procedure `void main()` and procedures it calls, without any assembly language, that will first input up to 32 characters from the keyboard to the 6812 (using `getchar()`), and will then jump to one of the procedures, given below, whose name is typed in (the names can be entered in either upper case or lower case, or a combination of both, but a space is represented as an underscore). The procedures: – `void start(), void step_up(), void step_down(), void recalibrate(),` and `void shut_down()` – just type out a message; for instance, *start()* will type out "Start Entered" on the host computer monitor. The `main()` procedure should generate the least number of bytes of object code possible and should run on HIWAVE. Although you do not have to use HIWAVE to answer this problem, you can use it without penalty, and it may help you get error-free results faster.

9. Suppose a string such as "SEE THE MEAT", "MEET A MAN", or "THESE NEAT TEAS MEET MATES" is stored in `char string[40];`. Using one-dimensional vector rather than linked list data structures to store the coding/decoding information:

a. Write a C or C++ procedure `encode()` to convert the ASCII `string` to Huffman code, as defined by the coding tree in Figure 2.6a, storing the code as a bit string, first bit as most significant bit of first element of `int code[16];`.

b. Write a C or C++ procedure `decode()` that decodes such a code in `int code[16]`, using the coding tree in Figure 2.6a, putting the ASCII string back as it was in `char string[40];`.

10. Repeat problem 9 for the Huffman coding tree in Figure 2.6b.

11. Write an initialization and four shortest C or C++ procedures – `void pstop(int)` push to top, `int pltop()` pull from top, `psbot(int)` push to bottom, and `int plbot()` pull from bottom, of a 10-element 16-bit word deque. The deque's buffer is `int deque[10]`. Use global `int` pointers, `top`, and `bottom`. (Note: the deque in §2.2.2 used indexes where this deque uses pointers. See also §2.3.3 for a queue that uses pointers.) Use global `char` variables for the size of the deque, `size`, and error flag `errors`, which is to remain cleared if there are no errors, and to be 1 if there are underflow or overflow errors. Note that C or C++ always initializes global variables to zero if not otherwise initialized. The procedures should manage the deque correctly as long as *errors* is zero. Procedures `pstop()` and `psbot()` pass by value, and procedures `pltop()` and `plbot()` pass by result.

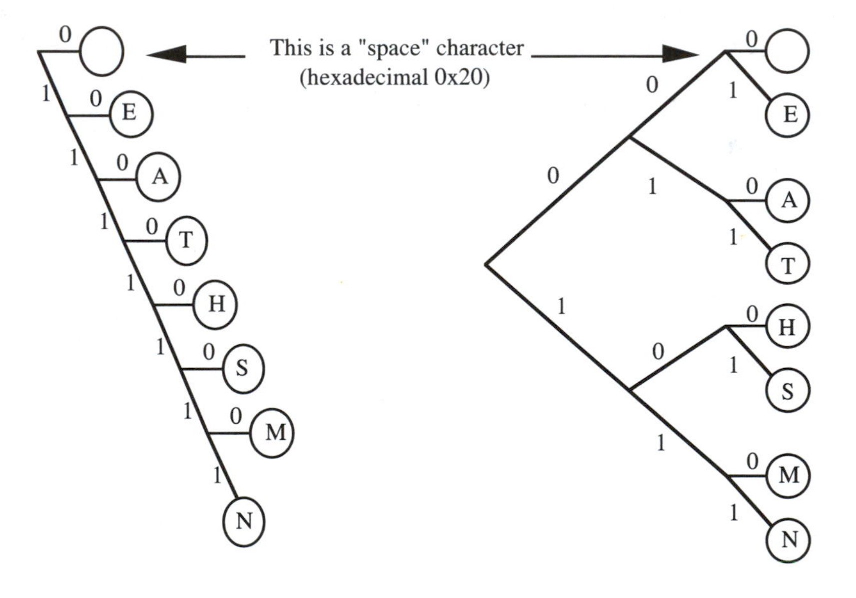

Figure 2.6. Other Huffman Codes

12. Write embedded assembly language in a C or C++ procedure `get(char *a, int i)` which moves `i` bytes following address `a` into a `char` global vector `v`, assuming `v` has a dimension larger than or equal to `i`. To achieve speed, use the `MOVB` and `DBNE` instructions. The call to this procedure, `get(s, n)`, is implemented as follows:

```
ldx s
pshx
ldd n
jsr get
leas 2,sp
```

13. Write a C or C++ procedure `monadic(int *a, int *result, int fun)` that executes any subroutine `fun()` in Chapter 1, problem 14, on 32-bit call-by-name argument `a`, returning the result in call-by-name argument `result`.

14. Write a C or C++ procedure `dyadic(int *a1, int *a2, int *result, int fun)` that uses an assembly-language subroutine `fun()` (described in Chapter 1, problem 30) to operate on 32-bit call-by-name arguments `a1` and `a2`, returning the result in call-by-name argument `result`.

15. Write a shortest C or C++ procedure `hexString(unsigned int n, char *s)` that runs in HIWAVE to convert an unsigned integer *n* into printable characters in `s` that represent it in hexadecimal so that `s[0]` is the ASCII code for the 1000s hex digit, `s[1]` is the code for the 100s hex digit, and so on. Suppress leading 0s by replacing them with blanks.

16. Write a shortest procedure `int inhex()` in C or C++ to input a four-digit hexadecimal number from the keyboard (the letters A through F may be uppercase or lowercase, typing any character other than '0'...'9', 'a'...'f', 'A'...'F', or entering more than four hexadecimal digits, terminates the input and starts the conversion) and convert it to a binary number, returning the converted binary number as an `unsigned int`. Although you do not have to use HIWAVE to answer this problem, you can use it without penalty, and it may help you get error-free results faster.

17. Write a shortest C or C++ program `int check(int base, int size, int range)` to write a checkerboard pattern in a vector of *size* $s = 2^n$ elements beginning at `base`, and then check to see that it is still there after it is completely written. It returns 1 if the vector is written and read back correctly, otherwise it returns 0. A checkerboard pattern is `range` $r = 2^k$ elements of 0s, followed by 2^k element of $FF, followed by 2^k elements of 0s, . . . for $k < n$, repeated throughout the vector. (This pattern is used to check dynamic memories for pattern sensitivity errors.)

18. Write a class `BitQueue` that is fully equivalent to the class `Queue` in §2.3.3, but that pushes, stores, and pulls 1-bit values; and all sizes are in bits rather than 16-bit words. The bits are stored in 16-bit *int* vectors allocated by the `allocate()` procedure.

19. Write a class `ShiftInt` that is fully equivalent to the class ***Queue*** in §2.3.3, but the constructor has an argument *n*, and function member `j = obj.shift(i);` shifts an `int` value `i` into a shift register of `n` `int`s, and shifts out an `int` value to `j`.

20. Write a class `ShiftChar` that is a derived class of the class `ShiftInt` in problem 19, where function member `j = shift(i);` shifts a `char` value `i` into a shift register of `n` `char`s and shifts out a `char` value to `j`. `ShiftChar` uses `ShiftInt`'s constructor.

21. Write a class `ShiftBit` that is fully equivalent to the class `ShiftInt` in problem 19, but that shifts 1-bit values, and all sizes are in bits rather than 16-bit words. The bits are stored in 16-bit *int* vectors allocated by the `allocate()` procedure.

22. Write a templated class `Deque` that is a derived class of templated class *Queue* and that implements a deque which can be pushed into and pulled from either end. The function members are `pstop()` push to top, `pltop()` pull from top, `psbot()` push to bottom, and `plbot()` pull from bottom. Use inherited data and function members wherever possible.

23. Write a templated class `IndexDeque` that is a derived class of templated class *Queue,* and that implements an indexable deque which can be pushed into and pulled from either end, and in which the `i`th member from the top or bottom can be read. The function members are `pstop()` push to top, `pltop()` pull from top, `psbot()` push to bottom, `plbot()` pull from bottom, `rdtop(i)` read the *i*th element from the top, and `rdbot(i)` read the *i*th element from the bottom of the deque. Function members `rdtop(i)` and `rdbot(i)` do not move the pointers. Use inherited data and function members wherever possible.

24. Write a templated class `Matrix` that implements matrix addition and multiplication for square matrixes (number of rows = number of columns). Overloaded operator + adds two matrixes resulting in a matrix; overloaded operator * multiplies two matrixes resulting in a matrix; and overloaded operators = and *cast* with overloaded operator `[]` writes or reads elements. For instance if *M* is an object of class `Matrix` then `M[i][j] = 5;` will write 5 into row `i,` column `j,` of matrix `M,` and `k = M[i][j];` will read row `i,` column `j,` of matrix `M` into `k.` `Matrix`'s constructor has an argument `size` that is stored as a data member *size,* and that allocates enough memory to hold a `size` by `size` matrix of elements of the template's data width, using a call to the procedure `allocate.`

25. Intervals can be used to calculate worst-case possibilities, for instance, in determining if an I/O device's setup and hold times are satisfied. An interval `<a,b>, a ≤ b,` is a range of real numbers between `a` and `b.` If `<a,b>` and `<c,d>` are intervals `A` and `B,` then the sum of `A` and `B` is the interval `<a + c, b + d>`, and the negative of `A` is `<-b, -a>.` Interval `A` contains interval `B` if every point in `A` is also in `B.` Write a templated class `Interval` having public overloaded operators + for adding two intervals resulting in an interval; - for negating an interval resulting in an interval; and an overloaded operator > returning a *char* value 1 if the left interval contains the right interval, and otherwise returning 0. If `A, B,` and `C` are objects of class `Interval,` the expression `A = B + C;` will add intervals `A` and `B` and put the result in `C, A = - B;` will put the negative of `A` into `B,` and the expression `if(A > B) i = 0;` will clear `i` if `A` contains `B.` The template allows for the values such as `a` or `b` to be `char, int,` or `long.` The class has a public variable `error` that is initially cleared, and set if an operation cannot be done or results in an overflow.

26. Write a templated class `Interval` having the operators of problem 25 and additional public overloaded operators `*` for multiplying two intervals to get an interval and `/` for dividing two intervals to get an interval, and also showing a procedure `sqrt(Interval)`, which is a friend of `Interval`, for taking the square root. Use the naive rule for multiplication, where all four terms are multiplied, and the lowest and highest of these terms are returned as the product interval, and assume there is already a procedure `long sqrt(long)` that you can use for obtaining the square root (do not write this procedure). If `A`, `B`, and `C` are of class `Interval`, the expression `A = B * C;` will multiply intervals `B` and `C` and put the result in `A`, `A = B/C` will divide `B` by `C` and put the result in `A`, and `A = sqrt(B);` will put the square root of `B` into `A`. Note that `a/b` is `a * (1/b)`, so the multiply operator can be used to implement the divide operator; `a - b` is `a + (-b)`, so the add and negate operators can be used to implement the subtract operator; and `4 * a` is `a + a + a + a`, so scalar multiplication can be done by addition. Also, `Interval` has a public data member `error` that can be set if we invert an interval containing 0 or get the square root of an interval containing a negative value. Finally, write a `main()` procedure that will initialize intervals `a` to <1,2>, `b` to <3,4>, and `c` to <5,6>, and then evaluate the result of the expression `( -b + sqrt( b * b - 4 * a * c ) ) / (a + a)`.

The MC68HC812A4 die.

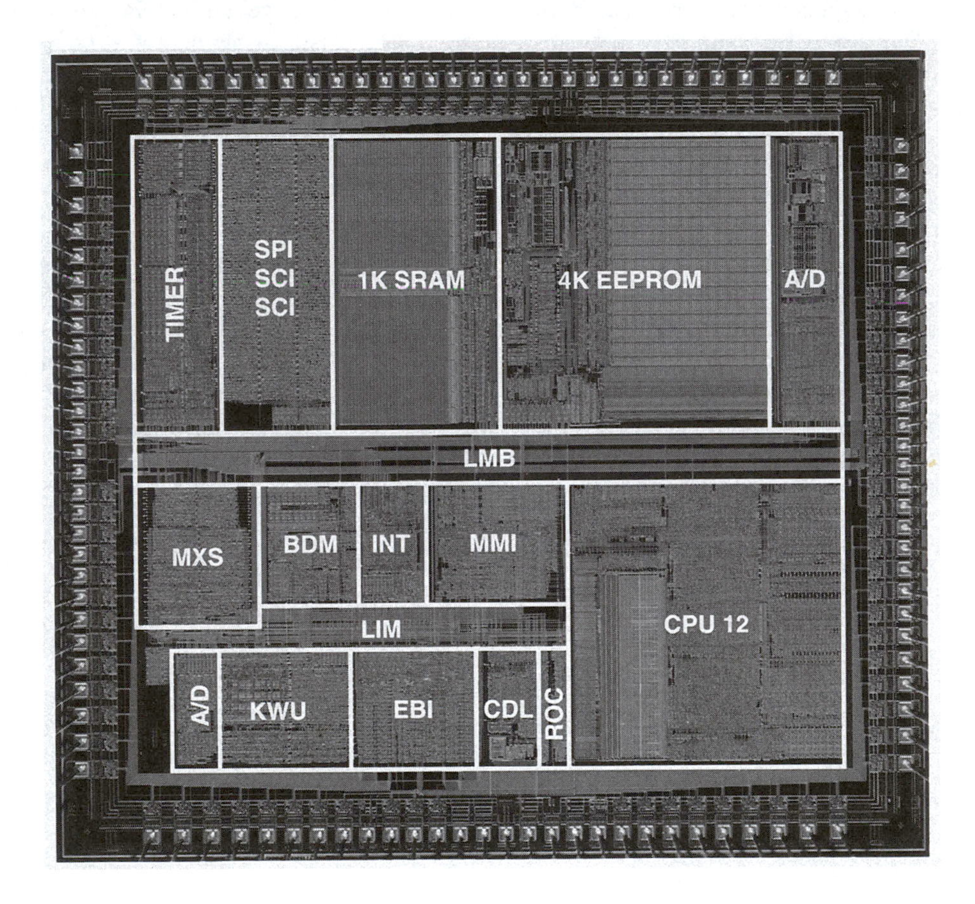

3

Bus Hardware and Signals

Understanding the data and address buses is critical, because they are at the heart of interfacing design. This chapter will discuss what a bus is, how data are put onto it, and how data from it is used. The chapter progresses logically, with the first section covering basic concepts in digital hardware, the next section using those concepts to describe the control signals on the bus, and the final section discussing the important issue of timing in the microprocessor bus.

The first section of this chapter is a condensed version of background material on computer realization (as opposed to architecture, organization, and software discussed in earlier chapters) needed to understand the remainder of the book. This led to the study of bus timing and control – very important to the design of interfaces. Its importance can be shown in the following experience. Microcomputer manufacturers have applications engineers who write notes on how to use the chips the companies manufacture and who answer those knotty questions that systems designers can't handle. The author had an opportunity to sit down with Charlie Melear, one of the very fine applications engineers at Motorola's plant, when the first edition of this book was written. Charlie noted that the two most common problems designers have are (1) improper control signals for the bus, whereby several bus drivers are given commands to drive the bus at the same time, and (2) failure to meet timing specifications for address and data buses. These problems, which will be covered in §3.2.2, remain. Even today, when much of the hardware is on a single chip and the designer isn't concerned about them, they reappear when I/O and memory chips are added to a single-chip microcontroller.

This chapter introduces a lot of terminology to provide background for later sections and enable you to read data sheets provided by the manufacturers. The terminology is close to that used in industry, and microprocessor notation conforms to that used in Motorola data sheets. However, some minor deviations have been introduced where constructs appear so often in this book that further notation is useful.

This chapter should provide enough background in computer organization for the remaining sections. After reading the chapter, you should be able to read a logic diagram or the data sheets describing microcomputers or their associated integrated circuits, and you also should have a fundamental knowledge of the signals and their timing on a typical microcomputer bus.

3.1 Digital Hardware

The basic notions and building blocks of digital hardware are presented in this section. While you have probably taken a course on digital hardware design that most likely emphasized minimization of logic gates, microcomputer interfacing requires an emphasis on buses. Therefore, this section focuses on the digital hardware that can be seen on a typical microcomputer bus. The first subsection provides clear definitions of terms used to describe signals and modules connected to a bus. The second subsection considers the kinds of modules you might see there.

3.1.1 Modules and Signals

Before the bus is explained, we need to discuss a few hardware concepts, such as the module and the signal. Since we are dealing in abstractions, we do not use concrete examples with units like electrons and fields.

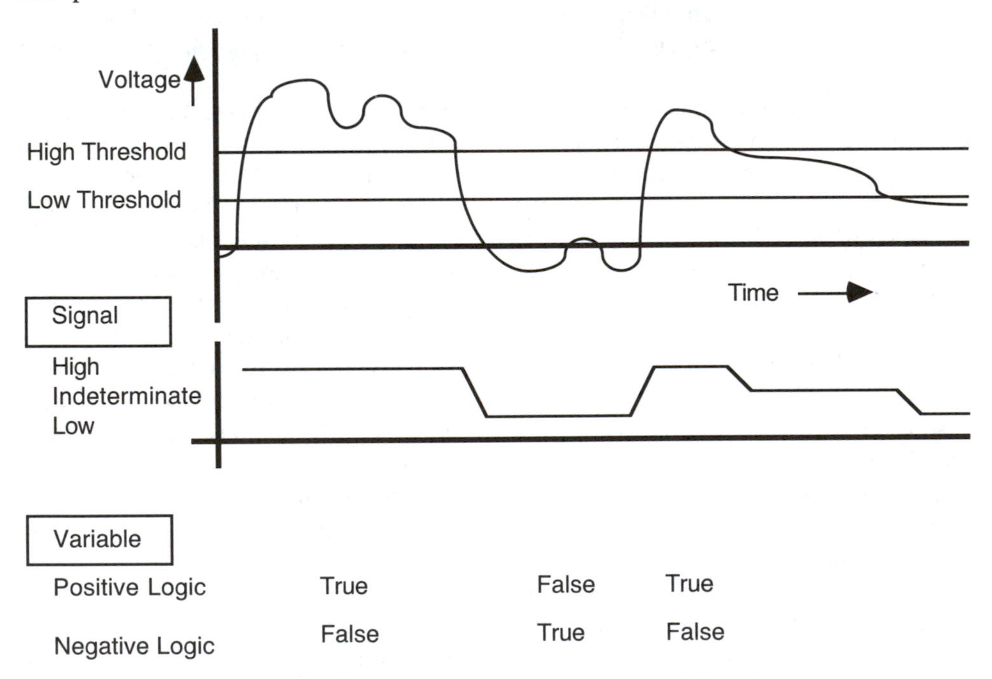

Figure 3.1. Voltage Waveforms, Signals, and Variables

One concept is the binary *signal*. (See Figure 3.1.) Although a signal is a voltage or a current, we think of it only as a *high* signal if the voltage or current is above a predefined threshold, or as a *low* signal if it is below another threshold. We will use the symbols H for high and L for low. A signal is *determinate* when we can know for sure whether it is high or low. Related to this concept, a *variable* is the information a signal carries, and has values *true* (T) and *false* (F). For example, a wire can carry a signal L, and being a variable called "ENABLE," it can have a value T to indicate that something is indeed enabled. The expression *assert a variable* means to make it true; *negate a variable* means make it false; and *complement a variable* means make it true if it was false or make it false if it was true. Two possible relations exist between signals and variables. In *positive logic*, a high signal represents a true variable and a low signal, a false variable. In *negative logic*, a high signal represents a false variable and a low signal, a true variable. Signals, which can be viewed on an oscilloscope or a logic analyzer, are preferred when someone, especially a technician, deals with actual hardware. Variables have a more conceptual significance and seem to be preferred by designers, especially in the early stages of design, and by programmers, especially when writing I/O software. Simply put, "true" and "false" are the 1 and 0 of the programmer, the architect, and the system designer; and "high" and "low" are the 1 and 0 of the technician and IC manufacturer. While nothing is wrong with using 1 and 0 where the meaning is clear, we use the words "true" and "false" when talking about software or system design and the words "high" and "low" when discussing the hardware realization, to be clear.

Two types of variables and their corresponding signals are important in hardware. A *memory variable* is capable of being made true or false and of retaining this value, but a *link variable* is true or false as a result of functions of other variables. A link variable is always some function of other variables (as the output of some gate). At a high level of abstraction, these variables operate in different dimensions; memory variables are used to convey information through time (at the same point in space), while link variables convey information through space (at the same point in time). Some transformations on hardware, like converting from a parallel to a serial adder, are nicely explained by this abstract view. For instance, one can convert a parallel adder into a serial adder by converting a link variable that passes the carry into a memory variable that saves the carry. Also, in a simulation program, we differentiate between the types because memory variables have to be initialized and link variables don't.

A *synchronous* signal is associated with a periodic variable (for example, a square wave) called a *clock*. The signal or variable is indeterminate except when the clock is asserted. Or, alternatively, the value of the signal is irrelevant except when the clock is asserted. Depending on the context, the signal is determinate either precisely when the clock changes from false to true or as long as the clock is true. The context depends on what picks up the signal and will be discussed when we study the flip-flop. This is so in the real world because of delays resulting from circuitry, noise, and transmission-line ringing. In our abstraction of the signal, we simply ignore the signal except when this clock is asserted, and we design the system so the clock is asserted only when we can guarantee the signal is determinate under worst-case conditions. Though there are asynchronous signals where there is no associated clock and the signals are supposed to be determinate at all times, most microprocessor signals are synchronous; thus in further discussions, we will assume all signals are synchronous. Then two *signals* are *equivalent* if they have the same (H or L) value whenever the clock is asserted.

The other basic idea is that of the *module*, which is a block of hardware with identifiable input, output, and memory variables. Input variables are the *input ports* and output variables are the *output ports*. Often, we are interested only in the behavior. Modules are *behaviorally equivalent* if, for equivalent values of the initial memory variables and equivalent sequences of values of input variables, they deliver equivalent sequences of values of output variables. Thus, we are concerned not about how they are constructed internally, what the precise voltages are, or what the signals are when the clock is not asserted, but only about what the signals are when the clock is asserted.

In §1.1.3, we introduced the idea of an integrated circuit (IC) to define the microprocessor. Now we further explore the concept. An integrated circuit is a module that is often contained in a *dual in-line package,* a surface-mount *thin-quad flat pack,* or a similar package. The pins are the input and output ports. Viewed from the top, one of the short edges has an indent or mark. The pins are numbered counterclockwise from this mark, starting with pin 1. Gates are defined in the next section, but will be used here to describe degrees of complexity of integrated circuits. A *small-scale integrated circuit,* or SSI, has in the order of 10 gates on one chip, a *medium-scale integrated circuit* (MSI) has about 100, a *large-scale integrated circuit* (LSI) has about 1,000, and a *very large scale integrated circuit* (VLSI) has more than 10,000. SSI and MSI circuits are commonly used to build up address decoders and some I/O modules in a microcomputer; LSI and VLSI are commonly used to implement 8- and 16-bit word microprocessors, 64K-bit and 128K-bit memory chips, and some complex I/O chips.

A *family* of integrated circuits is a collection of different types made with the same technology and having the same electrical characteristics so that they can be easily connected with others in the same family. Chips from different families can be interconnected, but this might require some careful study and design. The *low-power Schottky* or LS family, and the *complementary metal oxide semiconductor* or CMOS family, are often used with microprocessors. The LS family is used where higher speed is required, and the CMOS family, where lower power or higher immunity to noise is desired. The HCMOS family is a high-speed CMOS family particularly useful in 'A4 designs because it is fast enough for address decoding but requires very little power and can tolerate large variations in the power supply.

A block diagram was introduced at the beginning of §1.4. In block diagrams, names represent variables rather than signals, and functions like AND or OR represent functions on variables rather than signals. An AND function, for example, is one in which the output is T if all the inputs are T. Such conventions ignore details needed to build the module, so the module's behavior can be simply explained.

Logic diagrams (also called *schematics*) describe the realization of hardware to the level of detail needed to build it. In logic diagrams, modules are generally shown as rectangles, with input and output ports shown along the perimeter. Logic functions are generally defined for signals rather than variables (for example, an AND function is one whose output is H if its inputs are all H). It is common, and in fact desirable, to use many copies of the same module. The original module, here called the *type,* has a name, the *type name*. Especially when referring to one module copy among several, we give each copy a distinct *copy name.* The type name or copy name may be put in a logic diagram when the meaning is clear, or both may be put in the rectangle or over the left upper corner. Analogous to subroutines, inputs and outputs of the type name are *formal parameter names*, and inputs and outputs of the copy name are *actual parameter names*. Integrated circuits in particular are shown this way: formal parameters are shown inside a box representing the integrated circuit, and pin numbers and actual parameters are shown outside the rectangle for each connection that has to be made. Pins that don't have to be connected are not shown as connections to the module. (Figure 3.3 provides some examples of these conventions.)

Connections supplying power (positive supply voltage and ground) are usually not shown. They might be identified in a footnote, if necessary. In general, in LSI and VLSI N channel MOS chips such as microprocessors and I/O chips discussed in these notes,

Vss is the ground pin (0 volts) and Vcc or Vdd is usually +5 volts. You might remember this by a quotation improperly attributed to Winston Churchill: "*ground the SS* ." For SSI and MSI chips, the pin with the largest pin number is generally connected to +5 volts, while the pin catercorner from it is connected to ground. One should keep power and ground lines straight and wide to reduce inductance that causes ringing, and put a capacitor (.1 microfarad disc) between power and ground to isolate the ICs from each other. When one chip changes its power supply current, these *bypass capacitors* serve to prevent voltage fluctuations from affecting the voltage supplied to other chips, which might look like signals to them. Normally, such a capacitor is needed for four SSI chips or each LSI chip, but if the power and ground lines appear to have noise, more capacitors should be put between power and ground.

In connections to inner modules, negative logic is usually shown by a small bubble where the connection touches the rectangle. In inputs and outputs to the whole system described by the logic diagram, negative logic is shown by a bar over the variable's name. Ideally, if a link is in negative logic, all its connections to modules should have bubbles. However, since changing logic polarity effects an inversion of the variable, designers sometimes steal a free inverter this way; so if bubbles do not match at both ends, remember that the signal is unchanged, but the variable is inverted as it goes through the link.

A logic diagram should convey all the information needed to build a module, allowing only the exceptions we just discussed to reduce the clutter. Examples of logic diagrams appear throughout this book. An explanation of Figures 3.2 and 3.3, which must wait until the next section, should clarify these conventions.

3.1.2 Drivers, Registers, and Memories

This section describes the bus in terms of the D flip-flop and the bus driver. These devices serve to take data from the bus and to put data onto it. The memory – a collection of registers – is also introduced.

A *gate* is an elementary module with a single output, where the value of the output is a Boolean logic function of the values of the inputs. The output of a gate is generally a link variable. For example, a three-input NOR gate output is true if none of its inputs are true, otherwise it is false. The output is always determined in terms of its inputs. A *buffer* is a gate that has a more powerful output amplifier.

Your typical gate has an output stage that may be connected to up to f other inputs of gates of the same family (f is called the *fan-out*) and to no other output of a gate. If two outputs are connected to the same link, they may try to put opposite signals on the link, which will certainly be confusing to inputs on the link and may even damage the output stages. However, a *bus* or *buss* is a link to which more than two gate outputs are connected. The gates must have specially designed output amplifiers so that all but one output on a bus may be disabled. The gates are called *bus drivers.* An upper limit to the number of outputs that can be connected to a bus is called the *fan-in.* Bus drivers may also be buffers to provide higher power to drive the bus. And in these cases, the gate may be very simple, so that it has just one input, and the output is the complement of the input (inverting) or the same signal as the input (noninverting).

An *open collector gate* or open collector driver output can be connected to a *wire-OR* bus (the bus must have a *pull-up resistor* connected between it and the positive supply voltage). If any output should attempt to put out a low signal, the signal on the bus will be low. Only when all outputs attempt to put out a high signal will the output be high. Generally, the gate is a two-input AND gate, with inputs in positive logic and output in negative logic. Data, on one input, are put onto the bus whenever the other input is true. The other input acts as a positive-logic *enable*. When the enable is asserted, we say the driver is *enabled*. Since this bus is normally used in the negative-logic relationship, the value on the bus is the OR of the outputs, which is so common that the bus is called a wire-OR bus.

A *tristate gate* or tristate driver has an additional input, a *tristate enable*. When the tristate enable is asserted (the driver is enabled), the output amplifier forces the output signal high or low as directed by the gate logic. When the enable is not asserted, the output amplifier lets the output float. Two or more outputs of tristate gates may be connected to a *tristate bus*. The circuitry must be designed to ensure that no two gates are enabled at the same time, lest the problem with connecting outputs of ordinary gates arise. If no gates are enabled, the bus signal floats – it is subject to stray static and electromagnetic fields. In other words, it acts like an antenna.

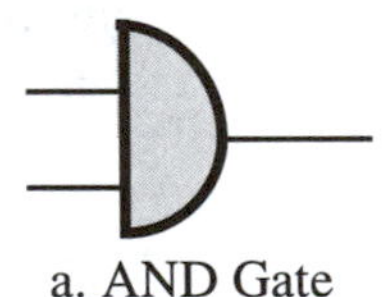

a. AND Gate

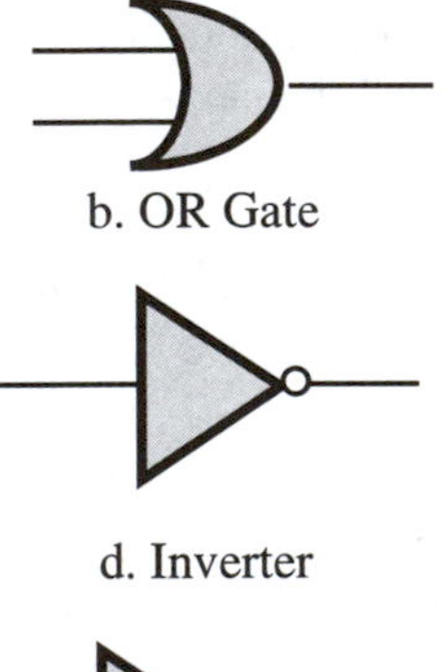

b. OR Gate

d. Inverter

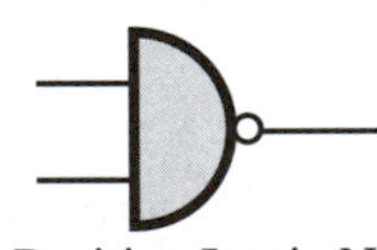

c. Positive Logic NAND Gate, or
AND Gate with Negative-Logic Output

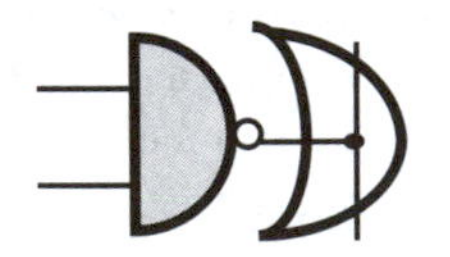

e. Open Collector AND Gate
Driving (Negative Logic) Wire-OR Bus

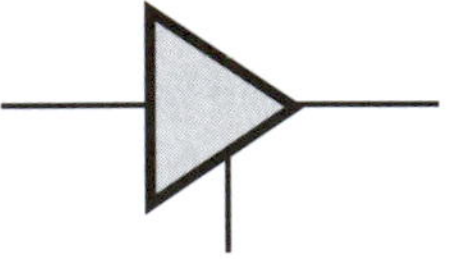

f. Noninverting Tristate Buffer

Figure 3.2. Some Common Gates

Gates are usually shown in logic diagrams as D-shaped symbols, the output on the round edge and inputs on the flat edge. (See Figure 3.2 for the positive-logic AND, NAND, and other gates.) Even though they are not shown using the aforementioned convention for modules, if they are in integrated circuits, the pin numbers are often shown next to all inputs and outputs.

Dynamic logic gates are implemented by passing charges (collections of electrons or holes) through switches; the charges have to be replenished, or they will discharge. Most gates use currents rather than charges and are not dynamic. Dynamic logic must be pulsed at a rate between a minimum and a maximum time, or it will not work; but dynamic logic gates are more compact than normal (static) logic gates.

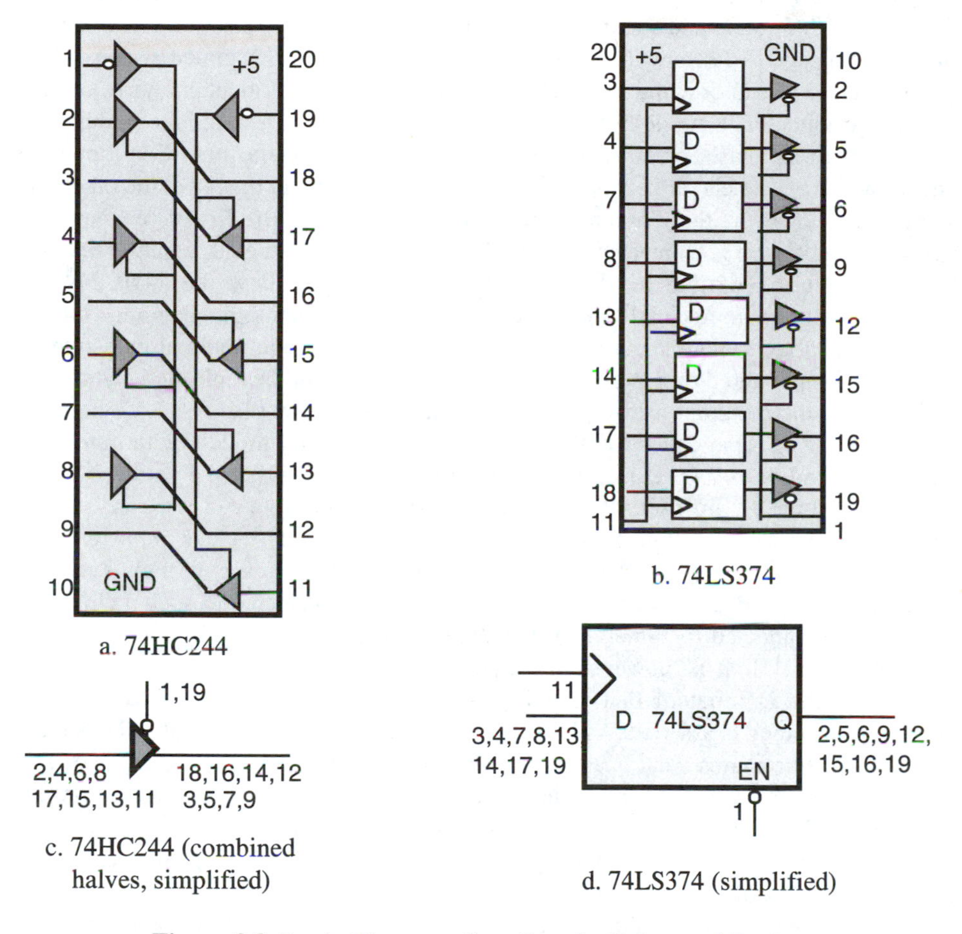

Figure 3.3. Logic Diagrams for a Popular Driver and Register

Gates are usually put into integrated circuits so that the total number of pins is 14 or 16, including two pins for positive supply voltage and ground. This yields, for instance, the quad two-input NAND gate, the 7400, which contains four two-input positive logic NAND gates. The 74HC00 is an HCMOS part with the pin configuration of the older 7400 TTL part. The 7404 has six inverters in a chip; it is called a hex inverter, so it is a good treat for Halloween (to invert hexes). A typical microprocessor uses an 8-bit-wide data bus, where eight identical and separate bus wires carry 1 bit of data on each wire. This has, in an IC, engendered octal bus drivers, with eight inverting or noninverting bus drivers that share common enables. The 74HC244 and 74HC240 are popular octal noninverting and inverting tristate bus driver integrated circuits. Figure

3.3a shows a logic diagram of the 74HC244 in which, to clearly show pin connections, the pins are placed along the perimeter of the module exactly as they appear on the dual in-line package. A positive 5-volt supply wire is connected to pin 20, and a ground wire to pin 10. If the signals on both pins 1 and 19 are low, the eight separate tristate gates will be enabled. For instance, the signal input to pin 2 will be amplified and output on pin 18. If pins 1 and 19 are high, the tristate amplifiers are not enabled, and pin outputs 18, 16, 14, 12, 3, 5, 7, and 9 are allowed to float. This kind of diagram is valuable in catalogs to most clearly show the inputs and outputs of gates in integrated circuits.

To save effort in drawing logic diagrams, if a number *n* of identical wires connect to identical modules, a single line is drawn with a slash through it and the number *n* is drawn next to the slash, or a list of *n* pins is written next to the line. Corresponding pins in the list at one end are connected to corresponding pins in the list at the other end. Commonly, however, the diagram is clear without showing the list of pin numbers. Also, if a single wire is connected to several pins, it is diagrammed as a single line, and the list of pins is written by the line. Figure 3.3c shows how the 74HC244 just discussed might be more clearly shown connecting to a bus in a logic diagram. Note the eight tristate drivers, their input and output links shown by one line and gate symbol. The list of 8 input and 8 output pins indicates the driver should be replicated eight times. Each input in the input list feeds the corresponding output in that list.

A *D flip-flop*, also called a (1-bit) latch, is an elementary module with *data input* D, *clock* C, and *output* Q. Q is always a memory variable having the value of the bit of data stored in the flip-flop. When the clock C is asserted (we call it a *clocked flip-flop*), the value of D is copied into the flip-flop memory. The clock input is rather confusing because it is really just a WRITE ENABLE. It sounds as though it must be the same as the microcomputer system clock. It may be connected to such a clock, but usually it is connected to something else, such as an output of a controller, which is discussed in §1.1.1. It is, however, the clock that is associated with the synchronous variable on the D input of that flip-flop, since the variable has to be determinate whenever this clock is asserted. As long as C is asserted, Q is made equal to D. As long as C is false, Q remains what it was. Note that when C is false, Q is the value of D at the moment when C changed from true to false. However, when C is asserted, the flip-flop behaves like a wire from D to Q, and Q changes as D changes. D flip-flops are used to hold data sent to them on the D inputs, so the data, even though long since gone from the D input, will still be available on the Q output.

A *D edge-triggered flip-flop* is like the D flip-flop, except that the data stored in it and available on the Q output are made equal to the D input only when the clock C changes from false to true. The clock causes the data to change (the flip-flop is clocked) in this very short time. A *D master slave flip-flop* (also called a dual-rank flip-flop) is a pair of D flip-flops where the D input to the second flip-flop is internally connected to the Q output of the first, and the second flip-flop's clock is the complement of the first flip-flop's clock. Though constructed differently, a D master slave flip-flop behaves the same as the D edge-triggered flip-flop. These two flip-flops have the following property: data on their Q output are always the former value of data in them at the time that new data are put into them. It is possible, therefore, to use the signal output from an edge-triggered flip-flop to feed data into the same or another edge-triggered flip-flop using the same clock, even while loading new data. This should not be attempted with D flip-flops because the output will be changing as it is being used to determine the value to be stored in the flip-flops that use the data. When a synchronous signal is input to a D edge-triggered flip-flop, the clock input to the flip-flop is associated with the signal, and the signal only has to be determinate when the clock changes from false to true.

In either type of flip-flop or in more complex devices that use flip-flops, the data have to be determinate (a stable high or a stable low signal) over a range of time when the data are being stored. For an edge-triggered or dual-rank flip-flop, the *setup time* is the time during which the data must be determinate before the clock edge. The *hold time* is the time after the clock edge during which the data must be determinate. For a latch, the setup time is the minimum time at the end of the period when the clock is true in which the data must be determinate; and the hold time is the minimum time just after that when the data must still be determinate. These times are usually specified for worst-case possibilities. If you satisfy the setup and hold times, the device can be expected to work as long as it is kept at a temperature and supplied with power voltages that are within specified limits. If you don't, it may work some of the time, but will probably fail according to Murphy's law, at the worst possible time.

In most integrated circuit D flip-flops or D edge-triggered flip-flops, the output Q is available along with its complement, which can be thought of as the output Q in negative logic. They often have inputs – set, which if asserted will assert Q, and reset, which if asserted will make Q false. Set and reset are often in negative logic; when not used, they should be connected to a false value or high signal. Other flip-flops such as set-reset flip-flops and JK edge-triggered flip-flops are commonly used in digital equipment, but we won't need them in the following discussions.

A *one-shot* is rather similar to the flip-flop. It has an input TRIG and an output Q, and has a resistor and capacitor connected to it. The output Q is normally false. When the input TRIG changes from false to true, the output becomes true and remains true for a period of time T that is fixed by the values of a resistor and a capacitor.

The use of 8-bit-wide data buses has engendered ICs that have four or eight flip-flops with common clock inputs and common clear inputs. If simple D flip-flops are used, the module is called a *latch;* if edge-triggered flip-flops are used, it is a *register.* Also, modules for binary number counting (*counters*) or shifting data in one direction (*shift registers*) may typically contain four or eight edge-triggered flip-flops. Note that even though a module may have additional capabilities, it can still be used without these capabilities. A counter or a shift register is sometimes used as a simple register. More interestingly, a latch can be used as a noninverting gate or using the complemented Q output as an inverter. This is done by tying the clock to true. The 74HC163 is a popular 4-bit binary counter; the 74HC164, 74HC165, and 74HC299 are common 8-bit shift registers; and the 74HC373 and 74HC374 are popular octal latches and registers, with built-in tristate drivers. The 74HC374 will be particularly useful in the following discussion of practical buses, since it contains a register to capture data from the bus as well as a tristate driver to put data onto the bus.

The following conventions are used to describe flip-flops in logic diagrams. The clock and D inputs are shown on the left of a square, the set on the top, the clear on the bottom, and the Q on the right. The letter D is put by the D input, but the other inputs need no letters. The clock of an edge-triggered flip-flop is denoted by a triangle just inside the jointure of that input. This triangle and the bubble outside the square describe the clocking. If neither appears the flip-flop is a D flip-flop that inputs data from D when the clock is high; if a bubble appears, it is a D flip-flop that inputs data when the clock is low; if a triangle appears, it is an edge-triggered D flip-flop that inputs data when the clock changes from low to high; and if both appear, it is an edge-triggered D flip-flop that inputs data when the clock input changes from high to low. This notation is quite useful because a lot of design errors are due to clocking flip-flops when the data is not ready to be input. If a signal is input to several flip-flops, they should all be clocked at the same time, when the signal will be determinate.

The logic diagram of the 74HC374 is shown in Figure 3.3b as it might appear in a catalog. Note that the common clock for all the edge-triggered D flip-flops on pin 11 makes them store data on their own D inputs when it rises from low to high. Note also that when the signal on pin 1 is low, the tristate drivers are all enabled, so the data in the flip-flops is output through them. Using this integrated circuit in a logic diagram, we might compact it using the bus conventions, as shown in Figure 3.3d.

An *(i,j) random access memory* (RAM) is a module with *i* rows and *j* columns of D flip-flops and an address port, an input port, and an output port. A *programmable read-only memory* (PROM) is like a RAM, but can only be read having been written at the factory. A row of the memory is available simultaneously and is usually referred to as a *word,* and the number *j* is called the *word width*. There is considerable ambiguity here, because a computer may think of its memory as having a word width, but the memory module itself may have a different word width, and it may be built from RAM integrated circuits having yet a different word width. So the word and the word width should be used in a manner that avoids this ambiguity. The output port outputs data read from a row of the flip-flops to a bus and usually has bus drivers built into it. Sometimes the input and output ports are combined. The address port is used to input the row number of the row to be read or written. A *memory cycle* is a time when the memory can write *j* bits from the input port into a row selected by the address port data, read *j* bits from a row selected by the address port data to the output port, or do nothing. If the memory reads data, the drivers on the output port are enabled. There are two common ways to indicate which of the three possible operations to do in a memory cycle. In one, two variables called *chip enable* (CE) and *read/not write* (R/W) indicate the possibilities; a do-nothing cycle is executed if CE is false, a read if CE and R/W are both asserted, and a write if CE is asserted but R/W is not. In the other, two variables, called *read enable* (RE) and *write enable* (WE), are used. When neither is asserted, nothing is done; when RE is asserted, a read is executed; and if WE is asserted, a write is executed. Normally, CE, RE, and WE are in negative logic. The *memory cycle time* is the time needed to complete a read or a write operation and be ready to execute another read or write. The *memory access time* is the time from the beginning of a memory cycle until the data read from a memory are determinate on the output, or the time when data to be written must be determinate on the input of the memory. A popular, fast (20-nanosecond access time) (4,4) RAM is the 74LS670. It has four input ports and four separate output ports; by having two different address ports it is actually able to simultaneously read a word selected by the read address port and to write a word selected by the write address port. A large (8K, 8) RAM is the 6264. It has a 13-bit address, eight input/output ports, and two Es and W variables that permit it to read or write any word in a memory cycle. A diagram of this chip appears in Figure 6.13.

The *programmable array logic* (PAL) chip has become readily available and is ideally suited to implementing microcomputer address decoders and other "glue" logic. (See Figure 3.4.)

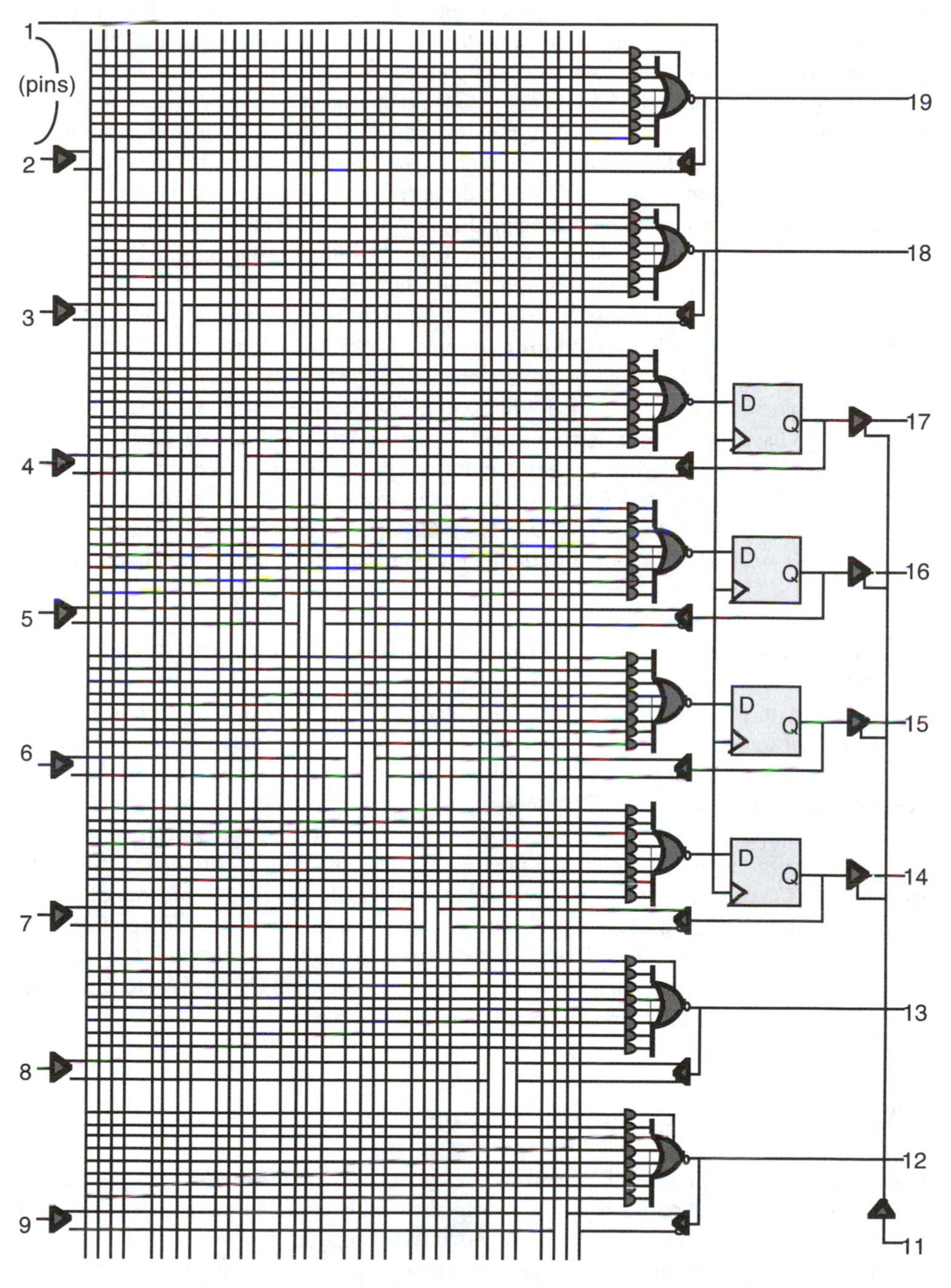

Figure 3.4. 16R4 PAL Used in Microcomputer Designs

A PAL is basically a collection of gates whose inputs are connected by fuses like a PROM. The second line from the top of Figure 3.4 represents a 32-input AND gate that feeds the tristate enable of a 7-input NOR gate, which in turn feeds pin 19. Each crossing line in this row represents a fuse, which, if left unblown, connects the column to this gate as an input; otherwise the column is disconnected, and a T is put into the AND gate. Each triangle-shaped gate with two outputs generates a signal and its complement and feeds two of the columns. The second line from the top can have any input from pins 2 to 9 or their complement, or the outputs on pins 12 to 19 or their complement, as inputs to the AND gate. For each possible input, the fuses are blown to select the input or its complement, or to ignore it. Thus, the designer can choose any AND of the 16 I/O variables or their complements as the signal controlling the tristate gate. Similarly, the next seven lines each feed an input to the NOR gate, so the output on pin 19 may be a Boolean "sum-of-products" of up to seven "products," each of which may be the AND of any I/O variable or its complement. This group of eight rows is basically replicated for each NOR gate. The middle four groups feed registers clocked by pin 1, and their outputs are put on pins 14 to 17 by tristate drivers enabled by pin 11. The registers can store a state of a sequential machine, which will be discussed further in §4.3.4. PALs such as the PAL16L8 have no registers and are suited to implementing address decoders and other collections of gates needed in a microcomputer system. There is now a rather large family of PALS having from zero to eight registers and one to eight inverted or noninverted outputs in a 20-pin DIP, and there also are 24-pin DIP PALs. These can be programmed to realize just about any simple function, such as an address decoder.

3.2 Address and Control Signals in 6812 Microcontrollers

One of the main problems designers face is how to control bus drivers so two of them will never try to drive the same bus at the same time. To approach this problem, the designer must be acquainted with control signals and the sequences of control signals generated by a microprocessor. This section is devoted to aspects of microprogramming and microcomputer instruction execution necessary for the comprehension and explanation of control signals on the microcomputer bus. The problem of controlling memory and I/O devices is first one of designing address decoders, and second, of timing the address, data, and control signals. The first subsection covers the decoding of address and control signals. The second subsection discusses timing requirements. With this discussion, you should understand how to interface memory and I/O devices to the buses, which is at the heart of the aforementioned problem.

3.2.1 Address and Control Timing

One common problem faced by interface designers is the problem of bus timing. To connect memory or I/O registers to the microprocessor, the actual timing requirements of the address bus and data bus have to be satisfied. When adding memory or I/O chips, one may have to analyze the timing requirements carefully. To build decoders, timing control signals must be ANDed with address signals. Therefore, we discuss them here. We discuss first the simpler, nonmultiplexed 'A4 bus and then the multiplexed 'B32 bus. The reader should study both timings, regardless of which system is to be used.

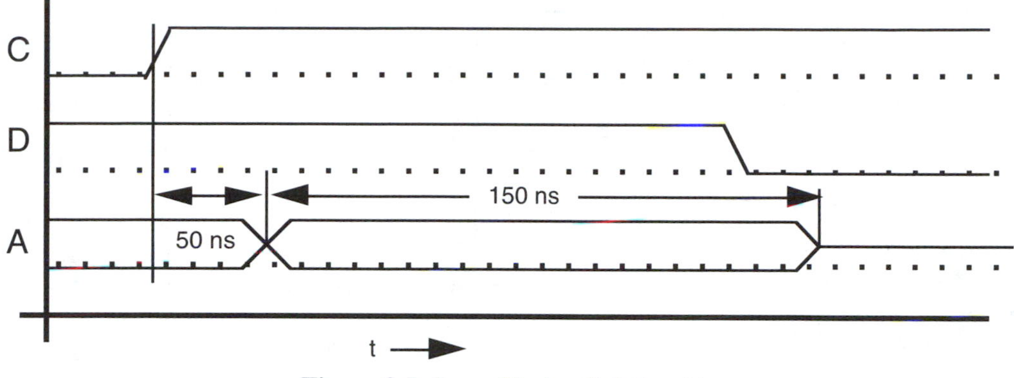

Figure 3.5. Some Timing Relationships

Timing diagrams are used to show the requirements. A timing diagram is like an oscilloscope trace of the signals, as is shown in Figure 3.5. For collections of variables, like the 16 address lines shown by the trace labeled A, two parallel lines indicate that any particular address line may be high or low but will remain there for the time interval where the lines are parallel. A crossing line indicates that any particular line can change at that time. A line in the middle of the high and low line levels indicates the output is floating because no drivers are enabled, or the output may be changing as it tries to reach a stable value. A line in the middle means the signal is indeterminate; it is not necessarily at half the voltage. (Motorola also uses a crosshatch pattern like a row of Xs to indicate that the signal is invalid but not tristated, while a line in the middle means the output is in the tristate open circuit mode on the device being discussed. That distinction is not made in this book, because both cases mean that the bus signal is indeterminate and cannot be used.) Timing is usually shown to scale, as on an oscilloscope, and requirements are indicated the way dimensions are shown on a blue print. On the left, the "dimension arrow" shows that addresses change 50 nanoseconds after C rises, and, in the middle, the "dimension arrow" shows that the address should be stable for at least 150 nanoseconds.

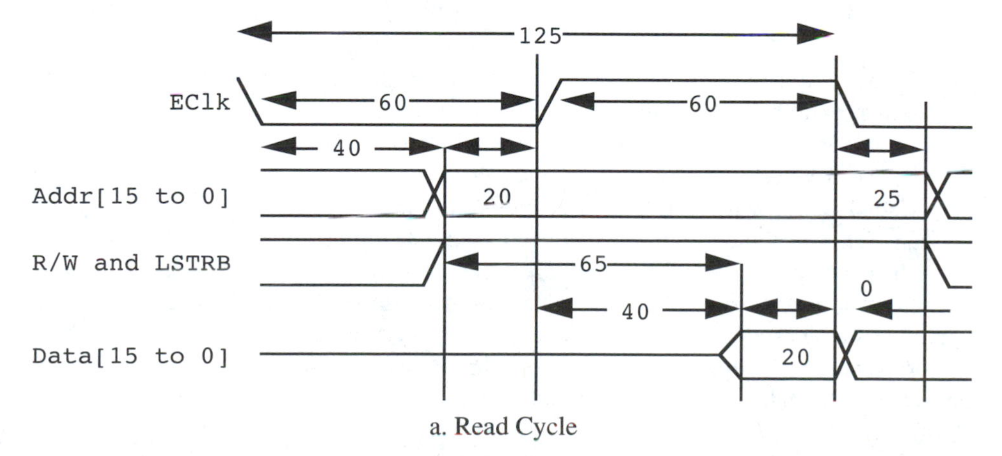

a. Read Cycle

Figure 3.6. Timing Relationships for the MC68HC812A4

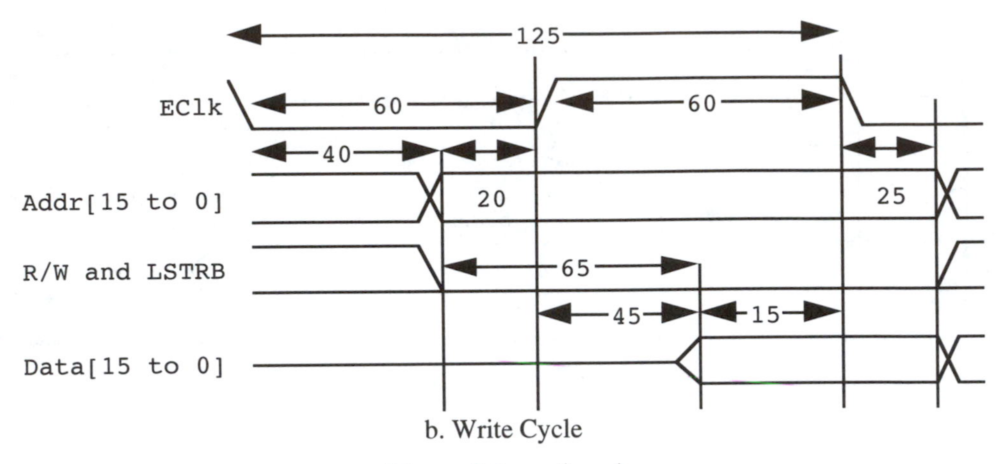

b. Write Cycle

Figure 3.6. continued

Pins otherwise used for ports A, B, and possibly C or D are used for the address and data bus. (Additional address bus expansion is considered in §6.1.) In the 'A4 *narrow mode,* pins otherwise used for port C are used for the data bus, and pins used for port E bits 2 and 4 are used for control signals *read/write* (R/W) and *E clock* (E). In the 'A4 *wide mode,* pins are used as in the narrow mode; in addition, pins otherwise used for port D are also used for the data bus, and a pin used for port E bit 3 is used for the *low strobe* LSTRB. The 'B32 time-multiplexes the address and data on port A and B pins.

A *memory cycle* is a period of time when the 6812 requests memory to read or write a word. The E clock is low in the first part and high in the second part of the memory cycle, and the address bus A[15 to 0] supplies the address of the data to be read or written as long as E is high. See Figure 3.6.

A memory cycle is either a *read cycle,* where data is read from a memory or input device, or a *write cycle,* where data is written into a memory or output device. In a read cycle, R/W is high during the cycle and the data bus – D[7 to 0] in the narrow mode, and D[15 to 0] in the wide mode – moves data from the memory or I/O device to the 'A4. In a write cycle, R/W is low during the cycle and the data bus – D[7 to 0] in the narrow mode and D[15 to 0] in the wide mode – moves data the opposite way.

We will first look at an example to show the principles of bus transfers, using approximate numbers for timing. (See Figure 3.6 for approximate timing relationships of the 'A4.) The timings represent worst-case numbers and do not necessarily add up; for instance, the E clock is low for at least 60 nanoseconds and high for at least 60 nanoseconds (adding up to 120 nanoseconds), but the cycle time may be no shorter than 125 nanoseconds. Other microprocessors have similar timing relationships.

The E clock can be an 8 MHz square wave. A memory cycle begins and ends when E falls from high to low. The address bus is indeterminate from the beginning of the cycle for 40 nanoseconds, which is 20 nanoseconds before E rises. This delay is due to the propagation of control signals to gate drivers and the propagation of the address through an internal bus to the external address bus. In a read cycle, the read/write signal, shown in trace R/W, remains high throughout the cycle, and the microprocessor expects valid data on the data bus for 20 nanoseconds before the falling edge of E, until it falls. These times are the setup and hold times of registers inside the 'A4. If any data line changes in this interval, the microprocessor may randomly input either an L or H on that line. The memory is responsible for providing valid and constant signals during this interval during the setup and hold time. In a write cycle, the R/W signal is guaranteed low at the time the address becomes stable and can rise 25 nanoseconds after the beginning of the next cycle. Due to delays in the path of the control signal and the delay through the bus driver between an internal bus and the data bus, the data to be written are put on the data bus and are guaranteed determinate 40 nanoseconds after the rising edge of E and remaining stable for 25 nanoseconds after E falls. These signals are available to control the memory. We note, however, that R/W does not have a rising edge whose timing can be depended on. R/W is *not a timing signal.* You cannot depend on it to satisfy setup and hold times. Similarly, address signals are not precisely aligned with the memory cycle. Such timing signals are often required and can be obtained by ORing the R/W with the inverted E clock or ANDing E with the AND of various address signals or their complements. Alternatively, the 'A4 chip select signals can be used; see §6.2.2.

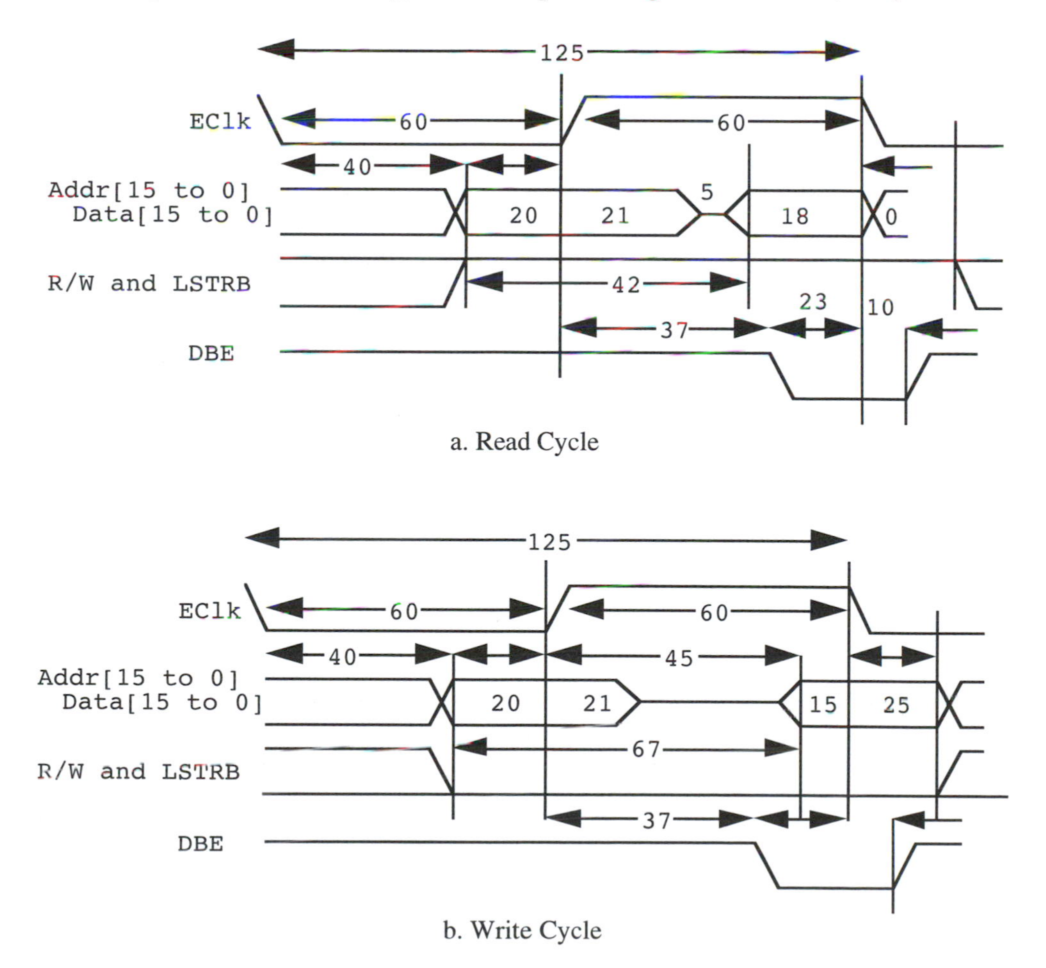

Figure 3.7. Timing Requirements of the Motorola MC68HC912B32

In the MC68HC912B32, to save pins, the port address is sent from the microprocessor on the same bus as the memory address. This bus replaces only PORTA and PORTB. DBE is high when the bus contains an address that is an I/O address or a memory address, and low when data are on the bus. See Figure 3.7. An external latch must be built to capture the address when DBE is T (1), and the address decoder should assert the enable low only when the E clock is T (1) and DBE is F (0).

In analyzing timing requirements, one compares the timing of two parts, such as the microprocessor and some memory to be connected, as we do in §6.5. The object is to verify whether data will be available at the receiver when needed. One should be aware that a specific chip may not meet all of its specifications. Some manufacturers just test a few chips from each batch, while others (such as Motorola) test each part for most specifications. A design in which some requirements are not satisfied may still work because some parts may surpass their specifications. In fact, if you are willing to take the time or pay the expense, you can *screen the parts* to find out which ones meet your tighter specifications. However, if the system fails because the design does not meet its parts' specifications, we blame the designer. If the design meets specifications but the system fails, we blame the part manufacturer or the part.

The key idea you need from this discussion is that the E signal has a falling edge which establishes the setup and hold timing for the read cycle, and the timing for determinate data during the write cycle. The R/W signal, which is often mistakenly used for write-cycle timing, has a rather sloppy rising edge, occurring well into the next cycle. The R/W signal is like the address signals in its timing. Therefore, as we build address decoders in Chapter 4 for I/O devices, we generally AND the complement of the E clock with address and R/W signals, or their inverses, to obtain the enables for I/O devices. Other aspects of bus timing will be further considered in Chapter 6.

3.2.2 Address and Control Signal Decoding

To define the problem of designing address decoders, we first describe the 'A4 address and data bus signals and the control signals associated with them. Figure 3.8 shows these signals for the two available external bus modes on both the 'A4 and 'B32. The narrow mode permits the attachment of 8-bit wide memory and I/O devices, while the wide mode permits attachment of 16-bit wide memory and I/O devices.

Table 3.1. Address Map for a Microcomputer

Address line	15	14	13	12	11	10	9	8	7	6	5	4	3	2	1	0
RAM	F	F	F	F	F	F	X	X	X	X	X	X	X	X	X	X
ROM	T	T	T	T	T	T	X	X	X	X	X	X	X	X	X	X
Input Device	T	F	F	F	F	F	F	F	F	F	F	F	F	F	F	T
Output Device	T	F	F	F	F	F	F	F	F	F	F	F	F	F	T	F

The design problem is to use the 'A4 address and control signals to enable each memory and I/O device when and only when it is supposed to read or write data. For a given system, the *address map* identifies all the memories and I/O devices used by the microcontroller and the range of addresses each device uses. Each memory or I/O device is listed on a line, and the address lines that must be 1 or true (T), must be 0 or false (F), or are not specified (X) are shown. See Table 3.1. If a device has 2^n memory words in it, then the low-order n address bits should be unspecified (X) because these bits are generally input to the device and decoded internally to select therein the word to be read or written. The device address ranges must be mutually exclusive; no two devices should be selected by any address (an exception called a shadowed output device is considered in §4.1.1). The selection of the addresses to be used for each device, which determines how the address map is written, can significantly affect the cost of the control logic, but while there is no exact theory on how to select the addresses to minimize the cost of the control circuit, most designers acquire an adequate skill through trial and error. Generally, a good design is achieved, nevertheless, if either the addresses for each device are evenly spaced in the range of addresses, or else if a number of devices are contiguously addressed.

The *address decoder* enables a device when it should read data in a read cycle, write data in a write cycle, or both. It is designed in two steps. First, an arbitrarily large negative-logic output AND gate (a NAND gate) is designed, that inputs address and control signals, some of which are to be inverted. Second, this generally unrealizable NAND gate is implemented in terms of available gates or other integrated circuits.

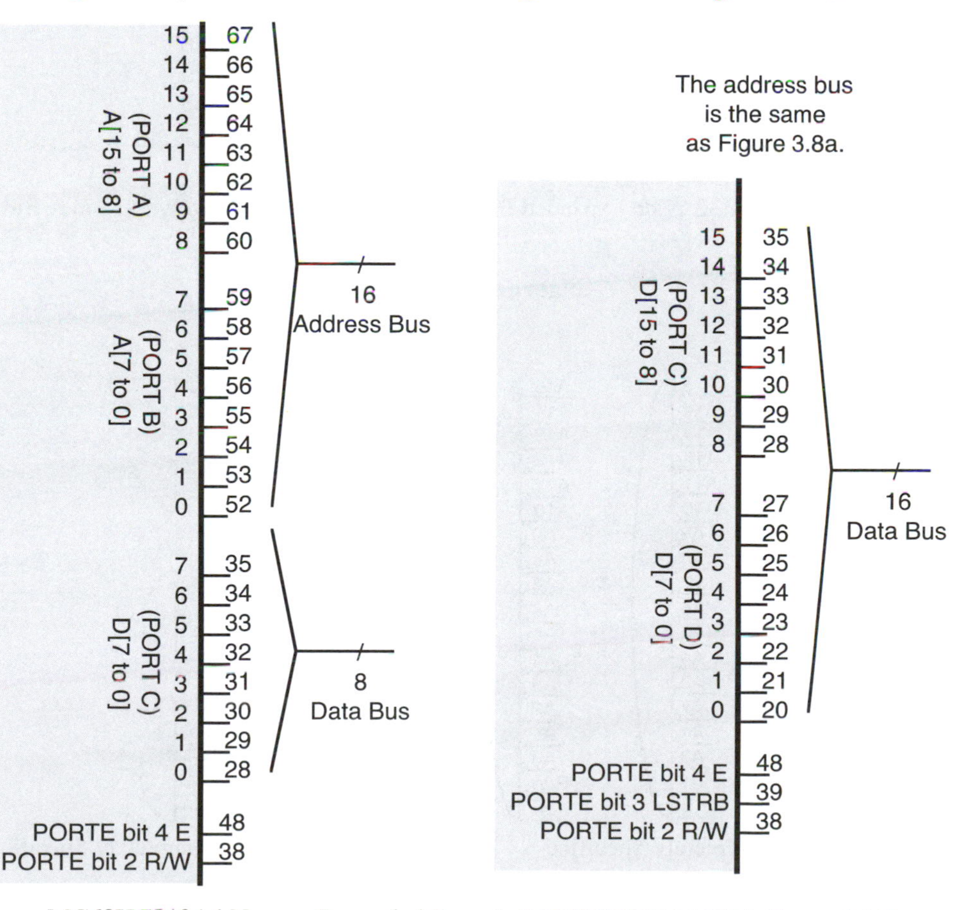

a. MC68HC812A4 Narrow Expanded Bus b. MC68HC812A4 Wide Expanded Bus

Figure 3.8. Address and Data Bus Signals

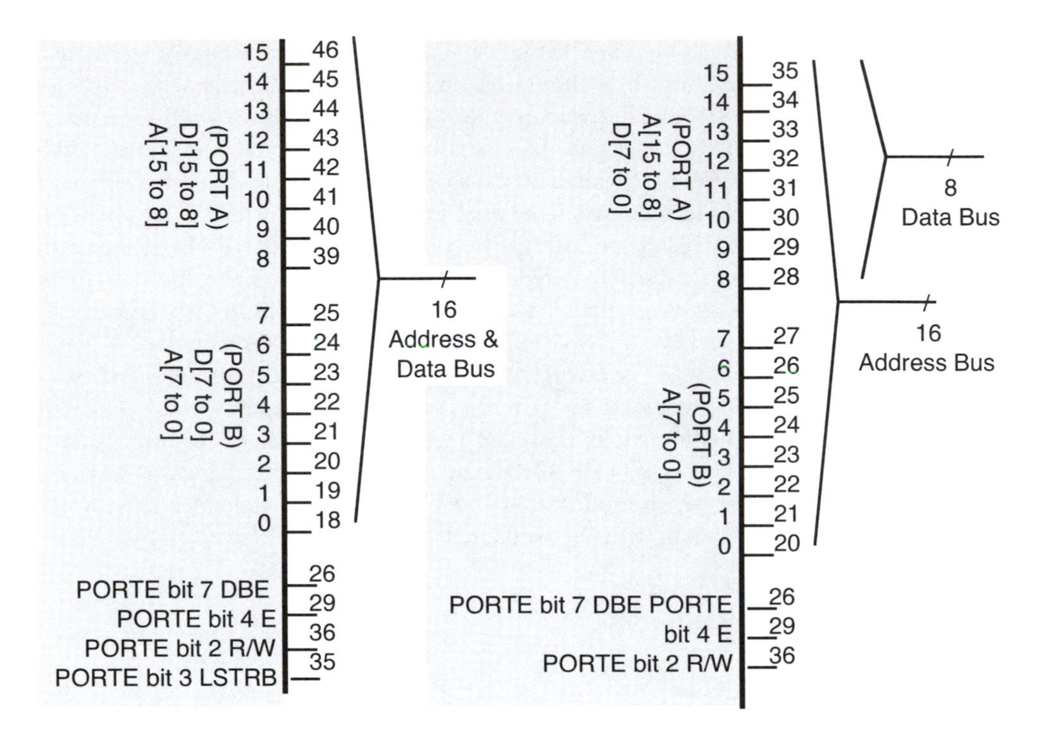

c. MC68HC912B32 Wide Expanded Bus d. MC68HC912B32 Narrow Expanded Bus

Figure 3.8. continued

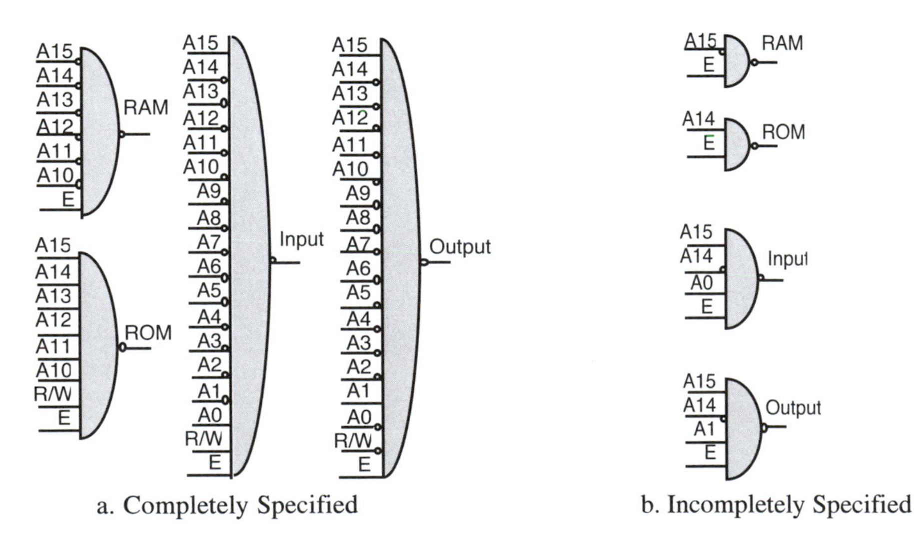

a. Completely Specified b. Incompletely Specified

Figure 3.9. Decoding for Table 3.1

When a reliable hardware system is needed, such as to debug software, *complete decoding* is indicated. Basically, every address line and control line must feed into the large NAND gate, unless it is supplied to the device to be used internally. Figure 3.9a shows the NAND gates needed to completely specify the decoding of the devices shown in Table 3.1 for the narrow mode. The E clock is ANDed into each gate because the addresses are only valid when E is true. R/W is not ANDed into a decoder gate if it is used inside the device.

To reduce hardware costs, *incomplete decoding* may be used. It assumes that the program will only use permissible addresses that are listed in the address map. The decoders use the least number of inputs to each gate such that no permissible address will enable a device other than the device specified in the address map. Consider the address map of Table 3.1 as you reduce the ROM's NAND gate. If A14 is deleted from the gate, then the ROM is enabled if the address is in the range 0xfc00 to 0xffff, which it should be, and *echo range* 0xbc00 to 0xbfff, which it shouldn't be. But the addresses 0xbc00 to 0xbfff are never used by any of the other devices, so they are not permissible. The program should never generate an address in this range. Therefore A14 is removed from the ROM's NAND gate. We extend this technique to delete further decoder inputs. The other gates are similarly reduced too. If only one memory module is used with the 'A4, then all inputs except the E clock could be eliminated. Incomplete decoding can eliminate much of the hardware used to decode addresses in a microcomputer. But it should only be used when the program can be trusted to avoid using impermissible addresses. The technique is especially useful for small microcomputers dedicated to execute a fixed program that has no software bugs which might use duplicate addresses.

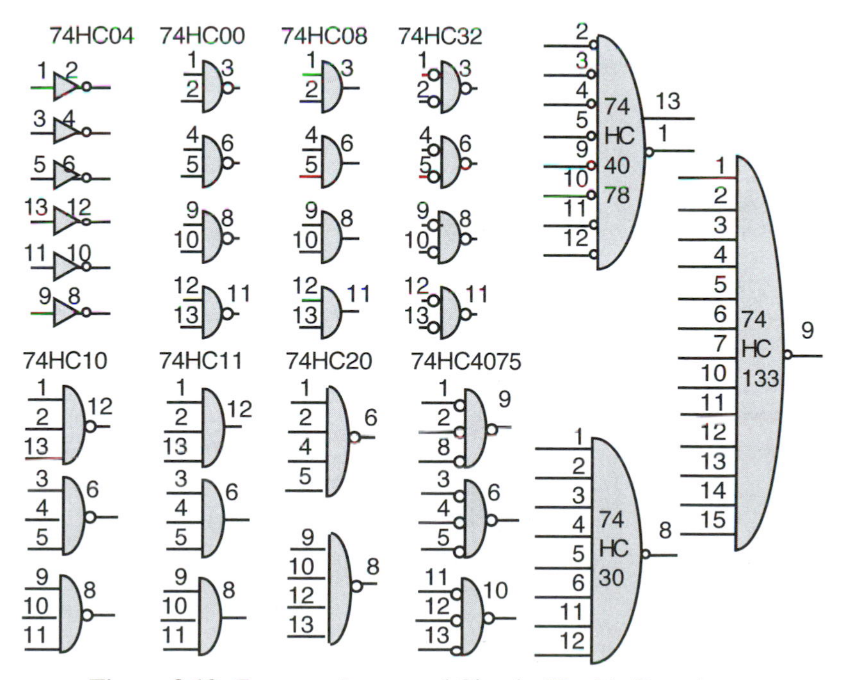

Figure 3.10. Common Integrated Circuits Used in Decoders

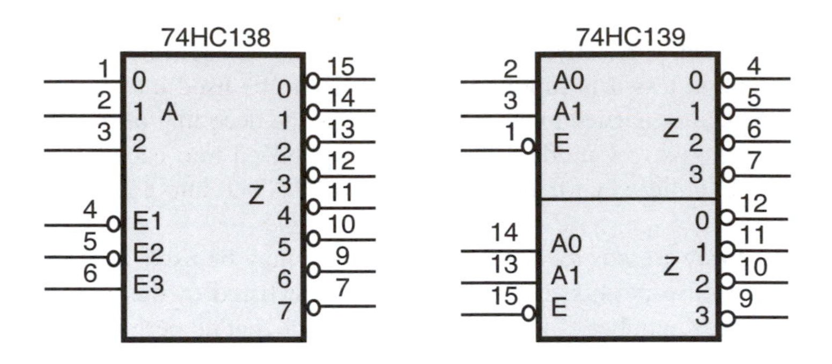

Figure 3.10. continued

The next step is to implement the decoder using existing integrated circuits. The real objective is to reduce hardware cost. However, as a reasonable objective for problems at the end of the chapter, we can restrict the selection of integrated circuits to those of Figure 3.10, defining the "best" solution as that using the least number of these chips. If two solutions use the same number of chips, we rather arbitrarily define the "best" to be the one using the least number of inputs.

Figure 3.10 shows common gates: the 74HC00 quad NAND gate, 74HC04 hex inverter, 74HC08 quad AND gate, 74HC10 triple NAND gate, 74HC11 triple AND gate, 74HC20 dual NAND gate, 74HC30 NAND gate, 74HC33 OR gate, 74HC133 NAND gate, 74HC4078 NOR gate, 74HC138 decoder, and 74HC139 dual decoder. The 74HC4078 provides two outputs that are complements of each other. The 74HC138 decoder asserts output Z[A] low if all three enables are asserted. Each half of the 74HC139 dual decoder asserts output Z[A] low if E is asserted.

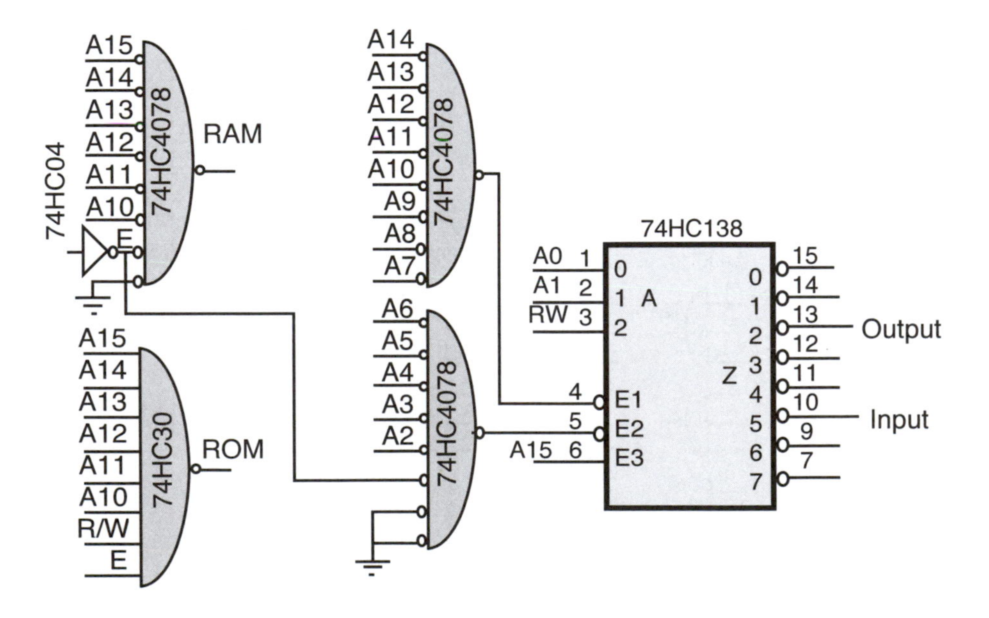

Figure 3.11. Logic Design of Minimal Complete Decoder

A minimal-logic design of the complete decoder is shown in Figure 3.11. Observe the shared gates used to implement two or more decoders. Also, note that when connecting gates to gates, bubbles indicating negative logic either appear on all ends of a line or do not appear on an end of any line. The reader is invited to design a minimal incomplete decoder for the address map shown in Table 3.1.

The wide mode has some additional decoding considerations with respect to the LSTRB signal. One may write into an 8-bit memory or output register using a `STAB` instruction or the equivalent, at an even or an odd address, or one may write into a 16-bit memory or output register using a `STD` instruction or the equivalent. An enable for an 8-bit memory handling the left, high-order byte of a 16-bit word should write only if a `STD` instruction is used or a `STAB` instruction uses an even address. This occurs when R/W and A0 are low. An enable for an 8-bit memory handling the right, low-order byte of a 16-bit word should write only if a `STD` instruction is used or a `STAB` instruction uses an odd address. This occurs when R/W and LSTRB are low. Reading is not generally a problem in the wide expanded mode, because regardless of whether 16 bits are read or 8 bits are read from an even or odd address, the 6812 will read 16 bits and, if necessary, use the correct 8-bit value out of the 16 bits that are read.

In §6.2, we resume the discussion of decoding. The 6812 has a Lite Systems Integration Module in it that decodes addresses and provides chip select signals. It can be used in lieu of, or in addition to, the SSI/MSI-based decoder discussed above.

3.3 Conclusions

The study of microcomputer data and address buses is critical because scanty knowledge in these areas leads to serious interface problems. Before getting on these buses, data inside the microprocessor are unobservable and useless for interfacing. But when data are on the bus, they are quite important in the design of interface circuitry. This chapter has discussed what address, data, and control signals look like on a typical microcomputer bus. You should now be able to read the data sheets and block or logic diagrams that describe the microprocessor and other modules connected to the bus. You should also be able to analyze the timing requirements on a bus. And, finally, you should have sufficient hardware background to understand the discussions of interface modules in the coming chapters.

If you found any difficulty with the discussion on hardware modules and signals, a number of fine books are available on logic design. We recommend *Fundamentals of Logic Design,* fourth edition, by C. H. Roth, PWS Publishing Co., because it is organized as a self-paced course. However, there are so many good texts in different writing styles that you may find another more suitable. Further details on the MC68HC812A4 can be obtained from the MC68HC812A4TS/D Motorola, 1996; §11 gives the timing specifications of the microcomputer. As noted earlier, we have not attempted to duplicate the diagrams and discussions in that book because we assume you will refer to it while reading this book; also, we present an alternative view of the subject so you can use either or both views. The final section in this chapter, however, has not been widely discussed in texts available before now. But several books on interfacing are currently being introduced, and this central problem should be discussed in any good book on interfacing.

Do You Know These Terms?

See page 36 for instructions.

signal
high
low
determinate
variable
TRUE
FALSE
assert a variable
negate a variable
complement a variable
positive logic
negative logic
memory variable
link variable
synchronous
clock
equivalent
module
input port
output port
behaviorally equivalent
dual in-line package
thin-quad flat pack
small scale integrated circuit (SSI)
medium scale integrated circuit (MSI)
large scale integrated circuit (LSI)
very large scale integrated circuit (VLSI)
family
low-power Schottky (LS)
complementary metal oxide semiconductor (CMOS)
logic diagram
type
type name
copy name
formal parameter name
actual parameter name
bypass capacitor
gate
buffer
fan-out
bus
buss
bus driver
fan-in
open collector gate
wire-OR
pull-up resistor
enable
enabled
tristate gate
tristate enable
tristate bus
dynamic logic
D flip-flop
data input
clock input
clocked flip-flop
D edge-triggered flip-flop
D master slave flip-flop
setup time
hold time
one-shot
latch
register
counter
shift register
random access memory (RAM)
programmable read-only memory (PROM)
word
word width
memory cycle
chip enable (CE)
read/not write (R/W)
read enable (RE)
write enable (WE)
memory access time
programmable array logic (PAL)
narrow mode
wide mode
E clock
low strobe
read cycle
write cycle
address map
screen the parts
address decoder
compete decoding
incomplete decoding
echo range

Problems

Problems 1 and 2 are paragraph correction problems. See the guidelines on page 38. Hardware designs should minimize cost (minimal number of chips, where actual chips are specified; and when the number of chips is the same, a minimal number of gates and then a minimal number of pin connections, unless otherwise noted). A logic diagram should describe a circuit in enough detail that one could use the diagram to build a circuit. When logic diagrams are requested, use bubbles to represent negative logic and gates representing high and low signals, and show pin numbers where applicable. A block diagram should describe a circuit in enough detail that one could write a program to use the block diagram. When block diagrams are presented, show variables and gates representing true and false values, and show the maximum detail you can, unless otherwise stated. (Note that a box with SYSTEM written inside it is a block diagram for any problem, but is not a good answer; give the maximum amount of detail in your answer that is possible with the information provided in the question.)

1.* A negative logic signal has a low signal representing a true variable. To negate a variable is to make it low. A synchronous variable is one that repeats itself periodically, like a clock. A family of integrated circuits is a collection of integrated circuits that have the same architecture. A block diagram describes the realization of some hardware to show exactly how to build it. In a block diagram, logic functions are in terms of true and false variables. Vss is normally +5 volts. We normally put .001-microfarad bypass capacitors across power and ground of each MSI chip or about every 4 SSI chips.

2.* A buffer is a gate whose output can be connected to the outputs of other buffers. Open collector drivers can be connected on a bus, called a wire-OR bus that ORs the outputs in positive logic. When a tristate bus driver is disabled, its outputs are pulled to 0 volts by the driver. A flip-flop is a module that copies the variable on the D input when the CLOCK input is high, and leaves the last value in it at other times. The setup time for a D edge-triggered flip-flop is the time the data must be stable before the edge occurs that clocks the data into the flip-flop. The word width of a microcomputer is the number of bits put into the accumulator during an LDAA instruction. The memory cycle time is from when the address is stable until data can be read from or written into the word addressed. Read-only memories store changing data in a typical microprocessor. A programmable array logic (PAL) chip is similar to a PROM, having fuses that are blown to program the device, and it is suitable for "glue" logic and address decoders.

3. Draw the integrated circuits of Figure 3.10 so that they use OR gates rather than AND gates, and put appropriate bubbles on inputs or outputs to get the correct function.

4. A 74CH133 chip being unavailable, show how to implement such a chip's function using the least number of 74HC04s and 74HC20s (see Figure 3.10).

5. A decoder chip being unavailable, show how to implement such a chip's function using the least number of SSI gates as indicated (see Figure 3.10).

a. Implement a 74HC138 using 74HC04s and 74HC20s.
b. Implement a 74HC139 using 74HC04s and 74HC10s.

6. The output signals of a gate are defined for each input signal by Table 3.2. What is the usual name for the logic function when inputs and outputs are considered variables if

Table 3.2. Outputs of a Gate

A	B	C
L	L	L
L	H	L
H	L	L
H	H	H

a. A, B, and C are positive-logic variables.

b. A and B are positive logic and C is negative logic.

c. A and B are negative logic and C is positive logic.

d. A, B, and C are negative-logic variables.

7. A 74HC74 dual D flip-flop stores two bits that can be changed by addresses on a 6812 address bus or by a separate signal. Show a logic diagram that makes the flip-flop have TF (10, left to right, as shown in Figure 3.12) if address 0x5A31 is presented on the 6812 address bus, makes it have FF (00) if address 0x4D21 is presented on the 6812 address bus, and makes it have 01 (FT) if an input signal CMPLT rises from low to high. (Hint: let CMPLT clock the shift register to change from TF to FT.)

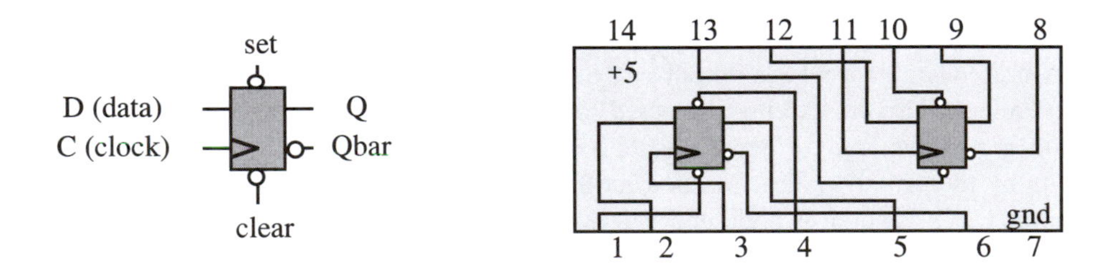

a. Graphical Notation for an Edge-Triggered Flip-Flop b. Pin Connections

Figure 3.12. A 74HC74

8. From the Motorola high-speed CMOS logic data book description of the 74HC163 (Figure 3.13a), determine which control signals among CLOCK, RESET, LOAD, ENABLE P, and ENABLE T are high, are low, or have a rising edge or falling edge, to

a. cause the data on pins 3, 4, 5, and 6 to be stored in its register.
b. cause data stored in its register to be incremented by 1.

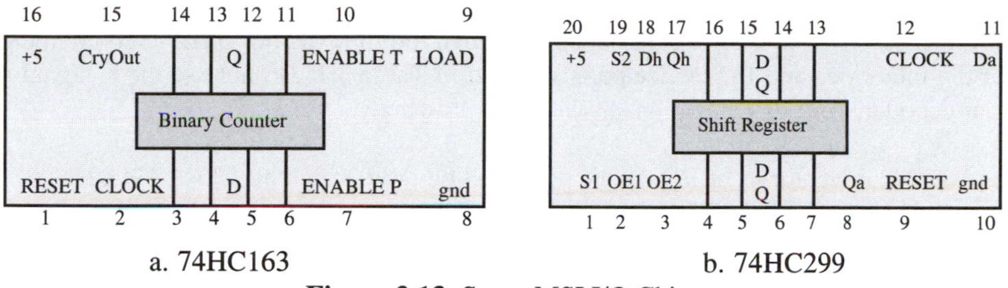

a. 74HC163 b. 74HC299

Figure 3.13. Some MSI I/O Chips

9. From the Motorola high-speed CMOS logic data book description of the 74HC299 (Figure 3.13b), determine which control signals among CLOCK, RESET, OE1, OE2, S1, and S2 have to be high, be low, or have a rising edge or falling edge to

a. cause data on pins 4, 5, 6, 7, 13, 14, 15, and 16 to be stored in its register.
b. cause data stored in its register to be output on pins 4, 5, 6, 7, 13, 14, 15, and 16.
c. cause data stored in the register to be shifted so that a bit shifts out pin 17 as a bit on pin 11 is shifted into the register.
d. cause data stored in the register to be shifted so that a bit shifts out pin 8 as a bit on pin 18 is shifted into the register.

10. Draw the block diagram of a completely specified decoder using arbitrarily large AND gates with appropriate bubble inputs and outputs, like Figure 3.9, for Table 3.3's memory map. Include R/W and E control signals in each decoder as needed.

Table 3.3. Another Address Map for a Microcomputer

Address line	15	14	13	12	11	10	9	8	7	6	5	4	3	2	1	0
RAM	F	F	F	F	T	F	X	X	X	X	X	X	X	X	X	X
ROM	T	T	T	T	X	X	X	X	X	X	X	X	X	X	X	X
Input Device	F	F	F	F	F	F	T	F	F	F	F	F	F	F	F	F
Output Device	F	F	F	F	F	F	T	T	F	F	F	F	F	F	F	F

11. Show a logic diagram of a minimum-cost complete decoder for the memory map of Table 3.3. Use only SSI chips from Figure 3.10.

12. Draw the block diagram of a complete decoder using arbitrarily large AND gates with appropriate bubble inputs and outputs, like Figure 3.9, that will select memory module 1 for addresses in the range 0x0000 to 0x1FFF, memory module 2 for addresses in the range 0xF000 to 0xFFFF, register 1 for address 0x4000, and register 2 for address 0x8000. Do not include R/W and E control signals in each decoder.

13. Show a logic diagram of a minimum-cost complete decoder for the memory map given in problem 12. Use only SSI chips from Figure 3.10.

14. Using just one 74HC10 (see Figure 3.10), show a logic diagram that can implement an incompletely specified decoder in an expanded multiplexed bus 6812 microcomputer for the memories and I/O device ports selected in Table 3.1. Do not use the E signal in your decoder. For this realization, answer the following.

a. What memories or I/O device ports will be written into by STAA $9002?
b. What memories or I/O device ports are stored into by STAA $FFFF?
c. What memories or I/O device ports are written into by STAA $80FF?
d. What memories or I/O device ports are written into by STAA $4000?
e. What five different addresses, other than $8001, access the I/O device port at $8001?

15. Using just one 74HC04 and one 74HC75, show a logic diagram that can implement the decoder for the memories and registers selected in Table 3.3, without decoding the E clock. For this realization, answer the following.

a. What memories or I/O device ports will be written into by STAA $300?
b. What memories or I/O device ports are stored into by STAA $FFFF?
c. What memories or I/O device ports are written into by STAA $812?
d. What memories or I/O device ports are written into by STAA $200?
e. What five different addresses, other than $300, access the input device?

16. Show the logic diagram of a minimum cost decoder, using a 74HC133 and a 74HC138, that provides enables for a set of eight registers addressed at locations %0111 1111 1111 1rrr, where bits rrr indicate which register is enabled.

17. A set of eight 8-byte memories are to be addressed at locations %1000 0000 00mm mxxx. The bits mmm will be FFF (000) to enable the first memory, . . . , and TTT (111) to enable the eighth memory. Bits xxx are sent to each memory and decoded internally to select one of eight words to be read or written. Show the logic diagram of a minimum cost decoder that provides these enables. Use the 74HC4078 and 74HC138.

18. Build an address decoder as discussed in §3.2.1, and determine the 6812 bus timing as discussed in §3.2.2. Using a 74HC04 chip and 74HC30 chip, design a decoder so that it recognizes addresses %1011 1x0x xxxx xxxx (asserts the decoder output low whenever an address appears with the indicated 1s and 0s). Show the oscilloscope traces of (1) the E clock, (2) R/W, and (3) decoder output. Then for inputs indicated below connected to the 74HC30 decoder chip, show the oscilloscope traces of the decoder output for (4) the inverted E clock only, (5) the R/W signal only, (6) both the inverted E clock and inverted R/W signal, and (7) both the inverted E clock and R/W signal. You should have seven sets of traces. Repeat this exercise and copy each of the tracings, to write location 0xbc00 instead of reading it. Finally, for the final decoder, determine and write down for what ranges of addresses the decoder will assert its output low. Show which pulse widths are approximately 60 nanoseconds and which are approximately 90 nanoseconds.

19. Intel's I/O devices have negative logic RD and WR instead of the R/W signal and E clock of the 6812. When RD is asserted low, the device reads a word, and when WR is asserted low, the device writes a word. Show the logic diagram for a minimum-cost circuit to generate RD and WR from R/W and E. Show the timing of the RD signal when the device is being read and of the WR signal when it is being written into.

20. A set of four Intel-style I/O devices are to be addressed at locations %0111 1111 1111 11rr; each device has an RD and a WR signal to enable reading and writing in it. The bits rr will be FF (00) to enable the first device, . . . , and TT (11) to enable the fourth device. Show the logic diagram of a minimum-cost decoder that provides these RD and WR signals. Use the 74HC133 and 74HC138.

Technological Arts' **Adapt812** is a modular implementation of the 68HC812A4, in single-chip mode, which includes all essential support circuitry for the microcontroller. A well designed connector scheme groups the dedicated I/O lines on one standard 50-pin connector, while routing the dual-purpose I/O lines to a second 50-pin connector, to form the address and data bus for use in expanded memory modes.

4

Parallel and Serial Input/Output

The first three chapters were compact surveys of material you need to know to study interface design. In the remainder of the book, we will have more expanded discussions and more opportunities to study interesting examples and work challenging problems. The material in these chapters is not intended to replace the data sheets provided by the manufacturers, nor do we intend to simply summarize them. If the reader wants the best description of the 'A4, 'B32, or any chip discussed at length in the book, data sheets supplied by the manufacturer should be consulted. The topics are organized around concepts rather than around subsystems because we consider the former more important in the long run. In the following chapters, we will concentrate on the principles and practices of designing interfaces for the 'A4 and 'B32.

The first section of this chapter discusses some terminology used in describing I/O ports and describes how to build and access generic parallel I/O ports. We then study the parallel ports in the 'A4 and 'B32. The third section introduces simple software used with parallel I/O ports. Indirect I/O is then discussed. Serial I/O devices, considered next, are particularly easy to connect to a computer because only a small number of wires are needed, and the devices are useful when the relatively slow operation of the serial I/O port is acceptable. These serial ports are called *synchronous* because a clock is used. Asynchronous serial ports are discussed in Chapter 9, where communications systems are described. Throughout these sections, we show how 6812 I/O devices can be accessed in C and how objects can be used to design I/O devices and their software.

Upon finishing this chapter, the reader should be able to design hardware and write software for simple parallel and serial input and output ports. Programs of around 100 lines to input data to a buffer, output data from a buffer, or control something using programmed or interpretive techniques should be easy to write and debug. The reader should understand the use of object-oriented programming for I/O device control. The reader should be able to write classes and use them effectively in debugging and maintaining I/O software. Moreover, the reader will be prepared to study the ports introduced in later chapters, which use parallel and serial I/O ports as major building blocks.

4.1 I/O Devices and Ports

We first consider the parallel port from the programmer's viewpoint (the I/O port's architecture). One question is whether I/O ports appear as words in primary memory, to be accessed using a pointer to memory, or as words in an architecturally different memory, to be accessed by different instructions. Another concern is where to place the port in the address space. A final aspect is whether the port can be read from or written in, or both. The "write-only memory" is usually only a topic for a computer scientist's joke collection, but an I/O port can be write-only; to understand why, you need to understand hardware design and cost. So we introduce I/O port hardware design and programming techniques to access the hardware. This section will also be useful in later sections that introduce the 6812 parallel ports and the software used with these ports.

From a designer's viewpoint, an *I/O device* is a subsystem of a computer that inputs or outputs data. I/O devices have ports. In simplified terms, a *port* is a "window" to the outside world through which a logically indivisible and atomic unit of data passes. An *input port* passes data into the computer and an *output port* passes data out of it. Data are moved to and from ports in the memory bus as words. Recall that a word has already been defined as a unit of data that is read from or written to memory in one memory cycle. A port can be a word, although it doesn't have to be one word, because the unit of data read or written in a memory cycle need not be an indivisible unit of data that passes into or out of the computer, as we shall see later in this section.

All three of the above terms can be hierarchical. An I/O device can be composed of I/O devices, since subsystems can be composed of smaller subsystems. A port can be composed of ports because a unit of data passed indivisibly at one time can be subdivided and an indivisible subunit can be passed at a different time. Even a word accessed in a memory cycle can be the same as two words accessed in two memory cycles elsewhere.

There are two major ways in a microcomputer to access I/O, relative to the address space in primary memory. In the first method, *isolated I/O*, the ports are read from by means of *input instructions,* such as IN 5. This kind of instruction would input a word from I/O port 5 into a data register. Similarly, *output instructions* like `OUT 3` would output a word from a data register to the third I/O port.

The second way is by *memory-mapped I/O,* in which the ports are read by means of LDAB instructions, or the equivalent, such as `LDAA`, `ADDB`, `ORB`, or `MOVB`. Memory-mapped I/O uses the data and address buses just as memory uses them. The microprocessor thinks it is reading or writing data in memory, but the I/O ports are designed to supply the data read or capture the data written at specific memory locations.

Memory-mapped I/O is more popular because most microcomputers have instructions that operate directly in memory, such as `INC 0x100` or `ROL 0x100`. If the program is in read-only memory, indexed addressing and indirect memory can be used to relocate memory-mapped I/O, while isolated I/O may not have this capability. The use of these instructions operating directly on (readable) output ports in memory-mapped I/O is very powerful; their use can also shorten programs that would otherwise need to bring the word into the data register, operate on it, and then output it. Finally, conventional C without embedded assembly language can access I/O ports using variable pointers or using constant address pointers. However, object-oriented C++ can make isolated I/O behave essentially like memory-mapped I/O.

We can accidentally write over an output port when we use memory-mapped I/O. Memory-mapped I/O can be protected, however, by a *lock.* The lock is an output port that the program can change. The lock's output is ANDed with address and other control signals to get the enable or clock signals for all other I/O ports. If the lock is F, no I/O ports can be read or written. Before reading an I/O port, the program has to store T in the lock; then store F in the lock after all I/O operations are complete – 6812 ports controlling EEPROM and flash memory programming are locked.

One of the most common faulty assumptions in port architecture is that I/O ports are 8 bits wide. For instance, in the 6812, the 1-byte-wide `LDAB` instructions are used in I/O programs in many texts. There are a large number of 8-bit I/O ports on I/O chips that are designed for 8-bit microcomputers. But 8 bits is not a fundamental width. In this book, where we emphasize fundamentals, we avoid that assumption. Of course, if the port is 8 bits wide, the `LDAB` instruction can be used – in C by accessing a variable of type *char.* There are also 16-bit ports. They can be read by `LDD` instructions, or as an *int* variable in C or C++. A port can be 1 bit wide; if so, a 1-bit input port is read in bit 7; reading it will set the N condition code bit, which a `BMI` instruction easily tests. Many ports read or write ASCII data. ASCII data is 7 bits wide, not 8 bits wide. If you read a 10-bit analog-to-digital converter's output, you should read a 10-bit port. Whatever your device needs, consider using a port of the right width.

4.1.1 Generic Port Architecture

As in the memory design to be presented in §6.5, generic parallel I/O device hardware, such as a tristate driver or a register, is enabled or clocked by an address decoder in the device that decodes the address on the address bus. The decoder can either be completely or incompletely specified, and built with SSI gates, decoders, or PALs. Generally, though, the 6812 must be read from or written into the port when a specific memory address is sent out and must not access it when any other address used by the program is sent out. (An exception, shadowed output, is discussed later in this section.)

The generic *input port* samples a signal when the microcomputer executes a `LDAB` or `LDD` instruction in memory-mapped I/O, and reads the sample into the 6812. Because most microcomputers use tristate bus drivers, the port must drive the sample of data onto the data bus exactly when the microprocessor executes a read command with this port's address. See Figure 4.1a for an 8-bit input device with an input port at 0x4000. Since this port address has many zeros, the use of negative-logic-input/output AND gates (positive-logic OR gates such as the 74HC4078) often reduces the decoder's cost.

The output port usually has to hold output data for an indefinite time, until the program changes it. The generic *basic output port* is therefore a latch or register that is capable of clocking data from the data bus whenever the microcomputer writes to a location in memory-mapped I/O. The D bus is connected to the D input of the register or latch, and the clock is connected to an address decoder so that the register is clocked when the microprocessor executes a `STAB` or `STD` instruction at the address selected for this port. See Figure 4.1b for an 8-bit output device with its port at 0x67ff. Since the port address has many 1s, the use of NAND gates such as the 74HC30 often reduces cost.

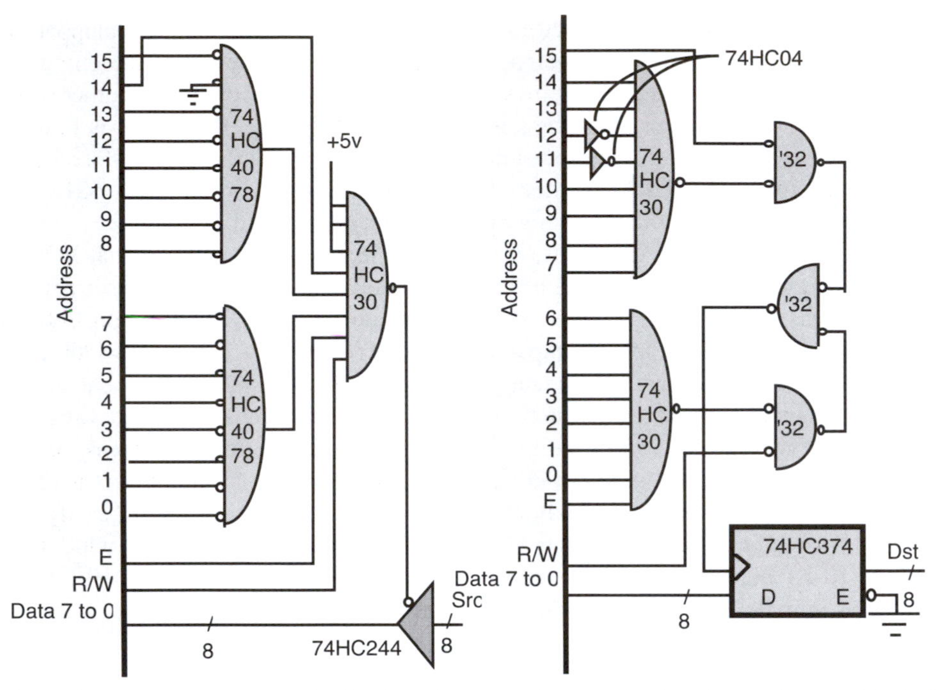

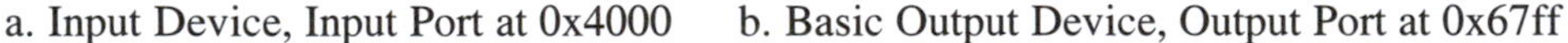
a. Input Device, Input Port at 0x4000 b. Basic Output Device, Output Port at 0x67ff

Figure 4.1. Logic Diagrams for I/O Devices for a Narrow Expanded Bus

The following discussion is oriented to 8-bit ports aligned to 8-bit word boundaries. A similar set of examples can be generated for 16-bit ports, and for ports that do not align with 8-bit or 16-bit words. We consider the design and use of a 16-bit port in problem 7. We defer consideration of odd-aligned ports to the end of this discussion.

An output port can be combined with an input port at the same address that inputs the data stored in the output port, to implement a more flexible but more costly generic *readable output port.* Figure 4.2 shows a readable output port that can be read from or written in at location 0x1000. The decoder shown therein should be implemented with a minimal number of available gates, but this design is left as problem 8 in this chapter.

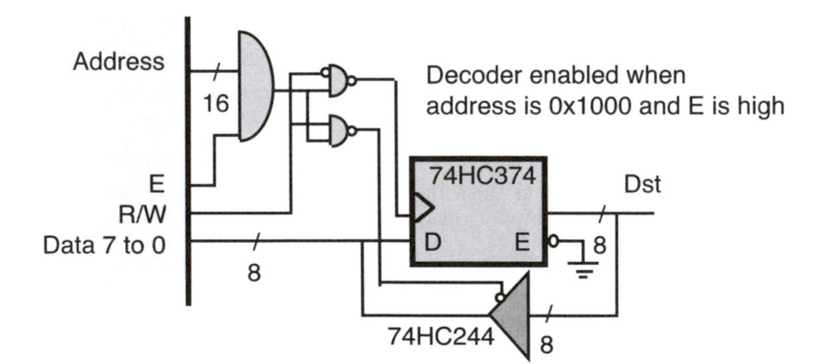

Figure 4.2. A Readable Output Port for a Narrow Expanded Bus

These generic ports can be accessed in assembly language. To read the data from an input port (Figure 4.1a) or readable output port (Figure 4.2), use the following instruction or its equivalent

```
LDAB    $4000
```

When LDAB recalls an 8-bit word from location $4000, the decoder responds, asserting the 74HC244's negative-logic tristate enable. At that time, the 74HC244 drives the data bus with its input data SRC, and the 6812 puts the data on the bus into accumulator B. Many instructions, including BRSET and BRCLR, also read from an input port. The following instruction writes into a basic output port (Figure 4.1b) or readable output port (Figure 4.2).

```
STAB    $67ff
```

When the STAB instruction memorizes an 8-bit word at location $67ff, the decoder responds, causing a rising edge on the 74HC374 clock at the end of the memory cycle when the E clock falls. The 6812 has put the contents of accumulator B onto the data bus, and the rising edge causes this data to be stored in the 74HC374. This data is available to the outside world until the 6812 writes new data into it. To read, modify, and write into a readable output shown in Figure 4.2, one can use the following instruction:

```
INC    $1000
```

When the INC instruction recalls an 8-bit word from location $1000, the bottom decoder output asserts the tristate gate's negative-logic tristate enable to input DST to the 6812's data bus. When the INC instruction memorizes an 8-bit word at location $1000, the top decoder output causes a rising edge on the 74HC374 clock at the end of the memory cycle when the E clock falls. The data from the 6812 data bus, which is the incremented value of DST, is written back into the 74HC374, which outputs this data until it is written into again. Besides INC, the instructions DEC, ASL, ASR, LSR, ROL, ROR, BSET, and BCLR read from, modify, and write into a readable output port.

These ports can be accessed in C or C++. The declaration or cast for an 8-bit port is usually *char* (if we want to test the sign bit) or *unsigned char* (if we want to prevent sign extension) and is further declared to be *volatile,* indicating that the data can change due to external activities, to prevent the compiler's optimizer from removing access to it. A constant, cast to a *volatile unsigned char* pointer, can be used to read a port. Or a *volatile unsigned char* global variable *in* can be forced to have an address 0x4000 by means of an "at" sign (@), as in *volatile unsigned char in@0x4000.* Finally, a *volatile unsigned char* pointer *inPtr* can be loaded with 0x4000. To read the data from an input port (Figure 4.1a) or readable output port (Figure 4.2) into a *char* variable *i,* use one of the following statements or their equivalent:

```
i = *(volatile unsigned char *)0x4000;
i = in;
i = * inPtr;
```

The statements `i = *(volatile unsigned char *)0x4000;` and `i = in;` (where `in` is 0x4000) generate `LDAB 0x4000`, and `i = *inptr;` generates a `LDAB 0,X`.

A constant, cast to a `volatile unsigned char` pointer, can write into an 8-bit port. Or a `volatile unsigned char` global variable `out` can be forced to have an address 0x67ff by means of an "at" sign (@), as in `volatile unsigned char out@0x67ff`. Finally, a `volatile unsigned char` pointer `outPtr` can be loaded with 0x67ff. To write *i* into a basic output port (Figure 4.1b) or readable output port (Figure 4.2), one can use any of the following statements:

```
*(volatile unsigned char *)0x67fff = i;
out = i;
*outPtr = i;
```

The statements `*(volatile unsigned char *)0x67fff = i;` and `out = i;` (where `out` is 0x67ff) generate `STAB 0x67ff`, and `*outPtr = i;` generates `STAB 0,X`.

A constant, cast to a `volatile unsigned char` pointer `portPtr`, can be used to read from, modify, and then write into an 8-bit readable output port. A `volatile unsigned char` global variable `port` can be forced to have address 0x1000 by means of an "at" sign (@), as in `volatile unsigned char in@0x1000`, or a `volatile unsigned char` pointer `outPtr` can be loaded with 0x1000; then, to increment a readable output as shown in Figure 4.2, one can use the following:

```
*(volatile unsigned char *)0x1000++;
port++;
(*portPtr)++;
```

The statements `(*(volatile unsigned char *)0x1000)++;` and `port++;` (where symbolic name `port` is 0x1000) generate `INC port`, and `(*portPtr)++;` generates `INC 0,X`. Similarly `port--; port |= i;` and `port &= i;` generally also access a readable output port with `DEC`, `BSET`, and `BCLR` read-modify-write instructions.

Notice that an instruction like `INC $1000`, generated by `port++;` fails to work on Figure 4.1b's basic output port, which is not a readable output port. The instruction reads garbage on the data bus, increments it, then writes "incremented garbage" into the output port. To use any read/modify/write instructions, build a more costly readable output port.

Alternatively, an output port can be at the same address as a word in RAM; writing at the address writes data in both the I/O port and the RAM and reading data reads the word in RAM. This technique is called *shadowed output.* This effect can be achieved, moreover, through software. If a basic output port is to be updated after being read, like a readable output port, a global variable can keep a duplicate of the data in the port so it can be read whenever the program needs to get the data last put into the port. In C or C++, for instance, using a pointer `outPtr` to a port, declare also a global `volatile unsigned char portValue;`, then write `*outPtr = portValue = i;` whenever we output to the port. Then `i = portvalue;` reads what is in the port. Also, `*outPtr = portValue++;` increments it, `*outPtr = portValue--;` decrements it, `*outPtr = (portValue |= i);` sets bits, and `*outPtr = (portValue &= i);` clears bits. If we consistently copy output data written into the port into `portValue;` we can read `portValue` when we want to read the port.

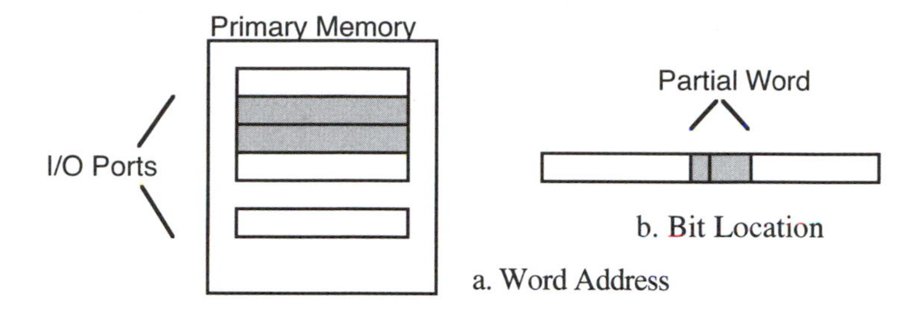

Figure 4.3. An Unusual I/O Port

A port may be only part of a word or of two adjacent words. See Figure 4.3. A "worst-case" 3-bit port, low bit of the 8-bit word at location 0x4000 and two high bits of the 8-bit word at 0x4001 (i.e., the middle 3 bits of the 16-bit word at 0x4000), occupies two consecutive 8-bit words in the memory map (Figure 4.3a) and parts of each word (Figure 4.3b). Generally, if an n-bit port ($n < 16$) is read as a 16-bit word, the other "garbage" bits should be stripped off, and if it is not left-aligned, logical shift instructions should align the port data. If such a port is written into, other ports that share word(s) written in order to write into this port must be read and then rewritten to be maintained. Consider an example of inputting from and outputting to a misaligned port.

Assuming that we use the declaration `int *ptr = (int*)0x4000` and that `0x4000` is the address of a pair of 8-bit words containing the 3-bit input port just discussed, the port is read into a variable `d` with `d = (*ptr >> 6) & 7`. The declaration, `(int*)`, causes reading or writing 16 bits at the address pointed to by `ptr`. The operator `>>` moves data from the port to the least significant bits of `d`, and the operator `&` removes data that is not in the port from the words read from memory; this generates the following:

```
0000086D ED00        LDY   0,X
0000086F C606        LDAB  #6
00000871 35          PSHY
00000872 160890      JSR   $0890
00000875 C407        ANDB  #7
00000877 87          CLRA
```

The subroutine at $890 shifts the data, in Y, accumulator B places to the left.

```
00000890 87          CLRA
00000891 D7          TSTB
00000892 B7C6        EXG   D,Y
00000894 2705        BEQ   *+7    ;abs = 089B
00000896 47          ASRA
00000897 56          RORB
00000898 03          DEY
00000899 26FB        BNE   *-3    ;abs = 0896
0000089B 3D          RTS
```

Omit the shift operator if the port is aligned so that the least significant bit of the port is the least significant bit of a word. The AND operator is omitted if the port consists of whole words. For instance, assuming pointer `ptr = (int*)0x4000` is the address of a 16-bit port, `d = *ptr;` will read a 16-bit port at `0x4000` into `d`.

Similarly, consider the case where those three bits are a readable output port within words that also contain other readable output ports that must be unchanged. Assume that pointer `ptr = (int*)0x1000` points to a 16-bit word containing the aforementioned 3-bit readable output port. Then `*ptr = *ptr & 0xFE3F | ( ( d << 6 ) & 0x1C0; );` puts `d` into the port, generating the following (without balancing the stack):

```
0000087B 13        EMUL
0000087C C4C0      ANDB  #192
0000087E 8401      ANDA  #1
00000880 3B        PSHD
00000881 EC82      LDD   2,SP
00000883 C43F      ANDB  #63
00000885 84FE      ANDA  #254
00000887 EA81      ORAB  1,SP
00000889 AA80      ORAA  0,SP
0000088B 6C00      STD   0,X
```

The expression `d << 6`, which is implemented by the `EMUL` instruction, moves the least significant bits of the data in `d` into position to put in the port; `& 0x1C0` removes parts of `d` that are not to be output; `*ptr & 0xFE3F` gets data in the words not in the port that must be unchanged. The OR instructions merge the data, in `d` to be written into the port, and the data, in the words not in the port together, putting this data into the words at 0x1000 and 0x1001. If the port is aligned so that the least significant bit of the port is the least significant bit of a word, the shift operator is omitted; and if the words being written into have no ports other than the port being written, the `*ptr & 0xFE3F |` may be omitted. For instance, assuming pointer `ptr = (int*)0x4000` is the address of a word, `*ptr = d;` will write 16-bit data in `d` into a 16-bit output port at location `ptr`. If the output port is not readable, but is a basic output port of Figure 4.1b, then a copy of the data, that is output to that and other ports in the words it writes into, must be kept in memory. It can be kept in `int` variable `portValue;` the statement `*ptr = portValue = portValue & 0xFE3F | ( ( d << 14 ) & 0x1C0 );` is not that much more complex than the statement for a readable output port.

The I/O port can also be accessed as an element of a vector. If an input port is at 0x4001 and `ptr` is 0x4000, then the expression `i = *(ptr + 1);` reads the port's data into `i`. Note that `ptr[1]` is the same as `*(ptr + 1)`, so this statement can be written `i = ptr[1];` which is often considered easier to understand. See also §4.4.5.1.

I/O ports can also be accessed as elements of a `struct`, which can be declared:

```
typedef struct F { volatile unsigned char A, B, int C; } F;
```

If we declare a pointer to this `struct; F *fptr = (F *)0x100;`, then if an input port is at 0x101, `d = fptr->B;` reads the port's data. Similarly, `fptr->B = d;` writes data into the port. Further, if we declare `F f@0x100;` then `d = f.B;` reads the port's data. Similarly, `f.B = d;`, writes data into the port. See also §4.4.5.4.

If the `struct` has bit fields, these statements generate code like that just shown for the misaligned port discussed under Figure 4.3, but are easier to understand.

```
typedef struct F { volatile unsigned char A:2, B:3, C:3, D:2; } F;
```

In HIWARE's C++ compiler, if in "advanced options" the setting for "code generation" has "bit field byte allocation" set to "most significant bit in byte first," bit fields are coded eight bits to a byte such that the left `struct` element corresponds to the leftmost bit. Otherwise, the right `struct` element corresponds to the leftmost bit (the default setting). But if a bit field were to overlap the boundary between bytes, as in Figure 4.3, the potentially overlapping bit field is left-aligned on the next byte boundary instead.

Some logic functions can be implemented in hardware upon writing to a port. The data can set bits in a *set port.* See Figure 4.4. A pattern of 1s and 0s is written by the processor via the data bus: wherever it writes a 1, the port bit is set; wherever it writes a 0, nothing is done. The data bit is ANDed with the decoded address enable, which is asserted only when a matching address appears; the E clock is high at the end of a memory cycle; and R/W is low because we are memorizing. Note that if the data bit and enable are true (1), a bit in the flip-flop at the bottom of this figure will be set. If a data bit is false (0), nothing is done.

Data can clear bits in a *clear port.* In such a port, writing a true (1) clears the port bit, and writing a false (0) does nothing. For instance, if a clear port at 0x4000 is to have bits 1 and 4 cleared, and `ptr` is `0x4000;`, the statement `*ptr = 0x12;` clears the two bits. The hardware is the same as in Figure 4.4, except that the clear input to the flip-flop is asserted low if the decoder enable is asserted and a data bit is true (1). Alternatively, wherever the processor writes a false (0), the port bit is cleared; wherever it writes a true (1), nothing is done. This hardware is the same as in Figure 4.4, except that the clear input to the flip-flop is asserted low if the decoder enable is asserted and a data bit is false (0).

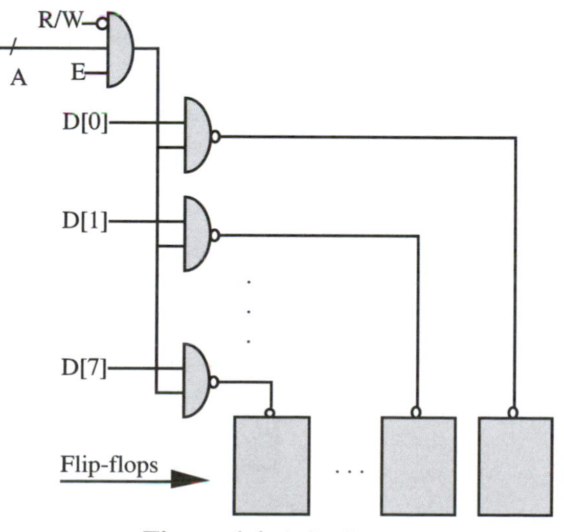

Figure 4.4. A Set Port

These ports are set or cleared by just writing a constant or variable into them. If a set port at the address 0x4000 is to have bits 0 and 3 set, and `ptr` is declared as above, the statement `*ptr = 9;` sets the bits. If an 8-bit clear port at 0x4000 is to have bits 1 and 2 cleared, and writing 0s clears a bit, we can use `*ptr = 0xf3;` or the statement `*ptr = ~6;` to clear the bits. Set ports and clear ports can be readable; if so, the data in the port are read without logic operations being done to them. A clear port is frequently used in devices that themselves set the port bit, and the programmer only clears the bit. An interrupt request, discussed in the next chapter, is set by the device and can be cleared by the processor writing to a clear port. C and some assembly languages have means to OR or to AND data to memory, so these port functions are redundant. Nevertheless, the 6812 timer flags are clear ports; writing a 1 in a bit position clears a timer flag bit there.

Finally, we introduce two techniques that use the address bus without the data bus to output data. These are the address trigger used in many Motorola I/O devices, and the address output port used in the 68000-based Atari 520ST.

In an *address trigger,* an instruction's recalling or memorizing at an address causes the address decoder to provide a pulse that can be used to trigger a one-shot, clear or set a flip-flop, or provide an output from the computer. Figure 4.5a shows an example of a trigger that produces a 60-ns pulse when location 0x1000 is read or written. An address trigger is, in effect, an output port in a word that requires no data bits in it. These instructions could simultaneously load or store data in another port even while the address triggers a one-shot in a manner similar to the shadow output port, or they might load garbage into the data register or store the data register into a nonexistent storage word (which does nothing). Some variations of an address trigger are the *read address trigger,* which produces a pulse only when a fetch or recall operation generates the address; a *write address trigger,* which generates a pulse only when the address is recognized in a memorize operation; and an *address trigger sequence,* which generates a pulse when two or more addresses appear in a sequence.

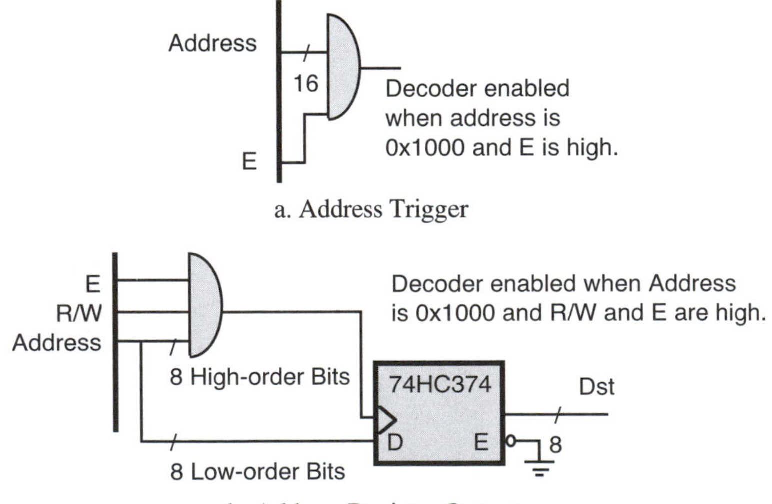

Figure 4.5. Address Output Techniques

An *address-register output* was used in the 68000-based Atari 520ST. The design of its extension ports supports only read-only memories, in order to plug games into the computer, and thus is suitable for use only for an input port. But output can be done. The high-order 8 bits of the address are decoded; the low-order 8 bits of the address are not decoded, as if a memory chip used these bits internally. Rather, these low-order 8 bits are input to the device's register. To use the port for 8-bit output, data are put on low-order 8 bits of the address while the device address is put on the high-order bits of the address bus. This is easily handled in C: let *a* be a dummy variable, `d` be `volatile unsigned char` data, and `ptr` point to the port; the low-order 8 bits of `ptr` are 0s, and `a = *(ptr+d)` puts the data in `d` into the register.

4.1.2 Generic Port Classes

Although a high-level language greatly simplifies I/O interfacing, objects further simplify this task through encapsulation, information hiding, polymorphism, inheritance, and operator overloading. This section shows how they can be used.

Objects can be used as part of the core of a program, as well as for I/O interfacing. The original idea of an object was to tie together a data structure with all the operations that can operate on it. For instance, we define a data structure such as a queue, and we tie to it the operations like push, pull, and so on. Further, and independently, the I/O software they use can be organized in terms of objects.

I/O devices can be encapsulated into objects, as discussed in §2.3.3. In effect, we will handle all access to parallel I/O devices using object function members. Since all these alternative techniques can now be implemented in the same way, replacing a device using one technique with a device using another technique is easy. We can separate the design into a part having to deal with algorithms and interpreting a data structure, and a part having to deal with the I/O, and test each part separately, or mix and match different alternatives of both parts of the design at will. We can make I/O operations *device-independent,* meaning that they are basically identically written at compile time in the main program, regardless of the implementation technique. By using different procedure arguments that *declare* or *bless* an I/O *object,* a different I/O device that uses the same techniques or one that uses different techniques can be substituted. When this can be done at run time, blessing pointers to objects, this is called *I/O redirection.*

Both device independence and I/O redirection are also key ideas of *device drivers* in operating systems. Device drivers appear to be better suited to complex I/O devices like terminals that may need to recognize special characters such as backspace, and disks that need directories and allocation bit maps. Objects appear better suited to simple I/O devices: parallel I/O ports, and A-to-D converters. These simpler devices can also use device independence and I/O redirection, but a device driver is much more complex.

We can define classes of objects for I/O devices in many ways, each having increased power and sophistication. We can get as sophisticated as a device driver. However, this sophistication also increases overhead. When a class has a function member with the same name as a function member of its base class, the class's function member overrides the base class's function member, as discussed in §2.3.3. However, the compiler generally loads both function members into memory, because it may be unable to determine if the base class's function members will ever get used. We must be cautious about what we put into classes and base classes or we will fill memory with a lot of unnecessary function members. (By the way, a similar kind of overhead appears when operating systems use device drivers.) Having tried a number of different approaches, we have concluded that the following rules should be used for I/O objects.

1. There will be a hierarchical library of classes designed for interchangeable I/O devices.

2. The root class's function members and instance variables will be defined for a template root class *Port;* all 8-bit, 16-bit, and 32-bit I/O devices are objects of this class or of its derived classes. There should be a constructor, virtual function members for input (`get`), output (`put`), and options (`option`), and possibly a destructor. There can also be virtual overloaded operators to cast, assign, and OR or AND into an object. The constructor initializes the I/O device as well as the object data members. One of the constructor's `int` parameters specifies the port address, which is copied into a data member `port`, to point to 8-bit, 16-bit, or 32-bit ports.

The root class `Port` and its data and function members are defined below. It will be redefined in §4.2.3 and again in §4.3.6 to include more capabilities, especially options. In this section we see how C++ I/O software uses templates and overloaded operators. We make all 8-bit ports objects of class `Port<char>;` 16-bit ports are objects of class `Port<int>.` Similarly, 32-bit ports can be objects of the class `Port<long>.` These ports are also often further templated as `unsigned` to inhibit sign extension.

```
template <class T> class Port{
    protected : T *port;                                        // data members
    public: Port(unsigned int a) { port = (T *)a; }             // constructor
    virtual void put(T data) { *port = data; }                  // output
    virtual T get(void){ return *port; }                        // input
    T operator = (T data) { put(data); return data; }           // assignment
    operator T () { return get(); }                             // cast
    virtual T operator |= (T data) {put(data |= get()); return data;}
    virtual T operator &= (T data) { put(data &= get()); return data; }
};
```

Consistent with C++ conventions, all input function members are `get,` and output function members are `put`. They simply use the pointer `port`, initialized by the constructor, to read data from or write to the port. An object of class `Port` can be declared, or a pointer `ptr` can be blessed for class `Port`, and then used as shown next. The rest of the program, here simply represented by the second-to-last line, is the same, regardless of which class the object is declared or blessed a member thereof.

```
void main(){unsigned char i;
   Port<char> port(0x1000);
   port.put(3); i=port.get();
}
```

```
void main(){unsigned char i;Port<char>*ptr;
   ptr = new Port<char>(0x1000);
   ptr->put(3); i=ptr->get();
}
```

Our classes are designed to be interchangeable, to effect device independence and I/O redirection. Whereas the above example switches classes at compile time, which is device independence, an object can also be blessed for different classes at run time, which is I/O redirection. In both cases, except for the declaration of the object or blessing of the object, the rest of the program will be unchanged.

```
void func(char mode){ unsigned char i; Port<char> *ptr;
    if(mode) ptr = new Port<char>(0x1000);
    else ptr = new Port<char>(0x2000);
    ptr->put(3); i = ptr->get();
}
```

The cast and assignment "=" operators can be overloaded for input or output operations throughout the remainder of the program, so they appear as I/O operations using the simple data types described in §4.1.1. Do not overlook the advantage of using these overloaded operators over using `put` and `get` function members; a program using simple input and output software as discussed in §4.1.1 can be converted to classes without modifying the main program, except for the inclusion of the class declaration and the constructor. Conversely, a program written using I/O classes can be changed back to one using simple data types without rewriting most of it.

Overloaded ORing `|=` and ANDing `&=` are useful for set and clear ports, as illustrated in the next two classes. A set port or clear port can be given its own class, a derived class that overrides `|=` and `&=`.

```
template <class T> class SetPort : public Port<T> {

    public: SetPort(int a) : Port(a) { }

    virtual T operator |= (T data) { put(data); return data|=get(); }
};

template <class T> class ClearPort : public Port<T> {

    public: ClearPort(int a) : Port(a) { }

    virtual T operator &= (T data) { put(~data); return data&=get(); }
};
```

Then if `port` is an object of class `Port` or `SetPort`, the expression `port |= i;` will OR the data `i` into the the data stored in a register inside the device. Similarly, clear ports directly handle `&=`. For instance, an I/O subsystem (a device) might be written and tested on an 'A4 using a clear port, and the object port will be of the class `ClearPort`. Then if it is implemented on a 'B32 using external hardware such as a 74HC74, which readily implements an ordinary output port, `port` can be of class `Port`, and the software will perform the ANDing operation. If `port` is an object of class `Port,` the software will AND the argument into the output port (which should be readable). But if the hardware ANDs the output data into its register, *port* should be of class `ClearPort,` which simply writes the data to the port, and the software will not interfere with the hardware operation that ANDs the data written into the register. A program written to clear a port's bits will work exactly the same way in both the 'A4 and 'B32 environments.

The advantage of object-oriented programming for I/O should be somewhat apparent from the preceding examples. However, object-oriented programming has additional useful features when one is designing a state-of-the-art microcomputer's I/O devices, as proposed by Grady Booch in his tutorial *Object-Oriented Computing.* Encapsulation is extended to include not only instance variables and methods, but also the I/O device and the digital, analog, and mechanical systems used for this I/O. An `object` is all these parts considered as a single unit. For instance, suppose you are designing an automobile controller. An object (Call it `PLUGS`) might be the spark plugs, their control hardware, and procedures. Having defined `PLUGS,` you call function members (for instance, `SetRate(10)` to `PLUGS`) rather like connecting wires between the hardware parts of these objects. The system takes shape in a clear intuitive way as the function members are defined. In top-down design, you can specify the arguments and the semantics of the methods that will be executed before you write them. In bottom-up design, the object `PLUGS` can be tested by a driver as a unit before it is connected to other objects.

An object can be replaced by another object, if the function calls are written the same way (polymorphism). If you replace your spark plug firing system with another, the whole old `PLUGS` object can be removed and a whole new `PLUGS1` object inserted. You can maintain a library of classes to construct new products by building on large pretested modules. Having several objects with different costs and performances, you can insert a customer-specified one in each unit. Factoring can be used to save design effort.

Factoring can be used in a different way to simplify the programming of rather complex 6812 I/O systems. In order to use the 6812's SPI module for external I/O devices, some basic routines, available in a library of classes, will be needed to initialize it, to repetitively exchange data with the device, or to exchange data with it only on the program's or the device's command. Then, as larger systems, such as `PLUGS,` are implemented that use the SPI, new classes can be defined as derived classes of these existing classes, to avoid rewriting the methods inherited from the classes in the library.

Putting these two notions together might produce an incompatible notion of factoring, but they actually appear to work synergetically. The hierarchy of classes at the root end can implement the factoring of routines needed to control a 6812's I/O system and prevent duplication of code. In this book, we will build up this infrastructure. For instance we will build an object for an I/O device that includes all the methods needed to initialize and use it. The leaf-ward part of the hierarchy can be used to add special functions to the basic I/O system to meet a specific application's requirements. For instance, an object for a robot controller might be coupled to the 6812 system by means of an RS-232 serial link as discussed in Chapter 9. The object `ROBOT` can be a member of a newly defined class `ROBOTDevice` that has additional methods, or can use the methods of its base class(es) to correctly and efficiently function calls sent to ***ROBOT*** or received from `ROBOT.` The control of `ROBOT` will be high-level, because all lower-level operations are invisible to the writer of the function calls (information hiding), which substantially reduces the design cost and improves system reliability.

4.1.3 Debugging Tools

Object-oriented programs for I/O devices, which separates I/O procedures from the rest of the program, can be debugged using techniques described in this section. An object driver can exercise the object, object stubs can replace the I/O device object, and function-member checking can make the function inform the designer of improper actions (at run time if the device is redirected, so the error isn't discovered until run time).

An *object driver* executes the object function members to test the passing of parameters between the object and the rest of the program and the passing of data between the object function members and the hardware. A simple output object driver, shown following, simply inverts the output pattern each time it executes an output operation. It is shown with an object `dataPort` declared a member of class `Port`.

```
Port<char> dataPort(0x1000); char pattern;
void main(){ do dataPort.put(pattern ^= 0xff); while(1); }
```

The expression `pattern ^= 0xff;` inverts the value of *pattern* and passes that value to the output function member *put* each time it executes. The output port should have a square wave on each bit, each bit having the same period. A slightly better driver for an output device simply increments the output pattern each time it executes an output operation. It is shown with an object pointer `ptr` blessed to make the object `ptr` points to a member of the class `Port`.

```
void main()
  {Port *ptr=new Port<char>(0x1000);do ptr->put(pattern ++);while(1);}
```

The output port should have a square wave on each bit, but each bit should have a period that is twice as long as its next-less significant bit. Using an oscilloscope and this output driver, a technician can expose hardware errors due to shorting outputs together or miswiring outputs. A simple input device's driver shown below uses an output port to provide data that are read through the input port and checked with the data that was sent. It is shown with an object pointer *ptr* blessed as a member of the class `Port`.

```
void main() { char error; Port outPort(0x2000);
     Port *ptr=new Port<char>(0x1000);
     do if(ptr->get(outPort.put(++pattern)!=pattern) error=1; while(1);
 }
```

Clearly, an easily written object driver can check the hardware without much effort. However, more complicated object drivers are needed when outputting arbitrary patterns causes undesirable effects – for instance, if the output is connected to hardware that acts catastrophically to some patterns. Also, one may not be able to rewire an input port so that during testing it is connected to an output port but during normal operation it is connected to something totally different. In these cases it is necessary to write specialized object drivers that are compatible with the environment of the I/O device.

An object can be blessed as a member of the *Stub* class to verify the program that uses an I/O device. A `Stub` class is defined below.

```
template <class T> class Stub : Port<T> {

    public: Stub(T * a) : Port<T>((int)a) { }               // constructor

    virtual T get(void){ return *port++; };                 // input

    virtual void put(T data) { *port++ = data; };           // output

    virtual T operator = (T data) {put(data);return data;}
};
```

Note that the input function member merely gets consecutive items from an input vector, presumably a vector initialized with constant values that are a useful input test pattern. The output function member merely puts consecutive items into consecutive elements of a vector, presumably a variable vector. This vector can be examined after the program stops. Examples of defining an object as a member of this `Stub` class in lieu of `Port` are shown below.

```
unsigned char outStream[5], inStream[] = { 1,2,3,4,5 };

void main(){ unsigned char i;
    Stub<unsigned char> in(inStream), out(outStream);
    out = 5; i = in;
}

void main(){ int i, j; Port<unsigned char> *inPtr, *outPtr;
    inPtr = new Stub<unsigned char>(inStream);
    outPtr = new Stub<unsigned char>(outStream);
    *outPtr = 5; i =*inPtr;
}
```

The *Stub* function members can also have function calls to output data in them to verify that they are executed. Constructor parameters can be output, or *put* function member parameters can be output to verify that the right data are being sent to them. However, calls to output data can slow down execution, which can interfere with debugging real-time programs. The use of input and output vectors as shown in the example above interferes less with real-time programming; that is why we recommend using these `Stub` members.

Another tool in debugging object-oriented I/O programs is to use *function-member checking*. Here, the class is expanded to include illegal calls to function members, which set `Port`'s data member *errors* when they are executed. Alternatively, *printf* can be used to indicate an error. These errors can detect when hardware is asked to do something it can't do, such as loading arbitrary data into a set port, and when the function member's parameters are illogical, such as when a pointer to a *char* is passed in place of a *char* that is expected by the function member. The example below illustrates the `SetPort` class with some function-member checking.

```
template <class T> class SetPort : Port { char errors;
    public: SetPort(long l) : Port<T>(l) { }                        // constructor
    virtual T operator |= (T data) {put(data); return data|=get();}
    virtual void put(T data) { errors = 1; };                       // illegal
    virtual T operator = (T data) { errors = 1; return data; }      // illegal
    virtual T operator &= (T data) { errors = 1; return data; }     // illegal
    virtual T operator |= (T *data) { errors = 1; return 0; }       // illegal
};
```

One can list every data type and operation that is illegal in this class, to let the compiler tell the programmer when an improper operation is requested. Further, new classes can be defined with checks for improper requests. We used `Port` for input ports as well as readable output ports. An input port could use a derived class of `Port` with a function member `put` that sets `errors` to indicate that the operation was not completed.

Normally, the use of function-member checking doesn't warrant the effort needed to define additional classes or function members; a programmer can simply not use the overloaded operator `&=` with `SetPort` or the function member `put` for input ports. We don't use function-member checking in most examples in this book so that they will be easier to understand. However, classes in a 6812 library should insert or remove function-member checking using conditional compiling, so it can be used to catch run-time errors. For instance, if `CHECKING` is defined, function-member checking is inserted, otherwise it is removed in the following procedure:

```
virtual T operator &= (T data) {
#ifdef CHECKING
    errors = 1; return data;
#else
    return 0;
#3ndif
}
```

While this discussion of debugging is specific to C++ object-oriented I/O interfacing, it can be adapted in part to conventional C programming by using `#define` statements that expand into different macros, depending on how the `#define` statements conditionally compile the program. For instance, if `debug` is `#defined,` then some message is printed, but if `debug` isn't `#defined,` the program reads a port's data.

```
#ifdef debug
#define inputPort (0 & printf("read data"));
#else
#define inputPort (*(unsigned char *)0x1000)
#endif
```

4.2 6812 Parallel Ports

The first two subsections independently describe the 'A4 and the 'B32 parallel ports. You can read the subsection that applies to the system you are using, without having to read the other. The last subsection describes an object-oriented class for parallel ports.

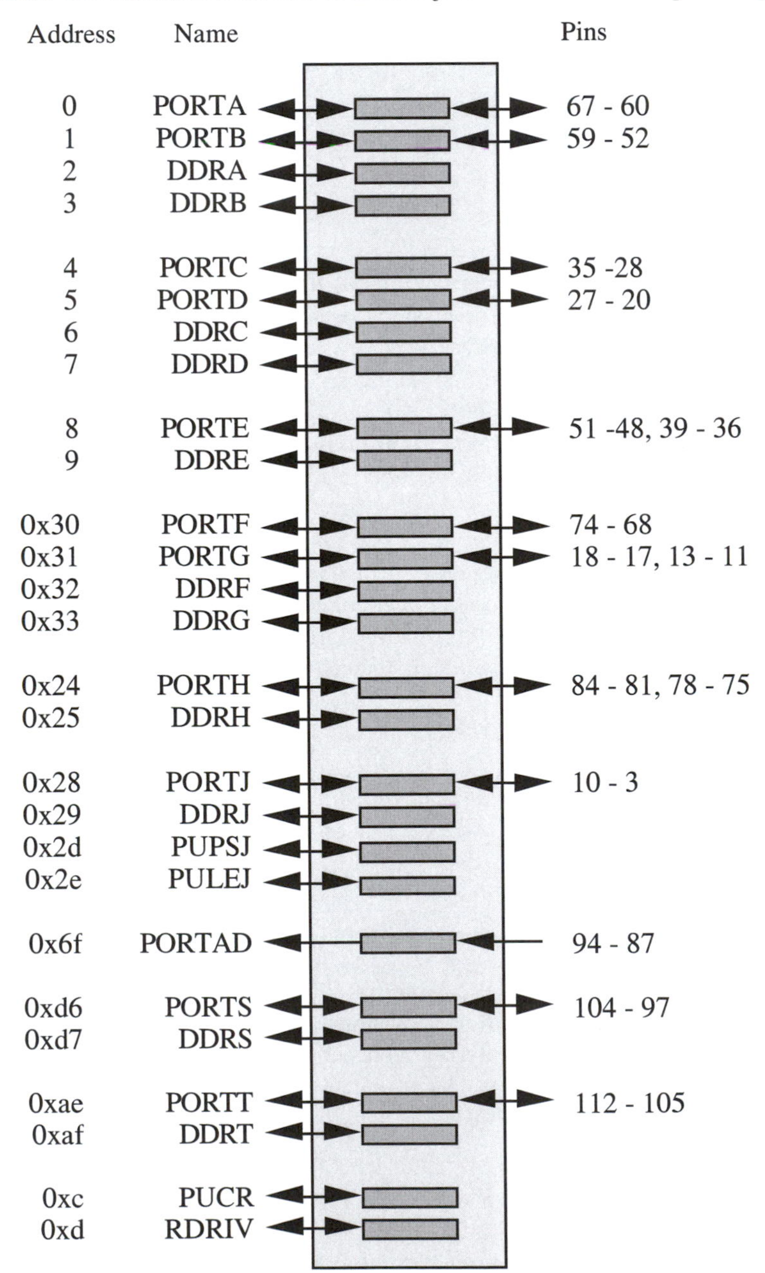

Figure 4.6. MC68HC812A4 Parallel I/O Ports

4.2.1 MC68HC812A4 Port Architecture

The 'A4 has a dozen parallel ports – A through H, J, S, T, and AD – which are quite similar to each other (see Figure 4.6). However, they are also extensively used for other functions – ports A and B for extended mode address bus, ports C and D for extended mode data bus, ports E, F, and G for extended mode bus control signals, port S for serial I/O, port T for timer I/O, and port AD for analog to digital I/O – that they are not always available for parallel I/O. Ports can be named in assembly language using EQU directives, or in C or C++ using `#define` or declaration statements. The type *volatile* means the data can be changed from outside, so the compiler will not optimize statements using it, and the @ symbol precedes the address of the port.

```
volatile unsigned char PORTA@0 PORTB@1, DDRA@2, DDRB@3, PORTC@4,
    PORTD@5, DDRC@6, DDRD@7, PORTE@8, DDRE@9, PORTF@0x30, PORTG@0x31,
    DDRF@0x32, DDRG@0x33, PORTH@0x24, DDRH@0x25, PORTJ@0x28, DDRJ@0x29,
    PORTAD@0x6f, PORTS@0xd6,DDRS@0xd7, PORTT@0xae, DDRT@0xaf,
    PUPSJ@0x2d, PULEJ@0x2e, PUCR@0xc, RDRIV@0xd;
volatile int PORTAB@0, DDRAB@2, PORTCD@4, DDRCD@6, PORTFG@0x30,
    DDRFG@0x32;
```

Pull-ups are needed if wire-or logic is used on a port pin; the pull-up can be provided by an external 10K resistor to +5 V, or by having the 'A4 pull up the pin. Each bit of port `PUCR` can attach a pull-up on one of the ports. See Figure 4.7. If `PUPH` has a T (1) stored in it, then all bits of `PORTH` have pull-ups, causing their lines to be pulled high if nothing is connected to them, otherwise all the bits of `PORTH` do not have pull-ups, causing their lines to float if nothing is connected to them. Similarly, `PUPG` pulls up all pins on `PORTG`, etc. The 'A4 can reduce the drive current on all port bits of a port in order to reduce power consumption or noise that interferes with A-to-D (analog-to-digital) conversion. Each bit of port `RDRIV` can reduce drive power on one of the ports. If `RDPJ` has a F (0) stored in it then all bits of `PORTJ` supply enough current to drive a TTL gate; otherwise, all `PORTJ` pins have reduced power, about 40% of this power. `RDPH` reduces power on `PORTH`, and so on.

Pull-ups or pull-downs can be connected on a line connected to any `PORTJ` pin. Pull-downs are useful if the desired default condition, when no device is driving the port pin, is low (0). Each bit of port `PULEJ`, if true (1), can cause a pull-up if the corresponding bit of `PUPSJ` is true (1) or can cause a pull-down if the corresponding bit of `PUPSJ` is false (0). If the bit of `PULEJ` is false, neither pull-up nor pull-down is used on the corresponding `PORTJ` pin. `PUPSJ` should be written into before `PULEJ`.

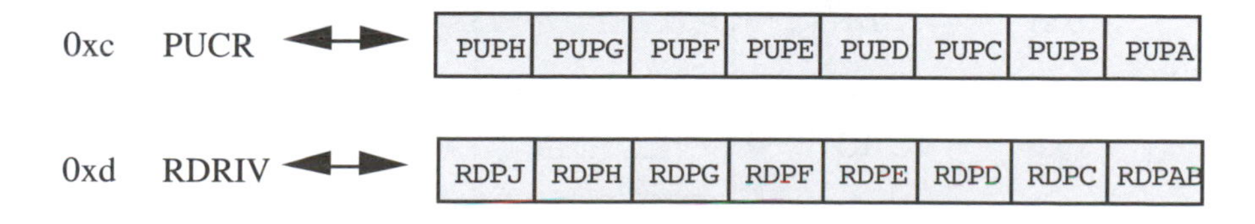

Figure 4.7. MC68HC812A4 Parallel I/O Control Ports

4.2.2 MC68HC912B32 Port Architecture

The 'B32 has eight similar parallel ports (Figure 4.8) – A, B, E, DLC, P, S, T, and AD. However, they are also extensively used for other functions – ports A and B for extended mode address and data buses, port E for bus control signals, port DLC for the BDLC module, port P for pulse width modulation, port S for serial I/O, port T for timer I/O, and port AD for analog to digital I/O – that they're not always available for parallel I/O.

These ports can be named in assembly language using EQU directives, or in C or C++ using `#define` or global declaration statements. The following declarations can be put in a header file which is incorporated into each program using `#include`s.

```
volatile unsigned char PORTA@0, PORTB@1, DDRA@2, DDRB@3, PORTE@8,
    DDRE@9, PORTP@0x56, DDRP@0x57, PORTAD@0x6f, PORTS@0xd6,
    DDRS@0xd7, PORTT@0xae, DDRT@0xaf, PUCR@0xc, RDRIV@0xd,
    PORTDLC@0xfe, DDRDLC@0xff;
volatile int PORTAB@0, DDRAB@2;
```

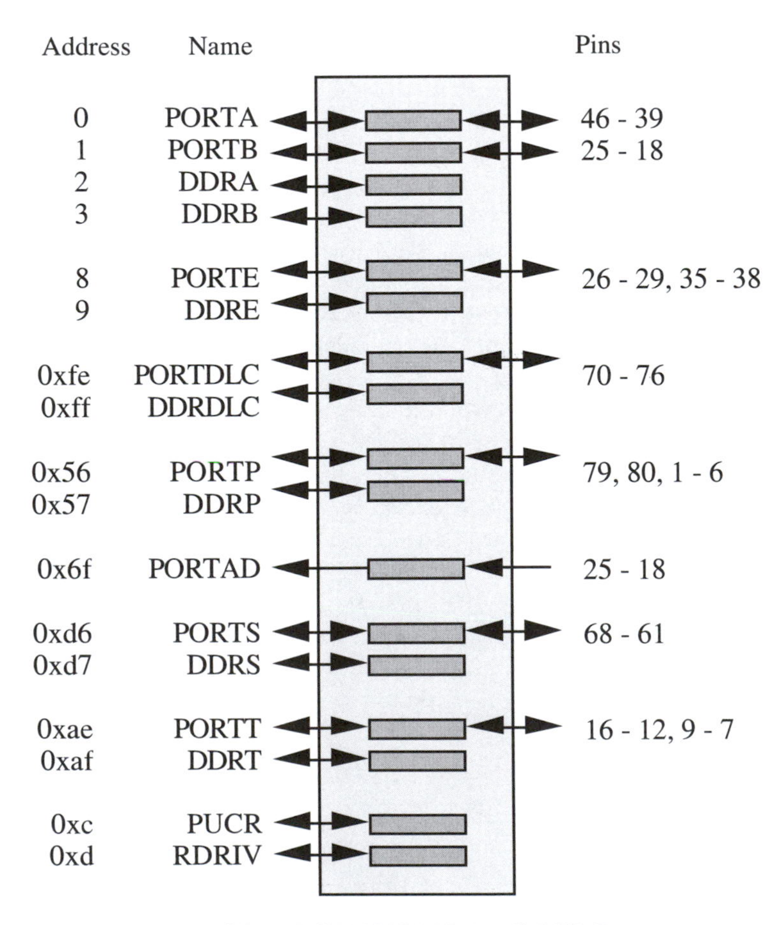

Figure 4.8. MC68HC912B32 Parallel I/O Ports

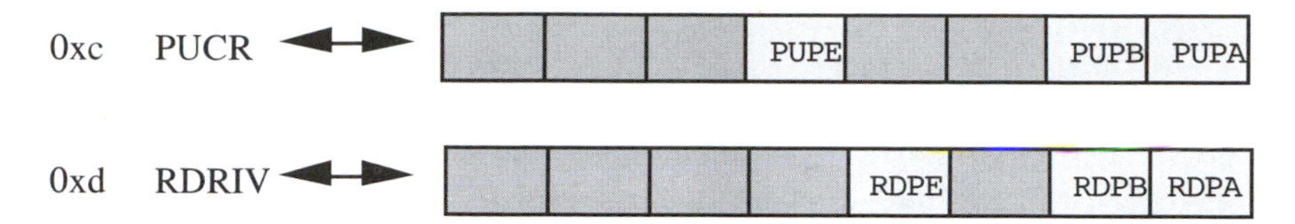

Figure 4.9. MC68HC912B32 Parallel I/O Control Ports

Pull-ups are needed if wire-or logic is used on a port pin; the pull-up can be provided by an external 10K resistor to +5 V, or by having the 'B32 pull up the pin. Each bit of port `PUCR` can attach a pull-up on one of the ports. See Figure 4.9. If `PUPA` has a T (1) stored in it, then all bits of `PORTA` have pull-ups, causing their lines to be pulled high if nothing is connected to them; otherwise, all the bits of `PORTA` do not have pull-ups, causing their lines to float if nothing is connected to them. Similarly `PUPB` pulls up all pins on `PORTB`, and so on. The 'B32 can reduce the drive current on all port bits of a port in order to reduce power consumption or noise that interferes with A-to-D conversion. Each bit of port `RDRIV` can reduce drive power on one of the ports. If `RDPA` has a F (0) stored in it, then all bits of `PORTA` supply enough current to drive a TTL gate, otherwise all `PORTA` pins have reduced power, about 40% of this power. `RDPB` reduces power on `PORTB`, and so on.

4.2.3 Programming of PORTA

This section illustrates assembly language and C programming techniques to access port A. Througout this chapter we use port A in our examples because port A is available in both the 'A4 and 'B32. So we introduce techniques to program port A here. But these techniques can be applied to any of the parallel ports in either the 'A4 or the 'B32.

All ports, except the analog port `PORTAD`, have a *direction port.* For port A, for each bit position, if the direction bit is F (0), as it is after reset, the port bit is an input, otherwise if the direction bit is T (1) the port bit is a readable output bit. The other ports and their direction port exhibit the same relationship, except `PORTAD`. A direction port is an example of a *control port,* which is an output port that controls the device but does not send data outside it. Writing the contents of a device's control ports is called the *initialization ritual.* This *configures* the device for its specific use.

We illustrate the use of `PORTA` in assembly language first, and in C or C++ after that. To make `PORTA` an output port, we can write in assembly language.

```
ldab    #$ff    ; generate all ones
stab    DDRA    ; put them in direction bits for output
```

Then, any time after that, to output accumulator B to `PORTA` we can write

```
stab    PORTA   ; output accumulator B
```

To make `PORTA` an input port, we can write

```
clr     DDRA    ; put zeros in direction bits for input
```

Then, any time after that, to input *PORTA* into accumulator B we can write

```
ldab    PORTA    ; read PORTA into accumulator B
```

It is possible to make some bits – for instance, the rightmost three bits – readable output bits and the remaining bits input bits, as follows:

```
ldab    #7       ; generate three 1s in rightmost bits
stab    DDRA     ; put them in direction bits for output
```

The instruction `stab` *PORTA* writes the rightmost three bits into the readable output port bits. The instruction `ldab` *PORTA* reads the left five bits as input port bits and the right three bits as readable output bits. A minor feature also occurs on writing the 8-bit word: the bits written where the direction is input are saved in a register in the device, and appear on the pins if later the pins are made readable output port bits.

The equivalent operations in C or C++ are shown below. To make *PORTA* an output port, we can write

```
DDRA = 0xff;
```

Note that *DDRA* is declared an *unsigned char* variable. Then, any time after that, to a *char* variable *i* to *PORTA*, write

```
PORTA = i;
```

Note that *PORTA* is declared an *unsigned char* variable. To make *PORTA* an input port, we can write

```
DDRA = 0;
```

Then, any time after that, to input *PORTA* into an *unsigned char* variable *i* we can write

```
i = PORTA;
```

Generally, the direction port is written into in assembly language or C before the port is used the first time, and need not be written into again. However, one can change the direction port from time to time, as shown in the IC tester example in a later section.

PORTA and *PORTB* together, and their direction ports *DDRA* and *DDRB* together, can be treated as a 16-bit port because they occupy consecutive locations. Therefore they can be read from or written into using `LDD` and `STD` instructions. These examples are considered in problem 9 at the end of the chapter. Furthermore, they can be accessed in C and C++ as shown below. To make *PORTA* and *PORTB* an output port, we can write

```
DDRAB = 0xffff;
```

Note that *DDRAB* is declared an *int* variable. Then, any time after that, to an *int* variable *i*, high byte to *PORTA* and low byte to *PORTB*, we can write

```
PORTAB = i;
```

Note that *PORTAB* is declared an *int* variable. To make *PORTA* and *PORTB*, an input port, we write

```
DDRAB = 0;
```

Then, any time after that, to input `PORTA` (as high byte) and `PORTB` (as low byte) into an `int` variable `i`, we can write

```
i = PORTAB;
```

In the 'A4, `PORTC` and `PORTD` can similarly be read and written as 16-bit data, and `PORTF` and `PORTG` can also be written this way. However, `PORTF` and `PORTG` have missing bits: `PORTF` bit 7 and `PORTG` bits 6 and 7 are not implemented.

4.2.4 A Class for Ports with Direction Control

The 'A4 and 'B32 parallel ports are symmetrically and consistently organized, which makes them well suited to being described by classes. Our class `Port` is redefined below to manage extra features of these ports, most of which have direction ports.

```
template<class T>class Port{ protected:unsigned char attr;T *port,value;
  public: Port(unsigned int a=0, unsigned char attr=0, T mask=0)
    { this->attr=attr;port=(T *)a;if(attr&0x60) option(1,mask);}
  virtual void put(T data) { *port = value = data; };
  virtual T get(void){if(attr&0x80) return value;else return *port;};
  virtual int option (int c = 0, int mask = 0) {
    switch(c){
      case 0: c = attr & 0x1f; attr &= 0xe0; return c;
      case 1:if(attr&0x20)port[2]=mask;if(attr&0x40)port[1]=mask} break;
      case 2: *(int *)0xc |= mask; break; // PUCR (hi) and RDRV (low)
      case 3: *(int *)0x2d|= mask; break; // PUPSJ (hi) and PULEJ (low)
    }
    return 0;
  }
  T operator = (T data) { put(data); return data; }
  operator T () { return get(); }
  virtual T operator|=(T data) { put(data|=get()); return data; }
  virtual T operator &= (T data) { put(data &= get()); return data; }
};
```

The constructor performs the initialization ritual for a parallel port. Its left argument can be conveniently initialized using this `enum`:

```
enum { aA = 0, aB = 1, aC = 4, aD = 5, aE = 8, aF = 0x30, aG = 0x31,
     aH = 0x24, aJ = 0x28, aS = 0xd6, aT = 0xae, aP = 0x56, aAD = 0x6f,
     aDLC = 0xfe, aAB = 0, aCD = 4, aFG = 0x30};
```

The constructor has two additional operands, which are zero if they are omitted, to initialize the object's `attr` and its direction port. Values from the *enum* statement

```
enum {wrOnly = 0x80, dirAt1 = 0x40, dirAt2 = 0x20};
```

can be ORed into the attribute argument. *wrOnly* should be ORed into it if the port is a basic output port that is write-only; it causes the *get* function to return the value that was last output by the `put` function. `dirAt1` should be ORed into it if the port has a direction port that is in the next location below the data port, and `dirAt2` should be ORed into it if the port has a direction port that is in a location two words below the data port; these values cause the direction port to be initialized using the constructor's third argument. Ports A, B, C, D, F, and G should use `dirAt2`, while all other ports except AD should use `dirAt1`. Note that ports AB, CD, and FG should use `dirAt1`.

The `get` and `put`, function members, and overloaded cast, assignment `=`, ORing `|=`, and ANDing `&=` operators essentially duplicate those of the previously defined `Port` class, except that `get` and `put` have been modified to handle a basic output port.

Devices generally have operations other than input and output. The `Port` class above needs a way to report errors and write into direction, `PUCR`, `RDRV`, `PUPSJ`, and `PULEJ` ports. All such special functions are implemented by an `option` function member, whose first operand (the "case") designates which operation is to be performed with the optional second parameter (the "input"), or what data are output from the function. Case 0 returns errors, just as the queue class used an `error` procedure, to ensure proper response to error conditions. The error value is stored in the rightmost bits of `attr`. Case 1 is reserved for putting the input into the direction register, case 2 puts the input into `PUCR` and `RDRV`, and case 3 into `PUPSJ` and `PULEJ`. Other numbers will be reserved for special functions used in other devices. This is done so that, if an option doesn't apply to a device, such as setting an SPI's "direction", the program compiles without errors and executes without crashing. However, in function-member checking, illegal cases are written to set the error indicator.

Case 2 ORs the input's rightmost byte into `PUCR` to enable pull-ups, and its leftmost byte into the `RDRV` to select low power outputs. `enum` values shown below, beginning with "u" enable pull-ups; for instance, `option(2,uA);` enables `PORTA` pull-ups. Values beginning with "p" reduce power; for instance, `option(2,pA);` reduces power on `PORTA`.

```
enum { uA = 0x100, uB = 0x200, uC = 0x400, uD = 0x800, uE = 0x1000,
    uF = 0x2000, uG = 0x4000, uH = 0x8000, uJ = 0, uS = 0, uT = 0,
    uP = 0, uDLC = 0, uAB = 0x300, uCD = 0xc00, uFG = 0x6000};
enum { pA = 1, pB = 1, pC = 2, pD = 4, pE = 8, pF = 16, pG = 32,
    pH = 64, pJ = 128, pS = 0, pT = 0, pP = 0, pDLC = 0, pAB = 0x01,
    pCD = 0x06, pFG = 0x18};
```

Case 3 ORs the input's rightmost byte into `PULEJ` to enable pull-ups or pull-downs, and its leftmost byte into the `PUPSJ` to select whether pull-ups (1) or pull-downs (0) will be used. For instance, to make `PORTJ` bit 2 pull up, execute `option(3,0x404);` and to make its bit 4 pull down, execute `option(3,0x10)`. Such values can be ORed together in a single call to `option` for either case.

For instance, an object for 'A4's 8-bit parallel `PORTJ` is initialized for readable output, configured for pull-ups and reduced power, and then used for output, as follows:

```
void main(){ char i;                              // local variable
    Port<char> port(aJ, dirAt1, 0xff); // declaration
    port.option(3,0xffff);                        // make all bits pull-up
    port.option(2,pJ);                            // reduce power
    port = 5; i = port;                           // write to port, read data written
}
```

Port's constructor's parameters are an address, attributes, and the direction port value initialize *port* and *attr*. First parameter *aJ* means the address for `PORTJ`. `Port`'s constructor then sets the device's direction port calling `option` with parameter 1, because we always set up the port's direction before the port is used; `option` can also be called later to change the device's direction. Function member `option` can be explicitly called right after the object is declared or blessed, to make `PORTJ` use pull-ups and to reduce power on all pins. Having declared or blessed a `Port` device, function members `get`, `put`, and `option`, and overloaded operators cast, assignment, `|=`, and `&=`, can be used with `Port`. Additional members will be added to *Port* in §4.3.6.

An object for 'B32's 16-bit parallel port AB can be blessed for readable output, configured for pull-ups and reduced power, and then used for output, as follows:

```
void main(){ int i; Port<int> *ptr;
    ptr = new Port<int>(aAB,dirAt1,0xffff);// constructor
    ptr->option(2, pAB|uAB );                  // for reduced power, pull-up
    *ptr = 5; i = *ptr;                        // Rest of program uses same code.
}
```

An object for an 8-bit basic output port at 0x200 can be declared and used for output, as follows:

```
void main(){ char c; Port<char> d(0x200, wrOnly); // constructor
    d = 5; c = d;                                  // Rest of program uses same code.
}
```

This class doesn't provide function-member checking, but that can be added as shown in problem 13 at the end of the chapter. Another interesting class in problem 15 is for one-bit input or output. The constructor gets an *id* whose low-order 3 bits designate a bit number and whose high-order bits designate a port (0 is `PORTA`, etc.).

The use of object-oriented programming simplifies the use of the 'A4 and 'B32 parallel ports. It is easy to design software for one port, and then change the software to another port if the hardware design has to be changed because the first port is needed for another device. This makes possible a library of programs that use a port with a block of hardware; a program and block of hardware can be taken from the library and easily modified to adapt it to another port. One merely has to change the constructor's arguments. Finally, the class function members are easy to test in a function-member driver, and they can be substituted for by a stub. Function-member checking, if it is implemented, also catches illegal use of devices if an inappropriate device is chosen for an application, or an inappropriate option is specified.

4.3 Input/Output Software

Software for input and output devices can be very simple or quite complex. In this section, we look at some of the simpler software. We show C programs to make the discussion concrete. The software to use a single input and a single output device to simulate (replace) a wire will be considered first because it provides an opportunity to microscopically examine what is happening in input and output instructions. We next discuss input to and output from a buffer by analogy to a movie. Programmed control of external mechanical and electrical systems is discussed next. We will discuss the control of a traffic light and introduce the idea of a delay loop used for timing. Then, in a more involved example, we'll discuss a table-driven traffic light controller and a linked-list interpreter, which implement a sequential machine. Finally we discuss an IC tester.

4.3.1 A Wire

The program *main()* following this discussion will move data from an 8-bit input port at *0x4000* (Figure 4.1a) to an 8-bit output port at *0x67ff* (Figure 4.1b) repetitively. This program only simulates eight wires, so it is not very useful. However, it illustrates the reading of input and writing of output ports in C. Observe the manner in which the addresses of the ports are set up. The cast *(char*)* is only needed if the C compiler checks types and objects to assigning integers to addresses. Alternatively, the addresses can be specified when the pointers are declared, as we will show in the next example. This program is worth running to see how data are sampled and how they are output, using a square-wave generator to create a pattern of input data and an oscilloscope to examine the output data. You may also wish to read the assembly-language code that is produced by the C compiler and count the number of memory cycles in the loop, counting also the number of cycles from when data are read to when they are output (the latency). Timing is hard to predict for all C compilers, and the best way to really determine it is to run the program and measure the timing.

```
void main() {
    char *src, *dst;
    src = (char*)0x4000; dst = (char*)0x67ff;
    do *dst = *src; while(1);
}
```

The assembly language generated for this C procedure is

```
00000869 CE4000      LDX   #16384
0000086C CD67FF      LDY   #26623
0000086F A600        LDAA  0,X
00000871 6A40        STAA  0,Y
00000873 20FA        BRA   *-4    ;abs = 086F
```

The 6812 actually reads the input port in the instruction LDAA 0,X, and then writes into the output port in STAA 0,X about 375 ns later. The loop executes in 1.125 μs. However, loop timing depends on the compiler and the programmer's style. Timing is best determined by measuring chip enable pulses on an oscilloscope.

4.3.2 A Movie

We may wish to input data to a buffer. The declaration *buffer[0x100]* creates a vector of length 0x100 bytes to receive data from an input port at 0x4000. Observe that the address of the input port is initialized in the declaration of the pointer.

```
void main(){unsigned char *src=(unsigned char*)0x4000,buffer[0x100], i;
      for(i =0; i < 0x100; i++) buffer[i] = *src;
}
```

The assembly language generated by this C `for` loop is shown below:

```
0000866 C7          CLRB
0000867 37          PSHB
0000868 87          CLRA
0000869 1981        LEAY  1,SP
000086B 19EE        LEAY  D,Y
000086D E600        LDAB  0,X
000086F 6B40        STAB  0,Y
0000871 33          PULB
0000872 52          INCB
0000873 C140        CMPB  #64
0000875 25F0        BCS   *-14    ;abs = 867
```

Finally we may wish to output data from a buffer. Observe that the address `pnt` of the buffer is initialized in the `for` loop statement, and is incremented in the `for` loop statement rather than in the third expression of the `for` statement, which is missing. Note the ease of indexing a vector or using a pointer in a `for` loop statement. This operation, emptying data from a buffer to an output port or filling a buffer with data read from an input port, is one of the most common of all I/O programming techniques. It can use either pointers or indexes to read from or write into the buffer. The programmer should try both approaches, because some architectures and compilers give more efficient results with one or the other approach.

```
void main() {
    char buffer[0x100], *pnt; volatile char *dst = (char*)0x67ff;
    for(pnt = buffer; pnt < buffer + 0x100; ) *dst = *pnt++;
}
```

The assembly language generated by this C `for` loop is

```
0000866 1A80        LEAX  0,SP
0000868 2004        BRA   *+6    ;abs = 86E
000086A A630        LDAA  1,X+
000086C 6A40        STAA  0,Y
000086E B754        TFR   X,D
0000870 1AF20100    LEAX  256,SP
0000874 3B          PSHD
0000875 AEB1        CPX   2,SP+
0000877 B745        TFR   D,X
0000879 22EF        BHI   *-15    ;abs = 86A
```

4.3.3 A Traffic Light Controller

Microcomputers are often used for *logic-timer control.* In this application, some mechanical or electrical equipment is controlled through simple logic involving inputs and memory variables, and by means of delay loops. (Numeric control, which uses A/D and D/A converters, is discussed in Chapter 7.) A traffic light controller is a simple example; light patterns are flashed on for a few seconds before the next set of lights is flashed on. Using *light-emitting diodes* (LEDs) instead of traffic lights, this controller provides a simple and illuminating laboratory experiment. Moreover, techniques used in this example extend to a broad class of controllers based on logic, timing, and little else.

In the following example, a *light pattern* is a collection of output variables that turns certain lights on and others off. (See Figure 4.10a.) Each bit of the output port LIGHTS turns on a pair of LEDs (see Figure 4.10b) if the bit is T. For example, if the north and south lights are paralleled, and the east and west lights are similarly paralleled, six variables are needed; if they are the rightmost 6 bits of a word, then TFFFFF would

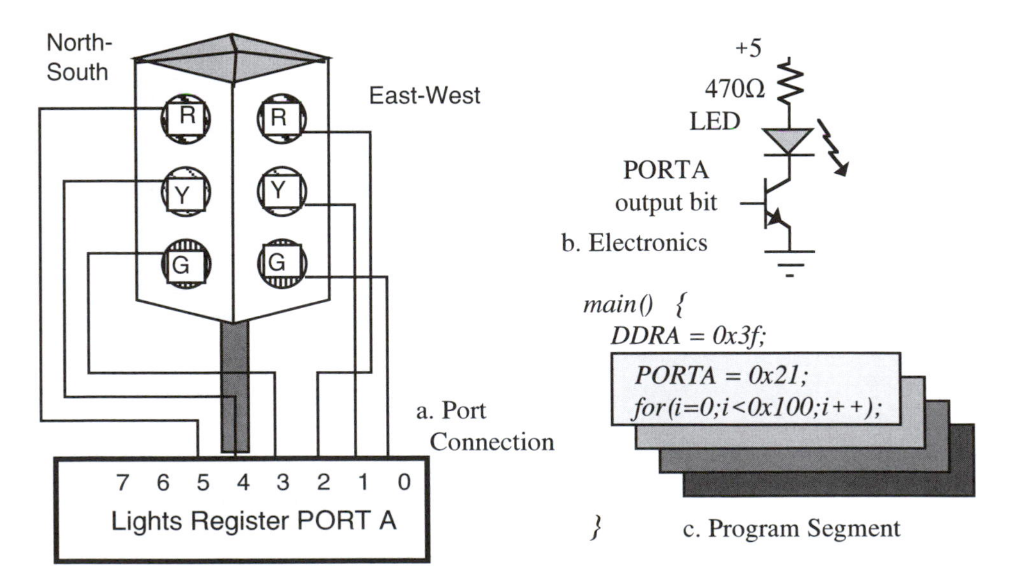

Figure 4.10. Traffic Light

turn on the red light, FTFFFF would turn on the yellow light, and FFTFFF would turn on the green light in the north and south lanes. FFFTFF, FFFFTF, and FFFFFT would similarly control the lights in the east and west lane. Then TFFFFT would turn on the red north and south and green east and west lights. We will assume that the 6812 output is connected so its right 6 bits control the lights as just described. The right 6 bits of DDRA are set to make these bits of `PORTA` outputs. The left 2 bits of the 8-bit output port need not be connected at all. Also, for further reference, TIME will be a binary number whose value is the number of seconds that a light pattern is to remain on. For example, the pair LIGHT = TFFFFT and TIME = 16 will put the red north and south and green east and west lights on for 16 seconds. Finally, a *sequence* of light patterns and associated times describes how the traffic light is controlled. This is an example of a *cycle,* a sequence that repeats itself forever.

In this technique, as the program in Figure 4.10c is executed it supplies multiple instances of immediate operands to the output port (as in `PORTA = 0x21;`) and immediate operands to control the duration of the light pattern. A loop such as `for(i=0;i<0x100;i++);` is called a *delay loop.* It is used to match the time of the external action with the time needed to complete the instruction. Delay loops are extensively used in I/O interface programs. The usual loop statement after the `for(;;)` and before the ending semicolon (Figure 2.3c) is missing because the control part of the `for` statement provides the required delay. The constant *0x100* that must be put in the statement to get a specific loop delay is hard to predict analytically and varies from one compiler to another, but it can be empirically determined.

A better way than programming a control sequence using immediate operands is an interpreter; it is usually recommended for most applications because it simplifies writing the control sequences and to store them in a small microcomputer memory. An *interpreter* is a program that reads values from a data structure such as a vector, a bit or character string, a list, or a linked-list structure to control something, like drill presses or traffic lights, or to execute interpretive high-level languages such as BASIC, LISP, or JAVA. You might want to scan §2.2 to review data structures before looking at interpreters. Table and linked-list interpreters are particularly useful in interface applications. The table interpreter is described first, then the linked-list interpreter is introduced by modifying the table interpreter.

A traffic light cycle might be described by Table 4.1. It can be stored in a table or array data structure. Recall from §2.2.1 that arrays can be stored in row-major order or column major order. C accesses arrays in row major order. The array has two columns – one to store the light pattern and the other to store the time the pattern is output – with one row for each pair. Consecutive rows are read to the output port and the delay loop.

Table 4.1. Traffic Light Sequence

LIGHT	TIME
TFFFFT	16
TFFFTF	4
FFTTFF	20
FTFTFF	4

```
void main() { unsigned char *lights, i, j, k, tbl[4][2];
    lights = (unsigned char *)0;       /* set up pointer to I/O port */
    lights[2] = 0x3f;                  /* initialize direction register for output*/
    tbl[0][0] = 0x21;                  /* initialize first light pattern */
    tbl[0][1] = 16;                    /* initialize first delay period */
    tbl[1][0] = 0x22;                  /* initialize second light pattern */
    tbl[1][1] = 4;                     /* initialize second delay period */
    tbl[2][0] = 0x0c;                  /* initialize third light pattern */
    tbl[2][1] = 20;                    /* initialize third delay period */
    tbl[3][0] = 0x14;                  /* initialize fourth light pattern */
    tbl[3][1] = 4;                     /* initialize fourth delay period */
    do                                 /* do the four-step sequence forever */
        for(i=0;i<4;i++){              /* do four steps */
            PORTA = tbl[i][0];         /* output a light pattern */
            for(j=0; j<tbl[i][1]; j++)          /* repeat following to delay */
                for(k=0;k <0xffff;k++) ;        /* delay a bulk amount */
        };
    while(1);
}
```

Structures can implement tables more efficiently than arrays can, so their columns can have different data types and sizes. You can use a pointer to point to the structure, as discussed next. We first introduce a very simple link mechanism using the index in an array, and then the use of pointer variables and structures to implement links.

4.3.4 A Sequential Machine

Linked-list interpreters strongly resemble sequential machines. We have learned that most engineers have little difficulty thinking about sequential machines, and that they can easily learn about linked-list interpreters by the way sequential machines are modeled by a linked-list interpreter. (Conversely, programmers find it easier to learn about sequential machines through their familiarity with linked-list structures and interpreters from this example.) Linked-list interpreters or sequential machines are powerful techniques used in sophisticated control systems, such as robot control. You should enjoy studying them, as you dream about building your own robot.

A *Mealy sequential machine* is a common model for (small) digital systems. While the model, described soon, is intuitive, if you want more information, consult almost any book on logic design, such as *Fundamentals of Logic Design,* by C. H. Roth, West Publishing Co.; Chapter 14 is especially helpful. The machine is conceptually simple and easy to implement in a microcomputer using a linked-list interpreter. Briefly, a Mealy sequential machine is a set S of internal states, a set I of input states, and a set O of output states. At any moment, the machine is in a *present internal state* and has an *input state* sent to it. As a function of this pair, it provides an *output state* and a *next internal state.* In the next time step, the next internal state is the present internal state.

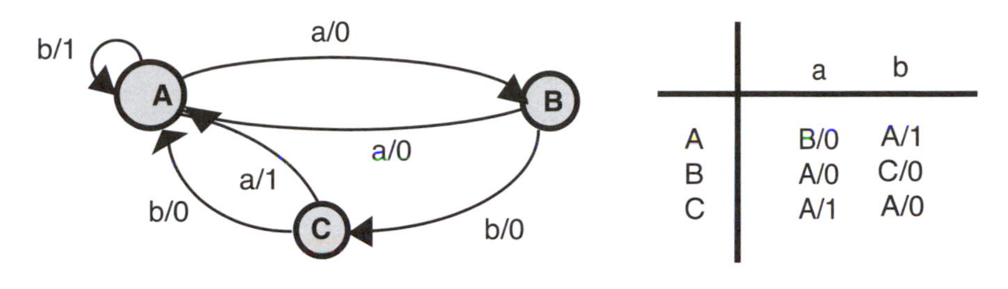

	a	b
A	B/0	A/1
B	A/0	C/0
C	A/1	A/0

a. Graph Representation

b. Table Representation

Figure 4.11. Mealy Sequential Machine

The Mealy sequential machine can be shown in graph or table form. (See Figures 4.11a and 4.11b for these forms for the following example.) Herein, the machine has internal states S = {A, B, C}, input states I = {a, b}, and output states O = {0, 1}. The graph shows internal states as nodes, and, for each input state, an arc from a node goes to the next internal state. Over the arc, the pair representing the input state/output state is written. In the table, each row describes an internal state and each column, an input state; the pair signifying the next internal state/output state is shown for each internal and output state. Herein, if the machine were in state A and received input a, it would output 0 and go to state B; if it received input b, it would output 1 and go back to state A.

Consider a simple example of a sequential machine operation. If the machine starts in internal state A and the input a arrives, it goes to state B and outputs a 0. In fact, if it starts in state A and receives the sequence abbaba of input states, it goes from internal state A through the internal state sequence BCABCA and generates output 000001.

The table representation can be stored in a microcomputer in a three-dimensional array in row-major order. The interpreter for it would read an input, presumably from the least significant bit of *PORTB*, and send the output to an output, least significant bit of *PORTA*. The input state a is the value 0x00, when read from the input port, and b is 0x01. The internal state is associated with the leftmost array index being read. If the initial internal state is A, then the program implements this by initializing an index to index 0 associated with state A. The table is interpreted by the following program.

```
enum{A , B, C};
unsigned char PORTA@0, PORTB@1, DDRA@2, DDRB @3;
char tbl[3][2][2] = {{{B,0},{A,1}},      /* initialize table */
                     {{A,0},{C,0}},
                     {{A,1},{A,0}}};
void main() { char i,j;
    DDRA = 1; DDRB = 0; i = 0;           /* set up initial state */
    while(1){                            /* interpret forever */
        j = PORTB & 1;                   /* get input state */
        PORTA = tbl[i][j][1];            /* output the output state */
        i = tbl[i][j][0];                /* read out next internal state */
    }
}
```

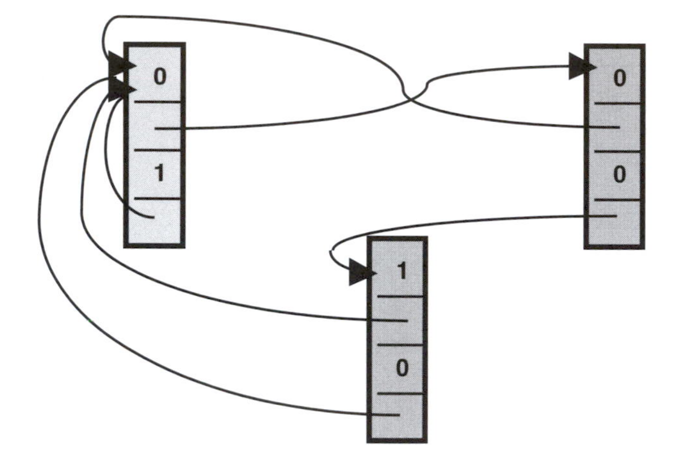

Figure 4.12. A Linked-List Structure

For low-end C compilers, two- or three-dimensional arrays are not implemented. However, the traffic light and sequential machine examples can be implemented with one-dimensional vectors. See the problems at the end of the chapter. To assist checking such a vector, it can be laid out like a higher-dimensional array; an example of this layout will be given when an object-oriented traffic light controller is discussed.

The program using structures is a bit more complex, but it is a more correct use of linked lists in C. In either program, an input port senses the input state and an output port provides the output state to some external system. We introduced the linked-list structure by comparing it to a row of the table. The structure is accessed (read from or written in) by a program, an interpreter. The key idea is that the next row to be interpreted is not the next lower row, but a row specified by reading one of the table's columns. For example, after interpreting the row for state A, if a b is entered, the row for A is interpreted again because the address read from a column of the table is this same row's address. This view of a list is intuitively simple. More formally, a *linked-list structure* is a collection of *blocks* having the same *template.* A block is a list like the row of the table and the template is like the column heading. Each block is composed of *elements* that conform to the template. Elements can be 1 bit to tens of bits wide. They may or may not correspond directly to memory words, but if they do, they are easier to use. In our example, the block (row) is composed of four elements: The first is an 8-bit element containing a next address, the second is a 16-bit output element, and the third and fourth elements are like the first and second. Addresses generally point to the block's first word, as in our example, and are loaded into the address register to access data in the block. Elements are accessed by using the offset in indexed addressing. Another block is selected by reloading the address register to point to that block's first word. Rather than describe blocks as rows of a table, we graphically show them, with arcs coming from address fields to the blocks they point to, as in Figure 4.12. Note the simple and direct relationship between Figure 4.12 and Figure 4.11a. This intuitive relationship can be used to describe any linked-list structure, and, without much effort, the graph can be translated into the equivalent table and stored in the microprocessor memory.

Linked lists generally have elements that are of different sizes. Also, pointers that are addresses to memory may be needed because they are not multiplied and added to compute memory addresses, as array indexes are, and thus are faster. Such linked lists should be stored as structures. Recall that in order to point to an element *e* of a structure *s*, we used *s.e* in earlier discussions. If a pointer is moved to different copies of a structure as the current internal state in the sequential machine, we can put it in the structure pointer variable `ptr`, and `(*ptr).e` is the element `e` of the structure pointed to by the pointer `ptr`. As a shorthand, the operator `->` is used; `ptr->e` is equivalent to `(*ptr).e`. It can now be rewritten:

```
void main(){
   struct state{struct state *next;char out;}A[2],B[2],C[2], *ptr = A;
   A[0].next = B ; A[0].out = 0 ; A[1].next = A ; A[1].out = 1;
   B[0].next = A ; B[0].out = 0 ; B[1].next = C ; B[1].out = 0;
   C[0].next = A ; C[0].out = 1 ; C[1].next = A ; C[1].out = 0; DDRB=1;
   while(1){ ptr += PORTA & 1; PORTB = ptr -> out; ptr = ptr -> next; }
}
```

Note that a data type, `struct state`, stores the next internal and the output states for a present internal state and input state combination. There is a vector for each internal state; `A[2]` has, for each input state, a pointer to a `struct state`, and the entire vector represents the internal state A. We have to initialize the structures in the program, not in the declaration, because locations of structures must be declared before they are used as entries in a structure. The initial internal state is initialized at the end of the declaration of the structure to be state A. In the `while` loop, the input number is read from the input port and added to the pointer *ptr*. If a "1" is read, *ptr* is moved from `A[0]` to `A[1]`, from `B[0]` to `B[1]`, or from `C[0]` to `C[1]`. Then the structure's element `out` is output, and the structure's element `next` is put in the pointer `ptr`.

The structure and pointer are very useful in I/O programming. An I/O device may have many ports of different sizes that can be described best by structures. The device's address can be put in a pointer `ptr`, and `ptr->port` will access the `port` element of it. Linked-list structures are especially useful for the storage of the control of sophisticated machines, robots, and so on. You can model some of the operations as a sequential machine first, then convert the sequential machine to a linked-list structure and write an interpreter for the table. You can also define the operations solely in terms of a linked list and its interpretive rules. Some of our hardware colleagues seem to prefer the sequential machine approach, but our software friends insist that the linked-list structure is much more intuitive. You may use whichever you prefer. They really are equivalent.

The interpreters are useful for logic-timer control. A table is a good way to represent a straight sequence of operations, such as the control for a drill press that drills holes in a plate at points specified by rows in the table. A linked-list interpreter is more flexible and can be used for sequences that change depending on inputs. Interpreters are useful in these ways for driving I/O ports. Their use, though, extends throughout computer applications, from database management to operating systems, compilers, and artificial intelligence.

4.3.5 An IC Tester

In this subsection we consider a design problem to be able to test standard 14-pin ICs, about 30% of the ones we use, at the behavior level. We want to be able to put an IC into a socket, then run a test program that will determine whether the IC provides the correct sequence of outputs for any sequence of inputs; but we are not testing the delays, the input and output electrical characteristics, or the setup, hold, rise, or fall times of signals. Such a tester could be used to check bargain mail-order house ICs.

In principle, there are two design strategies: top-down and bottom-up. In top-down design, you try to understand the problem thoroughly before you even start to think about the solution. This is not easy, because most microcomputer design problems are said to be "nasty"; that means it is hard to state the problem without stating one of its solutions. In bottom-up design, one has a solution – a component or a system – for which one tries to find a matching "problem." This is like a former late-night TV show character, Carnack the Magnificent. Carnack reads the answer to a question written inside an envelope, then he opens the envelope and reads the question. This is bottom-up design. We do it all the time. The answer is microcomputers; what was the question? Now, if you are an applications engineer for Zilog, you are paid to find uses for a chip made by Zilog. But a good design engineer *must* use top-down design!

We now design this IC tester top-down. We need 12 I/O bits to supply signals to all the pins and to examine the outputs for all the pins except power and ground. But the pins are not standard from chip to chip. Pin 1 may be an input in one chip and an output in another chip. An 'A4's or 'B32's ports A and B would be more suitable than the simple parallel I/O device because a line to these ports can be made an input or an output under control of software for different chips. Note that this is not always the case, and a simpler I/O device (a basic output device using a 74HC374 or an input device using a 74HC244) may be indicated if it is cheaper or uses up less board space. Assuming these ports are available, we choose them. We examine ports A and B.

We will configure the devices so the A data port will input or output data to the high-number pins, and the B data port the low-number pins. A rugged (ZIF) socket will be used for 14-pin ICs, with power and ground connections permanently wired to pins 14 and 7, and other pins connected to the port bits as shown in Figure 4.13a, making it impossible to connect a port's output pin to +5 V or ground, which may damage it. The user will plug a 14-pin IC into the 14-pin socket to test it. Another rugged (ZIF) 16-pin socket can be used for testing 16-pin dual in-line packages.

B0	1	14	+5
B1	2	13	A3
B2	3	12	A2
B3	4	11	A1
B4	5	10	A0
B5	6	9	B7
ground	7	8	B6

a. 14-Pin Socket

B0	1	16	+5
B1	2	15	A5
B2	3	14	A4
B3	4	13	A3
B4	5	12	A2
B5	6	11	A1
B6	7	10	A0
ground	8	9	B7

b. 16-Pin Socket

Figure 4.13. Connections for a Chip Tester

The general scheme for programming will be as follows. A direction pattern will be set up once, just before the chip is inserted, and a sequence of patterns will be tested, one at a time, to check out the chip. A pattern of T and F values will be put into the direction ports: an F if the corresponding pin is an output from the test IC (and an input to the ports), and a T if the corresponding pin is a chip input (output from the ports). Then a test pattern will be put in the data port to set up inputs to the IC under test wherever they are needed. The test pattern bits corresponding to the IC's output pins are, for the moment, "don't cares." Data will be read from the I/O ports, and the bits corresponding to the test chip's output pins will be examined. They will be compared against the bits that should be there. The bits corresponding to the input pins on the test chip are supposed to be exactly the bits that were output previously. The other bits of the pattern, which were don't cares, will now be coded to be the expected values output from the IC under test. Summarizing, if an IC's pin is a chip input, the corresponding port bit's direction bit is T, and its data bit in the test pattern is the value to be put on the test IC's input pin; otherwise, if the IC's pin is a chip output, the corresponding direction bit is F, and the data bit is the value that should be on the pin if the chip is good. The test sequences are read from a vector by a vector-driven interpreter.

Constants for chip testing are `#define`d; these 16-bit values are finally put into ports A (high byte) and B (low byte). From Figure 4.13a we construct the definitions:

```
#define pin1 1        #define pin5 0x10     #define pin10 0x100
#define pin2 2        #define pin6 0x20     #define pin11 0x200
#define pin3 4        #define pin8 0x40     #define pin12 0x400
#define pin4 8        #define pin9 0x80     #define pin13 0x800
```

We illustrate the general scheme by showing concretely how a quad-2 input NAND gate, the 74HC00, containing four independent gates, can be tested. Figure 4.14a is the truth table for one of the four gates in the 74HC00. We use the above definitions to construct a value that is the number of testing iterations, a value to be put in the direction ports, and values to be tested by the vector interpreter. From the 74HC00 chip pin connections shown in Figure 4.14b, we recognize that the truth table A value should be put on pins 1, 4, 10, and 13; the truth table B value must be put on pins 2, 5, 9, and 12; and the Z result will appear on pins 3, 6, 8, and 11. So we easily write the `#define` statements shown to the right of Figure 4.14.

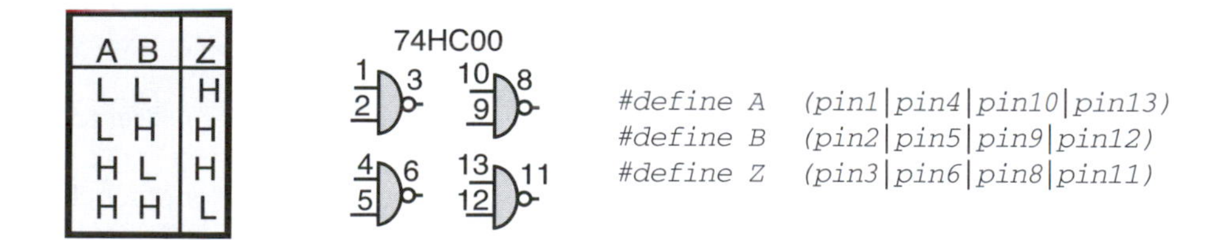

a. Truth Table b. Pin Connections

Figure 4.14. The 74HC00

These corresponding contents of the vector, which are actually evaluated to be `v[6] = {4, 0x264, 0x264, 0x6f6, 0xb6d, 0xd9b};` are used in the procedure `check()`. (The advantages of using `#define` statements can be appreciated if you try to construct these constants manually.) The first element of `v`, the number of iterations, is used in the `for` loop; the second element is used to initialize the direction port. The next element, corresponding to the top row of the truth table, has 1s exactly where Z appears because the truth table so indicates, so we initialize it to the value `Z`; the next element of `v`, corresponding to the second row of the truth table, has 1s exactly where `B` and `Z` appears, so we initialize it to the value `B|Z`, and so on. The program sets up `PORTA` and `PORTB` to be inputs where `v[1]` appears to be true (1), so their direction is initialized to `~v[1]`. Then the vector is read, element by element; the values of bits A and B are output, and the value returned is checked to see if it matches the element value. For a particular element, the vector value is output; wherever the direction bit is T the element's bit is output and wherever the direction bit is F the element's bit is ignored. The ports are read, and wherever the direction bit is T the element's bit is compared to the bit from the port. `v[1]` is a mask to check only the bits read back from the chip. The procedure `check()` returns 1 if the chip agrees with test inputs, and 0 if it fails to match the patterns in *v*.

```
#define PORTAB *((unsigned int*) 0)
#define DDRAB *((unsigned int*) 2)

unsigned int v[6]={ 4, Z, Z,B|Z, A|Z, A|B}; /* test vector for the 7400 */

int check(){ unsigned int i, bits;
    DDRAB = ~v[1]; /* initialization ritual for ports A and B */
    for(i = 0; i < v[0]; i++){ /* for all rows of truth table */
        PORTAB = bits = v[i + 2]; /* output to chip, save pattern for testing */
        if((bits&v[1])!=(PORTAB&v[1])) return 0; /* mismatch? exit with 0 */
    }
    return 1; /* if all match, return 1 */
}
```

The procedure above tests the 74HC00 chip. Other combinational logic chips can be tested in an almost identical manner, requiring only different vectors *v*. Chips with memory variables require more care in initialization and testing. Also, a thorough test of any chip can require a lot of patterns. If a combinational chip has a total of n input pins, then 2^n patterns must actually be tested.

A very powerful message of this example is the ability of high-level languages to abstract and simplify a design. By `#define` statements that are in turn defined in terms of other `#define` statements, we are able to utilize the ports of the 'A4 or 'B32 in a manner that is easy to understand and debug. It is easy to develop vectors for other chips like the 7408, which has a different truth table, or the 7404, which has a different pin configuration and a different truth table. It is easy to modify this program to use, for instance, ports C and D, or F and G, which have some missing bits. This modification is not unlike porting the program to a machine with a different I/O architecture. High-level languages simplify porting a program from one machine to another.

4.3.6 Object-oriented Vector Functions and Interpreters

In this subsection, we repeat this section's examples shown earlier to use object-oriented programming. The program *wire* illustrates simple input/output.

```
void main() { char i;
          Port<char>*dst; Port<char> *src;
          dst = new Port<char>(0x67ff);
          src = new Port<char>(0x4000);
          do *dst = i = *src; while(1);
}
```

Overloaded cast and assignment operators make this program look almost like the first wire example, but `*src` calls the overloaded cast operator that calls the `get` function member, and `*dst` calls the overloaded assignment operator that calls `put`. The local variable `i` is used herein to force the use of overloaded cast and assignment operators; otherwise `*dst = *src;` would call the copy constructor instead to make the object `dst` a copy of the object `src`. Objects have some overhead, especially in using virtual function members, but this overhead is often not a problem. But to simplify debugging, or change the entire program to work with 6812 parallel ports, we can just change `main`'s first three lines to bless the object as a member of the `Stub` or another class.

Earlier movie examples illustrate input and output from a buffer. This very important genus of I/O software will be handled by additional function members and overloaded operators. Functions to input (`get(T *v, int n)`) or output (`put(T *v, int n)`) whole vectors are useful because direct memory access (DMA) and floppy disk sector input and output can directly input or output whole vectors. It is better to pass them a vector rather than passing one byte at a time from a vector, then combining the bytes into a vector, to input or output the vector as a whole. In the event that your I/O operation is ever redirected to a DMA or disk, user-callable vector functions will allow efficient input or output of whole vectors. So even though your current design might not transfer whole vectors, using vector input and output functions and overloaded operators may make it possible at a later date, if the I/O should be redirected to a disk or DMA device, to take advantage of such devices. Besides, using these vector functions and operators factors the code used to step through the vectors into one place, in `Port` function members and overloaded operators, rather than having this code throughout the program. The vector `get` and `put` functions input and output whole vectors, without the application program managing each byte transfer.

These vector input and output functions are often called "raw I/O" because data are not interpreted as control characters. `Port` supports a minimal subset of conventional C++ *IoStream* functions. It uses the overloaded operators `>>` for "cooked I/O" input and `<<` for "cooked I/O" output of ASCII data, honoring control characters like carriage return. A string of `<<` operators provides a convenient output mechanism for terminals and keyboards similar to C's `printf`. Character input, `Port &operator >> (T &c)`, inputs one character and echoes it, using the object's *put* function. Character string input `Port &operator >> (T *b)` inputs a character line ending in a carriage return `'\r'`, allowing line editing responsive to backspace `'\b'` and delete line `'ctl x'`.

```
template <class T> class Port {
 protected:unsigned char attr, curPos; T *port,value, b[MAXCHARS];

 public:Port(unsigned int a=0, unsigned char attr=0, T mask=0)
  {port=(T *)a;curPos=echoPort=0;if(this->attr=attr<<5)option(1,mask);}

 int option (int c = 0, int mask = 0) {
   if(c == 0) { char i; i=attr & 0x1f; attr &= 0xe0; return i; }
   if(c == 1){if(attr&0x20)port[2]=mask; if(attr&0x40)port[1]=mask};
   if(c == 2) *(int *)0xc |= mask; break; // PUCR (hi) and RDRV (low)
   if(c == 3) *(int *)0x2d|= mask; break; // PUPSJ (hi) and PULEJ (low)
   return 0;
 }

 virtual void put(T data) { *port = value = data; };

 virtual T get(void){if(attr&0x80) return value; else return *port;};

 virtual void put(T *v, int n){ while(n--) put(*v++); };// vector output

 virtual void get(T *v, int n){ while(n--) *v++ = get();}; // vector input

 T operator = (T data) { put(data); return data; }

 operator T () { return get(); }

 virtual T operator|=(T data) { put(data|=get()); return data; }

 virtual T operator &= (T data) { put(data&=get()); return data; }

 Port &operator << (T c) { put(c); return *this; }

 Port &operator << (T *s) { while(*s) put(*s++); return *this; }

 Port &operator >> (T &c) {c = get(); put(c); return *this; }

 Port &operator >> (T *b) {
    do {value=get(); if(echoPort) echoPort->put(value);}
       while(value < '0');
    do { /* input alphanumerics, permit backspace and cancel line */
       if(value==0x08){if(curPos){backspace();curPos--;}}// bspc
       else if(value==0x18){while(curPos){backspace();curPos--; }}
       else { b[curPos++] = value;
           if(curPos >= MAXCHARS) {attr |= 1; return *this;}}
       value = get();if(value>='0') put(value);
    } while((value >= '0') || (value == 8) || (value == 0x18));
    curPos = 0; b[curPos++] = '\0'; return *this; /* null terminate */
 }

 virtual void backspace(void){ put('<');} // app. can override for device's bspc
};
```

The movie examples, exhibiting raw I/O, become very simple using vector function members, as shown below:

```
void main()
 { char b[0x10]; Port *ptr = new Port(0x4000); ptr->put(b, 0x10); }

void main()
 { char b[0x10]; Port *ptr = new Port(0x67ff); ptr->get(b, 0x10); }
```

Streamed output, exhibiting cooked I/O, simplifies formatting, as shown below:

```
void main(){ int x = 0x1234; Port<char> c(aH);
      c <<"x is "<<itoa(x)<<" or in hexadecimal: "<<htoa(x)<<'\r';
}
```

Single-character output is implemented by the segment `<< '\r'`. Since a single character `'/r'` appears to the right of the `<<` operator, the `Port` class's overloaded operator `Port &operator << (T c)` is used. A string of characters is output by a segment such as `<< " or in hexadecimal: "`. Since a character string `" or in hexadecimal: "` appears to the right of the `<<` operator, the `Port` class's overloaded operator `Port &operator << (T *s)` is used. Using procedures that return a character pointer to a null-terminated ASCII character string, such as `itoa(x)` and `htoa(x)`, numbers can be converted to character strings, and the `<<` operator can output the strings, as in `<< itoa(x)` or `<< htoa(x)`. Since a function returning a character string *itoa()* appears to the right of the `<<` operator, the `Port` class's overloaded operator `Port &operator << (T *s)` also is used to output the string returned by `itoa()`. A library can supply conversion routines such as `itoa` or `htoa`, or the programmer can write tailored conversion routines for special applications.

Cooked input echoes input characters so a user can see, on the output device, what he or she enters on the input device. String cooked output further handles special characters such as backspace and cancel (control-X, 0x18). The overloaded operator assembles all other characters, here loosely called "alphanumeric" characters, until a non-alphanumeric character is met again. Such collected characters can be passed to library or user-defined functions like *atoi* or *atoh* that convert the ASCII character string to a signed integer or a hexadecimal number corresponding to it. For instance,

```
void main(){char s1[MAXCHARS],s2[MAXCHARS],i;int x,y;Port<char> c(aB);
      c >> s1 >> i >> s2; x = atoi(s1); y = atoh(s2);
}
```

skips non-alphanumeric characters, then enters alphanumeric characters into `s1;` skips non-alphanumeric characters, then puts the next character into `i;` skips further non-alphanumeric characters, then enters alphanumeric characters into `s2,` until it encounters non-alphanumeric characters and a carriage return. `s1` is then converted as a signed decimal number, and `s2` is converted as a hexadecimal number. If the user types 123,a,456(c.r.) then `x` becomes 123, `i` becomes ASCII a, and `y` becomes 0x456.

The vector functions and overloaded operators `<<` and `>>` extend all the I/O class's capabilities, so they are included in the base class `Port`. However, all these functions and overloaded operators generate a significant amount of code, which HIWARE's current linker does not remove. Therefore, in the file `Port.c`, we have put conditional compilation preprocessor commands around most of these functions and overloaded operators to remove them when they are not needed. The user can `#define` a constant `USAGE` so as to compile only the functions and overloaded operators needed in an application. In future HIWARE linkers, which should be able to load only the functions actually used, these conditional compilation preprocessor commands won't be needed.

Interpreters such as the traffic light controller that interprets an array, discussed in §4.3.3, and the chip tester that interprets a vector of test patterns in §4.3.5, are essentially operations that can be applied to the data structures storing the patterns. The data structures and their (interpreter) operations can be encapsulated like arrays, and their operations are often encapsulated using objects. We will illustrate the traffic light controller object in the following example.

```
Port<char> P(idA,0x3f);

const char table[8] ={ 0x21, 16, 0x22, 4, 0xc,20, 0x14, 4 };

class traffic_table {
  char *tbl; short rows; char errors;
  public : char error; traffic_table(char *t, short size);
  void Execute(void), install(char, short);
} T(table, 4);

void main(){ if(!(T.error || P.option())) T.Execute(); }

traffic_table::traffic_table(char *t, short size){
  tbl = (char *)allocate(size << 1); rows = size; errors = 0;
  for(short i = 0; i < (rows << 1); i++) install(t[i], i);
}

void traffic_table::install(char data,short location){ // verify parameters
  if(((location & 1) == 1) &&
     !((((data & 7) == 1)||((data & 7) == 2)||((data & 7) == 4)) &&
     (((data&0x38)==8)||((data&0x38)==0x10)||((data&0x38)==0x20))))
       errors = 1;
     if(((location & 1)==0) && ((data < 4) && (data > 24))) errors=1;
     tbl[location] = data;
}

void traffic_table::Execute() { char i, j; long k;
  while(P.error() == 0) {
     for(i = 0; i < rows; i++) {
       P.put(j = tbl[i << 1]);
       for(j = 0; j < tbl[(i << 1) + 1]; j++)
         for(k = 0; k < 0x100; k++) ;
     }
  }
}
```

Observe that before *main* is executed, constructors of global objects *T* of class `traffic_table` and *P* of class `Port<char>` are executed. All `main` does is call *Execute* if no errors occur in the constructors. All `Execute` does is follow the procedure of the earlier traffic light example. But *T*'s constructor allocates room for a copy of the traffic light table from an external constant vector *table* (observe that both are vectors, but the constant global vector *table* is written spaced out to look like a two-dimensional array for ease of checking against the table provided to the user (Table 4.1). This use of an external vector and an internal vector illustrates the protection provided by object-oriented programming. The external global vector is an "initial" value that is copied into space provided by the `allocate` procedure into a "working copy" whose pointer, and therefore whose contents, are protected. Input data is verified by *install* to ensure that exactly one light is on in a north-south lane and in an east-west lane, and the delay time is reasonable. `install` could conceivably be used, after initialization is complete and the interpreter is running, to change a light pattern. The internal working copy of the traffic light pattern is changeable but protected against illegal patterns, while the global initial vector is really used just to set up this internal copy without having to use a lot of parameters to the constructor, or a lot of calls to a build function member. Such duplication of data structures is common in operating-system device drivers, where a constant data structure is used to initialize a working copy of the data structure, which permits modification but is protected against improper modification.

We conclude with an object-oriented example of the IC tester. We use the same `#define` statements as in the earlier IC tester example; it incorporates similar concepts to those used in the previous traffic light controller. Streamed output to an object *cout* indicates the test result. You should therefore study this last example on your own.

```
Port<int> P(idAB, 0);
unsigned int p00[6]={4, Z, Z, B|Z, A|Z, A|B};

class ICTest { unsigned int *pattern;
    ICTest(unsigned int *pattern) {
        this->pattern = (unsigned int *)allocate(2 + *pattern);
        for(short i=0;i<(2+*pattern);i++)this->pattern[i]=pattern[i];
    }
    public: int check(void);
} Test00(p00);

int ICTest::check(void){ register unsigned bits, i;
    P.option(1, pattern[1]); /* initialize ports A and B */
    for(i = 0; i < *pattern; i++){
        P = pattern[i + 2];
        if((P & pattern[1]) != (pattern[i + 2] & pattern[1])) return 0;
    }
    return 1; /* if all match, return 1 */
}

void main()
  { if(Test00.check())cout<<"good"; else cout<<"bad"; cout<< '\r';}
```

4.4 Input/Output Indirection

When studying single-chip microcomputers, we found it easy to use parallel ports on them to simulate the control signals on a memory bus, flipping them around in software. The kind of I/O considered up to now is analogous to direct memory addressing. The use of a parallel port to simulate a memory bus is like indirect addressing. Shift register-connected I/O is further indirection. Incidentally, a coprocessor actually is an I/O device that is read or written in microcode, which is one level below normal I/O, analogous to immediate addressing. In this section, we examine I/O indirection and examine some issues a designer should consider regarding I/O indirection. We will cover indirect I/O in the first subsection, followed by serial I/O, and will conclude with object-oriented programming and a discussion of design issues.

4.4.1 Indirect Input/Output

Up to now, the I/O device has been attached to the address and data buses. We shall call this *direct I/O.* We will show a direct I/O connection of a time-of-day chip, the 6818. The address, data, and control pins of an I/O device can be connected to a parallel I/O device's I/O port pins. One or more parallel ports are used to connect to an I/O chip's address, data, and control pins that are normally connected to the memory bus. Explicit bit setting and clearing instructions, often called *bit-banging,* can raise and lower the control signals for the I/O chip. Note that the 'A4 has 12 parallel ports, so it is exceptionally well suited to using some parallel ports to control another I/O device. We will show the advantages of this *indirect I/O* technique.

We want to keep track of the time of day, so we choose the MC6818A or MC146818A time-of-day clock chip to do this even when the microcomputer is turned off. Figure 4.15a shows the memory organization of the MC6818A. The current time is in locations 0 to 9, except for locations 1, 3, and 5, which hold an alarm time to generate an interrupt. Control ports at locations 0xA to 0xD allow different options. Locations 0xE to 0x3F are just some CMOS low-power RAM. After an initialization, the time may be loaded into locations 0 to 9, and then 0x8 is put into control port C to start the timekeeping. Locations 0 to 9 can be read after that to get the current time.

The MC6818A can be indirectly controlled through the 'A4's `PORTA` and `PORTB`, as shown in Figure 4.15d. M6818A control-signal timing, and address and data sequencing, are taken from Motorola data sheets. Figure 4.15b shows the write cycle and Figure 4.15c shows the read cycle. Control signals – address strobe `as`, data strobe `ds`, read-write `rw`, and chip select `cs` – are set high or low in the 'A4 `PORTB`, to write a word. Except for `ds`, they are initially high. We first raise `as` high, put the address into `PORTA`, make `cs` and `rw` low, drop `as` low, make `ds` high, put data to `PORTA`, drop `ds` low, and raise `cs`, `rw`, and `as` high. Reading is essentially the same, except that `rw` remains high and data are read from `PORTA`. Control signals are defined in the `enum` statement by having a 1 in the bit position through which they connect to `PORTB`.

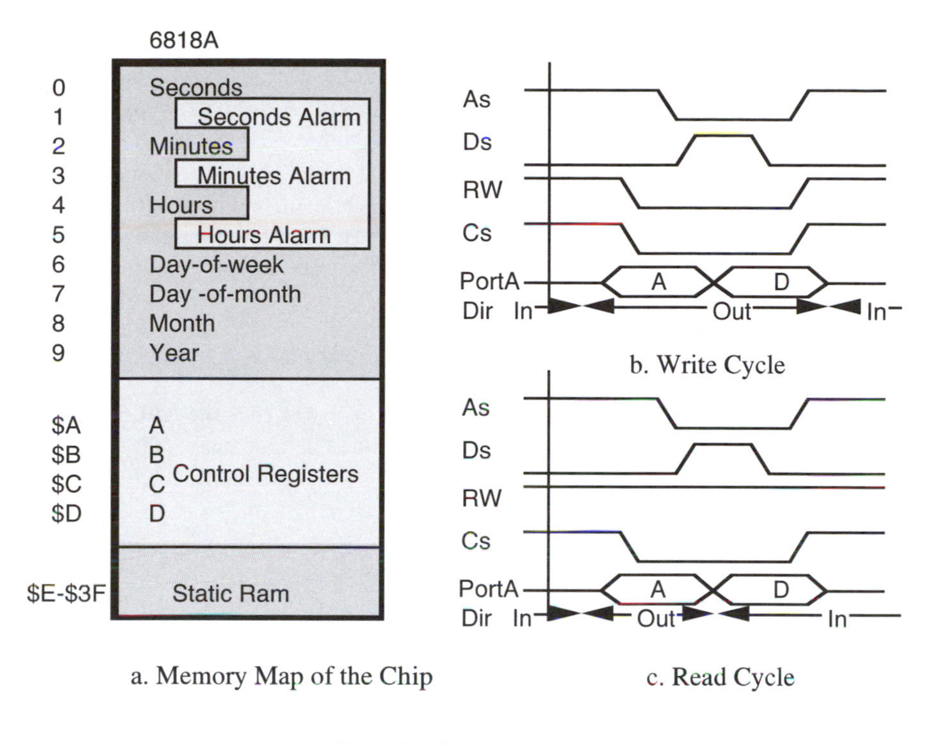

a. Memory Map of the Chip

c. Read Cycle

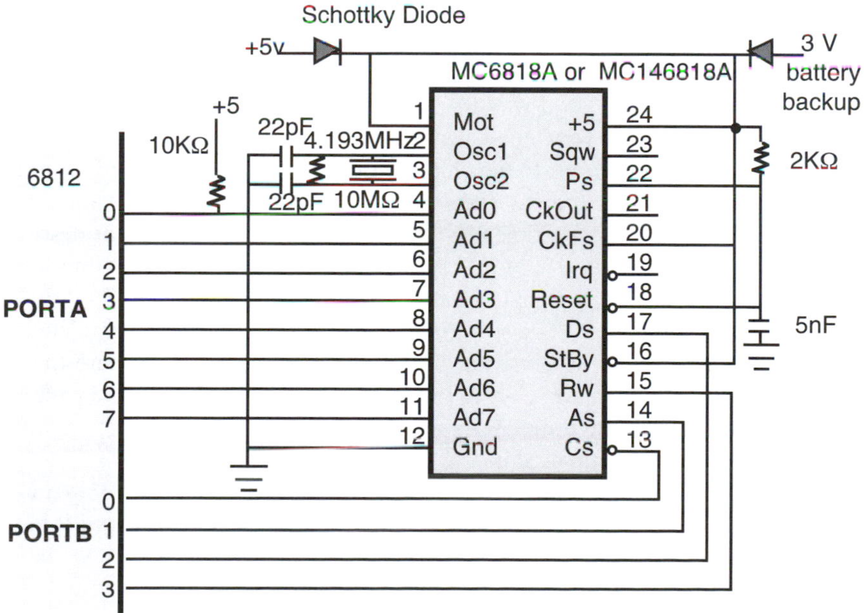

d. Connecting an MC6818A Time-of-Day Chip Using Indirect I/O

Figure 4.15. MC6818A Time-of-Day Chip

The C procedure `main` initializes port directions and initializes the 6818 control bits to their default states. The C procedure `outa` accesses the chip. Observe that `outa` rather tediously but methodically manipulates the MC68181A's control signals. A call `outa(d,6)` in indirect I/O writes *d* to location 6 in the MC6818A, which stores the day of the week. High-level language programs are easy to write. It is generally possible to write the procedure `outa` in assembly language, while the main program is in C, to regain some speed but keep most of the advantages of high-level languages. While the program shows how MC6818A memory can be written into, similar routines can read it.

```
enum { cs=1, as=2, ds=4, rw=8};

void main() {int yr,mo,dm,dw,hr,mn,se;
    DDRA = 0; /* make port A an input */ DDRB = 0xf; /* make port B output */
    PORTB = as+rw+cs;   /* initialize control bits to default state */
    outa(0x80,0xb); outa(0xf,0xa); outa(yr,9); outa(mo,8); outa(dm,7);
    outa(dw,6); outa(hr,4); outa(mn,2); outa(se,0); outa(8,0xb);
}

void outa(int d, char a) {
    DDRA = 0xff; /* make port A an output */
    PORTA = a; PORTB = as+cs; PORTB = as; /* output the address a */
    PORTB = 0; PORTB = ds; PORTA = d; PORTB = 0; PORTB = as;
    PORTB = as+rw; PORTB = as+rw+cs; DDRA = 0; /* make port A an input */
}
```

The main point of this section is the concept of indirect I/O, which we now elaborate on further. Besides being a good way to connect complex I/O devices to a single-chip computer, indirect I/O is a very good way to experiment with an I/O chip. The main advantage is that the connections to the chip are on the "other side" of an I/O port, rather than directly on the 6812's address and data buses. Therefore, if you short two wires together, the 6812 still works sufficiently to run a program. You have not destroyed the integrity of the microcomputer. You can then pin down the problem by single-stepping the program and watching the signals on the ports with a logic probe. There is no need for a logic analyzer. Indirect I/O is also a good way to implement some completed designs because it generally doesn't use external SSI chips; rather, it uses software to control a device. Indirect I/O is particularly easy to implement in the 'A4 because the address, data, and control buses in expanded mode are available as `PORTAB`, `PORTCD`, and `PORTE` parallel ports in the single-chip mode.

We used this technique to experiment with a floppy disk controller chip and a CRT controller chip set described in Chapter 10. We got these experiments to work in perhaps a quarter of the time it would have taken us using direct I/O. That experience induced us to write a whole section on this technique here in Chapter 4. There is a limitation to this approach. Recall from Chapter 3 that some chips use "dynamic" logic, which must be run at a minimum as well as a maximum clock speed. The use of indirect I/O may be too slow for the minimum clock speed required by dynamic logic chips. However, if the chip is not dynamic, this indirect I/O technique is very useful to interface to complex I/O chips.

4.4.2 LCD Interfacing

The liquid crystal display (LCD) has become the display device of choice for microcontrollers. An LCD features low power, full ASCII character displays of one to four lines, from 16 to 40 characters per line, and low cost. Many inexpensive LCD modules use the Hitachi HD44780 LCD controller chip. The LCDs of OPTREX's DMC series, which uses this Hitachi controller, can display a 16-column 1-row, a 16-column 2-row, a 20-column 1-row, a 20-column 2-row, a 20-column 4-row, or a 40-column 2-row ASCII message. Essentially, all displays use a standard interface that can be connected to 'A4's `PORTA,` as shown in Figure 4.16b.

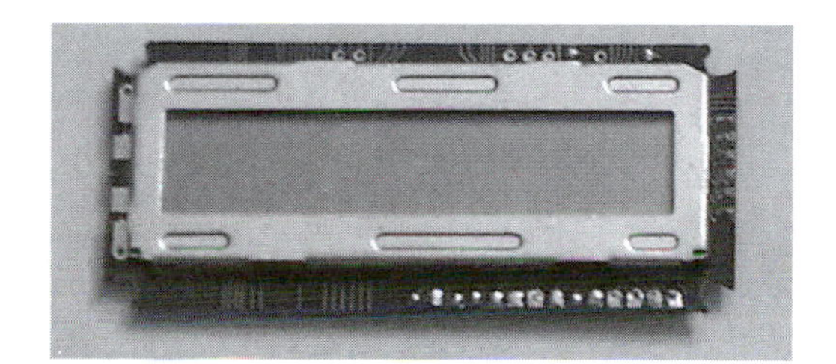

a. A 16-character, 1-line LCD Display

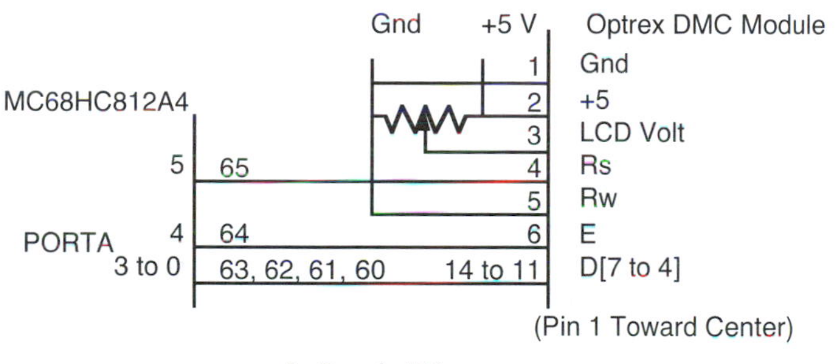

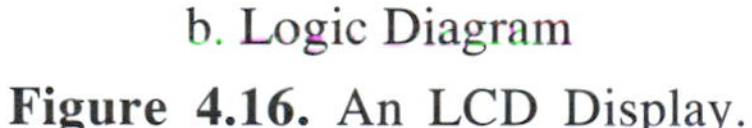

b. Logic Diagram

Figure 4.16. An LCD Display.

Table 4.2. LCD Commands

Binary Code	Command
`00000001`	Clear Display*
`0000001x`	Home Cursor*
`000001is`	Set Entry Mode
`00001dcb`	Set Display Mode
`0001srxx`	Shift Cursor/Display
`01aaaaaa`	Set Character Gen Addr
`1aaaaaaa`	Set Display Address

a . . . aa is an address
* needs 410 μs delay (otherwise, needs 10 μs)

	0	1
x	Don't care	Don't care
i	Autodecrement	Autoincrement
d	Display off	Display on
c	Cursor off	Cursor on
b	Blinking off	Blinking on
s	Shift cursor	Shift display
r	Shift left	Shift right
w	4-bit port	8-bit port
n	1-line display	2-line display
f	5 x 7 font	5 x 10 font

We show procedures for the 16-column 1-row display. *main*'s initialization ritual selects cursor blinking and movement. (See Table 4.2.) Its second line duplicates a command to configure its input port to 4 bits. The *put* procedure outputs a command or a character, using a delay loop to wait for the command's execution, and the *putStr* procedure outputs up to 16 characters. The constant *d410* puts a 410-μs delay and *d10* puts a 10-μs delay after the command is given. For an inexpensive 16-by-1 display, the cursor must be repositioned after outputting 8 characters, with the statement `if(i==7) put(0xc0,0,d10);`. To control other size displays, this statement is deleted.

```
enum { rs=0x20, e=0x10, d10=20, d410=800}; /* you might adjust d10, d410 */

void main(){ char i, j;
   DDRA= 0x3f; /* prepare PORTA for output */
   put(0x28, 0, d10); put(0x28, 0, d10); /* use 4-bit interface */
   put(6, 0, d10); /* set entry mode to autoincrement */
   put(0xe, 0, d10); /* set display mode: display and cursor on */
   put(1, 0, d410); /* clear display */
   putStr("Hello world,Hello world!"); /* print a message */
}

void put(char c, char a, int d) /* output high nibble, low nibble, then delay */
   { put4(((c >> 4) & 0xf) | a); put4((c & 0xf) | a); while(d) d-- ;
}

void put4(char c) {PORTA = c + e; PORTA = c;} /* display on falling e edge */

putStr(char *s) { int i;
   put(0x80, 0, d410); /* clear display */
   for(i = 0; *s; s++, i++) { /* output until null */
      put(*s, rs, d10); /* output a character */
      if(i == 7) put(0xc0, 0, d10); /* after 8th character, reposition cursor */
   }
}
```

4.4.3 Synchronous Serial Input/Output

Except when they come with a personal computer or are laid out inside a microcontroller chip, a parallel port and its address decoder take a lot of wiring to do a simple job. Just wire up an experiment using them, and you will understand our point. In production designs, they use up valuable pins and board space. Alternatively, a serial signal can be time-multiplexed to send 8 bits of data in eight successive time periods over one wire, rather than sending them in one time period over eight wires. This technique is limited to applications in which the slower transfer of serial data is acceptable, but a great many applications do not require a fast parallel I/O technique. Serial I/O is similar to indirect I/O, covered in §4.4.1, but uses yet another level of indirection, through a parallel I/O port and through a serial shift register, to the actual I/O device.

This subsection considers the serial I/O system that uses a clock signal in addition to the serial data signal; such systems are called synchronous. Asynchronous serial communication systems (Chapter 9) dispense with the clock signal. Relatively fast (4 megabits per second) synchronous serial systems are useful for communication between a microcomputer and serial I/O chips or between two or more microcomputers on the same printed circuit board, while asynchronous serial systems are better suited to slower (9600 bits per second), longer distance communications. We first examine some simple chips that are especially suited for synchronous serial I/O. We then consider the use of a parallel I/O port and software to communicate to these chips.

Although serial I/O can be implemented with any shift register, such as the 74HC164, 74HC165, 74HC166, and 74HC299, two chips – the 74HC595 parallel output shift register and the 74HC589 parallel input shift register – are of special value.

The 74HC589 is a shift register with an input port and a tristate driver on the serial output of the shift register. (See Figures 4.17b and 4.17d.) Data on the parallel input pins are transferred to the input port on the rising edge of the register clock RCLK. Those data are transferred to the shift register if the load signal LD is low. When LD is high, data in the shift register are shifted left on the rising edge of the shift clock SCLK and a bit is shifted in from IN, as in the 74HC595, but the data shifted out are available on the OUT pin only if the output enable EN is asserted low; otherwise it is tristated open.

74HC595

Pin	Signal	Signal	Pin
1	Q[1]	+5	16
2	Q[2]	Q[0]	15
3	Q[3]	IN	14
4	Q[4]	EN (o)	13
5	Q[5]	RCLK	12
6	Q[6]	SCLK	11
7	Q[7]	RST (o)	10
8	GND	OUT	9

a. Pin Connections for 74HC595

74HC589

Pin	Signal	Signal	Pin
1	D[1]	+5	16
2	D[2]	D[0]	15
3	D[3]	IN	14
4	D[4]	LD (o)	13
5	D[5]	RCLK	12
6	D[6]	SCLK	11
7	D[7]	EN (o)	10
8	GND	OUT	9

b. Pin Connections for 74HC589

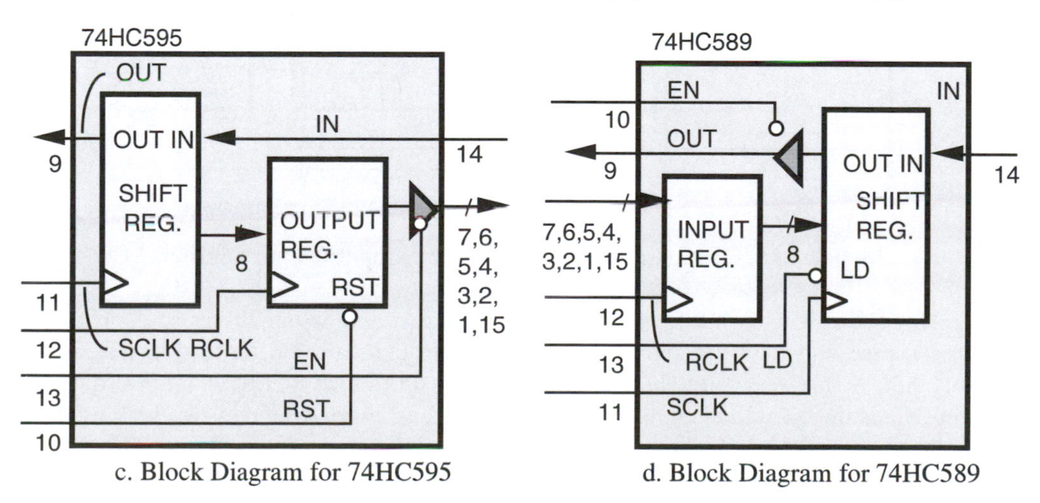

c. Block Diagram for 74HC595

d. Block Diagram for 74HC589

Figure 4.17. Simple Serial Input/Output Ports

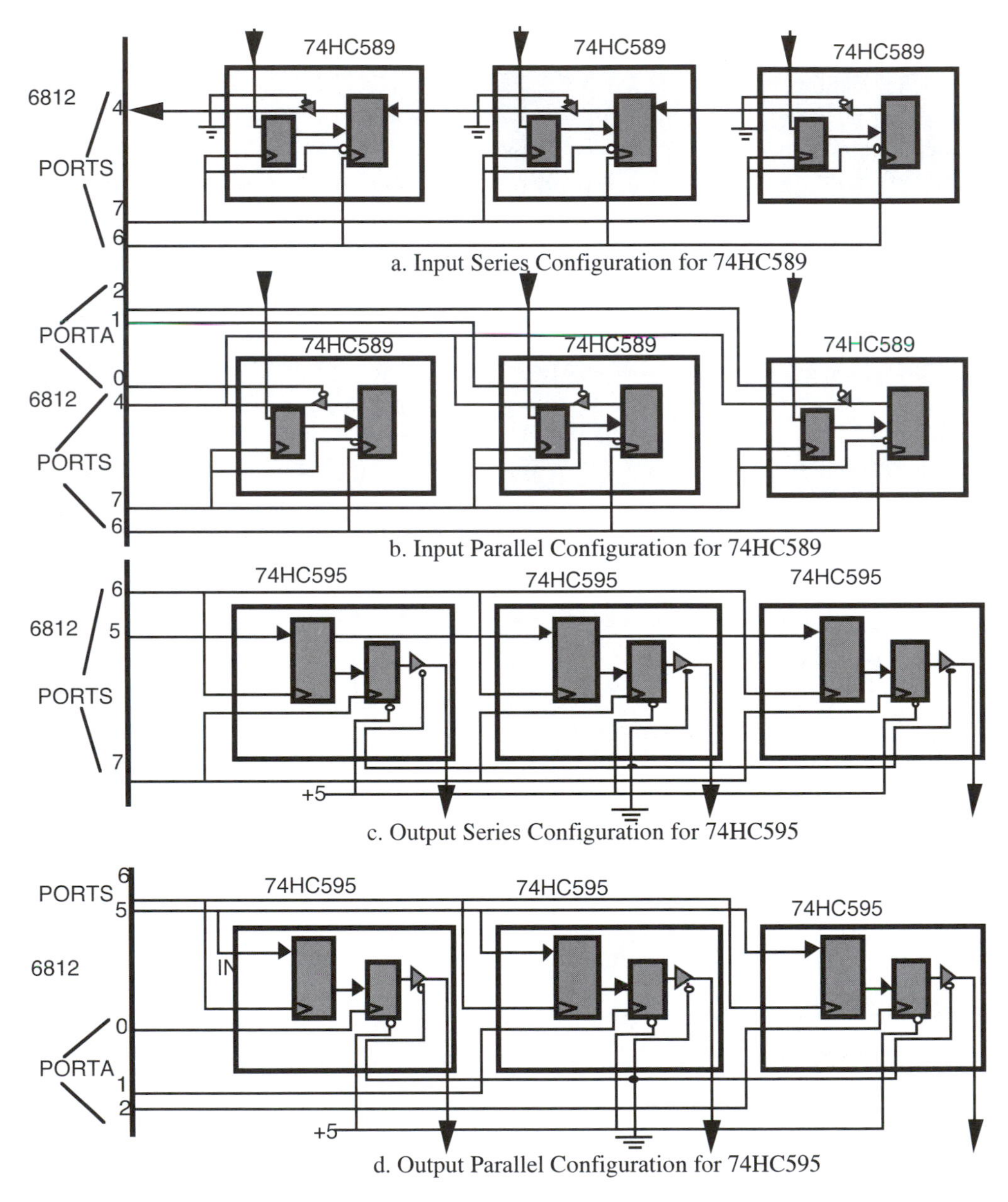

Figure 4.18. Configurations of Simple Serial Input/Output Registers

The 74HC595 is a shift register with an output port and tristate driver on the parallel outputs. (See Figures 4.17a and 4.17c.) We consider the shift register to shift left rather than right. A shift occurs on the rising edge of the shift clock SCLK. A bit is shifted in from IN and the bit shifted out is in OUT. On the rising edge of the register clock RCLK, the data in the shift register are transferred into the output port. If the output enable EN is asserted low, the data in the output port are available to the output pins; otherwise they are tristated open.

These chips can be connected in series or parallel configurations. (See Figure 4.18.) The 74HC589 can be connected in a series configuration to make a longer register, as we see in the 24-bit input port diagrammed in Figure 4.18a.

We will judiciously select `PORTS` and `PORTA` pins so that in the next discussion of the 6812 SPI we use the same pins for the same purposes – to run experiments with the SPI with the least amount of rewiring. In Figure 4.18a, The outputs OUT of each chip are connected to the inputs IN of the next chip to form a 24-bit shift register. Each 589's RCLK and LD pins are connected to `PORTS` bit 7 to clock the input ports and to load the shift registers together, and each chip's SCLK pins are connected to `PORTS` bit 6 to clock the shift registers together. The EN pins are connected to ground to enable the tristate drivers. In the software considered later, we pulse `PORTS` bit 7 twice to load the input registers at one time so as to get a consistent "snapshot" of the data; next we transfer this data into the shift register at one time by making LD low; and then, with LD high, we send 24 pulses on SCLK to shift the data into the 'A4.

The 74HC589 can be connected in a parallel configuration to make several separate input ports, as we see in the three 8-bit input ports of Figure 4.18b. Each chip's output OUT is connected to a common tristate bus line, and each chip's tristate enable EN is connected to different `PORTA` bits: 2 to 0, LD and RCLK connect to `PORTS` bit 7, and bits 4 and 6, are connected as in Figure 4.18a. Any of the input ports may be selected by asserting its tristate enable low, the others being negated high. Then a sequence similar to that discussed in the previous paragraph inputs the chip's data, using eight pulses on `PORTS` bit 6. While this configuration requires more output pins, software can choose any chip to read its data without first reading the other chip's data.

Figures 4.18c and 4.18d show the corresponding series and parallel configurations for the 74HC595. Reset can be connected to the 'A4 reset pin, which resets the system when it is turned on or when the user chooses; however, here it is merely connected to +5 V to negate it. The output enable EN is connected to ground to assert it. The series configuration makes a longer shift register. The parallel configuration makes separate ports that can output data by shifting the same data into each port but only pulsing the RCLK on one of them to transfer the data into the output register.

Series-parallel configurations, rather than simple series or simple parallel configurations, may be suited to some applications. The 74HC595 RCLK signals can come from the data source's logic, rather than from the microcomputer, to acquire data when the source is ready. The 74HC589 output enable EN can connect the output to a parallel data bus, so the output can be disabled when other outputs on that bus are enabled. These configurations suggest some obvious ways to connect serial ports.

Serial I/O chips can use parallel I/O port bits to control the lines to the chips using indirect I/O. We discuss the general principles after we consider this example: sending 24 bits of data to a series configured output, as shown in Figure 4.18c, following the flow chart in Figure 4.19. The outer loop of the procedure *serial_Out()* reads a word from a buffer and an inner loop shifts 1 bit at a time into `PORTS` bit 5, clocking `PORTS` bit 6 after each bit is sent, and then pulsing `PORTS` bit 7 to put the data into the output buffer register. Procedures for the other configurations in Figure 4.18 are similar to this one (see problem 24). The basic concept is that the individual signals needed to control the external chips can be manipulated by setting and clearing bits in parallel I/O ports. It is easy to write programs that will interface to serial I/O devices via a parallel I/O port.

```
void serialOut(unsigned char *s) {unsigned char i, j;
    DDRS = 0xe0;
    for(j = 0; j < 3; j++){
        for(i = 7; i >= 0; i--){
            PORTS = 0; /* make data and clock bits false (0) */
            if(0x80&s[j]) PORTS|=0x20;/* if msb is 1, make data true (1) */
            PORTS |= 0x40; PORTS &= ~ 0x40; /* pulse shift clock */
            s[j] <<= 1;/* shift data */
        }
    }
    PORTS |= 0x80 ; PORTS &= ~ 0x80; /* pulse output register clock */
}
```

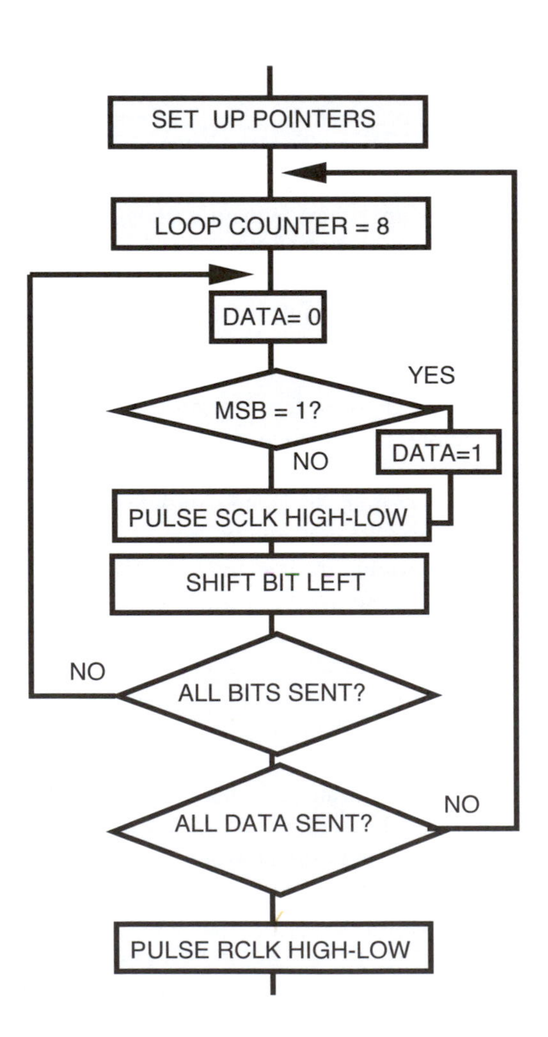

Figure 4.19. Flow Chart for Series Serial Data Output

The procedure above can be trivially modified to input data from three 74HC589s, connected as in Figure 4.18a. `PORTS` bit 7, normally high, has to be pulsed to move data from the 74HC589s' input pins into their parallel holding register on the rising edge before the shift procedure is executed, and is again pulsed low to transfer the data to the shift register; shifting, when `PORTS` bit 7 is high, puts that data serially into `PORTS` bit 4. Further, combining Figures 4.18a and 4.18c, three bytes of data can be output to the '595s as three bytes are input from the '589s at the same time.

Parallel input or output connections can be similarly implemented. Parallel output (Figure 4.18d) can get data to only one output register, without taking the time to shift data through all of them, as the serial connection requires. Initialize `PORTA` to output a high signal on bits 2 to 4 and, after shifting 8 bits, to pulse just one of them low. Parallel input (Figure 4.18b) similarly holds exactly one of `PORTA[2 to 4]` low while the shifting takes place (see `serialIn` below). Its parameter `a` is 0 to input a byte from the leftmost '589, is 1 to input from the middle, and is 2 to input from the right .

```
void serialIn(unsigned char a) { unsigned char i, value;
    DDRS = 0xec; DDRA = 7;
    PORTS &= ~ 0x80; PORTS |= 0x80 ; // clock data into first register
    PORTS &= ~ 0x80; PORTS |= 0x80 ; // load data into second register
    PORTA = ~( 1 << a ); // assert PORTA's ath bit low to enable ath '589
    for(i = 0; i < 7; i++){
        value <<= 1;/* shift data */
        if(PORTS & 0x10 ) value |= 1; /* get a bit from the port, insert it */
        PORTS |= 0x40; PORTS &= ~ 0x40; /* pulse the shift clock */
    }
}
```

The first line sets the direction to output control signals and input returned data. The next two lines pulse `PORTS` bit 7 to clock data into the '589's first register on its rising edge and then load the shift register when it is low. Then, when `PORTS` bit 7 is high, one of the tristate drivers is enabled, and then the data is shifted into the 6812.

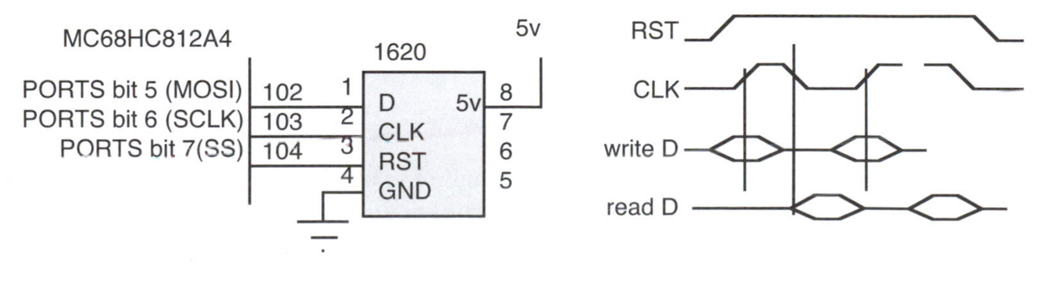

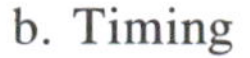

a. Logic Diagram b. Timing

Figure 4.20. Dallas Semiconductor 1620 Digital Thermometer

As an example of serial I/O, we consider a digital thermometer (see Figure 4.20). The Dallas Semiconductor 1620 has a CONFIG register and a TEMPERATURE register (among others), and uses a serial three-wire interface (see Figure 4.20). Data are shifted in and out, least significant bit first, on D (pin 1). Each message consists of sending an 8-bit command, optionally followed by sending or receiving 8 or 9 data bits. RST (pin 3) must be high from before the command is sent until when the data has been completely sent or completely received, and there must be a 5-ms delay between issuing commands, while RST is low. Temperature is measured about once per second.

The program below sends out a command to write 2 into the 8-bit 1620 CONFIG register to initialize it for temperature measurement; then it reads the 9-bit TEMPERATURE register, which is a 2's complement number, in units of 1/2 C°.

```
enum { D = 0x20, CLK = 0x40, RST = 0x80, WrCnfg = 0xc, RdTemp = 0xaa,
    start = 0xee } ;

void main() { int i;
    DDRS = 0xe0; PORTS = 0;
    PORTS|=RST;send(WrCnfg,8);send(2, 8);PORTS &=~RST; wait(2);// wait 5 ms
    PORTS|=RST; send(start,8); PORTS&=~RST; wait(20); // wait 1 s
    PORTS |= RST; send(RdTemp, 8); i = receive(9); PORTS &= ~RST;
}
void send(int d, char n) { char i;
    for(i = 0; i < n; i++){
        PORTS &= ~D; if(d & 1) PORTS |= D;
        PORTS |= CLK; PORTS &= ~CLK; d >>= 1;
    }
}
int receive(char n) { int d; char i;
    DDRS &= ~D;
    for(i = d = 0; i < n; i++)
        { d >>= 1; if(D & PORTS) d+=0x100; PORTS |= CLK; PORTS &= ~CLK;}
    DDRS |= D; if(n == 8) d >>= 1; return d;
}

void wait(int n){int i,j; for(j=0;j<n;j++) for(i=0; i<2105;i++ ) ; }
```

4.4.4 The 6812 SPI Module

The 'A4 and 'B32 were designed with the intent of exploiting serial modules like the 74HC589 and 74HC595. They incorporate a *serial peripheral interface* (SPI) that takes care of shifting serial data to and from the 'A4 or 'B32, essentially implementing the inner loop of `serialOut`, entirely in hardware. In this section we introduce the SPI module.

Figure 4.21 shows the SPI data, control, and status ports; `PORTS` and `DDRS` from Figures 4.6 and 4.8 are repeated here. We first consider ports needed to use the SPI for a procedure that is like `serialOut`, then the additional ports are discussed.

The enable port `SPE` must be set to use the SPI. Although the SPI module can also act as a slave to another computer, it is usually a master and we will consider it so until the end of this section. Thus master `SPE` and `MSTR` must be T. Input and output are determined by `DDRS`. When data are not being shifted, `PORTS` is output. Putting output data in the data port `SP0DR` starts the shifting. These bits are shifted out of master-out, slave-in (MOSI) `PORTS` bit 5, most significant bit first. The shift clock (SCLK), `PORTS` bit 6, clocks the internal and any external shift ports. Bits shifted in, most significant bit first, from master-in, slave-out (MISO) `PORTS` bit 4, can then be read from the data port `SP0DR`. The procedure `serialOut` shifts out the three byte vector *s*, using the series output configuration shown in Figure 4.18c.

Writing into `SP0DR` to begin the shifting is an example of a write address trigger. For input, a statement like `SP0DR = 0;` is needed to start a shift, even if no data need be output. The `SPIF` flag becomes set when all bits are shifted; it can be tested to wait for shifting to be done. In the next chapter this will be called a gadfly loop. This `SPIF` flag is an example of a *status port,* an input port that lets the programmer input information from the device, but this information is from the device itself, not from the outside world. This port tells the programmer that the SPI has completed its shift. Note that although the data shifted in is not needed in this example, nevertheless the statement `d = SP0DR;` is needed to clear `SPIF`. Reading `SP0DR` after reading `SP0SR` clears `SPIF`. This is an example of an address-trigger sequence.

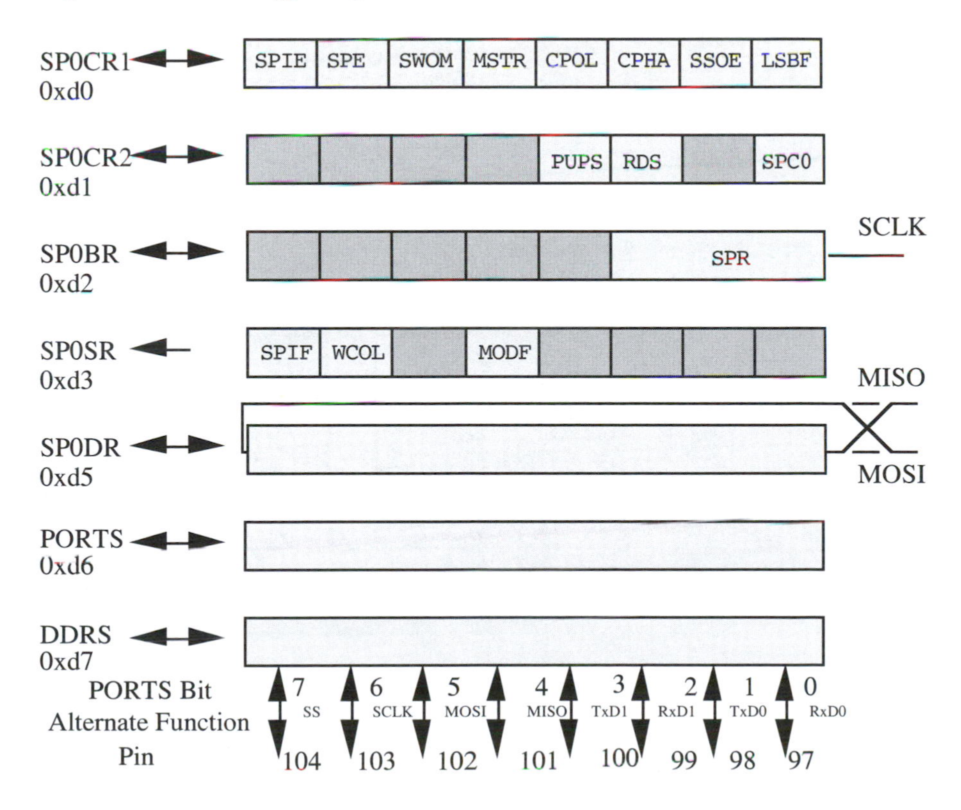

Figure 4.21. SPI Data, Control, and Status Ports

```
volatile unsigned char SP0CR1@0xd0, SP0CR2@0xd1, SP0BR@0xd2, SP0SR@0xd3,
    SP0DR@0xd5;
enum { SPIE = 0x80, SPE = 0x40, SWOM = 0x20, MSTR = 0x10, CPOL = 8};
enum { CPHA = 4, SSOE = 2, LSBF = 1, PUPS = 8, RDS = 4, SPC0 = 1};
enum { SPIF = 0x80, WCOL = 0x40, MODF = 0x10);
enum { SS = 0x80, SCLK = 0x40, MOSI = 0x20, MISO = 0x10};
void serialOut(unsigned char *s) {unsigned char j, d;
    DDRS = SS + SCLK + MOSI; /* make outputs for connections to the S.R. */
    SP0CR1 = SPE + MSTR; /* set up all bits in the control port at once */
    for(j = 0; j < 3; j++) /* transfer one byte at a time */
        { SP0DR=s[j];while(!(SP0SR&SPIF)) ; d=SP0DR;} /* move byte */
    PORTS |= SS ; PORTS &= ~ SS; /* pulse output port clock */
}
```

We now consider SPI ports useful in other applications. The SCLK bit rate is given by the `SPR` field of the `SP0BR` port. It is reset to 0, the default case, which results in a 4-MHz shift clock that the 74HC595 can handle. `SPR` can be set at any time to other values; if the E clock is 8 MHz, SCLK's frequency is $4/(2^{SPR})$. The `SWOM` bit determines whether the SPI's pins are wire-or and is usually F. The bits `CPOL` and `CPHA` determine the shape of the shift clock SCLK pulse. The value F, F (0, 0) is suitable for positive-edge clocked registers like the 74HC589 or 74HC595. Port `SSOE` modifies the master mode so that if `DDRS` bit 7 is true (1), `PORTS` bit 7 is asserted low only during shifting, and if `DDRS` bit 7 is false (0), asserting `PORTS` bit 7 low can cause an interrupt called a mode fault. `LSBF` permits data to be shifted least significant bit first. `PUPS` puts pull-ups on the high nibble of `PORTS`, and `RDS` reduces output power on these bits, in the same way the other ports have pull-ups and reduced power. Port `SPC0` permits MOSI to be used for bidirectional shifting of data for single-wire data transfers: if `DDRS` bit 5 is true (1), MISO is serial data output; otherwise, MISO is serial data input. MISO is then called master-out, master-in (MOMI).

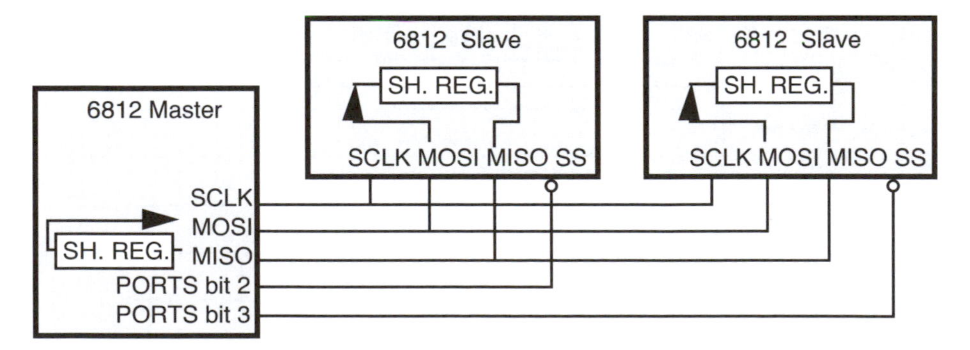

Figure 4.22. Multicomputer Communication System Using the SPI

The SPI interface can be used to communicate among several 6812s as shown in Figure 4.22. One 6812 is made a master, and all the others are slaves. In slaves, `PORTS` bit 7 is used as a slave select SS input. The `MOSI`, `MISO`, and `SCLK` pins are connected together. In contrast to the master, the slave clears its `SP0CR1` bit `MSTR` and `DDRS` bits 5, 6, and 7, and sets `DDRS` bit 4. As one SPI is initialized and inadvertently pulses its output port bits that input to the other SPIs, other SPIs may have to ignore the resulting spurious input byte. The master controls the slaves through some parallel output port like `PORTS` bits 2 and 3, by asserting exactly one slave SS input low. Then the SPI exchanges the data of the master SPI shift register with the data in the selected slave SPI shift register using a program like that just shown. To send data, the slave microcontroller writes its data into its `SP0DR` before the master writes its data into its `SP0DR`, which begins the shift operation. The slave's `SPIF` bit 7 is tested until it becomes asserted, indicating data have been exchanged. The slave reads the master's data from its `SP0DR`. This address trigger sequence – writing, testing, and reading – is needed even when a step doesn't seem to be needed, because this controls the slave's SPI just as it does the master's SPI. In addition to this master/slave relationship, several 6812s can communicate as equals (see Chapter 9, problem 7).

The serial port is a valuable alternative to the parallel port. It requires substantially fewer pins and wires. The 6812's SPI interface makes it easy to use these devices, but, with a modest amount of software, any parallel I/O port can be used to control them. However, a parallel port is required where speed is needed, because the serial port implemented with indirect I/O is considerably slower than §4.1's parallel port.

4.4.5 Accessing Devices Using Vectors and *structs*

The 6812 SPI device has a fair number of ports; such a collection of ports can be accessed using a vector or a `struct`, and `#define` or `enum` statements can be used to clearly name its ports. Vector notation is useful in accessing neighboring ports having the same width. *struct* notation is useful in accessing ports having different widths and also can be useful in accessing ports of the same width. Vector notation can also be used in simple compilers that do not support `structs` to handle ports that have different widths. We illustrate these concepts in following subsections by writing §4.4.4's `serialOut`.

4.4.5.1 Vector Access to Ports

A vector can be assigned to the locations of the SPI by the global declaration `char spi[8]@0xd0;`. Then the procedure `serialOut` can be rewritten

```
void serialOut(unsigned char *s) {unsigned char j, d;
    spi[7] = SS + SCLK + MOSI; *spi = SPE + MSTR;
    for(j = 0; j < 3; j++) {
        spi[5] = s[j] ; while(!(SP0SR & SPIF)) ; d = spi[5];
    }
    spi[6] |= SS ; spi[6] &= ~ SS;
}
```

4.4.5.2 Vector Pointer Access to Ports

Alternatively, a global or local variable pointer to the SPI ports *char *spi;* can be initialized in the first statement of *main*. Then, treating this pointer with offsets as a vector with indexes, we write into the SPI ports in the remaining lines.

```
void serialOut(unsigned char *s) {
     unsigned char j, d, *spi; spi = (unsigned char*)0xd0;
      spi[7] = SS + SCLK + MOSI; *spi = SPE + MSTR;
      for(j = 0; j < 3; j++) /* transfer one byte at a time */
           {spi[5]=s[j];while(!(SP0SR & SPIF)); d=spi[5];} /* move byte */
      spi[6] |= SS ; spi[6] &= ~ SS; /* pulse output port clock */
}
```

4.4.5.3 Using *#defines* to Name Ports

The above example illustrates the use of vectors in handling devices with a lot of identically wide ports. However, *#define* or *enum* statements can make access to these ports more self-documenting. (These appear in three columns to save space.)

```
#define SP0CR1 *spi        #define SP0CR2 spi[1]        #define SP0BR spi[2]
#define SP0SR spi[3]       #define SP0DR spi[5]         #define PORTS spi[6]
#define DDRS spi[7]
```

Then we write into the SPI ports in the remaining lines. Note that the procedure looks like the first *serialOut* in this subsection, except that the pointer *spi* must be initialized, but the assembly-language code actually uses this pointer to access the ports.

```
void serialOut(unsigned char *s) {
     unsigned char j, d, *spi; spi = (unsigned char*)0xd0;
      DDRS = SS + SCLK + MOSI; SP0CR1 = SPE + MSTR;
      for(j = 0; j < 3; j++)
           { SP0DR = s[j] ; while(!(SP0SR & SPIF)) ; d = SP0DR;}
      PORTS |= SS ; PORTS &= ~ SS; /* pulse output port clock */
}
```

4.4.5.4 *struct* Pointer Access to Ports

The `struct` SPI illustrates how `struct`s access ports of different widths, and our procedure `serialOut` shows how these `struct` members can be used. However, in HIWARE's compiler, the "advanced options" setting for "code generation" must have "bit field byte allocation" set to "most significant bit in byte first" for the `struct` members to correspond to ports in the same order in the declaration as they are in the data sheet.

```
typedef struct SPI {
   unsigned int SPIE:1, SPE:1, SWOM:1, MSTR:1, CPOL:1, CPHA:1, SSOE:1,
   LSBF:1, :4, PUPS:1, RDS:1, :1, SPC0:1, :5, SPR:3, SPIF:1,
   WCOL:1, :1, MODF:1;
} SPI;

void serialOut(unsigned char *s) {
   unsigned char j, d; SPI *spiPtr = (SPI *) 0xd0;
   DDRS = SS + SCLK + MOSI; spiPtr->SPE = 1; spiPtr->MSTR = 1;
   for(j = 0; j < 3; j++)
      { SP0DR = s[j] ; while(!spiPtr->SPIF) ; d = SP0DR;}
   PORTS |= SS ; PORTS &= ~ SS;
}
```

4.4.5.5 *struct* Access to Ports

Alternatively, if we globally declare the SPI using an @ to force its location, as in `SPI spi@0xd0;` we can write the same procedure

```
void serialOut(unsigned char *s) {unsigned char j, d;
   DDRS = SS + SCLK + MOSI; spi.SPE=1; spi.MSTR=1;
   for(j = 0; j < 3; j++)
      { SP0DR = s[j] ; while(!spi.SPIF) ; d = SP0DR;}
   PORTS |= SS ; PORTS &= ~SS;
}
```

Either a pointer is initialized to the port address, or else a globally defined `struct` is forced to be at a predefined location using the @ symbol in the global declaration. `struct`s can give self-documenting names to ports, especially where the ports have a variety of widths or even are identical-width ports otherwise described by vectors.

4.4.6 Indirect and Serial I/O Objects

Indirect and serial I/O using the SPI are suited to object-oriented programming. The classes `Indirect` and `SerialOut` shown below illustrate encapsulation, inheritance, overriding functions, and operator overloading. We discuss these two classes in this subsection. We also describe a class for an LCD utilizing `Port` class's `IoStream`.

Indirect I/O for the 6818 uses two parallel ports. This creates a minor dilemma: how do we define a class that might inherit function and data members from two classes? While C++ provides multiple inheritance to handle such cases, we couldn't find a simple way to express our design using multiple inheritance. Instead, we define one of these, the port through which address and data are time-multiplexed, as `Indirect`'s base class, which is `Port`. The other port, carrying control information, is a *Port* blessed inside `Indirect`'s constructor. The `put` and `get` function members execute the same algorithms as C procedures `outa()` and `ina()` described in §4.4.1.

```
class Indirect : Port<char> { public: char index; Port<char> *control;

   Indirect(char id, Port<char> *control) : Port<char>(id, 0)
      { this->control = control; *control = as + rw + cs; }

   void put(char value){
      *control = as + rw + cs; option(1,0xff); /* make an output */
      Port<char>::put(index);*control=as+cs;*control=as;/* output addr. */
      *control=0;*control=ds;Port<char>::put(value);*control=0;/* out data */
      *control = as + rw + cs; option(1,0); /* make this port an input */
   }

   char get(void){
      *control = as+rw+cs; option(1,0xff); /* make this port an output */
      Port<char>::put(index);*control=as+rw;*control=rw; /* out address */
      option(1,0); /* make this port an input */
      *control=rw+ds;value=Port<char>::get();*control=rw;/* input data */
      *control = rw + cs; *control = as + rw + cs; return value;
   }

   virtual char operator = (char data) { put(data); return data; }

   virtual Indirect &operator [] (char data){index=data;return *this;}
};
```

The overloaded index operator `[]` illustrates another C++ feature that is very well suited to indirect I/O. This overloaded operator is called whenever the compiler sees an index `[]` to the right of an object, as in `clock[0]`, whether the object and index are on the left or right of an assignment statement =. It executes the overloaded operator `[]` before it executes the overloaded cast or overloaded assignment operator. This overloaded operator stores what is inside the square brackets, `0` in our example, in object data member `index`. Then the following overloaded cast or assignment operator can use this saved value to supply an address, such as the address of a 6818 port or memory location.

`main()` declares the object *clock* to be of class `Indirect`. Its first parameter, `idJ`, is used in `Indirect`'s constructor's call to `Port`'s constructor, so `this` is an object through which data and addresses are passed to the 6818. *Indirect*'s constructor's second parameter is a pointer to an object of class `Port`; the object is blessed as `main` declares the object `clock`, to be used to output the control signals to the 6818.

```
void main() { Indirect clock(idJ, new Port<char>(idH, 0xf));
   char i, yr=1, mo=2, dm=3,d w=4, hr=5, mn=6, se=7;
   clock[0xb]=0x80;clock[0xa]=0xf;clock[9]=yr;clock[8]=mo;clock[7]=dm;
   clock[6]=dw;clock[4]=hr;clock[2]=mn;clock[0]=se;clock[0xb]=8;i=clock[0];
};
```

Using the overloaded index operator `[]` with overloaded assignment and cast operators, `main`, after initialization, accesses the 6818 ports using vector notation. The statement `clock[8] = mo;` puts local variable *mo* into 6818's location 8. Actually,

`[8]` calls the `Indirect` class's overloaded operator `[]`, which just stores `8` in the data member *index*. Then because an object is on the left of an assignment, `Indirect`'s overloaded operator = is called; it uses `index` to provide the address sent to the 6818A. Similarly, `i = clock[0];` uses `Indirect`'s overloaded `[]` operator to store 0 in member variable `index`. Then because an object is on the right of an assignment, `Indirect`'s overloaded *cast* operator is called; it uses `index` to send the 6818A address. The code looks like last section's access to a device's multiple ports using vectors. `#define` statements could make the code even more self-documenting.

```
const char bitPosition[8] = {~1,~2,~4,~8,~0x10,~0x20,~0x40,~0x80};

template <class T> class SerialOut : public Port<T> {

  public : SerialOut(int format, char baud) : Port (0xd0)
    { ((int *) port)=format; port[2]=baud; port[7]=0xe0; DDRA=0xff}

  virtual void put(T c) { unsigned char i, d;
    for(i = 0; i < sizeof(T); i++) {
      port[5]=c>>((sizeof(T)-i)*8);
      while(!(port[3]&0x80));
      d=port[5];c<<=8;
    }
    PORTA = ~ 0; /* pulse an output register clock */
  }

  virtual T get(void) { unsigned char i, d; T c;
    for(i = 0; i < sizeof(T); i++) {
      port[5] = d;
      while(!(port[3] & 0x80)) ;
      d = port[5];  c <<= 8;
    }
    PORTA = ~ 0; /* pulse an output register clock */ return c;
  }

  virtual int option (int c = 0, int mask = 0) {
    if(c == 0) return Port::option(0, m);
    else if((c >= 4) && (c < 7)){((char *)port)[c&3) = d; return 0; }
    else if((c >= 8) && (c < 0xb)) return ((char *)port)[c & 3);
  }

  virtual T operator = (T data) { put(data); return data; }

  virtual Port<T>operator [] (char data)
    { PORTA = bitPosition[data]; return *this; }

};

void main(){ SerialOut<char> sDevice(0x5003);char i=5;// example program
    for(i = 0; i < 5; i++) s[i] = sDevice[i];
    for(i = 0; i < 5; i++) sDevice[i] = s[i];
}
```

Class `SerialOut` shows again how the overloaded index operator `[]` can be used with objects, and how the function member `option` simplifies taking care of all the options if an object is redirected. The templated class `SerialOut` outputs an 8-bit `char`, a 16-bit `int`, or a 32-bit `long` variable to one of three groups of '595s, where each group is internally connected in series (Figure 4.18c) but the groups as a whole are connected in parallel (Figure 4.18d). *SerialOut*'s constructor's *int* argument initializes the SPI control ports. Its left 8 bits are put into *SP0CR1,* the next four bits go into *SP0CR2,* and the right four bits go into *SP0BR*. The class `SerialOut` has a `put` function member, which sends one byte out at a time through the SPI for as many bytes as there are in the template data type, and then, through `PORTA` it pulses the clock of the group that loads the shifted data into its output registers. The `option` function member with first argument 0, calls `Port`'s `option` function member (§4.3.6) to return errors. If the first argument is between 4 and 7, it writes the second argument into an SPI control port chosen by the two low-order bits of the first argument, and if the first argument is between 8 and 0xb, it reads an SPI control/status port similarly chosen. Using this `option` function member, if an object is redirected, inappropriate commands are safely ignored. The programmer can request any capability that the SPI can deliver.

The LCD can use C++'s `Port` class for formatted output. This class is

```
enum { rs = 0x20, e = 0x10, d10 = 20, d410 = 800};

class Lcd : public Port<char> { // use §4.3.6's PORT class as base

    void put4(char c) { Port<char>::put(c + e); Port<char>::put(c) }

    virtual void backspace() { put(0x10, 0); } // overrides Port's bspc

     public: Lcd(int a; unsigned char attr) : Port(a, attr, 0x3f) {
        put(0x28, 0, d10); put(0x28, 0, d10); put(6, 0, d10);
        put(0xe, 0, d10); put(1, 0, d410); col = 0; lcr = 1;
    }

    void put(char c, char a = rs, char d = d10) {
        put4(((c>>4)&0xf)+a); put4((c&0xf)+a);if(a)col++; while(d)d-- ;
    }

} cout;

void main()
    {cout<<"x is "<<itoa(x)<<" or in hexadecimal: "<<htoa(x)<<'\r';}
```

Port's overloaded operator << is used to output different data types as described in §4.3.6. But backspacing is taken care of by calling `Lcd`'s virtual `backspace()` function, which for this class, by shifting the cursor left, writing a space, and shifting the cursor left again, clears the character that the backspace removed, then moves the cursor back one position. This class can be used to output data on the LCD display. `main` displays the ASCII character stream "x is", prints the value of *x* in decimal, then displays the message "or in hexadecimal:" and displays the value of *x* in hexadecimal, using library or user-defined functions `itoa(x)` and `htoa(x)`.

4.5 A Designer's Selection of I/O Ports and Software

The parallel I/O device is the most flexible and common I/O device. When designing a parallel I/O device, the first step is to decide on the architecture of the port.

First we select the address. The address is often selected with an eye to minimizing the cost of address decoders.

Second, we select the data transfer mode of the port. One of the major design questions is whether the port should be directly or indirectly coupled to the microcontroller, or serially coupled through a three-wire interface.

Indirect I/O is a mode where one I/O device is used to provide the address, data, and control signals for another I/O device. Software emulates the microprocessor controller and generates its signals to the I/O device. It is generally an order of magnitude slower than direct I/O. But it is very useful when a parallel I/O device is available anyhow, such as in single-chip microcomputers like the 6812, or in personal computers that have a parallel port – often used for a printer. It is not necessary to attach devices to the address, data, and control buses within the computer; doing so might destroy the integrity of the computer and render debuggers inoperable.

Serial I/O, whether we use indirect I/O or the SPI, provides an obvious advantage over direct I/O by requiring fewer pins or wires. If the computer needs to be isolated from the external world, serial I/O uses fewer opto-isolators. Many manufacturers of specialty I/O devices, such as D-to-A converters, select this least expensive data-transfer mode, and three-wire interface devices are becoming widespread.

The main factor affecting the design decision is the speed of the I/O device. Sometimes the speed is dictated by the external system's speed, as when data are sent to or from a fast communication network using light pipes; and sometimes it is dictated by the process technology used to build the chip, as when dynamic logic is used, requiring a maximum time between events. Generally, the slower the required speed, the simpler the system. Many I/O devices are overdesigned with respect to speed. You should carefully determine the minimum allowable speed for the device and then choose the technique that fits that requirement. Then look at the system and determine if it has the needed mechanisms – a parallel port of sufficient width for indirect I/O, or a coprocessor mechanism for coprocessor-immediate I/O. We suspect that a lot of cases where indirect I/O is suitable and available are designed around direct I/O, which significantly increases their design and maintenance costs.

If indirect I/O is used, the next decision should be which parallel ports to use for determining address, data, and control signals for the device to be connected. If directly coupled, the decision is similar: which of the 12 parallel ports in the 6812 should be selected, or else what external port should be built. In the latter case, we need to consider whether an output port should be basic, readable, shadowed, or a set or clear port, or whether an address-trigger or address-register output is indicated. The availability of existing ports and the I/O port latency are the most important criteria. If the port is available on the 'A4 and is not needed by other devices, the first choice is to use one of these ports. If there is competition for their use, then external devices need to be considered.

If an 'A4 port is to be used, note that `PORTS` is generally used for SPI and SCI devices (§9.3.5), so their main attraction is for contrasting the operation using a parallel port to the same operation for using the SPI or SCI, and they are thus less desirable for general parallel port use. Similarly, `PORTT` will be shared with the counter/timer device (§8.1), so the main advantage to using these port bits is to compare a solution using a parallel port with a solution using the counter-timer device; they are thus also less desirable for general parallel port use. Again similarly, `PORTAD` is shared with the A-to-D converter devices (§7.5.3), and they are thus also less desirable for general parallel port use. Besides, `PORTAD` is input-only. Finally, `PORTE`, `PORTF`, and `PORTG` have less than 8 bits of general I/O and are used for special control signals. `PORTA` and `PORTB`, or `PORTC` and `PORTD`, suit a 16-bit port in a single-chip 'A4. `PORTA` and `PORTB` suit 1-bit or 8-bit ports since these ports are in the 'A4 and the 'B32. The choice of a first I/O port follows from the preceding discussion, but the selection of a fourth or fifth device usually becomes less clear. We can use less desirable ports for less critical devices, but we are lucky that the 'A4 and 'B32 have so many generally useful ports.

For external ports other than the serial three-wire interface, a basic output port is cheapest, but the software cannot read the data in it, so many *struct*s having bit fields, and explicit software equivalent to bit fields, won't work with basic output ports. A readable output port is most general, but also about twice as expensive as the basic output port. However, a RAM shadow, or software keeping a duplicate copy of the output data, can be used to avoid building a readable output port.

While some questions regarding the use of narrow or wide external data buses for external ports are yet to be considered (see Chapter 6), the basic issues of hardware implementation of different external direct ports and the software they use has been well covered in this chapter. The hardware for a basic output, readable output, input, or other external port can be implemented using simple TTL MSI chips. This chapter showed how to use SSI gate chips and decoders to implement the address decoder, and the popular 74HC244 and 74HC374 medium-scale integrated circuits to implement these ports. While other chips can be used, these chips are often used in printed circuit card microcomputers that are mass produced and intended for a wide variety of applications.

Besides hardware cost which govern the choices above, software costs are more often critical. Programming decisions generally affect clarity, which is reflected in the cost to write, maintain, and debug programs, and often affect static or dynamic efficiency. We now consider the designer's choice of software techniques to be used with I/O device.

A fundamental question is which language to use. Use assembly language for code that must operate quickly or with precise timing, and to use microcontroller features not available through the compiler, such as fuzzy logic instructions. However, programming in C or C++ is preferred over assembler-language programming because such code is about an order of magnitude cheaper to design and maintain. Because of its simplicity and wide acceptance, C is the language of choice for most I/O interfacing code. C provides operators that directly specify pointer usage, shifting, or masking, whereas other languages like PASCAL, BASIC, or JAVA miss some key parts that C provides. However, C++ offers object-oriented programming, which we discuss later.

I/O devices can be accessed with constant pointers, as in `d=(char*)0x4000;`; with variable pointers, as in `char ptr=(char*)0x4000; d=*ptr;`; using global variables positioned with embedded assembly-language origin statements; with vector indexing, as in `char ptr=(char*)0x4000; dp[0];`; and with pointers to structures, as in `spiPort->bd=2;`. We have shown several examples of each. Constant pointers seem to be useful where an I/O port is accessed once or twice. Global variables work well for most I/O devices. Vectors seem to be clearer when a device has a number of same-size ports. Structure pointer access is very useful for devices with many ports of assorted sizes. Finally, any long program can likely benefit from the intelligent use of `#define` statements to rename these constructs to be meaningful port names to provide self-documenting code.

Objects provide a mechanism to efficiently implement the capabilities of device drivers, such as device independence and I/O redirection. They achieve a major fraction of the capabilities of operating system I/O device drivers, but with a small fraction of the overhead. They provide I/O independence, which permits compile-time substitution, without changing the body of the code outside the I/O device objects, and I/O redirection, which permits the same flexibility at run time. They provide this independence and redirection by protecting functions and data so that there are no subtle interactions among I/O devices and the main procedure or its subroutines, other than the directly stated interactions in public function-member arguments and data members. Also, this ability to provide protection enhances documentation and maintainability, and thus reduces software design cost. One is less likely to confuse or abuse a variable that is bound to the functions that use it. If C++ is unavailable, its mechanisms can be emulated in C by enforcing appropriate conventions in symbolic names and function arguments.

Objects described in this chapter, in clearly marked and optional sections, provide more than just a way to show students who have had a course in C++ how object-oriented programming can be used in interface design. The fundamental ideas of object-oriented programming can elevate the programmer from mere software to full system design. By considering the object as encapsulating the function and data member, as well as the I/O port and external hardware, the designer can have predesigned and tested objects (functions, data, and external hardware) that can be inserted into an application as a unit, – plug-and-play – or be available in a library, to significantly reduce the design cost of a microcontroller-based system. Further uses of objects will appear in the next and following chapters. An outstanding article, "Object-Oriented Development" by Grady Booch, in the IEEE Computer Society Press tutorial *Object-Oriented Computing*, Gerald Peterson, Ed., vol. 2, led us to appreciate the use of objects in the design of embedded microcomputer systems. You might consult it for additional insights on this approach.

4.6 Conclusions

The interfacing of a microcomputer to almost any I/O system has been shown to be simple and flexible, using parallel and serial I/O devices. We studied different ways data can be passed through a port, into or out of a microcontroller. We saw some I/O software that moved data through a microcomputer, moved data into a buffer, and implemented a traffic light controller and IC tester using the simple I/O devices. Because timing is important to them, we studied the timing of such program segments. We studied indirect and serial I/O, which are especially attractive to the 'A4 and other microcontroller systems. Finally, we considered how the SPI can assist in serial I/O. We can use the same approach to designing an IC (or an I/O system) as we can for studying it, and thus develop an understanding of why it was designed as it was and how it might be used. In the remaining chapters, these techniques are extended to analog interfacing, counters, communications interfacing, display, and magnetic recording chips.

Do You Know These Terms?

See page 36 for instructions.

synchronous
I/O device
port
input port
output port
isolated I/O
input instructions
output instructions
memory-mapped I/O
lock
function-member checking
basic output port
readable output port
shadowed output
set port
clear port
address trigger
read address trigger
write address trigger
address-trigger sequence
address-register output
direction port
control port
initialization ritual
configure
device-independent
I/O redirection
Port
object
object driver
Stub
logic-timer
light pattern
sequence
cycle
delay loop
interpreter
Mealy sequential machine
present internal state
input state
output state
next internal state
linked-list structre
block
template
direct I/O
bit-banging
indirect port
status port

Problems

Problem 1 is a paragraph correction problem. See page 38 for guidelines. Guidelines for software problems are given on page 86, and for hardware problems, on page 115.

1*. A port is a subsystem that handles I/O. Memory-mapped I/O is used on the 6812 and is popular even on microcomputers that have isolated I/O, because it can use instructions that operate directly on memory and is more reliable in the face of a runaway stack than is isolated I/O. However, if a program error writes over I/O devices, a lock can be used to prevent the calamity. A basic output port is a tristate driver and a decoder; the decoder needs only to look at the address and the R/W line to see if the device is to be written into. An input port is a tristate driver and a decoder; the decoder needs only to look at the address and the R/W line to see if the device is to be read. A basic output port is a read-only port that cannot be written by the program. Therefore, the program should keep an extra copy of a word in such an output port if it wants to know what is in it. The data can be recorded automatically in memory by using an address trigger. Such output devices are commonly used because they are cheaper than readable output devices. An address output line uses a register to capture the low-order address bits when the high-order address bits match the decoder pattern.

2. A group of eight 1-bit input ports is to be addressed at locations 0x2C30 to 0x2C3F so they will be read in the sign bit position. Show the logic diagram of such a port, whose decoder is fully specified and whose input is a 74HC251 (Figure 4.23). This chip has tristate outputs Z and Z inverted. Use only a 74HC4078 and 74HC30 (Figure 3.10).

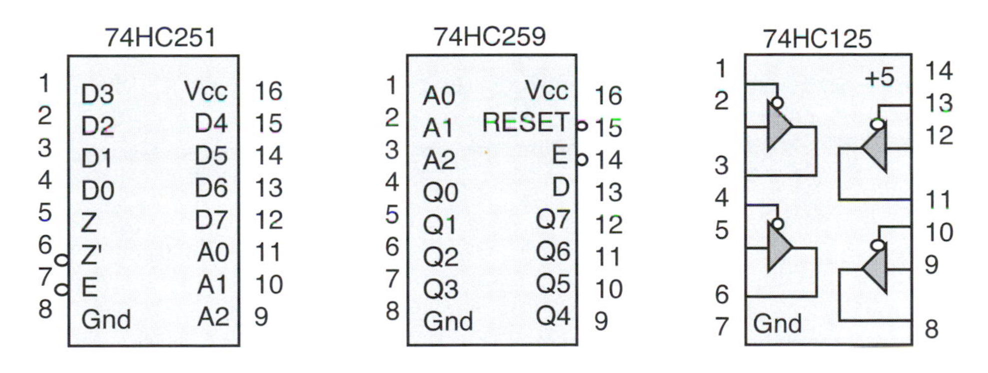

Figure 4.23. Some ICs for I/O

3. A group of eight 1-bit output ports is to be addressed at locations 0x73A0 to 0x73AF so they will write the sign bit of these words. Show the logic diagram of such a port, whose decoder is fully specified and whose output latches are in the 74HC259 addressable latch. Use only a 74HC259 (Figure 4.23), 74HC4078, and 74HC30 (Figure 3.10). (Note: This group of output ports can have the same address as a read/write memory using a shadow, so that when words are written in the memory, the sign bits appear in the outputs of the corresponding latch to be used in the outside world.)

4. Suppose a 1-bit input device using a 74HC125 (see Figure 4.23) inputs a signal A in the sign bit of location 0x2000

a. Show a logic diagram of the input device using incompletely specified decoding and chips from Figure 3.10 to implement the decoder. Decode all necessary address and control signals (E, R/W, etc.) but do not use chip selects. Assume that the program uses only addresses 0 to 0xFFFF (for RAM), 0x2000 (for this device), 0x6000 to 0x7FFF (for internal I/O), and 0xF000 to 0xFFFF (for ROM). Full credit is given to the design using the least number of chips.

b. A wave form is initially low when your program begins sampling it, then goes high and then low for the rest of the time. Write a C or C++ function *int pwdth()* to return the width of this positive pulse (in microseconds) as accurately as you can, assuming the E clock is 8 MHz.

c. What is part b's worst-case pulse-width measuring error, in microseconds?

5. Suppose a 1-bit output device using a 74HC74 (see Figure 3.11) outputs the least significant bit of location 0x2000. The output is to be a square wave.

a. Show a logic diagram of the output device using completely specified decoding, using chips from Figure 3.10 to implement the decode. Decode all 16 bits of the address and appropriate control signals (E, R/W, etc.), but do not use chip selects. Show all lines connected to +5 V or ground.

b. Write a self-initializing procedure *void squr(int n)* to generate a square wave with frequency in Hz given in *n* using delay loops. Assume the E clock is 8 MHz.

c. What is the lowest and highest frequency that part b can generate?

6. An output device having 16 output bits is addressed at location 0xD3A2. If a number $2n + 1$ is written into this location, the nth 1-bit latch is set, $0 \le n < 16$, and if a number $2n$ is written into this location, the nth 1-bit latch is cleared. Show a logic diagram of such a system of output latches, whose address decoder is fully specified and whose latches are in two 74HC259 addressable latches. Use only two 74HC259s, a 74HC04, a 74HC32, a 74HC4078, and a 74HC30. Show all chips and pin numbers.

7. Show the logical design of the decoder for Figure 4.2's readable output port, which can be read from or written in at location 0x1000. The decoder shown therein should be implemented with a minimal number of available gates using the chips in Figure 3.10. Show only the decoder, and not the output register or input tristate driver.

8. Show the logic diagram of a wide expanded-mode 16-bit readable output port at $200. The program will always read or write 16 bits, and never read or write 8 bits, using this port. However, do not show pin numbers on the data bus or on the other gates and flip-flops connected to the data bus (use vector notation to indicate a bus). Give assembler-language instructions to read this port, write data into it, and increment its value. Give C or C++ statements to read the data from it, write data into it, or increment it.

9. Write a single assembler-language instruction to write a value $1234 in `PORTA` and `PORTB`, and write another instruction to write a value $1234 in `DDRA` and `DDRB`.

10. Show the logic diagram of two ports, which are both addressed at location 0x8020. In one, a readable output port, the 7 least significant bits of the output word are readable; in the other port, an input port, the most significant bit read is input from the outside world. The address decoder is to be completely specified using a minimum of chips. Use a 74HC244 for the input port and a 74HC374 for the output port, and two 74HC4078s, a 74HC04, and a 74HC20 for the decoder. Show all chips and pin numbers and the 6812 address and data bus line numbers and control signals used.

11. Show the logic diagram of a readable set port and a clear port, which are both addressed at location 0xFF22. If a 1 is written in bits 7 to 0, the corresponding port bit is set, otherwise the data is unchanged; if a 0 is written in bits 15 to 8, port bits 7 to 0 are cleared, otherwise the data is unchanged. Data stored in the port must be available to the outside world. Reading 16 bits from 0xFF22 reads the port bits to data bus lines 7 to 0 only. The program will never read or write 8 bits using this port. The address decoder is to be completely specified using a minimum of chips. Use a 74HC244 for the input port, a 74HC08 and a 74HC32 for the set/clear logic, a 74HC374 for the output port, and a 74HC133, a 74HC04, a 74HC32, and a 74HC4078 for the decoder. Show all chips and pin numbers and the 6812 address bus, data bus, and control signals used.

12. A 16-bit input port at locations 0x6F3A is connected to a 16-bit serial-in parallel-out shift register to input, into a 16-bit-wide buffer, bit serial data shifted into the shift register. Bits are stored in each 16-bit word in the buffer, most significant bit first.

a. Show the logic diagram for this pair of ports using two 74HC244s and a pair of 8-bit serial-in, parallel-out shift registers that use 74HC164s. The address decoder is to be completely specified using a minimum of 74HC133s and 74HC4078s.

b. Write a (fastest) assembly-language program for an 8-MHz ’A4 to store these data into a buffer, using **`MOVW`** and **`DBNE`** instructions in a DO-loop. How many bits per second can it collect?

13. Write a templated class `Port` with function-member checking. In the constructor, if `CHECK` is defined, the *id* must be less than `idAD`, and in `option`, if `id` is `idAd`, then the direction must be input, otherwise `errors` is asserted to warn the user. If `CHECK` is not defined, the code is identical to `Port`’s in §4.2.2.

14. Write a templated class `Port` to access all 12 parallel ports as 8-bit ports, and ports A and B, C and D, and F and G as 16-bit ports, with function-member checking to check any possible error at run time. Use the following *enum* statement:

```
enum {iA, iB, iC, iD, iE, iF, iG, iH, iJ, iS, iT, iAD, iAB, iCD, iFG};
```

15. Write templated class `BitPort` to access any bit of any parallel port, derived from *Port* (problem 14). Its constructor’s arguments are `p, b,` and `d,` where `p` is the 8-bit parallel port, `b` is a bit number, and `d` is F (0) for input and T (1) for output.

16. Write the shortest possible self-initializing procedure `inbuf(char *a, char *p, int n)`, using assembly language embedded in C, to input *n* bytes of data from port *p* to a buffer at location *a,* which, like `main()` (§4.3.2), is the fastest in execution, and give the rate at which words can be input to the buffer for this procedure, assuming an 8-MHz CLK clock. Consider using `MOVB` with different addressing modes, and `DBNE`.

17. Design a traffic light controller that uses the power line frequency to time the lights and immediate operands to control the lights.

a. Show an I/O system logic diagram. A 60-Hz square wave is input in 0x4000 bit 7, using a 74HC125; and a 74HC374 outputs a 6-bit light pattern to control the lights, at location 0x4000, as in Figure 4.8. Use incompletely specified decoding, assuming the program uses only the addresses 0 to 0x7F, 0x4000 (for these ports), and 0xFF00 to 0xFFFF, using a minimum of 74HC04s, 74HC30s, and 74HC4078s.

b. Show a C or C++ procedure `void dly()` that tests the input to delay exactly 1/60 second. Assume the input has become high just before entry to this procedure.

c. Write a C or C++ procedure `void main(),` using immediate operands to control the light patterns. Use the program segment in part b to sequence the lights as in Table 4.1.

18. Write an array interpreter (like a linked-list interpreter) and an array `char a[][][]` as in the first `main` procedure in §4.3.4, to control Figure 4.8's traffic light. For internal state `s` and input state `x,` the array `a[s,x)[]` entries are next state, output state sent to `PORTA,` and time delay. Use a procedure `delay()` that delays 1 s. The main sequence is shown in Table 4.2. A late-night sequence has north-south lanes red and east-west lanes yellow, both on for 1 s and off for 1 s. A fire truck emergency sequence is the north-south lanes red while the others are green for 20 s, then the east-west yellow for 2 s while north-south is still red, after which the north-south are green and the others are red for 10 s, and then the main or late-night sequence is resumed with its first line. Transitions to new sequences exhibit no time delay and output the output state of the sequence being begun. `PORTB`'s input state is either 0 if neither late night nor emergency occurs, 1 if an emergency occurs, and 2 if it is late at night and no emergency occurs. The emergency input state can last only one state transition to begin the emergency sequence. The other input states last as long as the sequences are to be executed. *enum* the main sequence states as `M0, M1, M2,` and `M3,` having values 0 to 3; the late-night sequence as `N0` and `N1,` having values 4 and 5; and the emergency sequence as `E0, E1, and E2,` having values 6, 7, and 8.

19. Consider a vending machine controller. Its input port at location 0x1003 has value 0 if no coins are put into the machine, 1 if a nickel, 2 if a dime, 3 if a quarter, and 4 if the coin return button is pressed. The output port, at location 0x1003, will dispense a bottle of pop if the number 1 is output, a nickel if 2 is output, a dime if 3 is output, and a quarter if 4 is output. This vending machine will dispense a bottle of pop if 30 cents have been entered, will return the amount entered if the coin return button is pressed, and otherwise will keep track of the remaining amount of money entered (i.e., it will not return change if greater than 30 cents are put in). All control is done in software.

a. Show the logical design of the I/O hardware, assuming incompletely specified decoding if the program uses only addresses 0 to 0x80, 0x1003 (for these ports), and 0xFF00 to 0xFFFF, and the ports are only 3 bits wide. Use a minimum of 74HC04s and 74HC20s, and a 74HC244 and a 74HC374 for input and output port chips.

b. Show this controller's tabular and graphical sequential machine. Internal states S = {zero, five, ten, fifteen, twenty, twenty-five} will be the total accumulated money. Input states I = {B, N, D, Q, R} represents that no (blank) inputs are given; that a nickel, a dime, or a quarter are given; or that the coin return button has been pressed. Output states O = {b, p, n, d, q} will represent the fact that nothing (blank) is done, or a bottle of pop, a nickel, a dime, or a quarter, respectively, is to be returned. If multiple outputs are indicated, output a pop first, and go to the state that represents the amount of change left in the machine, and output the larger coin first. Assume the coin return button is pressed repeatedly to return all the coins.

c. Show a self-initializing procedure *void seqMch()* to implement this sequential machine by a linked-list interpreter, and show the linked list as a three-dimensional array, using an index as the link. Activate the solenoids for 0.1 s (assume procedure `void delay()` causes a 0.1 s delay), and then release the solenoids that dispense the bottles and the money. Guard against responding to an input and then checking it again before it has been removed. (Hint: Respond to an input only when it changes from that input back to the blank input.)

d. As an alternative to part c, show a self-initializing procedure `void algor()` to implement this sequential machine using arithmetic and conditional expressions in an algorithm. Use the same assumptions as in part c. Note the ease or difficulty of modifying part c's state machine, or the code in part d, when the cost of a soda is changed to 35 cents (don't jump to conclusions).

20. Design a keyless-entry module (KEM) for a car.

a. The KEM has five SPST switches (A to E); when any is pressed (only one can be pressed at a time) its corresponding binary number (A is 1, B is 2, C is 3, D is 4, E is 5) is input in negative logic to `PORTB`. For instance, if switch E is pressed, `PORTB` bits 0 and 2 are asserted low (representing TLT or binary number 5). Show a logic diagram of an input state encoder that uses a 74HC08 to OR the switch signals into port input bits (in negative logic) and switches with pull-up resistors.

b. The hardware state decoder of part a was found more costly than a software solution that inputs each key's signal directly to a `PORTB` pin (switch A to bit 0, B to 1, C to 2, D to 3, E to 4). (Do not draw the logic diagram of this new hardware circuit.) Write `char getkey()` that returns the currently pressed key's value (noKey returns 0, A returns 1, B is 2, C is 3, D is 4, E is 5) using this new hardware. (Hint: table-lookup may provide the shortest program, or you can give a shorter one.)

c. The KEM recognizes pressing $A^* C^* D^* B^*$, where A^* means `char getkey()` returns the value 1 (key A) one or more times, followed by one or more 0s (noKey). Give the Mealy sequential machine table that recognizes when the code is entered. Let the input state **A** represent that `char getkey()` returns the value A (1) one or more times, followed by one or more values noKey (0), etc. Input states **B**, **C**, **D**, and **E** are similarly defined. Let the internal state **a** represent that no keys have been recognized, that an illegal sequence has been input, or that an output pulse has been sent; **b** represent that the code A^* is recognized; **c** represent that the code $A^* C^*$ is recognized; and **d** represent that the code $A^* C^* D^*$ is recognized. Output state 1 indicates the whole code is recognized and a pulse should be generated; otherwise, the output state is 0. Show the Mealy sequential machine table.

d. Write a self-initializing procedure `void validate()` that produces a 500 ms positive pulse on `PORTA` bit 0 when the sequence is recognized, then continues to check for the sequence, forever. You must use the sequential machine interpreter model, rather than if-then or case statements. The procedure must store the state machine in a compact global `char` vector `v[4,5,2],` which you must show initialized in C, where the left index applies to the present internal state, the middle to the input state and the right to the output and next internal state. Use the `#define` statements to make your declaration more readable. Use a `for`-loop delay, assuming that constant `N` will be defined later to provide the correct pulse length.

21. Using §4.3.5's procedure `int check()`, write `#define` statements to generate patterns `A, B, ... , Z` and test vectors `unsigned int v[]` to test the

a. 74HC32 OR gate. b. 74HC04. c. 74HC10 NAND gate (see Figure 3.10).

d. 74HC138 decoder (see Figure 3.10). Follow this test procedure: for E1, E2, and E3 asserted, check for all combinations of A2, A1, A0; then for A2, A1, A0 all L (0) check for all combinations of E1, E2, and E3; but do not check one of the patterns already checked in the first sequence (15 tests, rather than 64).

e. 74HC74 (see Figure 3.11). Follow this test procedure:

1. Assert S, with clock and D low. Check that Q and Q-bar are set.
2. Clock a 0 into both flip-flops. Check if Q and Q-bar are clear.
3. Drop clock with data high. Check to see if both flip-flops remain cleared.
4. Clock a 1 into both flip-flops. Check to see if both flip-flops are set.
5. Clear both flip-flops asserting only R. Check to see if both flip-flops are clear.

22. Write a self-initializing procedure `int ina(a)` for Figure 4.15d, similar to `outa(a, d)` in §4.4.1, to return the 6818A's time of day, using indirect I/O. It should emulate Figure 4.15c's read cycle's timing diagram. However, if multiple outputs change at the same time, change `ds` first, `as` second, `rw` third, and `cs` last. (Disregard a possible timing problem, which reads the 6818A's time when it is changing.)

23. An LCD connects to `PORTA` as shown in Figure 4.16, but LCD signal Rw connects to `PORTA` bit 6 instead of to ground. When Rw and E are T (1) and Rs is F (0), `PORTA` bit 7 outputs a busy bit which is T (1) when the LCD is not ready for a new command, changing to F (0) when a new command can be given. Write the initialization and output procedures, and procedures to read the busy bit to synchronize writing to the LCD: `char get4(char a)` returns a nibble, `char get(char a)` returns a byte, `put4(char d, char a)` outputs a nibble, and `put(char d, char a)` outputs a byte – where `d` is data and `a` is the LCD register accessed (if `a` is 0, `Rs` is 0, otherwise `Rs` is 1).

24. Write self-initializing procedures for the other configurations in Figure 4.18, using `PORTS`, but not using the SPI. Assume address $a = 0$ selects the leftmost '589 or '595, 1 selects the middle '589 or '595, and 2 selects the right '589 or '595.

a. For Figure 4.18a, `char serialIn(char a)` returns a byte input from the *a*th '589.
b. For Figure 4.18a, `void serialIn(char *s)` puts 3 bytes input from the '589s in `s`.
c. For Figure 4.18b, `void serialIn(char *s)` puts 3 bytes input from the '589s in `s`.
d. For Figure 4.18c, `void serialOut(char c, char a)` outputs `c` to the `a`th '595.
e. For Figure 4.18d, `void serialOut(char c, char a)` outputs `c` to the `a`th '595.
f. For Figure 4.18d, `void serialOut(char *s)` outputs 3 bytes from `s` to the '595s.

25. The program at the end of §4.4.3 outputs a command to and inputs the measured temperature from the Dallas Semiconductor 1620. This chip has an output Tcom on pin 5, which can be used to control a heater or air conditioner, and registers TH for high and TL for low temperature limits. Tcom becomes asserted when the measured temperature exceeds TH and becomes negated when the measured temperature goes below TL. Use §4.4.3's procedures `send` and `receive`, in both parts of this problem.

a. TH is written with the command 0x01 and read with the command 0xa1. Write a procedure `int putHi(int t)` that outputs `t` to TH and reads it back, returning 1 if the value read back is not the same as the value output, and 0 otherwise.

b. TL is written with the command 0x02 and read with the command 0xa2. Write a procedure `int putLow(int t)` that outputs `t` to TL and reads it back, returning 1 if the value read back is not the same as the value output, and 0 otherwise.

26. Repeat problem 24 using the SPI to shift the data in and out, using §4.4.4's global variable ports and *enum*erated symbolic names. Data are shifted on the rising edge of SCLK, at 4 MHz, and SCLK is initially high.

27. Write the C or C++ procedure `void main()` to initialize the SPI for the 'A4's 8-MHz E clock and output on SCLK, MOSI, and SS, as follows:

a. Use §4.4.4's global variable port names and *enum*erated symbolic names to shift on the falling edge of SCLK, at 4 MHz, with SCLK initially high. The SPI outputs are open collector, and data are shifted least significant bit first.

b. Use a vector `char spi[8];` to access all SPI ports, to shift on the rising edge of SCLK, at 1 MHz, with SCLK initially high. SPI outputs have reduced power.

c. Use vector `char spi[8];` to access all SPI ports and give `#define`s equating `SP0CR1, SP0CR2, SP0BR, SP0SR, SP0DR, PORTS` and `DDRS;` to `spi`'s elements, to shift on the falling edge of SCLK (initially low) at 250 KHz.

d. Use §4.4.4's `struct SPI` to access SPI control ports 1 and 2, to shift on the rising edge of SCLK, with SCLK initially high.

e. Use a pointer `SpiPtr` to §4.4.4's `struct SPI` to access control ports 1 and 2, to shift on the rising edge of SCLK, with SCLK initially low.

The Motorola M68HC12B32EVB board can implement all the experiments and examples in this book, except those of Chapter 10. When used without another 6812 board, the debugger called DBUG_12 will use half of SRAM, permitting the other half to be used for an experiment.

5

Interrupts and Alternatives

The computer has to be synchronized with the fast or slow I/O device. The two main areas of concern are the amount of data that will be input or output and the type of error conditions that will arise in the I/O system. Given varying amounts of data and different I/O error conditions, we need to decide the appropriate action to be taken by the microcomputer program. This is studied in this chapter.

One of the most important problems in I/O systems design is timing. In §4.3.2, we saw how data can be put into a buffer from an input device. However, we ignored the problem of synchronizing with the source of the data so that we get a word from the source when it has a word to give us. I/O systems are often quite a bit slower, and occasionally a bit faster, than the computer. A typewriter may type a fast 30 characters per second, but the 6812 can send a character to be typed only once every 266,667 memory cycles. So the computer often waits a long time between outputting successive characters to be typed. Behold the mighty computer, able to invert a matrix in a single bound, waiting patiently to complete some tedious I/O operation. On the other hand, some I/O systems, such as hard disks, are so fast that a microcomputer may not take data from them fast enough. Recall that the time from when an I/O system requests service (such as to output a word) until it gets this service (such as having the word removed) is the latency. If the service is not completed within a maximum latency time, the data may be overwritten by new data and lost before the computer can store them.

Synchronization is the technique used to supply data to an output device when the device needs data, to get data from an input device when the device has some data available, or to respond promptly to an error if ever it occurs. Over ten techniques are used to match the speed of the I/O device to that of the microprocessor. Real-time synchronization is conceptually quite simple; in fact we have already written a real-time program in the previous chapter to synchronize to a traffic light. Gadfly synchronization requires a bit more hardware, but has advantages in speed and software simplicity. Gadfly was actually used in synchronizing the SPI. Three more-powerful interrupt synchronization techniques – polled, vectored, and real-time – require more hardware. Direct-memory access and context switching are faster synchronization mechanisms. Shuttle, indirect, time-multiplexed, and video memories can be used for very fast I/O devices. These synchronization techniques are also discussed in this chapter.

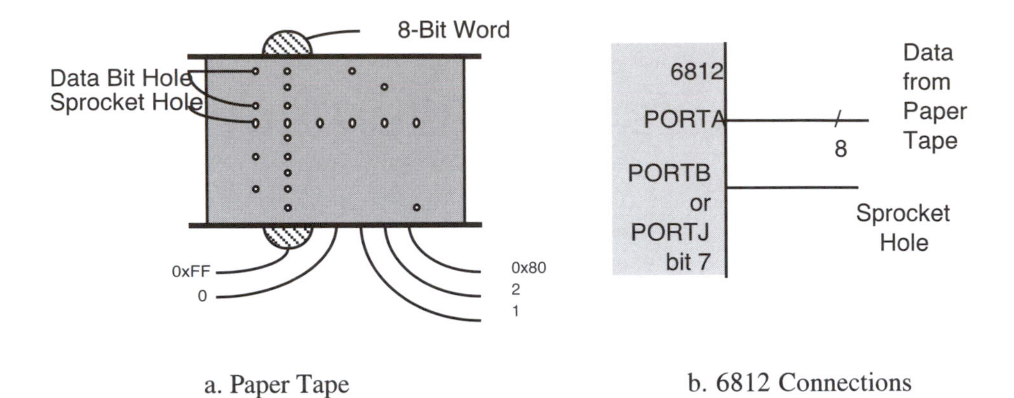

a. Paper Tape b. 6812 Connections

Figure 5.1. Paper-Tape Hardware

We first study the synchronization problem from the I/O device viewpoint, introducing *BUSY/DONE states* and terminology. An example of a paper-tape reader, illustrating the collection of data from a tape into a buffer, will be used to illustrate the different approaches to I/O synchronization. Paper tape (see Figure 5.1a) is used in environments like machine shops, whose dust and fumes are hostile to floppy disks.

Data from the data port can be read just as in the previous chapter and will be put into a buffer, as we now discuss. The pattern of holes across a one-inch-wide paper tape corresponds to a byte of data; in each position, a hole is a true value, and the absence of a hole is a false value. Optical or mechanical sensors over each hole position connected to the port pins signal an H (T) if the sensor is over a hole. We will read data from 6812's `PORTA` to realize this collection. At any time, the values of such a pattern of holes under the paper-tape head can be read by an instruction like *d = PORTA*. It can be put in a buffer by a statement like: `for(pnt=buffer; pnt<buffer+0x100; ) *(pnt++) = PORTA;`. However, in the last chapter we ignored, and in this chapter we focus on, the problem of getting the data at the right time, when the hardware presents it. The user can advance the paper manually. In this example, we have to read one byte of data from the pattern of holes when the sprocket hole sensor finds a sprocket hole.

A simple but general model (a Mealy sequential machine) of the device describes how a computer synchronizes with an I/O device so it can take data from it or send data to it. (See Figure 5.2a.) In this model, the device (or, equivalently, its object) has three states: *IDLE, BUSY,* and *DONE.* The device is in the IDLE state when no program is using it. When a program begins to use the device, the program puts it in the BUSY state. If the device is IDLE, it is free to be used, and, if BUSY, it is still busy doing its operation. When the device is through with its operation, it enters the DONE state. Often, DONE implies the device has some data in an output port that must be read by the program. The state transition from BUSY to DONE is associated with the availability of output from the device to the program. When the program reads this data, it puts the device into IDLE if it doesn't want to do any more operations, or into BUSY if it wants more operations done. An error condition may also put the device into DONE and should provide some way for the program to distinguish between a successfully completed operation and an error condition. If the program puts the device into BUSY or IDLE, it is called, respectively, *starting* or *stopping* the device. The device enters DONE by itself, *completing* the requested action. When the device is in DONE, the program can get the results of an operation and check to see if an error has occurred.

The IDLE state indicates the paper-tape reader is not in use. The user starts the paper-tape reader, putting it into the BUSY state. In that state, he or she pulls the paper tape until the next pattern is under the read head that reads a word from the tape. When the sprocket hole is under the tape reader or no paper is left, the reader reads a byte and enters the DONE state. The computer recognizes that when the reader is in the DONE state, data from the pattern should be read through the data port and put into the buffer in the next available location, or else an error condition might exist. Once read, if the program intends to read the next pattern because more words are needed to fill the buffer, it puts the reader back into the BUSY state to restart it. If another pattern should not be read because the buffer is full, the device goes to the IDLE state to stop or to ignore it. If an error condition is signaled, the device is left in the DONE state so it won't be used until the error is read and possibly fixed or reported. To read three bytes, the tape reader might pass through the following states: IDLE, BUSY, DONE, BUSY, DONE, BUSY, DONE, IDLE. (See Figure 5.2a.) Data are read each time the paper-tape reader goes into the DONE state. Note that there is a difference between the IDLE state and the DONE state. In the DONE state, some data in the input port are ready to be read, and the I/O device is requesting the computer to read them, or an error has rendered the device unusable; while in the IDLE state, nothing is happening, and nothing need be done.

In some I/O systems that do not return values or error messages back to the computer, however, DONE is indistinguishable from IDLE, so only two states are required. Consider a paper-tape punch. IDLE corresponds to when it is not in use. BUSY corresponds to when the program has put a byte out to it but the byte has not yet been completely punched. DONE corresponds to when the holes are punched, and another byte of data can be sent out. In this case, with no error conditions to examine in DONE, the DONE and IDLE states are indistinguishable, and we can say the device has just two states: IDLE and BUSY.

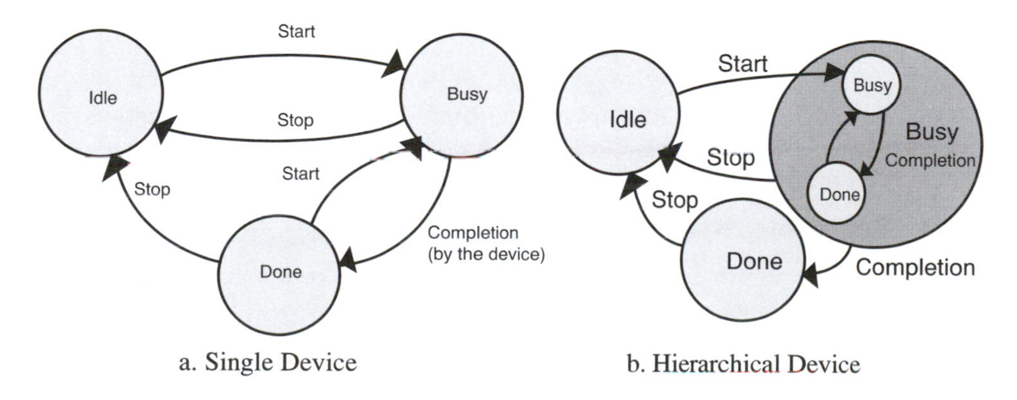

a. Single Device b. Hierarchical Device

Figure 5.2. State Diagram for I/O Devices

Figure 5.2b illustrates that the states can be hierarchically defined. If Figure 5.2a's BUSY/DONE states are for the arrival of a byte in a serial input device, the overall BUSY state of Figure 5.2a indicates that within a byte, bits have been requested but not all have arrived. Within this BUSY state, a lower level BUSY state (Figure 5.2b) is entered as a bit is requested; when the bit arrives, a lower level DONE state is entered. Thus, the overall BUSY state for the arrival of a byte is composed of lower-level BUSY and DONE states for the arrival of bits. The lower level state machine ping-pongs back and forth between the lower-level BUSY and DONE states until the whole byte has been serially input, while the higher-level state machine remains BUSY, then the higher-level DONE state is entered. Similarly, for the arrival of bytes described by the state machine in Figure 5.2a, there can be a higher-level state machine associated with the buffer into which the bytes are placed. This state machine is in the BUSY state as long as the buffer is dedicated for input but not filled with data. This state machine enters its DONE state when the entire buffer is filled and is ready to be used.

A typical microcomputer having several I/O devices has as many BUSY/DONE sequential machines, one for each device, and possibly a BUSY/DONE sequential machine for every buffer being emptied and filled. There exist product machines that can be used to describe interactions between devices. For instance, if a device has states i1, b1, and d1, and another device has states i2, b2, and d2, then the product machine has states i1i2, i1b2, i1d2, b1i2, b1b2, b1d2, d1i2, d1b2, and d1d2, where for instance i1b2 means that state machine 1 is in the IDLE and state machine 2 is in the BUSY state. In general, if there are n devices or buffers, there are 3^n states in their product machine. While many synchronization problems can be clearly defined for the simple state machine rather than the product machine, the product machine can define problems that occur when two or more devices or buffers interact, such that a condition arises only when each machine, and therefore the product machine, is in a specific state.

The general synchronization problem is to attend to a device or buffer when it enters its DONE state, in order to get input data, check for and respond to errors, or initiate an activity for future work. In this section, we consider real-time and gadfly synchronization techniques written in simple C and C++ object-oriented procedures.

5.1 Programmed Synchronization

There are several ways a microcomputer can synchronize with a slower I/O device, as discussed in this section. Two ways, real-time and gadfly synchronization, are programmed in the `main` procedure or in procedures it calls, and are studied here.

5.1.1 Real-time Synchronization

Real-time synchronization uses program timing delays to synchronize with the delays in the I/O system, to either equal or exceed the time a device is in BUSY. See Figure 5.3a. Two cases are (1) using a procedure's inherent delays, and (2) using a *wait* loop or other time-consuming statements to "pad" the processor's delay to match or exceed the I/O device's BUSY state time. If that happens to be the rate at which the paper tape is pulled, and the procedure is started so as to pick up the first pattern of holes, putting it into `buffer[0]` (a preposterous idea), the processor is synchronized with the device.

```
void main() {char buffer[0x100]; int i;
    for(i=0;i<0x100;i++) buffer[i] = PORTA;
}
```

To illustrate the second case, we insert a delay loop as we did in the traffic light application in §4.3.3. We can modify the procedure as follows:

```
void main() {char buffer[0x100]; int i, j;
    for(i = 0; i < 0x100; i++) { j = N; while(--j); buffer[i] = PORTA;}
}
```

The statically efficient delay loop `j = N; while(--j);` is empirically adjusted, by defining constant `N,` so that the loop executes during the BUSY state, which is related to the rate at which the paper tape is pulled. A longer delay can be implemented using nested loops, as in `i = N1; j = N2; while(--j) while(--i);`. If the procedure is started so as to pick up the first pattern of holes, putting it into `buffer[0]` (a nontrivial challenge), the processor is synchronized with the device. Because required program-statement timing delays already synchronize the processor to the tape reader, we are also using real-time synchronization. This program's assembly language is shown below.

```
0000086D CE0000     LDX   #0
00000870 CD1388     LDY   #5000
00000873 0436FD     DBNE  Y,*+0    ;abs = 0873
00000876 1980       LEAY  0,SP
00000878 B754       TFR   X,D
0000087A 19EE       LEAY  D,Y
0000087C 9600       LDAA  $00
0000087E 6A40       STAA  0,Y
00000880 08         INX
00000881 8E0100     CPX   #256
00000884 2DEA       BLT   *-20   ;abs = 0870
```

In real-time synchronization, the device has IDLE, BUSY, and DONE states, but the computer may have no way of reading them from the I/O device. Instead, it starts operations and keeps track of the expected time for the device to complete the operation. BUSY/DONE states can be recognized by the program segment being executed in synchronization with the device state. IDLE is any time before we begin reading the tape; the delay loop corresponds to BUSY; and DONE is when `buffer[i] = PORTA;` is executed. The device is started, and the time it takes to complete its BUSY state is matched by the time a program takes before it assumes the device is in DONE. While an exact match in timing is occasionally needed, usually the microcomputer must wait longer than the I/O device takes to complete its BUSY state. In fact, the program is usually timed for the longest possible time to complete an I/O operation.

Real-time synchronization is considered bad programming by almost all computer scientists. Dynamic memories can require refresh cycles, interrupts, and DMA cycles discussed in §5.4.1, which can occur at unpredictable times. A cache memory supplying instructions or data can speed up the execution of instructions, thus affecting delays based on their execution. If timing delays are implemented in high-level languages such as a `while` loop in C, the delay time can change when a later version of the compiler or operating system is used. It is difficult to provide a delay of a fixed time by means of delays inherent in instruction execution. The effort in writing the program may be the highest because of the difficulty of precisely tailoring the program to provide the required time delay, as well as being logically correct. This approach is sensitive to errors in the speed of the I/O system. If some mechanical components are not oiled, the I/O may be slower than what the program is made to handle. The program is therefore often timed to handle the worst possible situation and is the least responsive synchronization technique.

However, real-time synchronization requires the least hardware; it can be used with a basic input or output device, without the need for further hardware. It is a practical alternative in applications such as the traffic light controller discussed in §4.2.2 or a microcomputer that is dedicated to control a printer.

5.1.2 Gadfly Synchronization

The sprocket hole input can be sensed in the *gadfly* synchronization technique to pick up the data exactly when they are available. The program continually "asks" one or more devices what they are doing (such as by continually testing the sprocket hole sensor). This technique is named after the great philosopher, Socrates, who, in the Socratic method of teaching, kept asking the same question until he got the answer he wanted. Socrates was called the "gadfly of Athens" because he kept pestering the local politicians like a pesky little fly until they gave him the answer he wanted (regrettably, they also gave him some poison to drink). This bothering is usually implemented in a loop, called a *gadfly loop*, in which the microcomputer continually inputs the device state of one or more I/O systems until it detects DONE or an error condition in one of the systems. See Figure 5.3b. Gadfly synchronization is often called polled synchronization. However, polling means sampling different people with the same question, not bothering the same person with the same question. Polling is used in interrupt handlers, discussed later; in this text, we distinguish between a polling sequence and a gadfly loop.

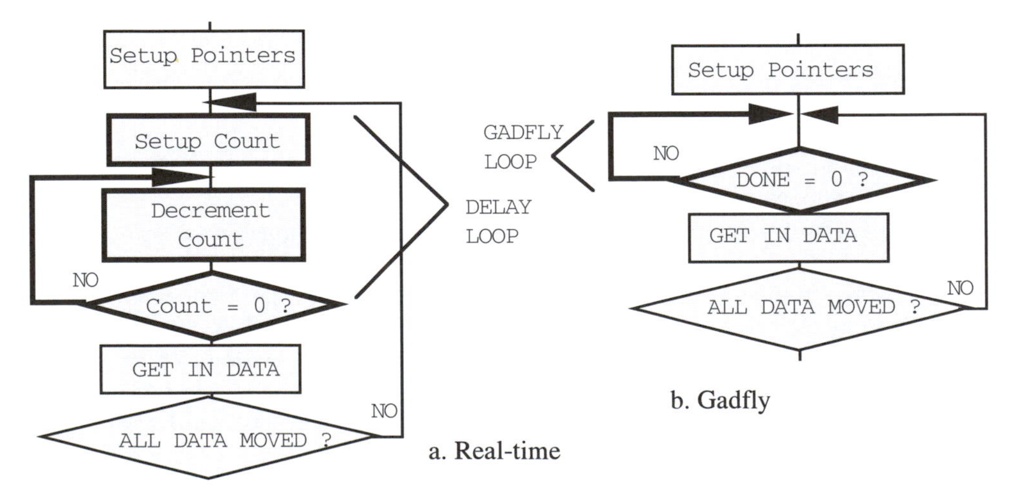

Figure 5.3. Flow charts for Programmed I/O

5.1.2.1 MC68HC812A4 Gadfly Synchronization

The 6812 'A4 has very simple key wakeup ports on ports D, H, and J that are very useful for gadfly and interrupt synchronization. (See Figure 5.4a.) They can also be used in monitoring keyboards, control switches, and position sensors. Additionally, `PORTT` has a flag register with similar characteristics (§8.3.1). All key wakeup ports detect falling edges, and `PORTJ` can detect rising edges on its inputs. Consider `PORTJ` bit 7. If `PORTJ` bit 7 is an input (`DDRJ` bit 7 is F) and `KPOLJ` bit 7 is F, then `KWIFJ` bit 7 sets upon a falling edge on `PORTJ` bit 7 (pin 10); if `KPOLJ` bit 7 is T, then `KWIFJ` bit 7 sets upon `PORTJ` bit 7's rising edge. `KWIFJ` bit 7 is cleared by writing a 1 into it (it is a clear port). A gadfly C procedure and its assembly language follows.

```
void main() {char buffer[0x100]; int i;
  for(DDRA = i = 0; i < 0x100; i++)
    {KWIFJ = 0x80; while (KWIFJ >= 0); buffer[i] = PORTA; }
}
```

```
0000086D C7          CLRB
0000086E 87          CLRA
0000086F 5B00        STAB  $00
00000871 B705        SEX   A,X
00000873 C680        LDAB  #128
00000875 5B2B        STAB  $2B
00000877 962B        LDAA  $2B
00000879 2AFC        BPL   *-2    ;abs = 0877
0000087B 1980        LEAY  0,SP
0000087D B754        TFR   X,D
0000087F 19EE        LEAY  D,Y
00000881 9600        LDAA  $00
00000883 6A40        STAA  0,Y
00000885 08          INX
00000886 8E0100      CPX   #256
00000889 2DE8        BLT   *-22   ;abs = 0873
```

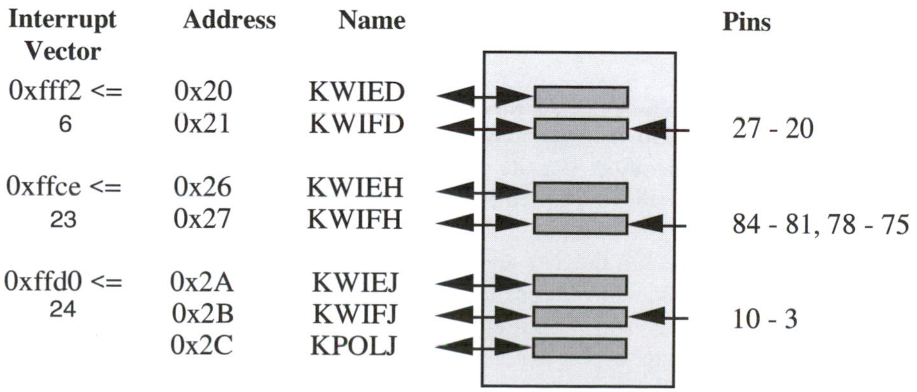

Interrupt Vector	Address	Name	Pins
0xfff2 <=	0x20	KWIED	
6	0x21	KWIFD	27 - 20
0xffce <=	0x26	KWIEH	
23	0x27	KWIFH	84 - 81, 78 - 75
0xffd0 <=	0x2A	KWIEJ	
24	0x2B	KWIFJ	10 - 3
	0x2C	KPOLJ	

Figure 5.4. Key Wakeup Ports for the MC68HC812A4

The gadfly loop, `while (KWIFJ >= 0);` waits until `PORTJ` bit 7 input falls, which is when a sprocket hole is about to pass from under its sensor. Data are picked up when this loop test fails, when the sprocket hole is still under its sensor.

5.1.2.2 MC68HC912B32 Gadfly Synchronization

The 6812 'B32 doesn't have built-in key wakeup hardware, but a signal such as the sprocket sensor can be sensed in a simple input, such as `PORTB` bit 7, for low and high values, as illustrated in the following program.

```
void main() {char buffer[0x100]; int i;
  for(DDRA = DDRB = i = 0; i < 0x100; i++)
     {while(PORTB&0x20); while( ! (PORTB & 0x20)); buffer[i]=PORTA; }
}
```

Gadfly loops `while(PORTB & 0x20); while( ! (PORTB & 0x20));` wait until `PORTB` bit 5 input falls and then rises, which is when a sprocket hole is under its sensor. Data are picked up then, when the sprocket hole is still under its sensor.

5.1.2.3 Gadfly Synchronization Characteristics

The programs above exhibit the IDLE, BUSY, and DONE states. IDLE is any time before we execute the procedure and DONE is when `buffer[i] = PORTA;` is executed, the same as in real-time synchronization. The device is BUSY when executing a gadfly loop. Gadfly synchronization exhibits lower latency than real-time synchronization, because the gadfly loop terminates exactly when the device is DONE, while the delay loop must delay for the worst-case time the device is BUSY. Real-time synchronization requires no hardware in addition to the basic I/O port, while gadfly synchronization requires hardware such as a key wakeup flag bit. Both techniques generally loop while the device is in BUSY, and that can be a long time for slow I/O devices. Interrupts, discussed later, can free the processor to do other things while it waits for DONE.

5.1.3 Handshaking

In addition to synchronization, a device may require a *handshake* signal to command the hardware to execute an operation. For the paper-tape reader, a handshake output bit R might be used to engage the motor pulling the paper tape; R would be asserted whenever the program decides to input the next byte. R can be a positive- or negative-logic-level signal, asserted until the data are input, or a positive- or negative-logic pulse asserted for about a microsecond. We consider the previous synchronization mechanisms and the handshake mechanisms in various combinations in subsequent examples. In each example, we use the paper-tape reader that was introduced earlier.

First, consider a 'B32 or 'A4 real-time paper-tape reader having data input on `PORTA,` handshaking on `PORTB` bit 0, and pulling high momentarily to request data.

```
void main() {char buffer[0x100]; int i, j;
    DDRB = 1; // arrange to output port bit 0 as a handshake
    for(i=0;i<0x100;i++) {
        PORTB = 1; PORTB = 0; // pulse the handshake signal
        j = N; while(--j); buffer[i] = PORTA; // synchronize and read port
    }
}
```

Note that, each time data are received, the handshake signal is high for about a microsecond before the real-time delay and the input operation.

Consider a 'B32 or 'A4 gadfly paper-tape reader that receives data on `PORTA`. We use `PORTB` bit 1 for synchronization, which falls when data can be read, and we use a handshake signal on `PORTB` bit 2, which becomes low to request data and returns high after the data is read. The handshake signal stays low until the data are transferred.

```
void main() {char buffer[0x100]; int i, j;
    PORTB = DDRB = 4; // arrange to output bit 2 as a handshake, and initialize it high
    for(i=0;i<0x100;i++) {
        PORTB = 0;            // begin handshake by asserting the signal low
        while( ! (PORTB & 2)); while(PORTB & 2); buffer[i] = PORTA;
        PORTB = 4;                        // end handshake by negating the signal
    }
}
```

Synchronization and handshake can be chosen independently of each other. Consider an 'A4 gadfly paper-tape reader as previously, synchronized with `KWIFJ` bit 7.

```
void main() {char buffer[0x100]; int i, j;
    PORTB = 0; DDRB = 4; // the handshake signal is output, and is initially low
    for(i=0;i<0x100;i++) {
        KWIFJ = 0x80; PORTB = 4;    // begin handshake by asserting the signal high
        while (KWIFJ >= 0); buffer[i] = PORTA;  // synchronize, read port
        PORTB = 0;                  // end handshake by negating the signal low
} }
```

5.1.4 Some Examples of Programmed I/O

We have already shown real-time programming with a traffic light controller, and gadfly synchronization with the SPI. Here we will provide three more illustrative examples of programmed I/O: generating infrared remote control signals, inputting magnetic card reader signals, and generating BSR X-10 signals.

The familiar infrared remote control controls TV sets and other home electronics. A rather large number of deliberately different control formats are used, so that commands given to one unit won't be inadvertently obeyed by another. A typical format sends data least significant bit first. For an F (0), `PORTA` bit 0 is H (1) for 350 μs; for T (1) it is H (1) for 700 μs. See Figure 5.5. An H (1) causes a 555 to generate a 38-KHz square-wave. See Figure 5.5c. The procedure `sendIr` sends 11 bits to the infrared LED.

a. Remote Control

b. Receiver Module

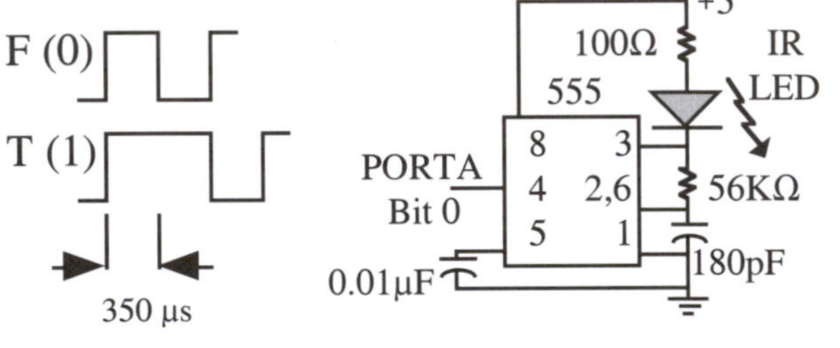

c. Infrared Pulses

d. Infrared LED Hardware

Figure 5.5. Infrared Control

```
void sendIr(int data) { char i;
   for(i = 0, DDRA = 1; i < 11; i++) {
      PORTA |= 1; if(data & 1)wait();
      wait(); PORTA &= ~1; wait(); data >>= 1;
   }
}

void wait() {int i = N; while(--i) ;} /* N is 350 * 8/3 to wait 350 ns */
```

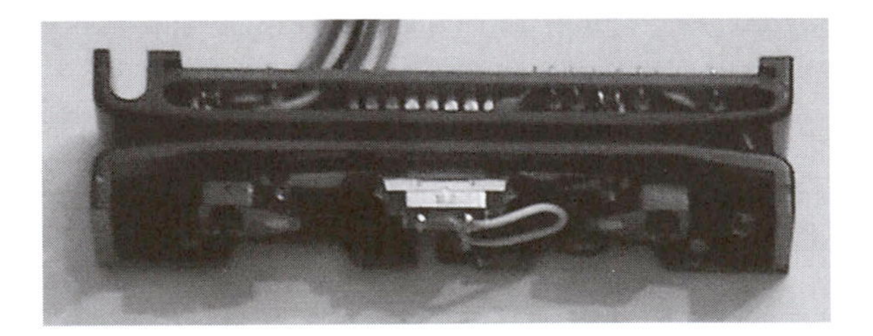

a. Hardware

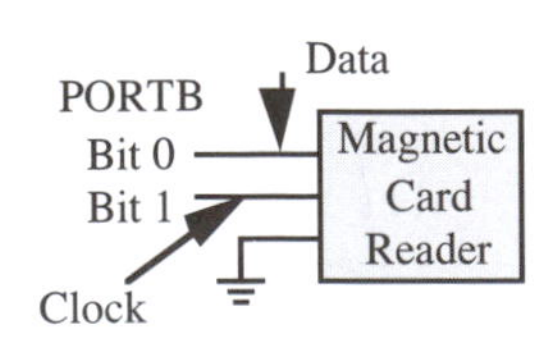

b. Pin Connections

Figure 5.6. Magnetic Card Reader

The also familiar magnetic credit card reader generates a falling-edge clock as it inputs data bits serially, most significant bit first. See Figure 5.6. The procedure `receiveMagCard` receives 16 bits serially, through `PORTB` bit 0.

```
int receiveMagCard (){ int i, data;
   for(i = 0, DDRB = 0; i < 16; i++) {
      while( ! (PORTB & 2)) ; while(PORTB & 2) ;
      data = (data << 1) | (PORTB & 1);
   }
   return data;
}
```

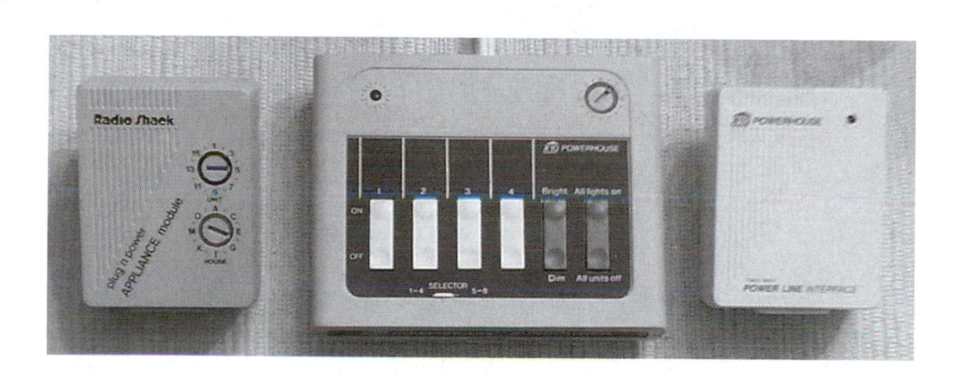

a. BSR Receiver b. BSR Controller c. TW523 Module

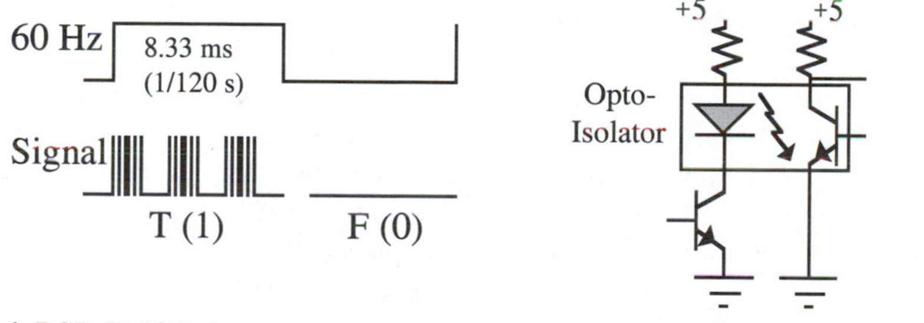

d. BSR X-10 Pulses e. BSR X-10 Transmitter

Figure 5.7. BSR X-10

An *X-10* receiver (Figure 5.7a) and controller (Figure 5.7b), originally designed by BSR Inc., and now distributed by Radio Shack, Sears, and other mail-order houses, controls lamps and appliances in the home via a signal sent over household 110-V, 60-Hz power wiring. One or more controllers and modules, each costing about $15, are plugged into power-wiring sockets, without the need for any other wiring. A controller sends commands as bursts of 100-KHz signals over the power wiring to the modules (Figure 5.7d). One bit is sent each half-cycle of the 60-Hz signal. An F (0) bit is no burst, and a T (1) bit is three 100 KHz bursts evenly spaced in the half cycle. The TW523 module (Figure 5.7c), uses three optical isolators (*opto-isolators*), described in Figure 5.7e. We will use it in §5.2.6.5. Here, we show how to generate a modulated signal using the capabilities of the 6812, essentially what is done inside the TW523. The procedure `sendBsr` sends 16 bits through `PORTB` bit 0, synchronized to `PORTB` bit 1's 60-Hz waveform.

```
void pulse(){ asm{
            ldaa #10 ; ten pulses
      L1:   bset PORTB,#1
            ldx #5 * 8 / 3
      L2:   dbne x,L2
            bclr PORTB,#1
            ldx #5 * 8 / 3
      L3:   dbne x,L3
            dbne a,L1
            ldx #(8000000 / (360*3)) - 20 * ( 5 * 8 / 3) ;
      L4:   dbne x,L4
}}

void sendBsr (int data) { int i;
  while(PORTB & 2) ; /* make sure we start on an edge */
  for(i = 0, DDRB = 1; i < 16; i++) {
    while( ! (PORTB & 2)) ; /* wait for next 1/2 cycle */
    if(data & 0x8000) { pulse(); pulse(); pulse(); } data <<= 1;
    while(PORTB & 2) ; /* wait for next 1/2 cycle */
    if(data & 0x8000) { pulse(); pulse(); pulse(); } data <<= 1;
}}

void main(){ sendBsr (0x55aa); }
```

`sendBsr` illustrates some interesting techniques. We wait for either edge of the 60-Hz square wave because bits are sent each half cycle. The top half and bottom half of `sendBsr` differ only in which edge the gadfly loop is waiting for. If a T (1) is sent, then the procedure `pulse` is executed three times; if a F (0) is sent, then `pulse` is not executed. The procedure `pulse` illustrates the use of embedded assembly language to produce 10 cycles of 100-KHz square wave and a small delay, which is written in assembly language since we are in it.

5.1.5 Object-oriented Classes for Programmed I/O

An object-oriented class `SyncPort` for real-time and gadfly synchronization mechanisms, derived from §4.1.2's generic `Port` class, is presented below. It is generalized to accommodate the handshake functions of the 6811 StrA and StrB pins and associated control logic, which are not implemented in hardware in the 6812. There are two parts to this general synchronization mechanism. In the first, an input signal edge indicates a device's transition from BUSY to DONE. The second part is an output signal, which can provide a pulse or a level handshake signal that prods the device to do something. We present the class and its function members, then we show subroutines it uses to take care of handshake and synchronization, and finally we show this class's uses.

```
template <class T> class SyncPort : public Port<T> {
    protected : long syncK, handK;

    public:SyncPort(int a,long syncP,long handP,int attr=0, T dir=0):
        Port(a, attr, dir) {
            orAt(syncK = syncP); if(handP & 0x8000) orAt(handP);
            orAtNext(handK = handP); // set direction for output
    }

    virtual T get(void){
        storeAt(syncK);             // clear flag
        handshakeBegin(handK);  // change level, or pulse the handshake signal
        synchronize(syncK);        // delay or gadfly loop
        T c = Port<T>::get();     // get data form input port
        handshakeEnd(handK);       // change level, or pulse the handshake signal
        return c;
    }

    virtual void put(T c) {
        storeAt(syncK);             // clear flag
        handshakeBegin(handK);  // change level, or pulse the handshake signal
        synchronize(syncK);        // delay or gadfly loop
        Port<T>::put(c);            // put data to output port
        handshakeEnd(handK);       // change level, or pulse the handshake signal
} };
```

To accommodate various synchronization and handshake options, we initialize two `long` words, `syncK` and `handK`, in `SyncPort`'s constructor; these `long` words take advantage of HIWARE's convention for passing a *long* word as a procedure argument, which puts the high 16 bits into index register X and the low 16 bits into accumulator D. The constructor saves its middle two *long* parameters in *long* data members `syncK` and `handK`, and passes the other parameters to its base class's constructor.

The `handK` parameter's high-order 16 bits are generally loaded into X and therefore contain the address of a handshake port such as `PORTA`, while the low-order 8 bits are generally loaded into accumulator B, which will be a bitmask that is stored into, ORed into, or complemented and ANDed into the port at address X. The bitmask will be all 0s, except that one bit will be 1. The remaining bits of the parameter, which are generally loaded into accumulator A, are used to control when the handshake signal is changed; setting the most significant bit put into accumulator A makes the handshake bit initially high, otherwise it is initially low. Setting the next bit causes a direction register at an address two locations higher than the handshake port address to be set; clearing that bit causes a direction register at the next higher address to be set. That bit should be set when ports A, B, C, D, F, or G are used for the handshake signal, and cleared for all other ports. The remaining bits are used each time `put` or `get` are called. Setting the bit loaded into accumulator A bit 5 causes the handshake signal to rise, setting the next bit loaded into accumulator A causes the handshake signal to fall, and setting the next bit causes the handshake signal to rise before the synchronization operation is performed. Setting the bit loaded into accumulator A bit 2 causes the handshake signal to rise, setting the next bit loaded into accumulator A causes the handshake signal to fall, and setting the least significant bit causes the handshake signal to rise after the synchronization operation is performed and the data is input or output. These operations are done by the assembly-language subroutines shown below:

```
#pragma NO_RETURN
void orAtNext(long handK){asm{ // to set direction to output
        bita   #$40    ; if bit 14 is set, increment twice
        beq    L0      ; for PORTs A to D, and F and G
        inx            ; otherwise increment once
  L0:   inx            ; In either case, fall through to orAt
}}

void orAt(long handK){asm{
        orab   0,x     ; set a bit at the port address
        stab   0,x
}}

#pragma NO_RETURN
void handshakeBegin(long handK){asm {
        lsra           ; called before synchronization is done, move middle
        lsra           ; three bits over to be tested in handshakeEnd
        lsra           ; to govern handshake before synchronization; fall through
}}

void handshakeEnd(long handK){asm {
        bita   #4      ; called before synchronization; bit 2 of third byte controls
        beq    L1      ; rise
        pshb
        orab   0,x     ; OR least significant byte with port data
        stab   0,x     ; put back
        pulb
  L1:   bita   #2
        beq    L2      ; bit 1 controls fall
        pshb
        comb           ; invert mask bits so 1 clears
        andb   0,x     ; AND with port data
        stab   0,x     ; put back
        pulb
  L2:   bita   #1
        beq    L3      ; bit 0 controls rise
        orab   0,y     ; OR with port data
        stab   0,y     ; put back
  L3:   ds.b   0       ; terminate
}}
```

The *long* data member *syncK* controls synchronization for real time, gadfly using a key wakeup port, gadfly using an input port bit, and time-sharing synchronization, which will be discussed in §5.3.2. Using it, this class accommodates all the synchronization techniques except for interrupts using a buffer or queue (and techniques for very fast devices discussed in §5.4). If the byte loaded into accumulator A is 0xff – bits 15 to 8 of the constant *syncK,* then its high 16 bits loaded into X are a key wakeup port address, such as *KWIFJ,* and its low byte, loaded into accumulator B, is the bitmask to test and clear the flag. If the byte loaded into accumulator A is 0xfe or 0xfd, then the high 16 bits are a synchronization port address, such as *PORTA,* and the low byte loaded into accumulator B is the bitmask to test that port. If the byte loaded into accumulator A is 0xfe, then data transfer waits for a falling edge of the bit that is 1 in the mask; and if 0xfd, then data transfer waits for a rising edge of this bit. If the byte loaded into accumulator A is 0xfc, and real-time interrupt synchronization is used (as will be discussed in §5.3.2) then a *sleep* procedure is called, with a sleep time that is the high 16 bits loaded into X. We discuss this further in §5.3.5. Finally, if the byte loaded into accumulator A is less than 0xfc, then real-time synchronization is used, delaying approximately *n* loops where a loop, which consists of a `dbne` instruction, executes in 0.375 ns if the 6812 uses an 8-MHz clock; *n*'s high 16 bits are the value in accumulator D minus 1, and the low 16 bits of *n* is in X. Conversely, to delay *n* loop times, make the constant *syncK* be [(*n* << 16) + 0x10000] | (*n* >> 16). The following subroutines implement the various synchronization techniques.

```
void storeAt(long syncK) { asm{
        ibne a,L1  ; used only for key wakeup synchronization, where a is 0xff
        stab 0,x   ; write 1 into the clear port (e.g., KWIFJ), to clear the flag
   L1:  ds.b 0
}
}

char testAt(long syncK) { asm{
        andb 0,x   ; check for 1 where the mask bit is 1
        rts
}
    return 0; // keep compiler happy
}

void orAtPrevious(long syncK){asm{
        orab  -1,x    ; to set the int. enable from the address of the flag port
        stab  -1,x    ; to enable key wakeup interrupts
}
}

void andAtPrevious(long syncK){asm{
        comb          ; to clear int. enable from the address of the flag port
        andb  -1,x    ; to disable key wakeup interrupts
        stab  -1,x
}
}
```

```
char testAtPrevious(asm{
          andb -1,x     ; check for 1 where the mask bit is 1
          rts
}
     return 0; // keep compiler happy
}

void synchronize(long syncK) { asm{
          cpd    #0xfc00; note: X is high 16 bits, D is low 16 bits of syncK
          bhi    L1       ; branch out for all but real-time delay synchronization
          beq    L7       ; 0xfc00xxxx is to synchronize using sleep()
   L0:    dbne   x,L0     ; real-time synchronization: wait for m to count down
          dbne   d,L0     ; xxxx is counted down first, the dddd is counted down.
          rts
   L7:    stx numTicks  ; store high 16 bits in a global variable, to pass it to sleep
   }
//   sleep(numTicks); return; // insert only if sleep is defined using multithreads
   asm{
   L1:    cmpa   #$0xfe ; other than real-time delay, X is location, b is bit pattern
          beq    L3       ; for aaaaffbb where third byte is 0xff
          bhi    L5       ; key wakeup gadfly using pattern bb at aaaa
   L2:    bitb   0,x      ; wait until flag is set
          beq    L2
          rts

   L3:    bitb   0,x      ; where third byte is 0xfe
          beq    L3       ; wait for bit to become high
   L4:    bitb   0,x      ; wait for bit to become low
          bne    L4
          rts

   L5:    bitb   0,x      ; where third byte is 0xfd
          bne    L5       ; wait for bit to become low
   L6:    bitb   0,x      ; wait for bit to become high
          beq    L6
}}
```

We now redo the examples in §5.1.3 to use the class `SyncPort`. First, consider the real-time paper-tape reader having data input on `PORTA`, using handshake signal on `PORTB` bit 0, which is pulsed high momentarily to request data. The constructor's first argument is the address of the data port, which is *aA*. Its second argument is the synchronization parameter loaded into `syncK`; real-time synchronization requires this to be the number of loops we desire to execute, rearranged to use the `synchronize` procedure shown earlier. We can use the expression `((n << 16) + 0x10000) | (n >> 16), (iB << 16)` developed earlier to generate it. The next argument is the handshake parameter loaded into `handK;`. Its most significant 16 bits are the address of the parallel port used, port B; its least significant 8 bits are the mask for the bit to be used for handshake; and the remaining byte designates the way this signal changes. Its direction register is two locations higher than its data port address, and it is pulsed high before synchronization, so its least significant 8 bits are binary 10110000 or 0xB0.

```
#define N 100 /* the number of dbne loops to be executed */
SyncPort p(aA,((N<<16) + 0x10000) | (N>>16), (aB << 16) | 0xB001);

void main() {char buffer[0x100]; p.get(b, 0x100) }
```

Note that since the class's function members do all the work, the program itself is trivially simple. Now consider our gadfly paper-tape reader whose data are received on `PORTA`, using `PORTB` bit 1 for synchronization, which pulses low before data can be read, and whose handshake signal on `PORTB` bit 2 is normally high, asserted low to input data, and returned high after the data is read. The constructor's first argument is the address of the data port which is `aA`. Its second argument is the synchronization parameter loaded into `syncK;`. Gadfly-time synchronization for `PORTB` bit 1's rising edge requires the most significant 16 bits be `PORTB`'s address, its least significant 8 bits be the mask for bit 1, 0x02, and the other bits be 0xfd. The next argument is the handshake parameter loaded into `handK`. Its most significant 16 bits are the address of the parallel port used, `PORTB`; its least significant 8 bits are the mask for bit 2, 0x04, to be used for handshake, and the other byte designates the way this signal changes. Its direction register is two locations higher than its data port address, it is initially high, and it becomes low before synchronization and goes high after data are recieved, so it is binary 11010100 or 0xD4. Note that this `main` is identical to the `main` procedure for real-time synchronization because the class's function members do the work.

```
SyncPort p(aA, (aB << 16) | 0xfd02, (aB << 16) | 0xD404) ;

void main() {char buffer[0x100]; p.get(b, 0x100); }
```

Finally, consider a gadfly paper-tape reader synchronizing using `KWIFJ` bit 7, and the same handshake as the previous example used. The constructor's first argument is the address of the data port, which is again *aA*. Its second argument is the synchronization parameter loaded into `syncK`. Gadfly-time synchronization for key wakeup J bit 7 requires the most significant 16 bits be `KWIFJ`'s address, 0x2B; its least significant 8 bits be the mask for bit 7, 0x80; and the other bits be 0xff. The next argument, the handshake parameter, is the same as in the previous example. The `main` procedure, of course, remains the same as in the other examples. Note how easy it is to switch from one port to another for data, synchronization, or handshaking.

```
SyncPort p(aA, 0x2BFF80, (aB << 16) | 0xD404) ;

void main() {char buffer[0x100]; p.get(b, 0x100) }
```

These examples suggest that a library of devices and their associated classes can be written and debugged in advance, and copied into the application late in the design. Object-oriented programming separates the I/O software from the main procedure to make such a substitution cleanly and correctly, without side effects.

5.2 Interrupt Synchronization

In this section, we consider interrupt hardware and software. *Interrupt* software can be tricky. At one time, some companies actually have a policy never to use interrupts, but instead use the gadfly technique. At the other extreme, some designers use interrupts just because they are readily available in microcomputers like 6812 systems. We advocate using interrupts when necessary but using simpler techniques whenever possible.

Interrupt techniques can be used to let the I/O system interrupt the processor when it is DONE, so the processor can be doing useful work until it is interrupted. Also, latency times resulting from interrupts can be less than latency times resulting from a large gadfly loop that tests many I/O devices, or a variation of the gadfly approach, whereby the computer executes a procedure, checks the I/O device, next executes another procedure, and then checks the devices, and so on – and that can be an important factor for fast I/O devices. Recall the basic idea of an interrupt from §1.2.3: a program P currently being executed can be stopped at any point; then a handler program D is executed to carry out some task requested by the device, and the program P is resumed. The device must have some logic to determine when it needs to have the processor execute the D program. P must execute the same way whenever and regardless whether D is executed. Hardware saves all the information that P needs to resume without error.

We first look at steps in an interrupt. Then we consider interrupt handlers, and the accommodation of critical sections. In the next subsections, details will be discussed as the multiple-interrupt case will be studied, using two techniques called polling and vectored interrupts. These will be simple extensions of the single-interrupt case.

5.2.1 Steps in an Interrupt

We now consider a 6812 microcontroller having just one interrupt, omitting some details in order to concentrate on the principles of the single interrupt. A complete subsection is provided to show how a key wakeup interrupt occurs in the 'A4 and another describes how an IRQ interrupt occurs in the 'B32. You need to read only one of these subsections appropriate to your equipment. However, if you study both mechanisms, you can generalize the concept of interrupt handling, which might apply differently in various machines that you may study and use later in your work.

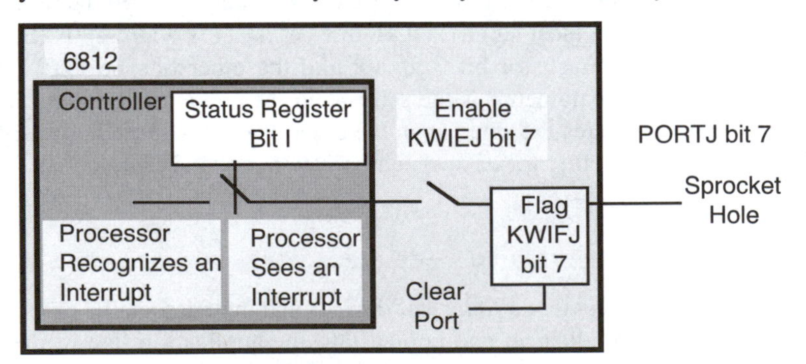

Figure 5.8. Key Wakeup Interrupt Request Path

5.2.1.1 Steps in an Interrupt in the MC68HC812A4

We consider a key wakeup interrupt in the 'A4 microcomputer. To use one of the key wakeup inputs for interrupt requests, the port's direction bit for that input must be clear. If the paper tape were pulled though the reader, the sprocket hole signal would be a train of 256 pulses that falls and then rises each time a hole is under the sensor. A falling edge causes a key wakeup interrupt. See Figure 5.8. The six-step sequence of actions leading to an interrupt and servicing it is outlined below.

1. When the external hardware determines it needs service either to move some data into or out of it or to report an error, we say that the device *requests an interrupt.* This occurs in the paper-tape reader when `PORTJ` bit 7 falls.

2. If the `PORTJ` bit 7 pin is an input (in `DDRJ` bit 7) and had been assigned (in `KWIEJ` bit 7) to sense interrupts, we say `KWIEJ` bit 7 *interrupts are enabled.*

3. If the microprocessor's condition code register's I bit is 0 we say the *microprocessor is enabled* for all non-XIRQ interrupts. When I is 1, the *microprocessor is masked* (or the *microprocessor is disabled*) for all non-XIRQ interrupts. If a signal from (any) device is sent to the controller, we say the *microprocessor sees a request,* or the *request is pending,* and an interrupt will occur, as described next. (The bit I is also controlled by hardware in the next step; also, XIRQ interrupts are handled differently.)

4. Most microcomputers cannot stop in the middle of an instruction. Therefore, if the microprocessor recognizes an interrupt, it *honors an interrupt* at the end of the current instruction. When the 6812 honors a key wakeup J interrupt, it saves the registers and the program counter on the stack, sets the condition code register I bit, and loads the 16-bit word at 0xffd0 into the program counter to process this interrupt. Importantly, condition code bit I is set after the former I was saved on the stack.

5. Location 0xffd0 contains the address of the key wakeup *handler.* The handler is like a "subroutine" that performs the work requested by the device. It may move a word between the device and a buffer, or it may report or fix up an error. One of a handler's critically important but easy-to-overlook functions is that it must explicitly remove the cause of the interrupt (by negating the interrupt request) unless the hardware does that for you automatically. This is done by writing 1 into bit 7 of the `KWIFJ` port.

6. When it is completed, the handler executes an `RTI` instruction; this restores the registers and program counter to resume the program where it left off.

The handler's address must be put in a 16-bit word in high memory where the interrupt mechanism reads it. Using the HIWARE C++ compiler (similar mechanisms will have to be used with other compilers), all we need to do is write `interrupt 23` before the procedure to be used as the key wakeup port J `handler`. This convention inserts the address of the handler into 0xffd0, and further ends the "procedure" with an `RTI`. Alternatively, the HIWARE linker can insert the address in high memory. The line

VECTOR ADDRESS 0xFFD0 handler

can be put at the end of the linker's `.prm` file. It puts the address of the key wakeup J handler into the word at 0xffd0. In order to use this technique, *handler* must be a C (not C++) procedure, preceded by the keyword *interrupt* so it ends in `RTI`, and must be declared *extern* so the linker will be able to find it. If you are using C++ write

```
extern "C" {extern interrupt void handler(void)
       {... (the body of the handler ) ... }}
```

However, this 16-bit word is in EEPROM in the 'A4, which is written using procedures given in §6.3. For DBUG12, its handler at that address jumps indirectly through an SRAM location 0xb10, so for it, execute `*(int*)0xb10=(int)handler;`, or, to use the interrupt number (23), execute `((int *)0xb3e}[-23]=(int)handler;` instead.

5.2.1.2 Steps in an Interrupt in the MC68HC912B32

We now consider a key wakeup interrupt in the 'B32. If the paper tape were pulled though the reader, the sprocket hole signal would be a train of 256 pulses, which falls and then rises each time a hole is under the sensor. This signal is input to the IRQ pin. A falling edge causes an IRQ interrupt. See Figure 5.9. The six-step sequence of actions that lead to an interrupt and that service it is outlined as follows.

1. When the external hardware determines it needs either to move some data into it or out of it or to report an error, we say a device *requests an interrupt.* This occurs in the paper-tape reader when the signal on the `IRQ` pin falls.

2. If *INTCR* bits 6 and 7 are set, we say `IRQ` interrupts are *enabled* (for falling edges).

3. If the microprocessor's condition code register's I bit is 0, we say the *microprocessor is enabled* for all non-XIRQ interrupts. When I is 1, the *microprocessor is masked* (or the *microprocessor is disabled*) for all non-XIRQ interrupts. If a signal from (any) device is sent to the controller, we say the *microprocessor sees a request,* or the *request is pending,* and an interrupt will occur, as described next. (The bit I is also controlled by hardware in the next step. Also, XIRQ interrupts are handled differently.)

4. Most microcomputers cannot stop in the middle of an instruction. Therefore, if the microprocessor recognizes an interrupt, it *honors an interrupt* at the end of the current instruction. When the 6812 honors an IRQ interrupt, it saves the registers and the program counter on the stack, sets the condition code register I bit, and loads the 16-bit word at 0xfff2 into the program counter to process this interrupt. Importantly, condition code bit I is set after the former I was saved on the stack.

5. Location 0xfff2 contains the address of the IRQ `handler`. The handler is like a subroutine that performs the work requested by the device. It may move a word between the device and a buffer, or it may report or fix up an error. An IRQ handler need not "remove the source of the interrupt"; this flip-flop clears automatically.

6. When it is completed, the handler executes an `RTI` instruction; this restores the registers and program counter to resume the program where it left off.

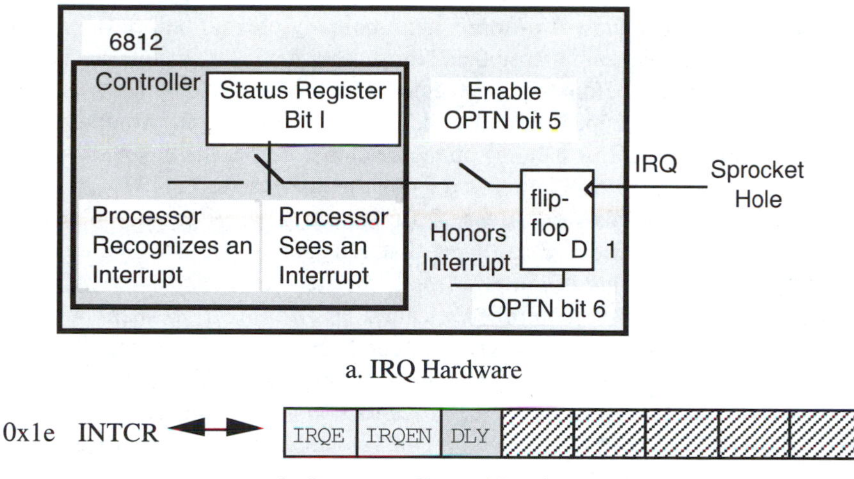

a. IRQ Hardware

b. Interrupt Control Register

Figure 5.9. IRQ Interrupt Request Path

In the 'B32, the handler's address must be put in a 16-bit word at 0xfff2 where the interrupt mechanism reads it. Using the HIWARE C++ compiler (similar mechanisms will have to be used with other compilers) all we need to do is write `interrupt 6` before the procedure to be used as the IRQ `handler`. This convention inserts the address of the handler into 0xfff2, and further ends the procedure with an `RTI`. Alternatively, the HIWARE linker can insert the address in high memory. The line

VECTOR ADDRESS 0xFFF2 handler

can be put at the end of the linker's .prm file. It puts the address of the IRQ handler into the word at 0xfff2. To use this technique, *handler* must be a C (not C++) procedure, preceded by the keyword *interrupt* so it ends in `RTI`, and must be declared *extern* so the linker will be able to find it. If you are using C++ write:

```
extern "C" {extern interrupt void handler(void)
      {... (the body of the handler ) ... }}
```

However, this 16-bit word is in flash memory in the 'B32, which is written using procedures given in §6.3. For DBUG12, its handler at that address jumps indirectly through an SRAM location 0xb32, so for it, execute `*(int*)0xb32 = (int)handler;` or to use the interrupt number (6), execute `((int *)0xb3e)[-6] = (int)handler;` instead.

5.2.1.3 Properties of Interrupt Synchronization

We stress that as soon as the 6812 honors an interrupt, it sets the I condition code to prevent repeatedly honoring the same interrupt. If it didn't, the handler's first instruction would be promptly interrupted – an infinite loop that will fill up the stack. Do not worry about clearing I because the preceding sequence's step 6 restores the condition code register, including the I bit, to its value before the interrupt was honored.

The IRQ interrupt request flip-flop is automatically cleared before the handler is executed. The 'A4 key wakeup interrupt, and most other device interrupts, needs to clear the source of the interrupt before `RTI` is executed. For similar devices, if `RTI` is executed and the source of the interrupt is not cleared, this same device will promptly interrupt the processor again and again – hanging up the machine. Before the key wakeup handler executes `RTI` or explicitly clears I, it *must* remove the interrupt source!

An interrupt request takes two steps. A flip-flop triggers if an edge occurs on its clock input. A switch in series with an input that can set this flag is called an *arm;* if it is closed the device is *armed,* and if opened, the device is *disarmed.* Any switch between the flip-flop and the controller is called an *enable;* if all such switches are closed the device is *enabled* and if any is opened, the device is *disabled.* Arming a device lets it record a request to request an interrupt, either immediately if it is enabled or later if it is disabled. Disarming a device prevents honoring an interrupt now or later. Disable an interrupt to postpone it, if you can't honor it now; but you may honor it when interrupts are enabled at a later time.

The programs above exhibits the IDLE, BUSY, and DONE states. IDLE is any time before the procedure executes, and DONE is when an interrupt occurs. The device is BUSY when waiting for an interrupt. Interrupt exhibits longer (worst-case) latency than gadfly synchronization because the instruction must be completed, the registers must be saved on the stack, and a few instructions in the handler must be executed before the device gets served (by reading data from the port or writing data to the port). The longest instruction, `EMACS`, might have just begun execution when an interrupt occurs, which adds 13 cycles to the latency. Further, if other devices use interrupts, a device's interrupt can be disabled for a long time while executing the other device's handler. Interrupt synchronization requires more hardware, such as a line (and pin) for the interrupt request and a gate (and flip-flop) to enable the interrupt. However, interrupt synchronization can avoid the loop while the device is in BUSY, and that can be a long time for slow I/O devices. This can free the processor to do other things while it waits for DONE.

5.2.2 Interrupt Handlers and Critical Sections

Three techniques are generally used in interrupt handlers: altering a global variable, writing into a buffer, and pushing into a queue. These are illustrated with the paper-tape example. We then discuss the problem of critical sections and how to manage them.

5.2.2.1 A Handler That Changes a Global Variable

First, we use a global variable *`char flag`*, which the main program gadflies upon. *`main`* inhibits interrupts using the `SEI` instruction. *`main`* then clears the buffer's index, and enables and clears the key wakeup interrupt. Interrupts are then enabled, using `CLI`. `SEI` and `CLI` take care of a critical section, as discussed at the end of this subsection. *`flag`* is cleared. Upon each falling edge, indicating the arrival of another pattern, an interrupt occurs and *`flag`* is incremented. Each time *`main`* sees *`flag`* nonzero, input data is transferred into the buffer. Then the key wakeup interrupt is disabled and we exit.

The first subsection shows a program whose interrupt handler sets a flag variable in the 'A4. Essentially the same program is written for the 'B32. The reader can study the one appropriate to his or her hardware, without having to read the other.

For the 'A4

```
char flag;

interrupt 23 void keyInt() { KWIFJ = 0x80; /* clear key */ flag++; };

void main() { char buffer[10], i;
        asm sei                 /* disable all non-XIRQ interrupts */
        KWIEJ =                 /* enable key wakeup interrupt */
        KWIFJ = 0x80;           /* clear flag */
        asm cli                 /* enable all non-XIRQ interrupts */
        for(DDRA = i = 0; i < 10; i++) {
                flag = 0;       /* prepare for next wait for interrupt */
                while(flag == 0) ;      /* wait for interrupt */
                buffer[i] = PORTA;      /* store data */
        }
        KWIEJ &= ~0x80;  /* disable key wakeup interrupt only */
}
```

The interrupt handler is coded in assembly language as follows.

```
0000086A C680          LDAB  #128
0000086C 5B2B          STAB  $2B
0000086E 720B40        INC   $0B40
00000871 0B            RTI
```

For the 'B32

```
char flag;

interrupt 6 void handler() { flag++; };

void main() { char buffer[10];
        asm sei                 /* disable all non-XIRQ interrupts */
        DDRA = 0;               /* make port A input */
        INTCR = 0xC0;           /* enable IRQ interrupt on falling edges */
        asm cli                 /* enable all non-XIRQ interrupts */
        for(unsigned char i = 0; i < 10; i++) {
                flag = 0;       /* prepare for next wait for interrupt */
                while(flag == 0) ; /* wait for interrupt */
                buffer[i] = PORTA; /* store data */
        }
        INTCR = 0; /* disable IRQ interrupt only */
}
```

The interrupt handler is coded in assembly language as follows.

```
0000086E 720B40        INC   $0B40
00000871 0B            RTI
```

5.2.2.2 A Handler That Fills or Empties a Buffer

This first technique really just transfers the synchronization from interrupt-based hardware to gadfly-based software. The second common technique lets the interrupt handler store data in the buffer. `main` will set up the buffer as in the first example, but the handler will fill it. `main` will wait for the buffer to be filled, gadflying on its index.

`main` sets up the index to the buffer and a key wakeup request interrupt as in the previous example. It then waits for the buffer's index to reach its end. We illustrate first the 'A4 interrupt and then the 'B32 interrupt mechanism.

For the 'A4

`PORTJ` bit 7's falling edge occurs each time another paper-tape sprocket hole comes under the reader. This causes the interrupt handler to be entered, which removes the source of the interrupt and reads the pattern into the next buffer element. When all data has arrived, the key wakeup interrupt is disabled. This gadfly loop waits for the buffer to be DONE, rather than for the arrival of each byte, as in the previous example.

```
unsigned char buffer[10], index;

 interrupt 23 handler() {
    KWIFJ = 0x80; /* clear flag */ buffer[index++]=PORTA; /* store data */
    if(index == 10) KWIEJ &= ~0x80;
  }

void main() {
    asm sei
    index = 0;                    /* set buffer pointer to beginning */
    KWIEJ = 0x80; KWIFJ = 0x80;   /* enable key wakeup int., clear flag */
    asm cli                       /* enable all non-XIRQ interrupts */
    while(index < 10) ;           /* wait for buffer to be done */
}
```

This interrupt handler is coded in assembly language as follows.

```
0000086A C680          LDAB  #128
0000086C 5B2B          STAB  $2B
0000086E B60B40        LDAA  $0B40
00000871 36            PSHA
00000872 42            INCA
```

```
00000873 7A0B40        STAA  $0B40
00000876 32            PULA
00000877 B705          SEX   A,X
00000879 9600          LDAA  $00
0000087B 6AE20B41      STAA  2881,X
0000087F B60B40        LDAA  $0B40
00000882 810A          CMPA  #10
00000884 2603          BNE   *+5    ;abs = 0889
00000886 4D2A80        BCLR  $2A,#128
00000889 0B            RTI
```

For the 'B32

The 'B32 mechanism is shown next; its assembly language is similar to the above. *IRQ*'s falling edge occurs each time another paper-tape sprocket hole comes under the reader. This causes the interrupt handler to be entered, which reads the pattern into the next buffer element. When all data has arrived, the key wakeup interrupt is disabled. This gadfly loop waits for the buffer to be DONE, rather than for the arrival of each byte.

```
unsigned char buffer[10], index;

 interrupt 6 handler() {
   buffer[index++]=PORTA; /* store data */ if(index==10) INTCR=0;}

void main() {
   asm sei
   index = 0;                    /* set buffer pointer to beginning */
   INTCR = 0xc0;                 /* enable int. */
   asm cli                       /* enable all non-XIRQ interrupts */
   while(index < 10) ;           /* wait for buffer to be done */
}
```

This interrupt handler is coded in assembly language as follows.

```
0000086E B60B40        LDAA  $0B40
00000871 36            PSHA
00000872 42            INCA
00000873 7A0B40        STAA  $0B40
00000876 32            PULA
00000877 B705          SEX   A,X
00000879 9600          LDAA  $00
0000087B 6AE20B41      STAA  2881,X
0000087F B60B40        LDAA  $0B40
00000882 810A          CMPA  #10
00000884 2603          BNE   *+5    ;abs = 0889
00000886 79001E        CLR   $1E
00000889 0B            RTI
```

5.2.2.3 A Handler That Uses a Queue for Input

The third technique's interrupt handler pushes data acquired from the input port into a queue until `main` pops the data to use it. Since the programs for the 'A4 and 'B32 are quite similar, we use this opportunity to show how a program can be maintained for two different applications or environments. This program uses a symbolic name *A4* to compile the program for the 'A4; otherwise, it is compiled for the 'B32. Note how conditional compilation can be used to simultaneously maintain software for two different applications or environments. Do not be confused between the queue, which is a temporary storage place to hold data from the handler to the main program, and the buffer which eventually stores the data. `main` just stores data into the buffer, but a typical program analyzes data, or uses it in some way, as it pops it from the queue.

```
unsigned char d[5], size, top, bot, error, buffer[10];

unsigned char pull() {
    if((size--) <= 0) {error = 1; return 0; } if (bot == 5) bot = 0;
    return( d[bot++] );
}
void push(unsigned char item) {
    if((size++) >= 5) { error = 1; return 0; } if (top == 5) top = 0;
    d[top++] = item;
}
unsigned char get() { while(!size) ; return pull(); }
#ifdef A4
interrupt 23 handler() {KWIFJ = 0x80; push(PORTA); } /* clear flag */
#else
interrupt 6 handler() { push(PORTA); }
#endif

void main() { unsigned char i;
    asm sei
#ifdef A4
    KWIEJ = 0x80; KWIFJ = 0x80;
#else
    INTCR = 0xc0;                         /* enable irq interrupt */
#endif
    asm cli /* set up interrupt */
    for(i = 0; (i < 10) && ! error; i++)
        buffer[i] = get(); /* gather data */

#ifdef A4
    KWIEJ &= ~0x80;                       /* disable key wakeup interrupt only */
#else
    INTCR = 0;                            /* disable irq interrupt only */
#endif

}
```

The interrupt handler, for the case where `A4` is `#define`d, is coded as

```
000008F4 C680        LDAB  #128
000008F6 5B2B        STAB  $2B
000008F8 D600        LDAB  $00
000008FA 07BD        BSR   *-65    ;abs = 08B9
000008FC 0B          RTI
```

While this handler appears to be simpler than the previous example's handler, it calls the `push` procedure. Its rather long `push` procedure is coded in assembly language as

```
000008B9 37          PSHB
000008BA F60805      LDAB  $0805
000008BD 37          PSHB
000008BE 52          INCB
000008BF 7B0805      STAB  $0805
000008C2 33          PULB
000008C3 C105        CMPB  #5
000008C5 2507        BCS   *+9    ;abs = 08CE
000008C7 C601        LDAB  #1
000008C9 7B0808      STAB  $0808
000008CC 201C        BRA   *+30    ;abs = 08EA
000008CE F60806      LDAB  $0806
000008D1 C105        CMPB  #5
000008D3 2603        BNE   *+5    ;abs = 08D8
000008D5 790806      CLR   $0806
000008D8 F60806      LDAB  $0806
000008DB 37          PSHB
000008DC 52          INCB
000008DD 7B0806      STAB  $0806
000008E0 33          PULB
000008E1 87          CLRA
000008E2 B745        TFR   D,X
000008E4 E680        LDAB  0,SP
000008E6 6BE20800    STAB  2048,X
000008EA 32          PULA
000008EB 3D          RTS
```

As in the two previous examples, `main` prepares the same IRQ or key wakeup interrupt. Then it calls the input procedure to get a byte at a time, and writes each byte into the buffer. However, in this example, the interrupt handler pushes the data obtained from `PORTA` onto the queue, using the `push` procedure. The `get` procedure gadflies as long as the input queue is empty; when the queue is nonempty, `get` pops an item, returning it to the caller; `main` then puts this data into its buffer. When the buffer is filled, `main` inhibits IRQ or key wakeup interrupts and the main procedure exits.

Observe how long the handler takes to run. As a general objective, we want to reduce latency time. Handlers disable interrupts throughout their execution, so they significantly increase worst-case latency for other devices. You should write handlers that are as fast as possible. Whenever you can do so, perform time-consuming operations in the main program, rather than the interrupt handler.

As in previous discussions of synchronization methods, we discuss the BUSY/DONE states of the devices using the interrupt synchronization method. The paper-tape reader is IDLE until after initialization sets it up and after `main` disables interrupts from the paper-tape reader just before it exits. It becomes BUSY after initialization. DONE is entered when an interrupt occurs and data are moved from `PORTA` to the input queue. After data are moved, the device generally returns to BUSY.

5.2.2.4 A Handler That Uses a Queue for Output

The output operation is quite similar to input in all previous cases except for the last case involving a queue. We leave the other cases as exercises for the reader, but we consider specifically our running paper-tape punch example because it illustrates two important points about interrupt software. Moreover, this example is carried out for the IRQ pin, which can be implemented in either the 'B32 or the 'A4.

The device generally causes the IRQ signal to fall if it is prepared to accept output data. The output interrupt should only be enabled when there is data to be output. Otherwise, an interrupt will occur and the handler will be entered, but there won't be anything to do because there won't be any data to output. Frustration!

In interrupt handlers using queues, the size of the output queue indicates whether or not there is data to be output. If the queue is empty, there is no data to be output and the output interrupt should be disabled; but if the queue is nonempty, there is data to be output, and the output interrupt should be enabled. The output interrupt initially is not enabled, but is enabled in `put` and disabled in the interrupt handler to follow this rule. (Observe, by contrast, that input interrupts are always left enabled.) Also a handshake, `PORTB` bit 6, is asserted exactly when the queue is nonempty, to advance the tape. We also have to terminate device use. The termination procedure should wait until the output queue is empty before it disables interrupts and terminates use of the device.

```
unsigned char d[5], size, top, bot, error, buffer[10];

unsigned char pull() {
    if((size--) <= 0) { error = 1; return 0; } if (bot == 5) bot = 0;
    return( d[bot++] );
}

void push(unsigned char item) {
    if((size++) >= 5) { error = 1; return; } if (top == 5) top = 0;
    d[top++] = item;
}

void put(char data) {
    while(size >= 5) ; push(data);
    if(size == 1) /* if queue just became nonempty */
        { INTCR = 0xc0; PORTB |= 0x40; }/* enable interrupt, turn on motor */
}
```

```
interrupt 6 void handler() {
    PORTA = pull();                    /* pull a byte and output it to the port */
    if(size == 0) INTCR = 0;           /* if queue empties, disable irq interrupts */
}

void main() { unsigned char i;         asm sei
    DDRA = 0xff; DDRB = 0x40; INTCR = 0; asm cli
    for(i = 0; (i < 10) && ! error; i++) put(buffer[i]); /* output data */
    while(INTCR) ; /* wait for queue to empty */
}
```

These procedures show some interrupt handler mechanisms, but these mechanisms are inefficient because gadflying on a global variable, such as a buffer index or a queue length, wastes time. Unless the microprocessor can do something else in the meantime, gadfly synchronization will be simpler, have lower latency, and be easier to debug.

5.2.2.5 Critical Sections

The correct management of critical sections is a very important aspect of interrupt software. A *critical section* is part of a procedure that if interrupted can cause incorrect results. For example, in a queue for an input device like a paper-tape reader, *size* is incremented in the handler when an item is pushed, and decremented in a procedure, called by *main,* when an element is pulled from the queue. *size* may be copied from its memory location to a register like accumulator B, decremented there, and put back. If an interrupt occurs between the time *size* is read until it is rewritten and the interrupt handler's *push* procedure increments *size*, then the interrupted program will cancel the change made by the handler. For instance, if *size* were initially 3, indicating there are 3 words in the output queue, and the *get* procedure pulled a word, it would decrement *size* to 2. But if, while *size* is effectively moved to accumulator B, an interrupt occurs and the interrupt handler saw a request to output a word for that interrupt request, it would push a word. It would increment the value of *size* that was in memory, changing it from 3 to 4. When the interrupted program that was decrementing *size* is resumed, it will write the number 2 in the memory variable *size.* But there are now 3 words in the queue, not 2. Subsequent queue-size checking will be faulty.

Another critical section occurs when a main program updates a global variable in parts, such as in the instruction sequence `CLR ALPHA` followed by `CLR ALPHA+1`, and the interrupt handler compares the 16-bit `ALPHA` with a constant. If the interrupt occurs after `CLR ALPHA` but before `CLR ALPHA+1`, the partially updated 16-bit `ALPHA` will be incorrectly compared to the constant in the handler.

The chances of a critical-section fault happening are actually very small. But if you believe in Murphy's law, such an error will occur at the worst possible time. Therefore, to write correct programs, you must avoid any possibility of such a critical-section error.

Critical-section errors can be correctly avoided by inhibiting interrupts in main program segments that change variables that are changed, read, or tested in an interrupt handler. Before a potential critical section is being executed, the condition code I bit is cleared; when it is left, the bit is restored to what it was before the critical section was entered. To set or clear I in a C procedure, assembly-language statements `SEI`, `CLI`,

TPA, or TAP are embedded in C. Alternatively, a memory-to-memory instruction such as CLR BETA or INC BETA, where BETA is an 8-bit variable, can avoid the possible critical-section error, because an instruction cannot be interrupted until it is completed.

The initialization of interrupt hardware and the variables, for instance counters and pointers for a queue, is often a critical section, although some study and considerable experience might be required to diagnose this problem. For instance, in the examples above, the initialization of key wakeup ports might accidentally cause an edge to set a flag flip-flop. Or the flip-flop might have been left set from prior use. The interrupt will then occur as soon as interrupts are enabled. But there is no data over the heads of the paper-tape reader, and garbage is read and entered into the buffer. This is why the initialization procedure writes 1 into the key flag register, to clear interrupt requests before interrupts are enabled. (Clearing a flag register with interrupt disabled is equivalent to having disarmed the device, as discussed earlier. This alternative is therefore called *software disarming* of the device.) Because this analysis is usually quite difficult, for any device using interrupts, we strongly advocate always disabling interrupts before, and enabling them after, an I/O device is initialized, and software-disarming the device.

5.2.3 Polled Interrupts

For multiple (polled) interrupts, the interrupt handler just finds out which interrupt request needs service. In this case, when the interrupt occurs, its handler, whose address is in 0xffd0, is executed. It *polls* the possible interrupts to see which one caused the interrupt. The polling program checks each possible interrupt, one at a time, in *priority order,* highest-priority interrupt request first, until it finds a device that requested service, executes that request, and clears that interrupt request.

5.2.3.1 Polled Interrupts in the MC68HC812A4

Consider an 'A4 having two paper-tape readers. The first reader's data and sprocket holes are on *PORTA* and *PORTJ* bit 7, respectively. The second reader's data and sprocket holes are on *PORTB* and *PORTJ* bit 6, respectively. See Figure 5.10. Either interrupt causes execution of the key wakeup J handler whose address is in 0xffd0.

```
unsigned char bufferA[10], bufferB[10], indexA, indexB;
interrupt 23 handler() {
    if(KWIFJ & KWIEJ & 0x80) /* if first flag is set and enabled */
        { KWIFJ = 0x80; bufferA[indexA++] = PORTA;} /* honor first int. */
    else if(KWIFJ & KWIEJ & 0x40) /* if second flag is set and enabled */
        { KWIFJ = 0x40; bufferB[indexB++] = PORTB;}/* honor second int. */
}
void main() { asm sei
    KWIEJ = 0xc0; KWIFJ = 0xc0; asm cli /* en. key wakeup int., clear flags */
    while((indexA < 10) && (indexB < 10)) ; /* wait for either buffer done */
    KWIEJ &= ~0xc0;                       /* disable key wakeup interrupts only */
}
```

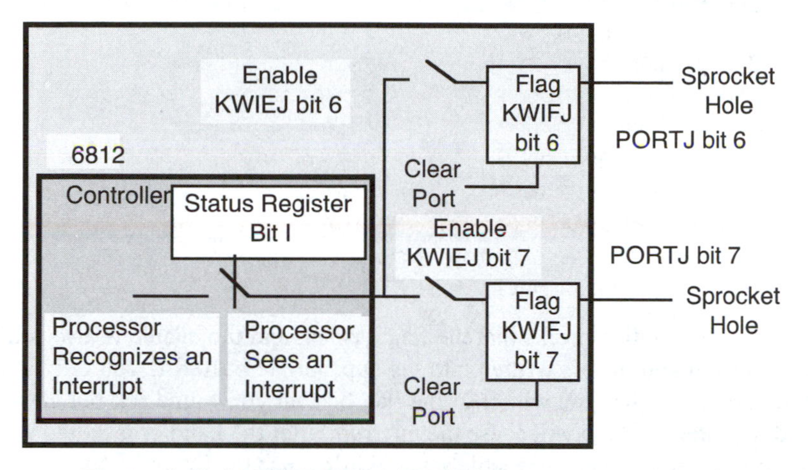

Figure 5.10. MC68HC812A4 Polled Interrupt Request Path

In the previous program, upon a key wakeup J interrupt, it is possible that *PORTJ* bit 7 or bit 6 saw a rising edge; we don't know. The key wakeup J handler determines whether input *PORTJ* bit 7 saw a rising edge, or input *PORTJ* bit 6 saw a rising edge. Once it finds a flag bit set, it removes the source by clearing the flag bit and writes data into the associated buffer. To make the program work sensibly, *main* terminates when either buffer is full. This interrupt handler is coded in assembly language as

```
0000086A 962B        LDAA  $2B
0000086C 942A        ANDA  $2A
0000086E 8480        ANDA  #128
00000870 B705        SEX   A,X
00000872 044517      TBEQ  X,*+26    ;abs = 088C
00000875 C680        LDAB  #128
00000877 5B2B        STAB  $2B
00000879 B60B54      LDAA  $0B54
0000087C 36          PSHA
0000087D 42          INCA
0000087E 7A0B54      STAA  $0B54
00000881 33          PULB
00000882 87          CLRA
00000883 B745        TFR   D,X
00000885 9600        LDAA  $00
00000887 6AE20B40    STAA  2880,X
0000088B 0B          RTI
0000088C 962B        LDAA  $2B
0000088E 942A        ANDA  $2A
00000890 8440        ANDA  #64
00000892 2716        BEQ   *+24    ;abs = 08AA
00000894 C640        LDAB  #64
00000896 5B2B        STAB  $2B
00000898 B60B55      LDAA  $0B55
```

```
0000089B 36          PSHA
0000089C 42          INCA
0000089D 7A0B55      STAA   $0B55
000008A0 33          PULB
000008A1 87          CLRA
000008A2 B745        TFR    D,X
000008A4 9601        LDAA   $01
000008A6 6AE20B4A    STAA   2890,X
000008AA 0B          RTI
```

Note that the handler checks that the flag has set, and the enable is also set, before the flag is cleared and data is written into the appropriate buffer. If you only check the flag bit, it is possible that you will clear the flag bit that you found set, but another flag bit caused the interrupt, and will cause the interrupt after the handler is exited with RTI. The flag bit that doesn't have its enable set couldn't cause the interrupt and shouldn't be serviced. There are even situations where this failing to service the interrupt making the request can hang up the machine, if the flag that is cleared continues to remain set. Therefore interrupt polling checks both the flag and the enable. This differs from a gadfly loop, which tests only the flag. Polling checks different bits than does a gadfly loop.

5.2.3.2 Polled Interrupts in the MC68HC912B32

Consider controlling two paper-tape readers in a 'B32 (Figure 5.11). We implement external logic to duplicate the 'A4's key wakeup logic. For each paper-tape reader, a 74CH74 flip-flop records when a rising edge occurs from its sprocket hole sensor, which, through the open drain 74HC01, can assert the IRQ line low. `PORTE` bits 7 to 5 connect to one reader and `PORTE` bits 4 to 2 to the other; bits 7 and 4 input the flip-flop value, bits 6 and 3 are the interrupt enables, and bits 5 and 2 clear the flip-flops. *INTCR* bit 6 enables IRQ interrupts. If enabled, either interrupt causes execution of the handler whose address is in 0xfff2. The handler examines these flip-flops to determine which paper-tape reader caused the interrupt; it clears its interrupt and reads its data port.

```
unsigned char bufferA[10], bufferB[10], indexA, indexB;

interrupt 6 handler() {
    if(PORTE & (PORTE << 1) & 0x80) /* if first FF enabled and set */
        { PORTE &= ~0x20; PORTE |= 0x20; bufferA[indexA++] = PORTA;}
    else if(PORTE & (PORTE << 1) & 0x10) /* if second FF enabled and set */
        { PORTE &= ~4; PORTE |= 4; bufferB[indexB++] = PORTB;}
}

void main() { asm sei
    INTCR = 0x40; /* enable IRQ int. when low */
    PORTE = DDRE = 0x6c; asm cli /* output high to avoid clearing FFs */
    while((indexA < 10) && (indexB < 10)) ; /* wait for either buffer done */
    PORTE = 0x24; /* disable tape reader interrupts only */
}
```

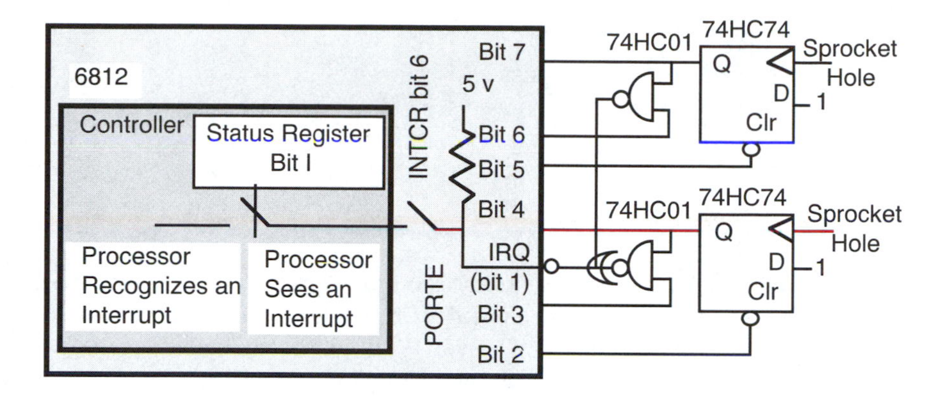

Figure 5.11. MC68HC912B32 Polled Interrupt Request Path

The last example illustrates the use of an IRQ line with external devices that request interrupts on either a 'B32 or an 'A4. Each such device has in it a request flip-flop, like the 74HC74 flip-flop, and an enable gate with open drain transistor, like the 74HC01, that can pull down the line when the flip-flop is set, a status port that reads the flip-flop directly, and a means to clear the interrupt request flip-flop. A number of devices can all be connected to this line. When any one device sets its request flip-flop in it, it pulls down the IRQ line, thus causing the interrupt handler whose address is in 0xfff2 to be executed. That handler polls the status registers and services the first device encountered that requested the interrupt. It clears that device's interrupt request flip-flop and performs its requested action.

5.2.3.3 Service Routines

In a more complex example, each different interrupt source has its own *service routine* that actually services the interrupt. This service routine is a conventional C or C++ procedure that is called from an interrupt handler, which may have local variables and may call other procedures, to satisfy the needs of the interrupting device. In this book, a service routine will obey the following convention, which makes it easier to write interrupt handlers. Each service routine, such as `srvc1()`, `srvc2()`, `srvc3()`, and so forth, will handle one source of interrupt on a device, and each source will have a flag or interrupt request register that can be individually polled and cleared. If a device has two or more sources, such as for normal interrupts and for errors, there will be two or more handlers for that device. Each service routine such as `srvc1()` will:

1. Test the flag and enable. If either are false, it returns a false value.

2. Otherwise, it performs the required operation, clears the source of the interrupt, and returns a true value.

The service routines can be tested in a C `if` condition that ORs the results of these service routines, as in the following handler having three service routines.

```
interrupt 23 handler(){ if(srvc1() || srvc2() || srvc3()) ; }
```

This interrupt handler is coded in assembly language as

```
00000882 07E6         BSR    *-24       ;abs = 086A
00000884 046407       TBNE   D,*+10     ;abs = 088E
00000887 07E9         BSR    *-21       ;abs = 0872
00000889 046402       TBNE   D,*+5      ;abs = 088E
0000088C 07EC         BSR    *-18       ;abs = 087A
0000088E 0B           RTI
```

This statement will call the service routines from left to right in the expression, until one of them returns a true value. When one does return true, it is not necessary to test any other service routines because any value ORed with a true value is true, so the condition must be true and the remaining tests should be skipped. In fact, C and C++ syntax requires that once a true value is found, the other values must not be tested. This rule for evaluation is not an optional rule for optimization, but is required by the language, so you can trust its being executed this way.

While this `if` condition looks very unusual, it produces good code. The highest-priority service routines are written on the left to be checked first. However, highest-priority devices can hog the system, preventing servicing of lower priority devices.

5.2.3.4 Round-robin Handlers

An alternative scheme is called a *round-robin* priority scheme. Here, the polling program is arranged as an infinite program loop. When the ith interrupt request in the priority order gets an interrupt and is serviced, the $i + 1^{th}$ interrupt request assumes the highest priority; so whenever the next interrupt occurs, the polling program starts checking the $i + 1^{th}$ interrupt request first. A round-robin polling handler is shown below, and its flow chart is given in Figure 5.12. A global variable `entry` determines which interrupt request bit to test first. It is set by the last interrupt that was honored, so that the interrupt request below it will be tested first when the next interrupt occurs.

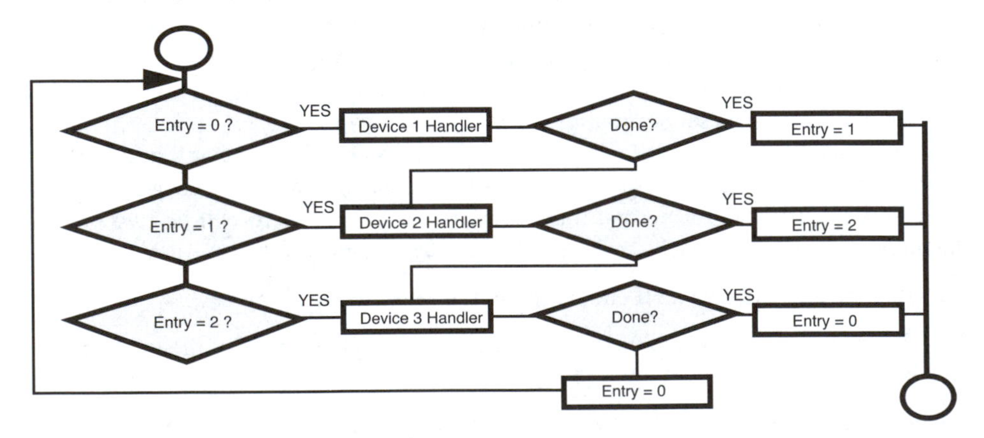

Figure 5.12. Flow Chart for Round-robin Interrupt Polling

```
interrupt 23 handler(){
    L: switch(entry){
        case 0: if(srvc1()) { entry = 1; break; }
        case 1: if(srvc2()) { entry = 2; break; }
        case 2: if(srvc3()) { entry = 0; break; }
        entry = 0; goto L;
    }
}
```

This interrupt handler is coded in assembly language as

```
0000088F F60B40       LDAB  $0B40
00000892 C102         CMPB  #2
00000894 222D         BHI   srvc3__Fv    ;abs = 08C3
00000896 87           CLRA
00000897 1608CA       JSR   srvc2__Fv
0000089A 03           DEY
0000089B 2703         BEQ   *+5    ;abs = 08A0
0000089D 0E1907C9     BRSET -7,X,#7,*-55    ;abs = 086A
000008A1 044406       TBEQ  D,*+9    ;abs = 08AA
000008A4 C601         LDAB  #1
000008A6 7B0B40       STAB  $0B40
000008A9 0B           RTI
000008AA 07C6         BSR   *-56    ;abs = 0872
000008AC 044406       TBEQ  D,*+9    ;abs = 08B5
000008AF C602         LDAB  #2
000008B1 7B0B40       STAB  $0B40
000008B4 0B           RTI
000008B5 07C3         BSR   *-59    ;abs = 087A
000008B7 044404       TBEQ  D,*+7    ;abs = 08BE
000008BA 790B40       CLR   $0B40
000008BD 0B           RTI
000008BE 790B40       CLR   $0B40
000008C1 20CC         BRA   *-50    ;abs = 088F
000008C3 0B           RTI
```

While a *goto* statement in this handler looks quite unusual, it also produces good code. Polling in the same priority order is useful when some interrupt requests clearly need service faster than others. Democratic round-robin priority is especially useful if some interrupt request tends to hog the use of the computer by frequently requesting interrupts.

The polling method and priority ordering affect latency. Priority polling provides shortest latency to the prior device, and longest latency to the least prior device, while round-robin priority provides average latency to each device. Aside from this, the method and ordering only affect performance when, in the product state machine, several devices change from BUSY to DONE in a short time (the latency of the first device to change state).

5.2.4 Vectored Interrupts

The previous example shows how multiple interrupts can be handled by polling them in the interrupt handler. Polling may take too much time for some interrupt requests that need service quickly. The *vectored interrupt* technique replaces the interrupt handler software by a hardware mechanism, so that the interrupt request handler is entered almost as soon as the device requests an interrupt. Table 5.1 gives the interrupt vectors for the 'A4; those of the 'B32 are almost identical, differing only for key wakeup and SCI1.

5.2.4.1 Vectored Interrupts in the MC68HC812A4

Again, consider having two paper-tape readers for the 'A4. The first reader's data and sprocket holes are sensed on `PORTA` and `PORTJ` bit 7. The second reader's data is still sensed on `PORTB`, but sprocket holes are now on `PORTH` bit 7. See Figure 5.13. The first interrupt causes execution of the handler whose address is in 0xffd0, but the second interrupt causes execution of the handler whose address is in 0xffce. The key wakeup J's handler's address is in 0xffd0 as usual, but the address of key wakeup H's handler is in location 0xffce, as shown in Table 5.1. The C program for the 'A4 follows shortly.

Table 5.1. Interrupt Vectors in the 6812

Name	Book §	Int #	Interrupt vector	DBug12 vector
Key wakeup H	5.2.4.1	24	0xFFCE,CF	0xB0E,0F
Key wakeup J	5.2.1.1	23	0xFFD0, D1	0xB10,11
AtoD	7.5.3	22	0xFFD2,D3	0xB12,13
SCI1	9.3.5	21	0xFFD4,D5	0xB14,15
SCI0	9.3.5	20	0xFFD6,D7	0xB16,17
SPI serial transfer complete	5.2.5.2	19	0xFFD8,D9	0xB18,19
Pulse accumulator input edge	8.3.4	18	0xFFDA,DB	0xB1A,1B
Pulse accumulator overflow	8.3.4	17	0xFFDC,DD	0xB1C,1D
Timer overflow	8.1	16	0xFFDE,DF	0xB1E,1F
Timer channel 7	8.2.1	15	0xFFE0,E1	0xB20,21
Timer channel 6	8.2.1	14	0xFFE2,E3	0xB22,23
Timer channel 5	8.2.1	13	0xFFE4,E5	0xB24,25
Timer channel 4	8.2.1	12	0xFFE6,E7	0xB26,27
Timer channel 3	8.2.1	11	0xFFE8,E9	0xB28,29
Timer channel 2	8.2.1	10	0xFFEA,EB	0xB2A,2B
Timer channel 1	8.2.1	9	0xFFEC,ED	0xB2C,2D
Timer channel 0	8.2.1	8	0xFFEE,EF	0xB2E,2F
Real-time interrupt	5.3.1	7	0xFFF0,F1	0xB30,31
IRQ or key wakeup D	5.2.1.1	6	0xFFF2,F3	0xB32,33
XIRQ	5.2.4.2	5	0xFFF4,F5	0xB34,35
SWI	5.3.5	4	0xFFF6,F7	0xB36,37
Unimplemented instruction	5.3.5	3	0xFFF8,F9	0xB38,39
COP failure		2	0xFFFA,FB	0xB3A,3B
Clock failure		1	0xFFFC,FD	0xB3C,3D
Reset	6.7	0	0xFFFE,FF	0xB3E,3F

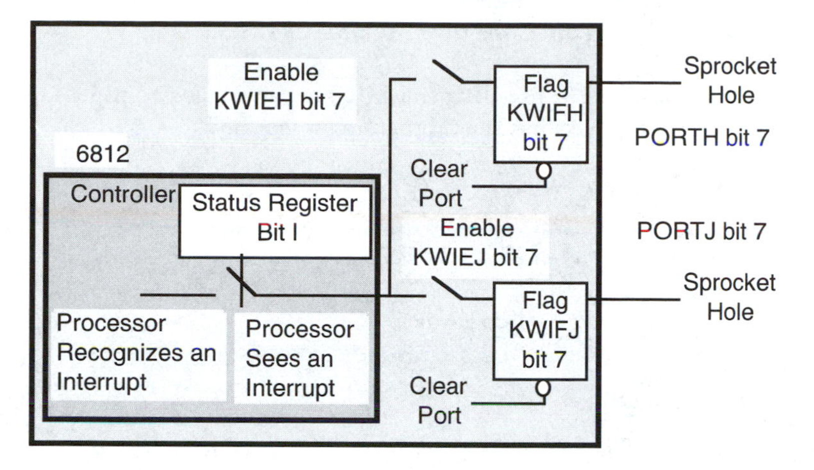

Figure 5.13. MC68HC812A4 Vector Interrupt Request Path

The 'A4 exhibits the vector interrupt well; the 'B32 has a pair of separately vectored interrupts, IRQ and XIRQ. The XIRQ interrupt is not an ideal I/O interrupt because, without external hardware, it cannot be disabled during critical sections or when the device is no longer needed. The XIRQ interrupt is designed as a "panic button," which can be used in debugging to stop execution when a program "goes wild." Nevertheless, we will illustrate vectored interrupts for 'B32 users, using IRQ and XIRQ. By using external hardware (Figure 5.14), the disadvantages of XIRQ can be overcome.

When an interrupt is honored in step three (§5.2.1), the 6812 uses an *interrupt vector,* which gives the address of the handler, in a manner discussed below. The 6812's interrupt vectors are in high memory, as shown in Table 5.1.

```
unsigned char bufferA[10], bufferB[10], indexA, indexB;

interrupt 23 void handler1() { KWIFJ=0x80; bufferA[indexA++] = PORTA;}

interrupt 24 void handler2() { KWIFH= 0x80; bufferB[indexB++] = PORTB;}

void main() { asm sei
      KWIEH = 0x80; KWIFH = 0x80; /* enable key wakeup H, J int., clear flags */
      KWIEJ = 0x80; KWIFJ = 0x80; asm cli  /* enable all non-XIRQ interrupts */
      while((indexA < 10) && (indexB < 10) ; /* wait for either buffer done */
      KWIEH &= ~0x80; KWIEJ &= ~0x80;      /* disable key wakeup interrupts only */
}
```

Note the similarities between polled and vectored interrupts using key wakeup flags. Assembler language for these interrupt handlers are essentially a pair of copies of the assembly language of §5.2.2.2.

A falling edge on `PORTJ` pin 7 causes `handler1` to be executed immediately. A falling edge on `PORTH` pin 7 causes `handler2` to be executed immediately. There is no need for polling. The 6812 "vectors" directly to the handler, reducing latency time.

5.2.4.2 Vectored Interrupts in the MC68HC912B32

A similar program for the 'B32 uses IRQ and XIRQ to handle the two paper-tape readers through different interrupt vectors. The C program for the 'B32 follows.

```
unsigned char bufferA[10], bufferB[10], indexA, indexB;

interrupt 5 handler1() // XIRQ handler
    { PORTE &= ~0x20; PORTE |= 0x20; bufferA[indexA++] = PORTA; }
interrupt 6 handler2() // IRQ handler
    { PORTE &= ~4; PORTE |= 4; bufferB[indexB++] = PORTB; }

void main() { asm sei
    INTCR = 0xc0; /* enable IRQ interrupts of falling edge */
    PORTE = DDRE = 0x64; andcc #0xaf /*avoid clearing FFs, en. both int.*/
    while((indexA < 10) && (indexB < 10)) ; /* wait for either buffer done */
    PORTE = 0x24; INTCR = 0; /* disable both tape reader interrupts */
}
```

Assembler language for these interrupt handlers is essentially a pair of copies of the assembly language of §5.2.2.2.

The first reader's data and sprocket holes are sensed on `PORTA`, but sprocket holes are now indicated by `XIRQ`. The second reader's data is still sensed on `PORTB`, and sprocket holes are on `IRQ`. See Figure 5.14. The first interrupt causes execution of the handler whose address is in 0xfff4; the second interrupt causes execution of the handler whose address is in 0xfff2, as shown in Table 5.1.

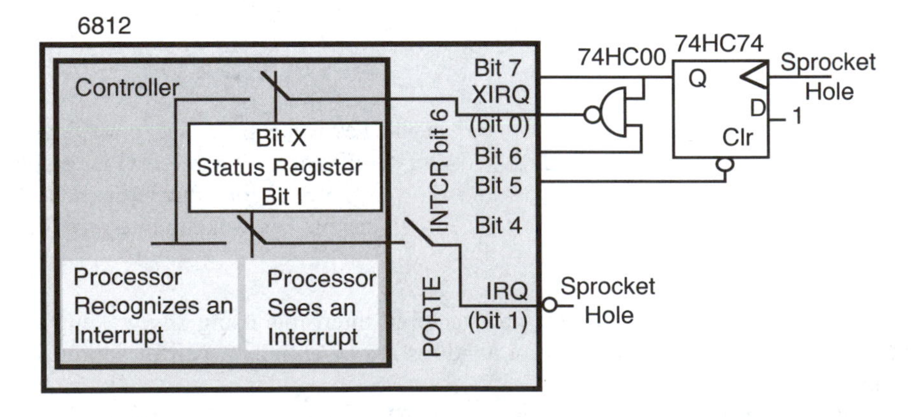

Figure 5.14. MC68HC912B32 Vector Interrupt Request Path

5.2.4.3 Vectored Interrupts for Other Devices

As Table 5.1 shows, the 6812 has a lot of interrupt vectors. The two right columns show the location of the vectors for hardware and Dbug12. For instance, key wakeup J's vector is at 0xffd0 and 0xffd1, while for Dbug12 it is at 0xb10 and 0xb11. The number to their left is the `interrupt` number used at the begining of a HIWARE interrupt handler. For instance, key wakeup J handler is designated `interrupt 23`. A section number appears in the second column where more details are given about the device and its interrupt request. For instance, key wakeup J is described in §5.2.1.

The main point of this section is that by using vectored interrupts, because the addresses of different handlers are at different locations and there is no software polling routine to go through, the specific interrupt request handler is executed without the delay of a polling routine. In effect, the polling routine is executed very quickly in hardware, and the winning handler is jumped to right after the registers are saved on the stack.

5.2.5 Examples of Interrupt Synchronization

The interrupt is useful in managing asynchronous requests and repetitive tasks. This section illustrates the use of interrupts in responding to key requests, inputting and outputting SPI data, and reading bar-code and X-10 signals.

5.2.5.1 Keyboard Handling

The 'A4's key wakeup ports, `PORTD`, `PORTH`, and `PORTJ`, respond well to sensors and push buttons, beginning procedures whenever such a sensor detects a situation needing correction or when a button is pressed requesting an action. Key contacts occur asynchronously, at random and unpredictable times, and are therefore handled by interrupts. Mechanical switches "bounce." A common debouncing technique is shown here. Finally, multiple keys have to be scanned; a simple scanning technique is given. That several sensors or push buttons might request actions in short order, while the microcontroller is taking care of one of them, is resolved in §5.2.6.

When mechanical switches or sensors close, a metal contact rebounds when it hits a metal plate, causing multiple closed/open events. If each transition from open to closed causes an edge, each edge causes an interrupt, and each interrupt initiates an action, then one physical closure of a switch or sensor may initiate multiple actions. This problem of *contact bounce* is mechanically addressed by using bounceless contacts, such as opto-electronic sensors or mercury-wetted contacts; by electrically debouncing the switch using analog techniques such as putting a capacitor across it, using digital techniques such as a set-clear flip-flop, or by computational accounting for multiple bounces. The most common technique used in microcontrollers is, upon detecting an apparent switch closure, seeing if it remains closed for 5 ms before recognizing the closure. This delay permits most switches time to stop bouncing. This *wait-and-see* technique has some disadvantages, but provides more than adequate performance at the least cost and complexity of all the techniques discussed above. An example of this technique is illustrated for a single switch, a linear select, and a coincident select or matrix keyboard.

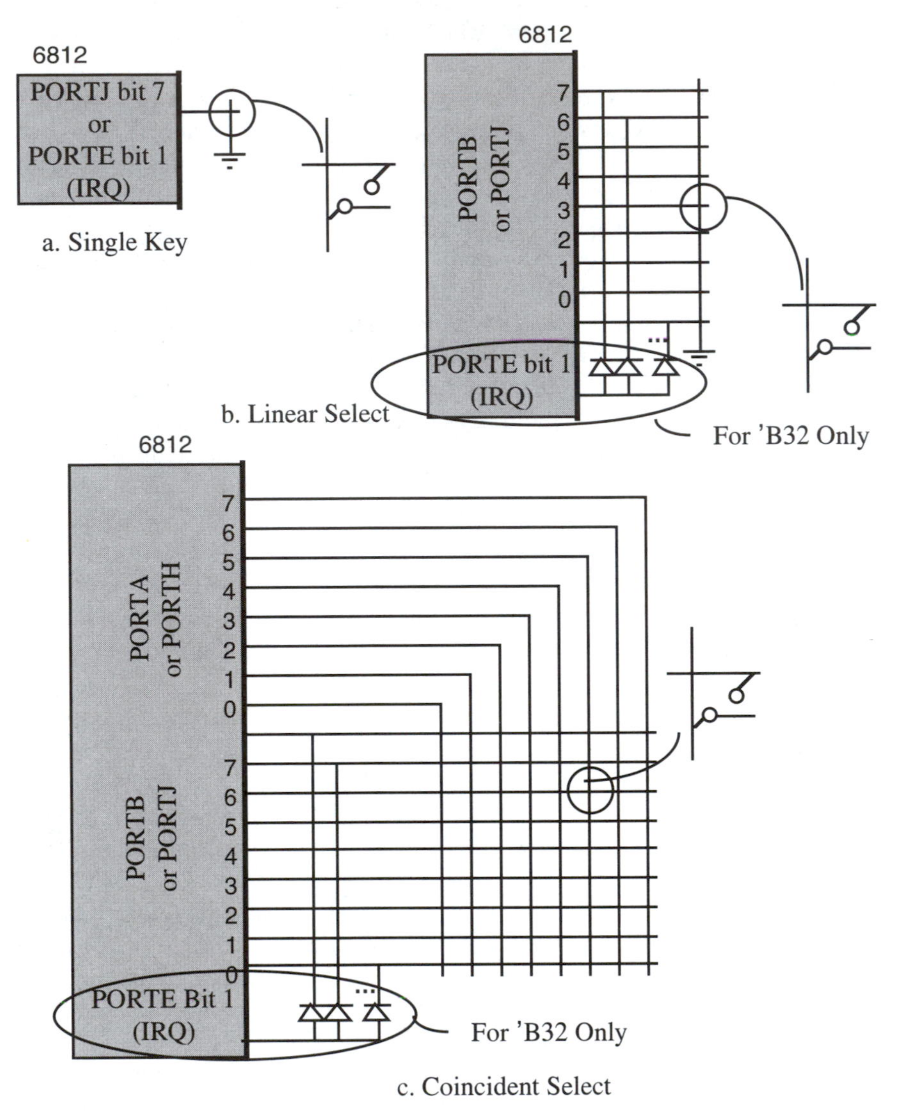

Figure 5.15. Keys and Keyboards

The key wakeup hardware of the 'A4 is exceptionally suited to interface to keys and keyboards. The 'B32, lacking this hardware, can use the IRQ signal, with external diodes for linear and coincident select keyboards, to interface to keyboards.

Figure 5.15a illustrates a single switch or sensor connected to an 'A4's `PORTJ` bit 7. This bit can be configured to use an internal pull-up so that if the switch is open, the pin voltage is high. When the switch is closed, the pin voltage falls, and a falling edge can generate an interrupt. In this short program, when `PORTJ` bit 7 sees a falling edge, it generates a key wakeup J interrupt; after 5 ms, if the switch remains closed and the input remains low, the handler puts a T (1) on `PORTJ` bit 7. The constant `N` is selected to provide a 5-ms delay, in this and other keyboard examples.

```
void interrupt 23 handler() {
    KWIFJ = 0x80; for(i = 0; i < N; i++) ;
    if( ! (PORTJ & 0x80)) PORTB |= 1;
}

void main() { asm sei
    DDRB = 1; KWIEJ = KWIFJ = PULJ = 0x80;
    asm cli
}
```

Figure 5.15b illustrates a single switch or sensor connected to 'B32's `PORTE` bit 1, which is the IRQ signal. This bit has an internal pull-up so that if the switch is open, the pin voltage is high. When the switch is closed, the pin voltage falls, and a falling edge can generate an interrupt. In this short program, when `PORTE` bit 1 sees a falling edge, it generates an IRQ interrupt; after 5 ms, if the switch remains closed and the input remains low, the handler puts a T (1) on `PORTE` bit 2. The program to assert `PORTE` bit 2 when a key is pressed is shown.

```
interrupt 6 void handler()
    { int i = N; while(--i); if( ! (PORTE & 2)) PORTB |= 1; };

void main() { asm sei        /* disable all non-XIRQ interrupts */
    INTCR = 0xc0; DDRB = 1; asm cli
}
```

In an 'A4, a *linear select keyboard* is, say, 8 switches, each connected between a `PORTJ` pin, configured as an input with pull-ups, and ground. See Figure 5.15b. Assume only one switch will close at any time. When any switch is closed, a falling edge can generate an interrupt. In the program below, a falling edge on any `PORTJ` pin generates a key wakeup J interrupt. In the handler, after 5 ms, if the switch where a falling edge occurred, remains closed and the input remains low, a procedure is executed, responsive to the switch; for instance, if `PORTJ` bit 0 falls, `p0` is executed, and so on.

```
void interrupt 23 handler() { char key; int i;
    key = KWIFJ; /* get edge */ KWIFJ = key; /* remove source(s) of interrupt */
    for(i = 0; i < N; i++) ; /* wait 5 ms */
    key &= ~PORTJ; /* AND negated port data after 5 ms */
    asm cli   /* permit interrupts: these procedures could execute for a long time */
    if(key& 0x01) p0(); if(key& 0x02) p1(); if(key& 0x04) p2();
    if(key& 0x08) p3(); if(key& 0x10) p4(); if(key& 0x20) p5();
    if(key& 0x40) p6(); if(key& 0x80) p7();
}

void main() { asm sei
    DDRJ = 0; KWIEJ = KWIFJ = 0xff; PULEJ = PUPSJ = 0xff; asm cli
}
```

In a 'B32, a linear select keyboard is, say, 8 switches, each connected between a `PORTB` pin, configured as an input with pull-ups, and ground; and through a diode, each switch can assert IRQ low. See Figure 5.15b. If a switch is closed, a falling edge generates an IRQ interrupt. After 5 ms, the handler causes a procedure to be executed, responsive to the switch if it is still closed; for instance, if `PORTB` bit 0 is low, procedure *p0* is executed, and so on. This procedure does not handle anomalous cases where one key causes an interrupt and another is low 5 ms later, while the second key might be bouncing. However, such anomalous cases do not occur often enough to warrant more sophisticated hardware or software.

```
void interrupt 6 handler() { char key; int i;
   key = ~PORTB; /* get keys */ for(i = 0; i < N; i++) ; /* wait 5 ms */
   key &= ~PORTB; /* AND negated port data after 5 ms */
   asm cli   /* permit interrupts: these procedures could execute for a long time */
   if(key&0x01)p0();if(key&0x02)p1();if(key&0x04)p2();if(key&0x08) p3();
   if(key&0x10)p4();if(key&0x20)p5();if(key&0x40)p6();if(key&0x80)p7();
}

void main() { asm sei
   PUCR = 2; INTCR = 0xc0; DDRB = 0; asm cli
}
```

An ’A4 *matrix*, or *coincident select, keyboard* is connected as (say eight) rows of wires and (say eight) columns of wires. A ’B32 matrix keyboard can be connected similarly to ports A and B, with diodes from the rows to the IRQ pin. See Figure 5.15c. Although electrically connected in a two-dimensional array, they can be physically positioned suited to an application such as a terminal keyboard, for instance. In the ’A4, rows are connected to input `PORTJ`, having pull-ups. Columns are connected to output `PORTH`. The main program initially makes `PORTH` an output, whose pins are all low. Then, if any one key is pressed in any row and in any column, a `PORTJ` input falls and a key wakeup J interrupt occurs. After being debounced in a manner similar to the linear select key, the key’s row and column number are catenated into a key code, which is pushed into a queue, to be pulled later. The ’B32 implementation is similar.

```
void interrupt 23 handler() { char fallingEdge, keys[8], r, c; int i;
  fallingEdge = ~KWIFJ; KWIFJ = fallingEdge /* get edge, remove int */
   for(i = 0; i < 8; i++) /* scan 8 columns */
      {PORTH = ~(1 << i); keys[i] = fallingEdge & ~PORTJ; }/* save col. */
   for(i = 0; i < N; i++) ; /* wait 5 ms */
   for(i = 0; i < 8; i++){ /* scan 8 columns again */
      PORTH = ~(1 << i); c = keys[i] & ~PORTJ; /* get data for that column */
      for(j = 0, r = 1; j < 8; j++, r <<= 1;)
          { if ( c & r ) { push( ( i << 3) + j ); i = j = 8;}}
  }
  PORTH = 0; /* all columns low to interrupt if any key is pressed */
}

void main() { asm sei
   KWIEJ = KWIFJ = PULJ = DDRH = 0xff; PORTH = 0;
   PULEJ = PUPSJ = 0xff; asm cli
}
```

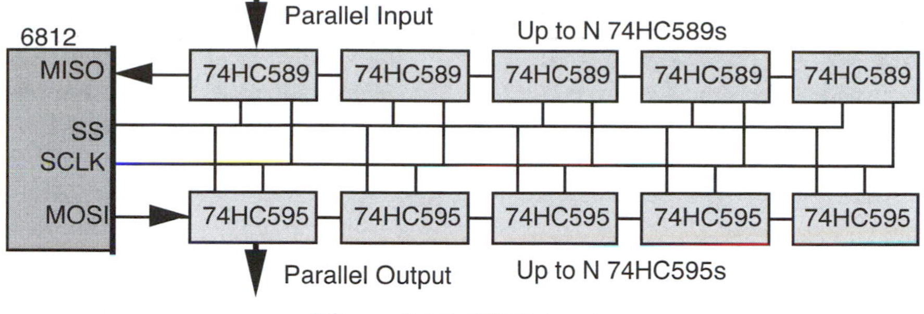

Figure 5.16. SPI Network

Coincident select uses the same strategy as linear select, but columns are scanned one at a time, and their first-sampled values are saved in a vector `keys`. These `keys`, ANDed with the same values scanned 5 ms later, indicate a key closure. If a key is found low then, its column number concatenated with its row number is a key code that is pushed onto a queue. The user can later pull the key value from the queue.

5.2.5.2 Interrupts for SPI Systems

The SPI can input and output serial data, as has been discussed in §4.4.4. In this example (see Figure 5.16), the SPI repetitively inputs and outputs multiple bytes of data using interrupts. The 6812 MOSI pin feeds a series of '595s, `MISO` collects from a series of '589s, SCLK feeds each chip's shift clock, and SS feeds each chip's RCLK and LD, as discussed in §4.4.3. Global vectors `inBuffer` and `outBuffer` hold data received from the '589s and sent to the '595s respectively, and a global `index` is used to access the vector elements in the interrupt handler. `main` initializes the SPI as before, but the SPI interrupt enable, `SPIE`, is set to permit interrupts. `SP0DR` is written to start the SPI. The shift clock rate is set low to give time for `main` to do other work. Each time eight bits have been shifted in and out, an SPI interrupt occurs. An address register sequence, reading `SPIF` and `SP0DR`, clears the interrupt and writes `SP0DR`'s data into `inBuffer`. Data from `outBuffer` is written into `SP0DR`, which also starts the SPI again. If `index` attains its maximum, it is cleared and the SS pin is pulsed low twice to transfer each 589's data between shift register and output register, and each 595's data between input register and shift register. `main` can now read from `inBuffer` and write data into `outBuffer`; the SPI moves this data in and out of the computer.

```
interrupt 19 void handler() { char dummy;
  dummy = SPIF; inBuffer[index] = SP0DR; SP0DR = outBuffer[index++];
  if(index == 5)
     {index=0;PORTS &= ~ SS; PORTS |= SS; PORTS &= ~ SS;PORTS |= SS;}
}

void main() { asm sei
  PORTS = SS; DDRS = SS + SCLK + MOSI; SP0CR1 = SPE + MSTR + SPIE;
  SP0CR1 = 4; /* low baud rate for more work */ SP0DR = 0; /* start */ asm cli
}
```

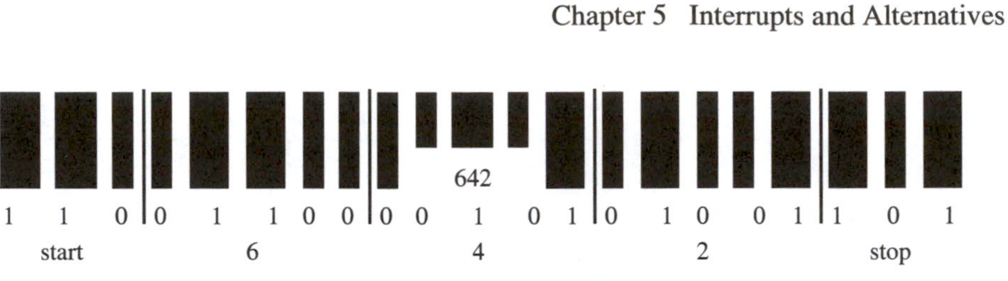

Figure 5.17. Bar Code

5.2.5.3 Histograms and Histories for the MC68HC812A4

Our next example illustrates the use of key wakeup interrupts to measure or record edge events. A *Universal Product Code* (UPC) is put on most packages to identify the package contents, and a *bar-code reader* is used to scan these codes. See Figure 5.17. A T (1) is a wide bar, and an F (0) is a narrow bar. The space between bars is the width of the narrow bar. The width of a wide bar is equal to that of two narrow bars. See §5.2.6.4. As a reader wipes over the code, bar widths are converted into pulse widths.

The program below stores the widths of a sequence of these pulse in a global vector *pulses,* which `main` will examine and decode. The reader's pulse output is input on both `PORTJ` bit 0, which is configured for rising edge interrupts, and `PORTB` bit 0, which is configured for falling edge interrupts. The 6812 has a 16-bit free-running counter `TCNT` at 0x84, which runs if bit 7 of `TSCR` at 0x86 is set. The handlers read the time of an edge from `TCNT`. Upon a falling edge at the beginning of a bar, in `handler1`, *TCNT* is saved in *start.* Upon a rising edge at the end of a bar, in `handler2`, the difference `TCNT` minus `start` is saved in `pulses`. The 'B32 doesn't have key wakeup hardware. Both the 'A4 and 'B32, however, can use the counter-timer to do so. Rather than jury-rig an example here for the 'B32, we refer to §8.3.2 to show how the 'B32 can measure pulse widths. The following program is for the 'A4 only.

```
int index, start, pulses[N];

interrupt 24 void handler1() { KWIFH = 1; start = TCNT; }

interrupt 23 void handler2() {
    KWIFJ=1;pulses[index++]=TCNT-start; if(index==N) KWIEH=KWIEJ=0;
}

void main(){ asm sei
    TSCR = 0x80; KWIEH = KWIEJ = PULEJ = KPOLJ = 1; asm cli
    do ; while(KWIEH);
}
```

The next example illustrates the use of key wakeup interrupts to measure or record *X-10*-coded bit patterns. Recall from §5.1.3 and Figure 5.7d that the X-10 T (1) is three bursts of 100-KHz signals, and F (0) is no burst of such signals. An amplitude modulation detector will output T (1) if a burst is present, and F (0) if no burst is present. Due to noise on the power line or to misalignment, the number of bursts could be off by one (or more). We count the number of bursts in one-half of the 60-Hz period. If about zero pulses occur (0 or 1 pulse) then the signal is F (0), but if about three pulses occur (two or more) then the signal is T (1). The detector produces three pulses if the X-10 code is T (1), or no pulses if it is F (0), and is connected to key wakeup H bit 0; its handler increments *pulseCount*. The 60-Hz squared-up waveform is connected to key wakeup J bit 0; its handler writes a T (1) into *bits[index]* if two or more pulses arrive, and writes F (0) if one or zero pulses arrive. (As in the previous discussion of X-10 codes, we strongly recommend that you use the TW523 module that isolates the 110-V system from your microcontroller and you.

```
int start, histogram[N];

interrupt 24 void handler1() { KWIFH = 1; start = TCNT; }

interrupt 23 void handler2() { unsigned int width;
    KWIFJ = 1; width = TCNT - start;
    if(width < N) histogram[width]++;
    if(histogram[0] == 100) KWIEH = KWIEJ = 0; // "bucket 0" full
}

void main(){ asm sei
    TSCR = 0x80; KWIEH = KWIEJ = PULEJ = KPOLJ = 1; asm cli
    do ; while KWIEH;    // wait here until terminated when "bucket 0" full
}
```

5.2.6 Object-oriented Classes for Interrupts

Interrupts can be very nicely handled using objects. For input, the class `IQFPort` (input queue flag port) and for output, the class `OQFPort` (output queue flag port), both shown in subsections below, are derived from `Port` and use class `Queue` from §2.3.3 and subroutines from §5.1.5. We have avoided making either `IQFPort` or `OQFPort` a derived class of `SyncPort` to reduce target memory requirements.

5.2.6.1 An IQFPort Class

The class `IQFPort` (input queue flag port) is a simple input class that uses a queue. We here illustrate this class and consider an application to our familiar paper-tape reader.

The constructor and handshake mechanisms are very similar, and the constructor's handshake argument and synchronization argument (when key wakeup flags are used) have the same meaning as for the `SyncPort` class. However, if the constructor's fourth argument is nonzero, an object of the `Queue` class is created. If the low-order 5 bits of the sixth argument is nonzero, they are used as an interrupt number; and an interrupt vector is inserted for Dbug12, taking the fifth argument as the handler address.

```
template <class T> class IQFPort : public Port<T> {

    protected : long syncK, handK; Queue<T> *Q;
```

```
    public:IQFPort (int a,long syncWord,long handWord,unsigned int qS,
          int hR, int attr=0, T dir=0) : Port(a, attr, dir){ char i;
            if(handWord & 0x8000) orAt(handWord);
            orAtNext(handK = handWord); orAt(syncK = syncWord);
            if(qS) Q = new Queue<T>(qS); else Q = 0;
            if(i = attr & 0x1f) ((int *)0xb3e)[-i] = hR; /* for dbug12 */
    }

    virtual T get(void) {
        if( ! Q->size ) handshakeBegin(handK); while( ! Q->size ) ;
        T data = Q->pull(); attr |= Q->error(); return data;
    }

    virtual char service(void) {
        if( ! (testAt(syncK) & testAt(syncK))) return 0;
        Q->push(Port<T>::get()); attr |= Q->error();
        if(Q->size==1 ) handshakeEnd(handK);storeAt(syncK);return 1;
    }
};
```

The interrupt handler service routine reads the port, pushing the result on the queue. The *get* function pulls the word. Handshaking isn't usually used with input interrupt, because data is provided to the device asynchronously, but is provided here such that if *get* is executed when the queue is empty, a handshake signal can be pulsed, or asserted until the interrupt handler responds to put data in the queue, to request data.

A paper-tape reader is a straightforward application of the class `IQFPort` without the need for a derived class. This paper-tape reader is shown below. The declaration of the object sets the data port to be `PORTA`, using port J bit 7 for synchronization on the falling edge, and port B bit 2 for handshake, which is asserted low when `get` is entered, and negates high after the interrupt brings in the data.

```
IQFPort<char> p(aA,0x2BFF80, (aB << 16)|0xD404,10,(int)handler, 23);

void interrupt 23 handler(){ p.service(); }

void main(){ char buffer[0x100]; p.get(buffer, 0x100); }
```

The constructor's first parameter is passed to *Port*'s constructor where it identifies the data port `id`. The next two parameters inititalize the `syncK` and `handK` data members to control synchronization and handshaking. The constructor's next parameter gives the queue size unless it is zero. The interrupt service routine `service` is written for not only single interrupts but also priority-ordered and round-robin polled interrupts. It therefore tests the flag and enable ports, returning a value of 1 if the interrupt was serviced, or a 0 if this device did not request an interrupt.

When an interrupt occurs, `handler` calls `service`, which uses `device`'s data members. `service` checks the flag and associated enable registers as discussed in §5.2.3. If this device requested the interrupt, data are read from this object's port and pushed onto its queue. The `main` procedure `get`s data by pulling it from the object's queue, but if the queue is empty, the `get` procedure gadflies until it is nonempty.

5.2.6.2 An OQFPort Class

For output, the class *OQFPort* (output queue flag port) illustrates additional complexities due to turning off interrupts when the output queue is empty. It is derived from *Port* and uses class *Queue* from §2.3.3 and subroutines from §5.1.5. The *put* function pushes some data. The interrupt handler service routine pulls data from the queue, which it writes to the port. Handshaking is often needed, such that if *put* is executed when the queue becomes nonempty, a handshake signal can be pulsed, or asserted until the interrupt handler responds to pull data from the queue.

```
template <class T> class OQFPort : public Port<T> {
    protected : long syncK, handK; Queue<T> *Q;

    public : OQFPort (int a, long sW, long hW, unsigned int qS,
        int hR,int attr=0, T dir=0) : Port(a, attr, dir){ char i;
            orAtNext(handK = handWord);
            if(handWord & 0x8000) orAt(handWord);
            orAt(syncK = syncWord);
            if(qS) Q = new Queue<T>(qS); else Q = 0;
            if(i = attr & 0x1f) ((int *)0xb3e)[-i] = hR; /* for dbug12 */
    }

    virtual void put(T data) {
        while(Q->size>=Q->maxSize);Q->push(value=data);attr|=Q->error();
        if(Q->size==1) { handshakeBegin(handK); orAtPrevious(syncK); }
     }

    virtual char service(void) {
        if( ! (testAt(syncK) & testAt(syncK))) return 0;
        Port<T>::put(Q->pull());
        if(! Q->size) { handshakeEnd(handK); andAtPrevious(syncK); }
        return 1;
    }

    ~OQFPort(void) { char i;
        i = Q->size;
        while(Q->size) ;if(i)handshakeEnd(handK); andAtPrevious(syncK);
    }
};
```

When *put* is called, it pushes its argument onto the output queue and enables interrupts if the queue becomes nonempty. When an interrupt occurs. The *handler* calls *service* which uses *device*'s data members. *service* checks the flag register and its associated enable register in the position of the synchronization mask, as in the discussion of polled interrupts in §5.2.3. If this device requested the interrupt, data is pulled from the object's output queue and output to its port. If the output queue empties, the output interrupt is disabled.

5.2.6.3 Polling IQFPort and OQFPort Classes

The classes `IQFPort` and `OQFPort` are suited to polled interrupts as described in §5.2.3. The interrupt handlers, such as `handler,` are individually written to establish the polling method and priorities of all the devices that are to be polled, rather like the constructors individually called for each class. For instance, if objects *O1, O2,* and *O3* are polled in that priority order for a key J wakeup handler, the following handler is

```
interrupt 23 handler(){if(O1.service()||O2.service()||O3.service());}
```

Then only the constructor declarations or blessings and the interrupt handlers have to be rewritten if an object is declared or blessed to be of different classes.

5.2.6.4 Bar-code Class

We now expand on our example of a bar-code reader (Figure 5.17) to include decoding the bar-code frame for an 'A4. Only decimal digits are encoded, each digit being a "two-of-five" code shown by the vector `cnvt`. The code for digit `i` is `cnvt[i]`. For instance, the code for digit 0 is 6. A `frame,` which is a sequence of representations of digits, is preceded by a start code (110) and followed by a stop code (101). The reader scans the wand from start to stop, possibly with varying speed. The initial width of a narrow bar is one-fifth the sum of the widths of the first three pulses, which is equal to the widths of two wide and one narrow bar. Thereafter, if a narrow bar is detected, its width is recalculated to be the average of the old and new narrow bar widths. Once five bars are counted (`bitCount` is 7), the pattern `bits` is searched in the constant vector *cnvt* for a matching pattern. When a digit is found, `get` returns it to the user.

```
char buffer[0x100], cnvt[10] = { 6,0x11,9,0x18,5,0x14,0xc,3,0x12,0xa };

interrupt 23 void handler() { device.service(); }

class barCode:public IQFPort<int>{int start,narrow; char bitCount,bits;
   public:barCode(void):IQFPort(aH,0,0,-1,10,(int)handler,23){asm sei
       pulPwr(uH); TSCR = 0x80; start = TCNT; bitCount = 0; asm cli
   }

   virtual int get(void) { char bit, i;
       do{
           int code = IQFPort<int>::get(); /* wait, then pull it */
           if( bitCount==0)narrow=code;else if(bitCount==1)narrow+=code;
           else if( bitCount == 2 ) { narrow += code; narrow /= 5; }
           else {
                bits=(bits<< )|(bit=code>(narrow + (narrow >> 1)));
                if(! bit) narrow = (narrow + code) >> 1;
                if(bitCount == 7) { /* if five bars, convert digit, reset count */
                    for(i=0; i<10;i++)if(cnvt[i]==(bits&0x1f)) break;
```

```
                    bitCount = 2;
                }
            }
        } while (bitCount++ != 2);
        return i;
    }

    virtual int service(void) { /* service rtn saves bar widths in input queue */
        if( ! (KWIFJ & 1) ) return 0; KPOLJ ^= 1; KWIFJ = 1;
        if(KPOLJ & 1) start = TCNT; else Q->push(TCNT - start);
        return 1;
} };

barCode device; char c;

void main(){ c = device.get(); }
```

5.2.6.5 An X-10 Class

We extend our implementation (§5.1.4) of an X-10 decoder (Figure 5.7), as an object of class *X-10* for the 'A4. X-10 modules are identified by a house code from "A" to "P," and a unit code from 1 to 16, by two 16-position rotary switches on the module. A 16-position rotary switches on the controller selects its house code. In a house, all controller and module house codes are generally the same, and each X-10 module in the house is assigned a different unit code. The controller buttons select a unit number and give commands. For instance, by pressing the buttons "1" and "on," you can turn on module 1. Pressing the "1" button sends a frame, and pressing the "on" button sends a frame. Each frame is 22 bits, which is sent twice for reliability; a frame (Figure 5.18c) consists of a start pattern (0xe), an 8-bit house code, an 8-bit "second" code, and a 2-bit type code that is TF (10) or FT (01). Type code FT indicates that the second code is a unit code, while Type code TF indicates that the second code is a command.

Although two frames are sent over the power line modulated by a 100-KHz carrier, the TW523's inputs and outputs (Figure 5.18a) transmit just one demodulated frame to the microcontroller and from the microcontroller, and one bit is transmitted each 60-Hz half-cycle. The TW523 provides a squared-up 60-Hz wave and a `signal`. Each side of Figure 5.7e's opto-isolators have separate ground and +5 V supplies, so 110-V power is not applied to the 6812. The TW523 module safely isolates power from the computer and user. *X-10*'s service routine is entered once every half-cycle, upon either edge of the 60-Hz squared-up signal on TW523's pin 1. This is connected to `PORTJ` bit 2, which is enabled to cause a key wakeup J interrupt. Each time the interrupt is handled, the KPOLJ bit 1 is inverted to interrupt upon the next rising or falling edge. If a T (1) occurs, an edge appears on TW523's pin 3's signal, which connects to `PORTJ` bit 1. This sets KWIFJ bit 1. This input bit is shifted leftward into a long word *InFrame* to collect a 22-bit frame (Figure 5.18c). If, due to a false start, the first four bits found are not start pattern 0xe, *inCount* resets to 0. When *inCount* is 22, a full frame has been collected; *Frame* is pushed onto a stack, to be popped by `get` when the program requests this value, and *inCount* is reset to begin looking for a next message.

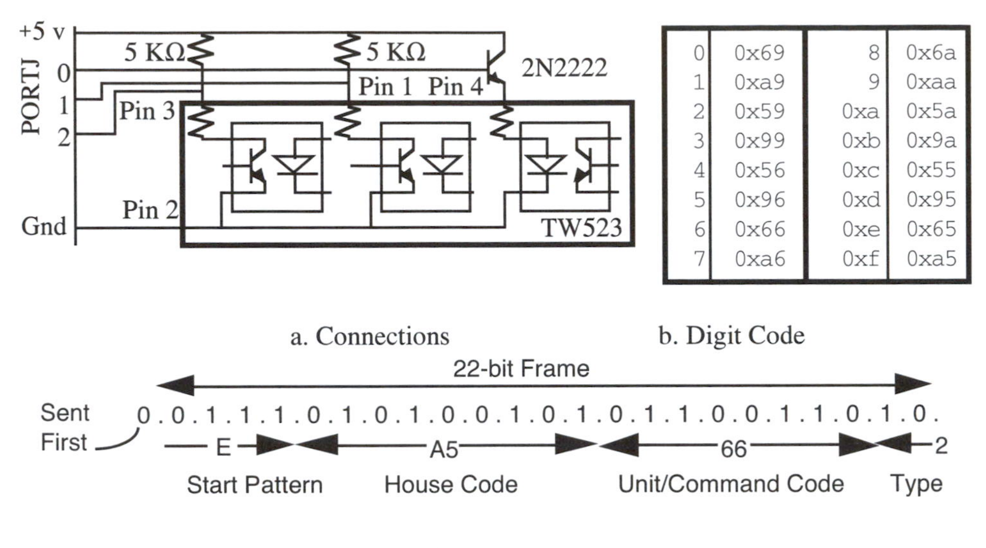

0	0x69	8	0x6a
1	0xa9	9	0xaa
2	0x59	0xa	0x5a
3	0x99	0xb	0x9a
4	0x56	0xc	0x55
5	0x96	0xd	0x95
6	0x66	0xe	0x65
7	0xa6	0xf	0xa5

c. The Code Frame

Figure 5.18. X-10 Frame

Function members *get* inputs X-10 commands. It pops an input frame from the queue, and decodes the data to deliver a command to the calling routine. The frame type is first analyzed. Then the house code bits are extracted, and function member *decode* converts it into a binary number. It uses a *for* loop to implement a linear search of the conversion vector *cnvt* (see Figure 5.18b). The *i*th element of this vector has the code for house *i* (house "A" is house binary number 0, house "B" is house binary number 1, etc.) In Figure 5.18c, the house code 0xa5 is for house 0xf which is house "P." The function member *decode* also converts the unit number or the command code. If the frame type was for a unit, this pattern is the unit number to select that unit, otherwise it is the command to be given to preselected units. The command values are 2 (on), 4 (all on), 6 (bright), 0xa (off), 0xc (all off), and 0xe (dim). In Figure 5.18c, 0x66 represents the command "bright." When the pattern in Figure 5.18c is received, *get* returns the value 6 to indicate it is a command to "brighten" the lights (of units previously selected) in house "P". Function member *get* returns their values in a struct *CMD*.

```
unsigned char inCount; long InFrame, OutFrame;

interrupt 23 void handler(void){
   InFrame<<=1; if(KWIFJ&4){InFrame++; if(!inCount)inCount=1;}KPOLJ^=2;
   KWIFJ = 6; /* clear key wakeup flag bits 1 and 2 */
   if(inCount==4){if((InFrame&0xf0!=0xe)inCount=InFrame=0;else inCount++;}
   if(inCount == 22){ push(InFrame); inCount = InFrame = 0; }
   if(inCount) ; inCount++;
   if(OutFrame & 0x8000) PORTJ |= 1; else PORTJ &= ~1; OutFrame << 1;
}}
```

```
const char cnvt[16] = /* table for get function member to decode bit sequences */
{0x69,0xa9,0x59,0x99,0x56,0x96,0x66,0xa6,0x6a,0xaa,0x5a,0x9a,0x55,0x95,
   0x65,0xa5 };

typedef struct CMD {char house, unit, command; } CMD;

class X10 : public IQFPort<int>{
   long Frame; char bitCount; public : char house, unit, command;

   public: X10 (void) : IQFPort(aH, 0, 0, 10, (int)handler, 23)
       { KWIEJ |= 4; KWIFJ = 6; DDRJ = 1; }

   unsigned char decode(unsigned char code) { unsigned char i;
       for(i = 0; i <= 16; i++) if((code & 0xf) == cnvt[i]) break;
       if(i > 15) error = 1; return i;
   }

   void get(CMD &result) { char bits; int i; long Frame;
       Frame=pull();if((Frame&3)==1) uni =decode((Frame & 0x3fc) >> 2 );
       else if((Frame&3)==2)command=decode((Frame&0x3fc)>>2);else error=1;
       house = decode(Frame >> 10);
       result.unit=unit; result.house=house; result.command=command;
   }

   void put( unsigned char house,unsigned char x,unsigned char type) {
       do ; while (OutFrame);
       OutFrame = 0x3800|(cnvt[house] << 10)|(cnvt[x] << 2)| type;
   }

   ~X10() { KWIEJ &= ~2; } /* disable interrupts */
} *ptr;

void main(){ CMD j; X10 device; ptr = &device ; device.get(j); }
```

Function member `put` constructs a frame and outputs X-10 commands. It waits until all T (1) bits in `OutFrame` are shifted out so that it will not stop a pattern that is being shifted, then it fills the fields of `OutFrame` with its parameters. The resulting pattern appears like the pattern in Figure 5.18c. A frame is sent by the key wakeup J handler by outputting, each 120th second, the most significant bit of `OutFrame` on `PORTJ` bit 0, which connects to the TW523's pin 4, and shifting `OutFrame` left 1 bit.

Both of the preceding examples illustrate that operations on data can be done in the interrupt handler service routine or in function members `get` or `put`. Function members are a better place to execute complex procedures. The handler should do the minimum amount of work to reduce latency for devices whose interrupts are deferred. The bar-code service routine merely saves the pulse width on a queue. The `X10` service routine merely collects the frame and pushes the undecoded 22-bit frame onto a queue. This strategy is used in many of the following examples.

5.3 Time-Sharing

The 6812 has a *real-time interrupt* (*RTI*) timer that can cause interrupts every 1.024 ms to every 65.5 ms Its interrupt handler can be used to synchronize I/O transfers to or from the outside world. In this section, we use it in a simple way and then in a manner similar to time-sharing, and show examples of time-sharing applications.

5.3.1 Real-time Interrupts

Simple RTI programs use the real-time interrupt enable `RTIE, RTICTL` bit 7; the real-time interrupt flag `RTIF, RTIFLG` bit 7; and `RTICTL`'s low nibble, real-time interrupt rate `RTR`. See Figure 5.19. `RTIF` is set every 512 ms/`RTR,` unless `RTR` is 0. If `RTIF` is T (1), and if `RTIE` is T (1), the interrupt vectors through 0xfff0. `RTIF` is cleared by writing 1 into it (it is a clear port).

After each time-out the RTI handler is executed. The handler can directly input or output data, or it can set a global variable bit, which another program continually checks (in a gadfly loop), that causes data to be input or output. The handler must remove the source of this interrupt.

The second technique can be used to implement a more accurate traffic light controller. The global variable `TimeUp` is set in the handler. We modify traffic light controller's `main` to time out the execution of the interpreter using variable `TimeUp`.

```
const unsigned char tbl[4][2] =
    {{0x21, 4}, {0x22, 1}, {0x0c, 6}, {0x14, 2}};

volatile unsigned char count;

interrupt 7 void handler(){ RTIFLG = RTIE; count--; }

void main(){ unsigned char i,j; asm sei
   DDRA = 0x3f; count = 15; RTICTL = RTIE + T65; asm cli
    do {
         for(i = 0; i < 4; i++){
              PORTA = tbl[i][0]; /* output pattern */
              count = 15 * tbl[i][1]; /* 15 ticks/s */
              do ; while(count);
         }
     } while (1);
}
```

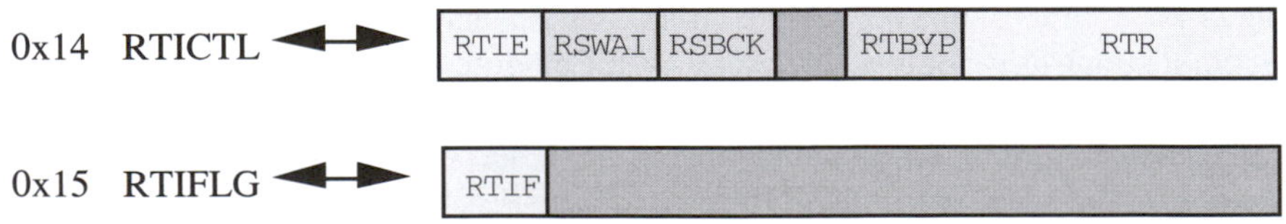

	RTIE	RSWAI	RSBCK		RTBYP	RTR
0x14 RTICTL	RTIE	RSWAI	RSBCK		RTBYP	RTR
0x15 RTIFLG	RTIF					

Figure 5.19. Periodic Interrupt Device

main is generally executing its gadfly `while` loop, because *count* is nonzero. Once each 32.7 ms, the handler decrements *count* each tick time-out. The `while` expression delays approximately `tbl[row][1]` seconds. Note that in place of a wait loop that uses possibly uncertain real-time program delays to time the lights, this example gadflies on the *count* variable synchronized to the more accurate RTI clock. The RTI clock is directly coupled to the crystal oscillator, and its time period is independent of software variability such as interrupts or changes in versions of the compiler, as well as hardware irregularities such as DMA and DRAM refreshing. This real-time interrupt version is thus much more accurate.

This example can be extended to many applications that require synchronization to very slow external devices. However, the 6812 yet remains tied up in a gadfly loop. In the next section, we will permit the 6812 to perform other functions while it is waiting to synchronize to a very slow I/O device.

5.3.2 Multithread Scheduling

A *multithread scheduling* technique, which is a primitive form of task scheduling done in a multitasking, multiuser operating system, would permit some other work to be done while this routine waits for real-time interrupts. We will generate real-time interrupts once every *tick,* where a tick is about 1/122th second (8.196 ms), rather than once every third of a second. This tick time is fairly commonly used because faster tick times consume too much time switching between threads, while slower tick times make the threads run erratically in time, in the perception of human users. We maintain three different threads, where a *thread* is a part of the program that is independent of other threads and that can be executed to do useful work. A thread can be executed for one tick, and then another thread might be executed for one tick, and so on. We can put a thread to *sleep* for a number of ticks; the other threads will be able to execute without competition from a sleeping thread. Rather than gadflying when waiting while an I/O device is BUSY, after using part of a tick to initiate the I/O operation, we can sleep for the remaining number of ticks until the I/O operation is completed and the DONE state is entered. The 6812 can do other useful work during those sleeping ticks; it does not spend all its time in a gadfly loop.

Multithread uses two simple ideas. First, the processor is tricked when returning from interrupts into restoring another thread's registers to run it. Second, a simple, effective way chooses the next thread to be executed: essentially, the thread waiting the longest to run is chosen to run. These two ideas are discussed now.

The information about a thread is maintained by a `struct`. Three threads can be maintained by a *struct* vector `threads[3]`. Members of each `struct` keep track of its `sleepTime, priority,` and `age` as well as the location of the thread's *stack.*

```
struct THREAD {
     volatile int sleepTime; unsigned char priority, age; char *stack;
} threads[3], *thisThread;
char *stackptr, nThreads;
```

The machine state needed for execution of a thread is frozen when the RTI interrupt occurs, by saving all its registers on the stack. Several threads will have their registers saved this way. The RTI `handler` shown below, which calls the procedure `findNext`, saving registers for one thread, may end by restoring these registers, but it may restore the registers for a different thread than the one that was just saved. That way a different thread can execute for the next tick.

```
enum { RTIE = 0x80, T8 = 0x04, RTIF = 0x80};

interrupt 7 void handler(){
    asm sts stackptr
    thisThread->stack = stackptr; /* save stack pointer for the int. thread*/
    thisThread->age = thisThread->priority; /* update this thread's age */

    findNext(); /* determine which thread runs in the next tick time */

    stackptr = thisThread->stack; /* restore stk pointer for max age's thread */
    asm lds stackptr /* move SP to new thread's saved stack, then return */
}

void findNext(){ unsigned char i, j; THREAD *p;

    for (i = j = 0; i < 3; i++) { /* try each thread */

        p = &threads[i]; /* p is thread pointer */

        if ( p->sleepTime ) /* if sleeping */
            { if(p->sleepTime != 0xffff) --p->sleepTime; }

        else { /* if not sleeping */
            if(RTIFLG && (p->age != 0xff)) p->age++;
            if(p->age >= j) { j = p->age; thisThread=p;} /* find max */
        }

        if(i == (nThreads - 1)) { /* if at end of search */
            RTIFLG = RTIF; /* remove interrupt */
            if(j == 0) { /* if no awake threads, restart search after next RTI */
                while ( ! (RTIFLG & RTIF)) ; i = 0xff; /* srch all thrds */
            }
        }
    }
}
```

The procedure `findNext` searches all the threads, deciding which thread will be executed, by its `sleepTime`, `priority`, and `age` variables. If `sleepTime` is nonzero, `sleepTime` is decremented each tick; otherwise, a thread will be executed from among all the nonsleeping threads `(sleepTime is 0)`. A thread can be made to sleep *N* ticks by making `sleepTime` equal to *N*. A thread goes to sleep if `sleepTime` is made nonzero. Among nonsleeping threads, a thread with the oldest *age* will be executed. (`sleepTime = 0xffff` will be used in the next section.) If no thread is "awake" so that the "oldest age" *j* remains zero after all threads are examined, a gadfly loop waits for the next RTI time-out, and the search of the threads is repeated. The *age* of each nonexecuting thread is incremented (up to 0xff), but when a thread is executed, its `age` is reset to its `priority`.

The `priority` is essentially *age*'s initial value. Nonsleeping threads share processing time in some sense "proportional" to their priority values. If all threads have the same `priority`, they will share processing time equally when they are not sleeping. If a thread has a much lower `priority` than all the other threads, it will execute when all the other threads are sleeping, or will execute very infrequently when they are not; such a thread is called a *background thread.* If a thread has a much higher `priority` than all the other threads, it will hog processing time; such a thread is called a `high-priority` *thread.*

The `sleep` procedure that follows gadflies while `sleepTime` is nonzero until an RTI interrupt selects another thread to run. When this thread runs again after waiting `d` tick times, its `sleepTime` becomes zero and it will exit the procedure.

```
void sleep(int d)
    { thisThread ->sleepTime = d; while(thisThread->sleepTime);}
```

A thread is started by the procedure `startThread`, which sets up the thread's execution entry point and its priority and age. The hardware stack is initialized to appear like the thread's stack just as the thread enters the RTI `handler`. This prepares the thread to run in the first tick time after being started, just as it will run in later tick times after it has been interrupted. However, just one thread, thread 0, continues to run as it starts other threads. Its stack is not initialized in `startThread`. Its stack is saved normally, when it is interrupted by the RTI device.

```
void startThread(int f,unsigned charp,int s){char *stack,t= nThreads++;
    threads[t].sleepTime=0; threads[t].age=threads[t].priority = p;

    if(t){ /* except for dummy thread 0, make stack space and initialize it */
        threads[t].stack = stack = (char*)malloc(s) + s - 9;
        stack[0] = 0; *(int *)(stack + 7) = f; /* initialize stack */
    }
}
```

The `restartThread` procedure following rewrites the program counter, and other thread variables, of an existing thread. It can be used after a thread is made to sleep "permanently" by setting its `sleepTime` to the maximim value (0xffff) with no other way, in order to awaken it to execute other functions.

```
void restartThread(int f,unsigned char p,unsigned char t) {char *stack;
   threads[t].sleepTime = 0; threads[t].age = threads[t].priority = p;
   stack = threads[t].stack; stack[0] = 0; *(int *)(stack + 7) = f;
}
```

Threads `main1` and `main2`, listed below, "flash" an LED on `PORTA` bit 4, at a rate of 1/2 Hz and an LED on `PORTA` bit 3 at 1 Hz. These threads sleep, and awaken to flash the LEDs from time to time.

```
void main1(){ do { PORTA ^= 0x10; sleep(122); } while(1); }

void main2(){ do { PORTA ^= 8; sleep( 61); } while(1); }
```

`main` initializes RTI with a 8.192 ms period and calls `startThread` to initialize its own thread as thread 0 and to initialize threads 1 and 2. The running thread, thread 0, calls `main0`. The other threads run `main1` and `main2` when they start executing, because the addresses of `main1` and `main2` are on these threads' stacks.

```
void main() { asm sei
    DDRA = 0xff; RITCTL = RTIE + T8;
    startThread (0, 0, 0); thisThread = &threads[0]; /* start threads */
    startThread ((int)main1, 50, 64); startThread ((int)main2, 50, 64);
    *(int *)0xb30 = (int) handler; /* insert dbug12 vector */ asm cli
    main0(); /* begin executing one of the threads */
}
```

5.3.3 Threads for Time-sharing

We now show threads that do some useful work while efficiently synchronizing to slow-speed devices. We will reexamine the keyboard controller and traffic light controller, and present a technique to support an alarm clock and a Centronics parallel printer device.

The human operator being a slow-speed device, keyboard scanning can be synchronized using time-sharing. Using this technique, newly pressed keys can be recognized even though other keys are pressed. Our earlier keyboard technique (§5.2.5) assumes that only one key is pressed at a time. If we allow two keys to be pressed simultaneously – as is often done by proficient keyboard users who press another key before releasing the first key – the program does not see an edge and does not generate an interrupt, if the newly pressed key is in the same row. Any technique that correctly recognizes any newly pressed key, even though $n - 1$ keys are still down, is said to exhibit *n-key rollover*. Two-key rollover is a common and useful feature that can be achieved using our improved technique, but for n greater than 2, one must avoid sneak paths through the keys. Sneak paths appear when three keys at three corners of a rectangle are pushed, and the fourth key seems to have been pushed because current can take a circuitous path through the pressed keys when the fourth key is being sensed. This can be prevented by putting diodes in series with each switch to block this sneak path current, but the solution is rather expensive and n-key rollover is not so useful.

A *2-key rollover* that uses a queue to record keystrokes, even while a previously pressed key may have to be recorded or responded to, can be appended to the multithread scheduler of §5.3.2. By sampling the keys about every 5 ms, a newly pressed key will be detected even though another key remains pressed. The time-sharing RTI `handler` calls the procedure `service` below, once each tick time.

```
char keys[8], previousKeys[8];

void service() { char r, c, i; // this service routine is called from handler
   for(i = 0; i < 8; i++){ /* scan 8 columns after a tick time is over */
      PORTA = ~(1 << i); /* put a low on just one column */
      c = keys[i] & ~previousKeys[i] & ~PORTB; /* analyze new key */
      if(c) for(j = 0, r = 1; j < 8; j++, r <<= 1;) /* scan the word */
         { if ( c & r ) { push(( i << 3) + j); thread[1].sleepTime=0;}
      previousKeys[i] = keys[i]; /* save older copy of keys to detect an edge */
      keys[i] = ~PORTB; /* save port data for next tick time */
}}
```

At the end of `service`, a first key pattern is copied to `keys`. After an 8-ms tick is executed, `handler` is again entered, which calls `service`. The first part of `service` is now executed, where the key pattern is compared to the key pattern 8 ms ago in `keys`, to determine if the key has been debounced. Moreover, a key pattern from 16 ms back is kept in `previousKeys`. This enables us to detect a falling edge of a debounced key. If a key input was high 16 ms ago, low 8 ms ago, and low now, the key has just been pressed, so push its code onto a queue. When a code is pushed onto the queue, a thread waiting for the data, thread 1 in this example, is awakened.

We illustrate a thread that uses key inputs. A `main1` program, which can substitute for §5.3.2's `main0`, `main1` or `main2`, can pull key codes from the queue where the above procedure `service` pushes them. `main1` sleeps indefinitely until a key is detected, where `service` awakens it. As in §5.2.5's linear select example, each key causes a different procedure to be executed; key code 0 causes `p0()` to be executed, etc.

```
void main1() { do {
      sleep(0xffff); // sleep until service() awakens thread to process data
      switch(Q.pull()) { // get the key code, execute its associated procedure
         case 0: p0(); break; // there can be 64 such procedures, each of
         case 1: p1(); break; // which is activated by one of the keys
      }
   } while(1); } // loop forever
```

Clearly this keyboard example is an improvement over the previous keyboard software. This technique doesn't waste time in a wait loop to get a couple of samples of the key signals to debounce the switch and establish a leading edge. Instead, it lets other threads run the computer, using up a tick time to wait and see if the key is still pressed. It further can handle n-key rollover if diodes are put in series with the switches. `main1` simply awaits key inputs, sleeping forever `(thread[1].sleepTime = 0xffff)`. It is awakened when `service` gets something for it to do `(thread[1].sleepTime = 0)`. This "wait until" scheduling mechanism is used by the Macintosh operating system; it waits on a next event queue to tell it what to do next.

We next look at our familiar traffic light controller's use of a multithread scheduler. Compared to the example in §5.3.1, this example puts the thread running the traffic light to sleep a number of tick times indicated by the time a light is to be left on.

```
void main1() { char i, tbl[4][2];
    DDRA = 0x3f;
    tbl[0][0]=0x21; tbl[0][1]=16; tbl[1][0]=0x22; tbl[1][1]=4;
    tbl[2][0]=0x0c; tbl[2][1]=20; tbl[3][0]=0x14; tbl[3][1]=4;
    do { for(i=0;i<4;i++){
            PORTA = tbl[i][0]; sleep(tbl[i][1] * 122);
        }
    } while(1);
}
```

Note that sleeping for a specific number of ticks is a "wait for" elapsed time scheduling mechanism. You can use it to make a thread wait for a rather long time until it is next able to do something. While it is waiting, other threads can use the computer without competition from this thread. However, the scheduler will not be able to guarantee that the thread will execute when it awakens. When it awakens, it only competes for time slices, along with all other nonsleeping threads. The one with the largest age will be given the use of the time slice.

a. View of Connector Seen from Printer

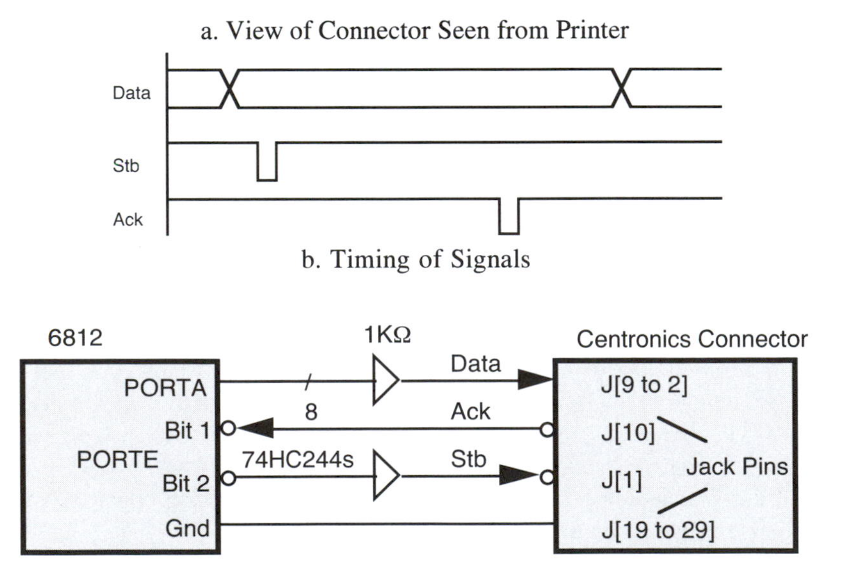

c. Connections

Figure 5.20. Centronics Parallel Printer Port

An alarm clock can be implemented using a thread. Numbers corresponding to events to be executed are stored in a vector *procedures*, in the order they are to be executed, and a vector *times* stores in element *i* the number of ticks from *procedure[i-1]* to *procedure[i]*. The procedure *main1* will execute *procedure[k]* after *times[0] + times[1] + ... times[k]* ticks have occurred.

```
int times[10], procedures[10];

void main2() { char i; do {
    sleep(times[i]);
    switch(procedures[i++]){
        case 0:p0();break;
        case 1:p1();break;
    }
} while(1); }
```

We illustrate a thread using gadfly and real-time interrupt synchronization for the Centronics printer example (Figure 5.20). Figure 5.20a shows the Centronics parallel printer connector. A character is printed by putting its ASCII code on the data lines, and asserting Stb low for at least 1 μs. When the printer accepts the character, it pulses Ack low (Figure 5.20b). We connect the printer data lines to *PORTA*, connect Ack to *PORTE* bit 1, and connect Stb to *PORTE* bit 2.

A personal computer mechanical printer usually has a buffer in it and can quickly put the character in the buffer, but if the buffer is full of data, the printer must wait milliseconds for a character to be mechanically printed before it has more room and can store the incoming character in the buffer. Thus, the time from Stb to Ack will be a few microseconds if the buffer is not full, or a few milliseconds if it is full.

Different response times indicate use of different synchronization mechanisms. Fast response, when the printer's buffer is not full, indicates gadfly synchronization, but the slow response, when it is full, indicates real-time interrupt.

We can use both gadfly and real-time interrupt synchronization. After writing data to *PORTA*, we produce a negative pulse on *PORTE* bit 2, which pulses Stb. This should cause the printer to assert Ack low, to put a falling edge on *PORTE* bit 1. Recalling that the printer might respond quickly or slowly, we check *PORTE* bit 1 right after Stb is pulsed, in a timed gadfly loop. Interrupts are disabled while this bit is checked, for if they were enabled, the interrupt handler would be entered before the key wakeup bit would be checked. If the printer responded quickly and this bit is set, then we re-enable interrupts and return. Otherwise, because we anticipate a long wait, in the procedure *sleep*, the current thread's *sleepTime* is set to make the thread sleep "forever" when the next tick occurs and the current tick is wasted using a gadfly loop on *sleepTime*. When the key wakeup interrupt occurs, its handler clears the thread's *sleepTime*, thus waking up the thread. Sleeping forever avoids the possibility that decrementing *sleepTime* will restart the thread when the printer has not responded. Of course this will hang up the thread if the printer is not on and does not respond. The user is supposed to recognize this and fix it.

```
interrupt 6 void handler(){threads[1].sleepTime=0;}/* awaken main1's put */
```

```
void put(char data) { unsigned char i;
    PORTA = data; PORTE &= ~4; PORTE |= 4; asm sei
    for(i = 0; i < 32; i++) if(~PORTE & 2) break; /* wait for Ack to fall */
    if(i < 32) { /* if Ack returns within the loop above, we are DONE */
        asm cli
        return; /* permit interrupts */
    }
    asm cli
    sleep(0xffff); /* otherwise sleep until interrupt */
}

void main2() { char i, string[10];  asm sei
    DDRE = PORTE = 4; DDRA = 0xff; INTCR = 0xc0; asm cli
    for(i = 0; i < 10; i++) put(string[i]); INTCR = 0; sleep(0xffff);
}
```

5.3.4 An Efficient Time Scheduler

The previous two sections have introduced a thread scheduler and its use in synchronizing slow I/O devices. This section refines the scheduler to make it more useful in microcontrollers having a very small amount of RAM. The key problem with the previous scheduler is that each thread had its own stack, whose buffers had to be large enough to accommodate the worst-case growth of its stack. This weakness makes such a scheduler too inefficient for a microcontroller with a small memory. A second problem is that once a thread is initialized but is no longer needed, its stack and structure can't be recovered to run other programs.

To circumvent these difficulties, the program uses a *shuffle stack.* The thread's stacks are shuffled when a new thread is selected to run, and the selected thread is put on top of the hardware stack. Other threads' stacks are shuffled to lower places in the hardware stack, and they have no space above them to expand. However, when a thread is run, its stack is on the top of the hardware stack and has room to expand as return addresses, operands, and local arguments are saved on it and as interrupts occur.

One subtle problem with the shuffle stack is that the stack can move whenever interrupts are disabled. Therefore, a pointer to a local vector should be recomputed, each time it is used directly from the stack pointer. Indexes to local vectors should also be used carefully. Pointers and indexes to local variables may be used in critical sections.

This program is presented in its entirety below, so you can see all of it. The *struct* THREAD is the same as used before. Many procedures are essentially identical to those used before, such as `main`, `main0`, `main1`, `main2`, `sleep`, *and* `findNext`. The procedure `startThread` has no RTS instruction, so it falls through to *sleep,* which falls through to the handler. The procedure `adjustStack` is called from the RTI `handler`, which is essentially the same as the previous RTI handler. `adjustStack` shuffles the stacks to make the selected thread's stack be the top stack. Its first `for` loop looks for the bottom of the selected thread's stack. Its next `for` loop adjusts the addresses of the top of each stack. Then the heap, pointed to by `free_ptr`, is used to temporarily hold the selected procedure's stack while it is moved in the assembly-language segment.

```
extern void main1(void), main2(void);
#define STACK 4
#define nThreads 3

char *stack,*local,keys[8],previousKeys[8]; Queue<char> Q(10);

struct THREAD{
    volatile int sleepTime; unsigned char priority, age,*stack;
} threads[nThreads], *thisThread;

void findNext(){ unsigned char i, j; THREAD *p;
    for(i = j = 0; i < nThreads; i++) { /* try each thread */
        p = &threads[i]; {/* p is thread pointer */
        if ( p->sleepTime ) /* if sleeping */
            if(p->sleepTime!=0xffff) /* 0xffff means sleep forever */
                --p->sleepTime; /* if not sleeping forever, decrement sleepTime */
        }
        else { /* if not sleeping */
            if(RTIFLG&&(p->age!=0xff))p->age++;/* increment age */
            if(p->age >= j) {
                j = p->age; thisThread = p; /* find max age */
            }
        }
        if(i == (nThreads - 1)) { /* if at end of search */
            RTIFLG = RTIF; /* remove interrupt */
            if(j == 0) { /* if no awake threads (did not change j) */
                i = 0xff;/* end for loop by setting loop counter to maximum */
                while ( ! (RTIFLG & RTIF)) ;/* restart search after next RTI */
            }
        }
    }
}

void adjustStacks(){ unsigned char i; int j;
    for (i = 0, stack = (char *)ENDSTACK; i < nThreads; i++) {
        if((threads[i].stack >
                thisThread->stack)&&(threads[i].stack<stack))
                    stack = threads[i].stack; /* get lowest */
    }

    for (j = stack - thisThread->stack, i = 0; i < nThreads; i++) {
        if(threads[i].stack < thisThread->stack)
            threads[i].stack += j; /* adjust pointers */
    }
```

```
asm{
    ldy thisThread         ; get pointer to selected thread
    ldy STACK,y            ; point to low address end of new thread's stack
    ldx free_ptr           ; get end of heap
L1: movb    1,y+,1,x+      ; move a byte from the stack to the heap
    cpy stack              ; at end of new thread's stack area?
    blo L1                 ; no, get more
    stx free_ptr           ; save high end of heap

    ldx thisThread         ; note: Y is at high end of this thread's stack
    ldx STACK,x            ; point to low end of new thread's stack
L2: movb    1,-x,1,-y      ; move a byte on the stack to close up the gap
    cpx local              ; all bytes in stack moved?
    bne L2
    ldx free_ptr           ; get end of heap
L3: movb    1,-x,1,-y      ; move a byte on the heap back onto the stack
    cpy local              ; all bytes moved from heap to stack?
    bne L3
    stx free_ptr           ; restore heap pointer for malloc
}}

void restartThread( int f, unsigned char t ) { char *stack;
    threads[t].sleepTime = 0;
    threads[t].age = threads[t].priority;
    stack = threads[t].stack;
    stack[0] = 0;
    *(int *)(stack + 7) = f;
}

#pragma NO_EXIT
void startThread(int f, unsigned char p) {
    stack = (char *)thisThread; thisThread = &threads[nThreads++];
    thisThread->age = thisThread->priority = p;

asm{
    clra                   ; this procedure builds the stack for a new thread
    clrb                   ; D is cleared; it becomes the argument for sleep( )
    ldy stack              ; get former thread
    beq L1                 ; if no former thread, don't save regs; else make stack
    ldx 2,sp               ; get location of function f to X
    leas -6,sp             ; make room on stack for former thread's saved D, X, and Y
    pshb                   ; cc register: clear interrupt bits
    sts STACK,y            ; note: y is loaded at the beginning of this procedure
    pshx                   ; save X as return address for new thread's stack
L1: ds.b    0              ; fall through to sleep, with argument time = 0
}}
```

```
void keyCheck() { char fallingEdge, r, c, i, j;
    for(i = 0; i < 8; i++){ /* scan 8 columns after a tick time is over */
        PORTA = ~(1 << i); /* put a low on just one column */
        c = keys[i] & ~previousKeys[i] & ~PORTB; /* analyze for new key */
        if(c) for(j = 0, r = 1; j < 8; j++, r <<= 1) /* scan the word */
            if (c & r) { Q.push(( i << 3) + j);threads[1].sleepTime=0;}
        previousKeys[i] = keys[i]; /* save older copy of keys to detect an edge */
        keys[i] = fallingEdge & ~PORTB; /* save port data for next tick time */
    }
}

#pragma NO_EXIT
void sleep(int time) {
    thisThread->sleepTime = time;
asm{
    leas -6,sp ; build stack for thread: make room for saving D, X, and Y
    pshc       ; condition code register (note: return address is on stack)
    sei
}}

interrupt 7 void rtiHandler(){
    asm sts local
    thisThread->age = thisThread->priority; thisThread->stack = local;
    keyCheck(); /* check keys for any pressed */
    findNext(); /* determine which thread runs in the next tick time */
    if(thisThread->stack != local) adjustStacks(); /* shuffle stacks */
}

#pragma NO_EXIT
void main() { asm sei
     PORTA = DDRA = 0xff; /* use PORTA for testing */
    RITCTL = RTIE + T8;/* start real-time interrupt */
    startThread ((int)main1, 50);/* fork thread 1 */
    startThread ((int)main2, 50);/* fork thread 2 */
    /* fall through to thread 0's procedure */

}
/* main0( ) is put here */
```

The procedure `restartThread` is used to cause an existing sleeping thread, selected by its second argument, to begin execution of a procedure provided by its first argument. Whereas the previous examples of `main1`, and so on, were infinite loops that did not return to the calling program, we can now have a procedure `main1` that ends by sleeping forever, without the possibility that an interrupt will awaken it. Such a thread can be restarted at another location using the procedure `restartThread`.

5.3.5 Special Instructions for Time-sharing

The 6812 has two instructions, *WAI* and *STOP,* that are used to manage interrupts. Though we introduced these in §1.2.3, we postponed a serious discussion of them until now because they are strongly related to interrupts. `WAI` is an improved kind of wait loop, and STOP can be a fast gadfly loop. We also have the `SWI` instruction and illegal instructions; these are considered again now that interrupts have been studied.

`WAI` waits for an interrupt, in lieu of the infinite loop at the end of `main()` in §5.3.2. A background thread can have a statement `asm WAI`. Once the `WAI` instruction is executed, the 6812 goes to "sleep," to be "awakened" by a reset or by an IRQ or XIRQ interrupt. However, `WAI` is optimized to reduce latency time. As soon as `WAI` is executed, it pushes all the registers on the stack as if honoring an interrupt. When an interrupt occurs, it does not have to save these registers. `WAI` can be used to reduce the latency to about five memory cycles. `WAI` stops the 6812 core. Although the MC68HC812A4 requires only 35 mA of 5-V supply current, `WAI` reduces this to half the amount (if the serial interfaces SCI, SPI, and timer are not running).

Another instruction, `STOP`, stops the oscillator and the entire microcomputer, reducing the supply current to a mere 300 μA. However, setting the S condition code bit disables execution of the `STOP` instruction. A background thread can be written:

```
void main0() { do asm { STOP}; while(1)}}
```

If no threads are awake, it will be executed to shut down the microcontroller.

If INTCR (0x1e) bit 5 is 1, after `STOP`, about a millisecond may be needed for the oscillator to restart so the processor can resume the next operation. Otherwise when the X condition code bit is set and XIRQ interrupts are therefore disabled, `STOP` is, in effect, an optimized gadfly loop. When the XIRQ line becomes asserted low, the stack pointer is repositioned to effectively remove the saved registers, which takes two clock cycles; the instruction following the `STOP` instruction is then executed next.

`SWI` has been discussed in §1.2.3. Here we observe that the condition code interrupt mask bit I is set exactly as if IRQ or XIRQ lines caused a hardware interrupt.

Many opcodes are not implemented in the 6812 instruction set. To catch bugs in programs, these unimplemented instructions result in an SWI-like operation, jumping to the location specified in 0xFFF8 and 0xFFF9. However, a debugger can use some of these unimplemented instructions as "hooks." Opcodes from 0x1830 to 0x1839 can perform functions `exit`, `compile-time breakpoint`, `getchar`, `gets`, `getdec`, `puts`, `gethex`, `putchar`, and `printf`. The hook `exit` stops execution, to resume at the beginning of the program; `compile-time breakpoint` stops execution, to resume after the next instruction; `getchar` gets a character, `gets` gets a string, `getdec` gets a decimal number, and `gethex` gets a hexadecimal number from the host's keyboard. The hook `putchar` puts a character on the screen; `puts` writes a string, and `printf` writes a string with characters, decimal and hexadecimal numbers, on the display. This last hook merely copies the `printf` format string and arguments from target memory to the personal computer memory, and then outputs using its `printf` procedure. While we did this successfully in a debugger UTBUG that ran on a Macintosh, it is not available in the HIWARE debugger that we are now using. But that might change someday.

5.3.6 Object-oriented Classes for Time-sharing

Object-oriented programming provides protection for each thread, and also sleep capability in place of delay or gadfly loops that lets the microcontroller perform useful work while waiting for an I/O operation. These concepts are covered in this section.

To use objects for programs like §5.3.2's `main1` and `main2`, we define a class `thread` that contains data members `sleepTime`, `age`, `priority`, and `stack`, used for time-slicing, as well as functions and data members common to two or more threads, and we generally define a derived class such as `thread1` for each thread unless it is completely identical to another thread. The `thread` class will have a function `main` that will be overridden in each derived class. The function `main` of class `thread1` will contain the starting procedure for thread 1. The `main` procedure that is started after reset, as `main` of §5.3.2 was started, will call class `thread1`'s constructor, which allocates its stack and initializes its data members. The real-time interrupt will start each thread's `main` function in turn, and when the thread's *age* is largest, run if for a time tick.

Object-oriented threads provide protection and polymorphism. Each thread has its own scope of names for function and data members; these can be declared private to protect them from other threads, and names can be reused. True global variables can be used to share information among threads and between interrupt handlers and threads.

Object-oriented classes of a thread can use `sleepTime` in lieu of delay or gadfly loops, if synchronization parameter bits 15 to 0 are 0xfc00 (`sleepTime` is loaded with bits 31 to 16). In the earlier class `IQFPort`, if the input queue empty, `get` gadflies on the queue size until an interrupt pushes some data, then `get` pulls this data from the queue. Using a synchronization parameter 0xfffffc00, `get` instead sleeps indefinitely until an interrupt pushes some data and wakes it up; then `get` pulls this data from the queue. These remarks also apply to the class `OQFPort` with the function `put`. A synchronization parameter 0x20fc00 makes it sleep for 20 tick times, which is assumed to be the time for the paper-tape punch to output the byte. These objects can be declared.

```
IQFPort<char> ppr(aA,0xFFFFFC02, (aB << 16)|0xB001, 10, (int)handler1);
OQFPort<char>
    ppp(aC,0x20FC08,(aB << 16)|0xB004,10, (int)handler2, dirAt2, 0xff);
```

A class `Pipe` is useful in linking a thread with another thread in a `pipeline`. The following templated class is a simple but effective *pipe*.

```
template <class T> class Pipe : public Port<T> {
    Queue <T>*Q; public : THREAD *thread;
    Pipe(unsigned char size) : Port(0){Q = new Queue<T>(size);}
    virtual void put(T data) {
        while(Q->size >= Q->maxSize) sleep(0xffff);
        Q->push(data);
        if(thread) thread->sleepTime = 0; thread = 0;
    }
```

```
    virtual T get(void) { T v;
        while(Q->size == 0) { thread = thisThread; sleep(0xffff); }
        v = Q->pull(); attr |= Q->error(); return v;
    }
    virtual T operator = (T data) { put(data); return data; } // assig.
    operator T () { return get(); }; // cast
};
```

The procedure `main1`, executing in `threads[1]`, outputs data to the procedure `main2`, which inputs it, using objects `ppr` and `ppp`, as shown below:

```
Pipe<char> pipe(10);

void main1() { char c; do { pipe = c = ppr; } while(1); }

void main2(){ char c; ppp = c = pipe; }
```

The procedure `main1`, executing in `threads[1]`, "outputs" data to the procedure `main2`, executing in `threads[2]`, where it appears as an "input." (The local variable *c* is used to force the use of assignment and cast operators. If it is not used, the compiler attempts to make one object a copy of the other object.) Since `pipe` is declared as a global object, its constructor is called before `main1` or `main2` are called. `main1`'s overloaded assignment operator calls `Pipe`'s `put` function member, and `main2`'s overloaded `cast` operator calls `Pipe`'s `get` function member. The queue holds `main1`'s output data until `main2` is ready to use it. One thread "outputs" to the pipe, while the other "inputs" from the pipe, as if the pipe were an I/O device. However, the pipe is merely a queue that holds "output" data until it is "input" to the other thread.

5.4 Fast Synchronization Mechanisms

In the previous section, we discussed the synchronization mechanisms used for slower I/O devices. There are seven mechanisms used for faster devices. These are direct memory access, context switching, coprocessing, and shuttle, indirect, time-multiplexed, and video memory. They are briefly outlined in the last section.

The first two subsections discuss three I/O synchronization techniques that are faster than interrupts. *Direct memory access* (DMA) is a well-known technique whereby an I/O device gets access to memory directly, without having the microprocessor in between. By this direct path, a word input through a device can be stored in memory, or a word from memory can be output through a device, on the device's request. It is also possible for a word in memory to be moved to another place in memory using direct memory access. The second technique, *context switching,* is actually a more general type of DMA. The *context* of a processor is its set of accumulators and other registers (as Texas Instruments uses the term) and the instruction set of the processor. To switch context means to logically disconnect the existing set of registers – bringing in a new set to be used in their place – or to use a different instruction set, or both. Finally, memory can be connected to the computer or to the I/O device at different times. These three techniques are now studied.

5.4.1 Direct Memory Access

One of the fastest ways to input data to a buffer is direct memory access. Compared to techniques discussed earlier, this technique requires considerably more hardware and is considerably faster. A *DMA channel* is the additional logic needed to move data to or from an I/O device. In DMA, a word is moved from the device to a memory in a *DMA transfer.* The device requests transferring a word to or from memory; the microprocessor CPU, which may be in the middle of an operation, simply stops what it is doing for one to five memory cycles and releases control of the address and data buses to its memory by disabling the tristate drivers that drive these buses; the I/O system including the DMA channel is then expected to use those cycles to transfer words from its input port to a memory location. Successive words are moved this way into or from a buffer.

The DMA device has a DESTINATION address, a COUNT, and a DONE status bit. The DESTINATION and COUNT registers are initialized before DMA begins, and are incremented and decremented, respectively, as each word is moved.

Two DMA techniques are generally available. An internal I/O device or an external I/O device can use DMA. If an external I/O device wishes to input or output data, it causes an edge on a pin control signal, which *steals a memory cycle* to transfer one word in *cycle steal mode,* or it asserts a level that halts the microprocessor to transfer one or more words in *burst mode,*as long as the level remains asserted. This sequence of events happens when a DMA request is made to input a byte using cycle stealing.

1. A falling edge occurs on an input pin, signaling the availability of data.
2. In the DMA controller, a request is made and granted, and a memory cycle is stolen from the processor.
3. The controller signals the I/O device to read a byte from a port.
4. The byte is written into memory using a DESTINATION address.
5. The COUNT value is decremented. If it becomes 0, the DONE status bit is set. The program gadflies or interrupts on this DONE bit, and resumes when it is set.

There is a two-level BUSY/DONE state associated with DMA, as with any I/O transfer that fills or empties a buffer, as discussed at the beginning of this chapter. The low-level BUSY/DONE state is associated with the transfer of single words. BUSY is when a word is requested from an input device and has not been input, or is sent to an output device and has not been fully output (the hardware is punching the paper in the paper-tape example). The high-level BUSY/DONE state is associated with the transfer of the buffer. BUSY is when the buffer is being written into from an input device and has not been completely filled, or the buffer is being read from into an input device and has not been completely emptied. The DMA channel synchronizes to the low-level BUSY/DONE state to move words into or out of the I/O device. The computer can synchronize with the high-level BUSY/DONE state in the ways discussed so far. A real-time synchronization would have the processor do some program or execute a wait loop until enough time has elapsed for the buffer to be filled or emptied. Gadfly synchronization was used in the example given above. An interrupt could be used to indicate that the buffer is full.

I/O used in high-level languages is often *buffered* or *cached.* For input, a buffer or *cache* is maintained and filled with more data than are needed. In *lazy buffer management,* the buffer is filled with data only when some data input is requested, but more data is put into the buffer than is requested in order to take data from the buffer, rather than from the input device, when some more data is needed later. In *eager buffer management,* the buffer is filled with data before some data input is requested, so that it will be in the buffer when it is requested. This technique makes the I/O device faster.

Finally, DMA can be used to synchronize to the high-level BUSY/DONE state; a kind of DMA^2. In larger computers such as an IBM mainframe, such a pair of DMA channels is called an *I/O channel.* In an I/O channel, a second DMA that synchronizes the high-level BUSY/DONE state of the first DMA channel will refill the COUNT, SOURCE, DESTINATION, and CONTROL of the first DMA channel that moves words synchronizing to the low-level BUSY/DONE state. Thus, after one buffer is filled or emptied by the first DMA channel, the second DMA channel sets up the first DMA channel so the next buffer is set up to be filled or emptied. The second DMA channel's buffer is conceptually a program called the *I/O channel program.* This channel itself has BUSY/DONE states. BUSY occurs when some, but not all, buffers have been moved, and DONE occurs when all buffers have been moved. How can this BUSY/DONE state synchronizing the high-level BUSY/DONE states be synchronized itself? Here we go again. It can be synchronized using real-time, gadfly, interrupt, or DMA synchronization. However, DMA^3 is not very useful; DMA would not be used to reload the COUNT, SOURCE, DESTINATION, and CONTROL of the second DMA channel.

Direct memory access requires more hardware and may restrict the choice of some hardware used in I/O systems. The DMA channel must be added to the system, and the other I/O chips should be selected to cooperate with it. However, the amount of software can be less than with other techniques because all the software does is initialize some of the ports and then wait for the data to be moved. The main attraction of DMA is that the data can be moved during a memory cycle or two anytime, without waiting for the M68340 to use software to move the data.

5.4.2 Context Switching

An interesting variation to DMA, uniquely attractive because it is inexpensive, is to use two or more microprocessors on the same address and data buses. See Figure 5.21. One runs the main program. This one stops when a device requests service, as if a DMA request were being honored, and another microprocessor starts. When the first stops, it releases control over the address and data buses, which are common to all the microprocessors and to memory and I/O, so the second can use them. The second microprocessor, which then can execute the interrupt request handler, is started more quickly because the registers in the first are saved merely by freezing them in place rather than saving them on a stack. The registers in the second could already contain the values needed by the interrupt request handler, so they would not need to be initialized. DMA using a DMA chip is restricted to just inputting a word into, or outputting a word from, a buffer; whereas the second microprocessor can execute any software routine after obtaining direct memory access from the first microprocessor.

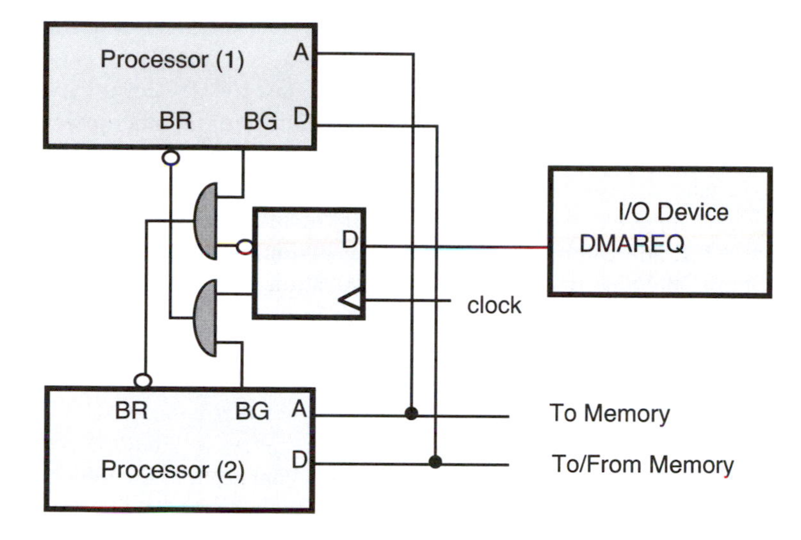

Figure 5.21. Connections for Context Switching

A complex operation is easy to do with context switching. While ordinary DMA cannot do this operation, context switching is faster than interrupt synchronization because not only are the registers not saved, but also they remain in the second processor, so they usually don't have to be initialized each time an interrupt occurs.

Finally, any set of microcomputers having DMA capability can be used in this manner; the one operating the main program need not be the same model as the one handling a device. This means you can put a new microprocessor in your old microcomputer. The old microprocessor is turned on to run programs left over from earlier days, and the new microprocessor is turned on to execute the new and better programs. This is an alternative to simulation or emulation in microprogramming. It is better because the best machine to emulate itself is usually the machine itself. And putting two microprocessors in the same microcomputer has hardly an impact on the system's cost.

A *coprocessor,* such as the floating-point 68881, essentially uses the same concept as context switching. Whenever the main processor detects a floating-point add that should be executed in the 68881, it gives up the bus to the coprocessor, which does one instruction, and then the main processor resumes decoding the next instruction. A one-instruction context switch makes the main processor and coprocessor appear to be part of a single processor having both the main processor's and the coprocessor's instructions. While coprocessors can be designed to handle I/O, they usually handle data computation.

Though this technique is not used often by designers because they are not familiar with it, it is useful for microcomputers because the added cost for a microprocessor is so small and the speed and flexibility gained are the equivalent of somewhere between those attained by true DMA and vectored interrupt, a quality that is often just what is required.

5.4.3 Memory Buffer Synchronization

The last techniques we will consider that synchronize fast I/O devices involve their use of memory, which is not restricted by memory conflicts with the microprocessor. One technique uses a completely separate and possibly faster memory, called a *shuttle memory.* A variant of it uses an I/O device to access memory, like indirect I/O, and is called an *indirect memory.* Another uses the same memory as the microprocessor, but this memory is fast and can be *time-multiplexed,* giving time slices to the I/O device. In a sense, these techniques solve the synchronization problem by avoiding it – by decoupling the microprocessor from the I/O via a memory that can be completely controlled by the I/O device.

Figure 5.22a shows a shuttle memory. The multiplexer connects the 16 address lines that go into the shuttle buffer and the 16 data lines that connect the buffer to the microprocessor or the I/O device. The buffer memory is shuttled between the microprocessor and the I/O device. When the buffer is connected to the I/O device, it has total and unrestricted use of the shuttle memory buffer whenever I/O operations take place. The microprocessor can access its primary memory and I/O at this time without conflict with the I/O's access to its shuttle memory because they are separate from the shuttle memory used by the I/O device. The multiplexer switches are both in the lower position at that time. Then, when the microprocessor wishes to get the data in the shuttle memory, the multiplexer switches are put in the upper position, and the microprocessor has access to the shuttle memory just as it has to its own primary memory. The buffer appears in the memory address space of the microprocessor. The microprocessor can load and store data in the shuttle memory. The synchronization problem is solved by avoiding it. Synchronization is required as data are moved to and from the I/O device from and to the shuttle memory; but the buffer memory is wholly controlled by the I/O device, so that it is not too difficult. The microprocessor can move data to and from the shuttle memory at leisure. It can even tolerate the delays that result from handling an interrupt at any time, when it moves data from one location in its memory to another location in it. There is no need for synchronization in that operation.

We built a parallel computer called TRAC, which used shuttle memories. The shuttle memories were connected to one processor or another processor. Once connected to a processor, the shuttle memory behaved like local memory and did not experience memory contention. Caches were simple to use with this variation of a shared memory. In I/O devices discussed in this book, the shuttle memory similarly removes the problem of memory contention from the synchronization problem.

A variation of a shuttle memory uses a parallel I/O device like 6812 ports in place of the multiplexer. (See Figure 5.22b.) The external port pins of an I/O device connect to the address and data buses of the memory. The processor writes an address to a parallel output port and then reads (or writes) the data to (or from) another port to access the memory. The only way to read or write in the buffer is to send addresses to, and data to or from, the I/O device, just as we accessed an indirect I/O device in §4.4.1, so we call it indirect memory. The M6818A's RAM from addresses 0xE to 0x3F is an indirect memory. Indirect buffer memory is completely separate from the microprocessor primary memory. When the (fast) memory is not controlled by the microprocessor through the I/O ports, it can be completely controlled by the I/O device, so it can synchronize to fast

I/O devices. Only the memory-mapped parallel I/O device takes up memory space in the primary memory, whereas the shuttle memory technique has the whole shuttle memory in the primary memory address space when the processor accesses it. But to access the buffer memory, you use slow subroutines as you do in indirect I/O.

Indirect memory using the MCM6264D-45 8K-by-8 chip is easily implemented on the 'A4. Tristate drivers (74HC244s) connect the external device to the memory when the 'A4 is not accessing it, so when they are not used, all 'A4 port bits are made inputs to allow the 74HC244s to access the memory signals. The 'A4 reads the memory, following the timing diagram in Figure 5.23a, by making PORTA and PORTB outputs; making E1, G, and W high, outputting the high byte of the address to PORTA, and the low byte to PORTB, asserting E1 low, asserting G low, reading the data from PORTC, negating E1 high, and negating G high. Writing, following the timing diagram in Figure 5.23b, is done by making PORTA and PORTB outputs, making E1, G, and W high, outputting the high byte of the address to PORTA and the low byte to PORTB, asserting W low, asserting E1 low, making PORTC output, writing the data to PORTC, negating E1 high, and negating W high. The hardware is connected as shown in Figure 5.23c. Only connections to the 'A4 are shown.

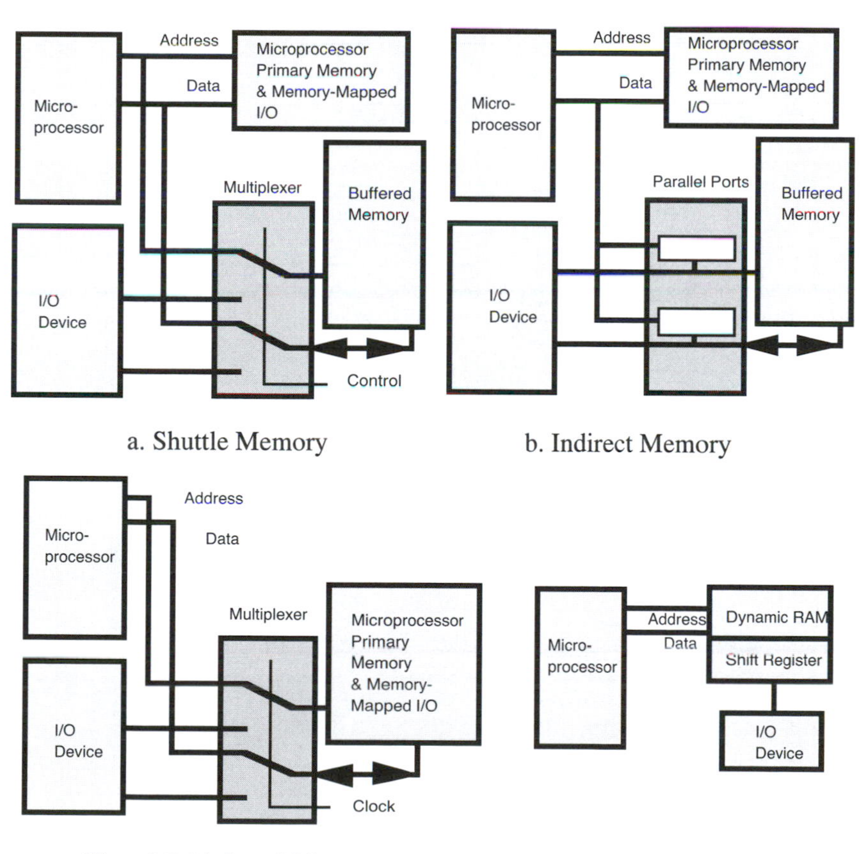

a. Shuttle Memory

b. Indirect Memory

c. Time-Multiplexed Memory

d. Video RAM

Figure 5.22. Fast Synchronization Mechanisms Using Memory Organizations

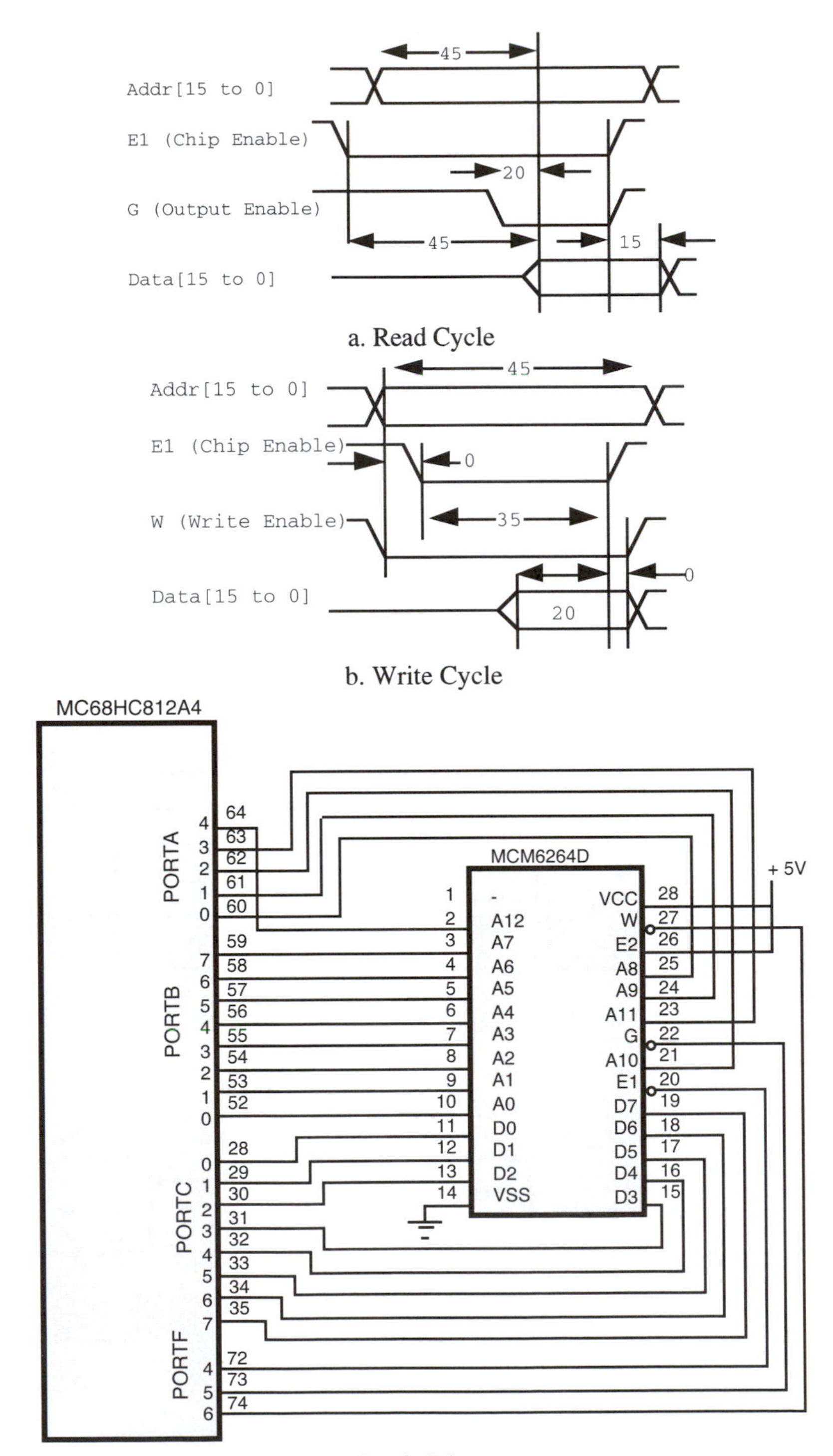

c. Logic Diagram

Figure 5.23. Indirect Memory Using an MCM6264D-45

```
enum {E1 = 0x10, G = 0x20, W = 0x40};

char get(int a) { char data;
    DDRA = DDRB = 0xff; DDRE = E1 + G + W; PORTA = a >> 8; PORTB = a;
    PORTE &= ~E1; PORTE &= ~G; data = PORTC; PORTE |= E1; PORTE |= G;
    DDRA = DDRB = DDRE = DDRC = 0; return data;
}

void put(int a, char data) {
    DDRA = DDRB = 0xff; DDRE = E1 + G + W; PORTA = a >> 8; PORTB = a;
    PORTE &= ~W; PORTE &= ~E1; DDRC = 0xff; PORTC = data;
    PORTE |= E1; PORTE |= W; DDRA = DDRB = DDRE = DDRC = 0;
}
```

A very similar mechanism uses the same memory for the primary memory and the buffer memory, but that memory is twice as fast as is necessary for the processor. (See Figure 5.23c.) In one processor memory cycle, the memory executes two memory cycles – one for the processor and one for the I/O device. The multiplexer is switched to the I/O device (for the first half of the memory cycle) and to the processor (for the last half of the memory cycle) to time-multiplex the memory. The I/O device always gets one memory cycle all to itself because the processor only uses the other memory cycle.

The time-multiplexed memory uses the same memory as the microprocessor, but this memory is twice as fast; the processor gets one time slice and then the I/O device gets one time slice, in an endless cycle. It is obviously less costly than the shuttle and indirect memories because a single large memory is used rather than two smaller memories. Its operation is very similar to DMA. In fact, it is sometimes called *transparent DMA*. However, the memory must be twice as fast as the processor, and the I/O device must synchronize to the processor (CLK) clock in this technique. The shuttle and indirect memories are more costly; however, a very fast (40-ns cycle time) memory can be used in the buffer and run asynchronously at full speed when accessed by the I/O device, and run at about the speed of the CLK clock when the processor accesses it. All three techniques provide for faster synchronization to the I/O device than the techniques discussed in the previous subsection. They can transfer data on every memory cycle without handshaking with the processor to acquire memory or use the processor. They find considerable use in CRT, hard-disk, and fast-communication I/O devices.

Finally, a *video RAM* or *VRAM* (Figure 5.22d) is a dynamic memory in which a row can be read from DRAM into a shift register, or written into DRAM from a shift register, in one memory cycle. The shift register can then shift data into or out of an I/O device. The TMS48C121 is a (128K, 8) DRAM with a (512, 8) shift register, which can shift data into or out of an I/O device at 30 ns per byte.

One of the main points of this section is that extra hardware can be added to meet greater synchronization demands met in fast I/O devices. While DMA is popular, it is actually not the fastest technique because handshaking with the microprocessor and the cycle time of the main memory slow it down. Shuttle or indirect memories that use fast static RAMs can be significantly faster than DMA. Moreover, for all of these techniques, the controlling software can usually be quite slow, and thus can be coded in C without loss of performance compared to programs coded in assembly language.

5.5 Conclusions

We have discussed over ten alternatives for solving the synchronization problem. Each has some advantages and some disadvantages. Figure 5.24 summarizes the techniques presented in this chapter.

Real-time synchronization uses the least hardware and is practical if an inexpensive microcomputer has nothing to do but time out an I/O operation. However, it can be difficult to program. Gadfly programs are easier to write, but they require that the hardware provide an indication of DONE. Also, a computer generally cannot do anything else when it is in a gadfly loop, so this is as inefficient as real-time synchronization. Real-time interrupt synchronization provides for long delays, giving up the processor to other threads or processes. Real-time and real-time interrupt synchronization are synchronous in that the external timing is determined by processor timing. Gadfly is asynchronous in that the I/O device's timing is not synchronized to the processor's timing, but the processor locks onto its timing when data are to be transferred.

The interrupt-polling technique and the vectored interrupt technique require more hardware to request service from the processor. The 6812 provides for autovectored and for external vectored interrupts. They are useful when the device needs service in a shorter time. However, the tendency to use them just because they are available should be avoided. Although interrupt polling only requires an interrupt bus line from device to processor, if the gadfly approach is exclusively used, this line invites the mayhem of an unrecognizable interrupt should a software error rewrite the control port in the device. Also, the interrupt technique can be used together with the gadfly technique. With the gadfly technique, the interrupts are all disabled by setting the status port current priority, as in the OR #0x700,SR instruction, or by clearing control bit 0 in an M6821 device. Then the program can loop as it tests the device, without fear of being pulled out by an interrupt. When utilized together, gadfly is used for careful, individual stepping of an I/O system; interrupt is used for automatic, rapid feeding of data.

The DMA technique is useful for fast devices that require low latency. This technique can only store data in a buffer or read data from a buffer. DMA^2 (the I/O channel) can restart a DMA transfer a little bit faster than simple DMA can. A variation of DMA, context switching, is almost as fast and flexible as the interrupt technique. A coprocessor uses a similar mechanism, and although it is generally used to execute data computation, it could be used for I/O. Shuttle, indirect, and time-multiplexed memories can be used for the fastest devices. What is somewhat surprising is that DMA, which requires a fair amount of extra and expensive hardware, actually is most desirable for a rather limited range of synchronization timing. Indirect and shuttle memories can be used for much faster synchronization, and context switching for slightly slower synchronization.

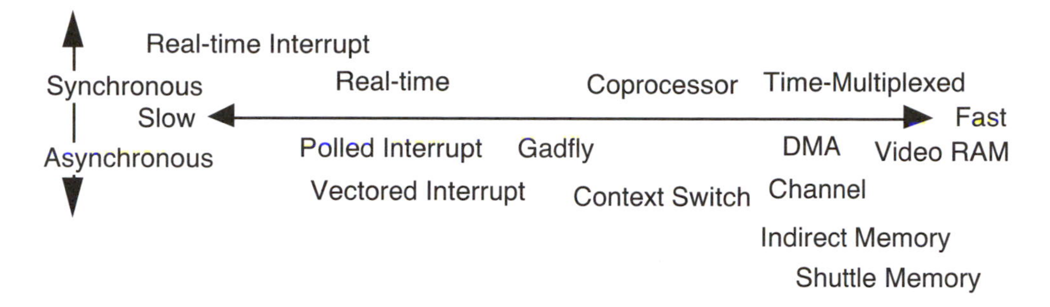

Figure 5.24. Synchronization Mechanisms Summarized

We have discussed eight alternatives for solving the synchronization problem. Each has some advantages and some disadvantages.

Real-time synchronization uses the least hardware and is practical if an inexpensive microcomputer has nothing to do but time out an I/O operation. However, it can be difficult to program. Gadfly programs are easier to write but require that the hardware provide an indication of the DONE state. Also, a computer cannot do anything else when it is in a gadfly loop, so this is as inefficient as real-time synchronization.

The interrupt-polling technique and the vectored interrupt technique require more hardware to request service from the processor. They are useful when the device needs service in a shorter time. However, the tendency to use them just because they are available should be avoided. Except when the processor has enough interrupt request lines to handle all interrupts, to provide the device handler's address, vectored interrupt may require an extra chip, which is a significant cost. Although interrupt polling only require a bus line from device to processor, if the gadfly approach is exclusively used, this line invites the mayhem of an unrecognizable interrupt should a software error rewrite the control register in the device. Also, the interrupt technique can be used together with the gadfly technique. With the gadfly technique, the interrupts are all disabled by setting the I condition code, as in the SEI instruction, or by clearing control bit 0 in the M6821 device. Then the program can loop as it tests the device, without fear of being pulled out by an interrupt. Commonly, gadfly is used for careful, individual stepping of an I/O system, and interrupt is useful for automatic and rapid feeding of the data.

The DMA technique is useful for fast devices that require low latency. This technique can only store data in a buffer or read data from a buffer. A variation of it, context switching, is almost as fast and flexible as the interrupt technique. Isolated memories and time-multiplexed memories can be used for the fastest devices.

Given these various techniques, the designer can pick one that suits the application. This chapter has shown how simple and flexible these techniques are.

Do You Know These Terms?

See page 36 for instructions.

synchronization
IDLE
BUSY
DONE
starting
stopping
completing
real-time synchronization
gadfly
gadfly loop
handshake
opto-isolator
signal
interrupt
device requests an interrupt
interrupts are enabled
microprocessor is enabled
microprocessor is masked
microprocessor is disabled
non-XIRQ
microprocessor sees a request
request is pending
honors an interrupt
handler
arm
armed
disarmed
enable
disabled
critical section
software disarming
interrupt handler
polls
priority order
service routine
round-robin
vectored interrupt
interrupt vector
contact bounce
wait-and-see
X-10
linear-select keyboard
coincident-select keyboard
Universal Product Code (UPC)
bar-code reader
real-time interrupt timer
multithread scheduling
tick
thread
sleep
sleepTime
priority
age
background thread
high-priority thread
n-key rollover
shuffle stack
WAI
STOP
exit
compile-time breakpoint
IoStreams
pipe
direct memory access
context switching
context
DMA channel
DMA transfer
release control
steal a memory cycle
cycle steal mode
burst mode
buffered I/O
cached I/O
cache
lazy buffer management
eager buffer management
I/O channel
I/O channel program
shuttle memory
indirect memory
time-multiplexed memory
transparent DMA

Problems

Problem 1 is a paragraph correction problem. See page 38 for guidelines. Guidelines for software problems are given on page 86, and for hardware problems, on page 115.

1.* Synchronization is used to coordinate a computer to an input/output device. The device has BUSY, completion, and DONE states. The BUSY state is when data can be given to it or taken from it. The device puts itself into the DONE state when it has completed the action requested by the computer. A paper-tape punch, by analogy to the paper-tape reader, is in the IDLE state when it is not in use; in the BUSY state when it is punching a pattern that corresponds to the word that was output just before the DONE state was entered; and in the DONE state when the pattern has been punched. The BUSY and IDLE states are indistinguishable in an output device like this one, unless error conditions are to be recognized (in the IDLE state). An address trigger will generate a pulse whenever an address is generated. Its output should never be asserted if the E clock is high. Address triggers are often used to start a device or to indicate completion by the device.

2. Write a C program that punches paper tape using real-time synchronization. Analogous to the latter procedure `main()` in §5.1.1, data are output through `PORTA` at a rate determined by the empirically evaluated constant `N`.

3. Write a hand-coded assembly-language program that punches paper tape using real-time synchronization. Analogous to the last program in §5.1.1, data are output through `PORTA` at a rate determined by the empirically evaluated constant `N`.

4. Write a `main()` procedure that punches paper tape using gadfly synchronization. Analogous to `main()` in §5.1.2, data are output through `PORTA` when bit 7 of `PORTB` falls, until the next time it falls. Use a key wakeup flag to detect the edge.

5. Write a hand-coded assembly-language program PUNCH that punches paper tape using gadfly synchronization. Analogous to the last program in §5.1.2, data are output through `PORTA` when bit 7 of `PORTB` falls, until the next time it falls. Use a key wakeup flag to detect the edge.

6. The LED signal (Figure 5.5a) can be fully generated by the 6812 without using a 555 timer chip; when the output of `PORTA` bit 0 is H (1) the LED is lit. To send a T (1) the LED should be pulsed at a rate of 38 KHz for 700 μs and be off for 350 μs. To send an F (0) the LED should be pulsed at a rate of 38 KHz for 350 μs and be off for 350 μs. Show a self-initializing procedure `void sendIr(int data)` that sends the least significant 11 bits of argument `data` through the infrared LED.

7. The BSR X-10 controller signal (Figure 5.7d) can be generated by the 6812 controlling a 555 timer chip (Figure 5.5c); when the output of `PORTB` bit 0 is H (1) the 555 generates a 100-KHz pulse train on the 110-V line; when this output is low, the

555 does not generate a pulse train. Rewrite the procedure `sendBsr` to send 16 bits through `PORTB` bit 0, synchronized to the 60-Hz waveform input on `PORTB` bit 1. A T (1) should be sent as a burst of 100-KHz pulses for 1/1080 second repeated each 1/360 second for 3 bursts after each edge of the 60-Hz waveform. An F (0), by comparison, should be sent as no burst for 1/120 seconds. Show a self-initializing procedure `void sendBsr (int data}` to send the 16-bit data.

8. Write a derived templated class `basicOutGadfly` of class `gadfly` that, like a basic output port, saves the last value sent out when `put` is called, so that when a `get` function member is called, it returns this saved value.

9. Write a derived templated class `SyncPort2` of class `SyncPort` providing a handshake output bit as well as a key wakeup input bit, using the same port and bit mapping convention of the input flag bit of `SyncPort`. Its constructor's argument list `(char id, unsigned char dir, char flag1, char flag2)` inputs or outputs 8-bit or 16-bit data on a port designated by `id` with direction `dir`, using input flag port `flag1`, and a 1-bit output port `flag2`. The new function member added to class `SyncPort2` is the function member `void handshake(char c);` that outputs the least significant bit of `c` to the output bit designated by `flag2`.

10.* The real-time synchronization technique times the duration of external actions using the microcomputer E clock as a timing reference and the program counter as a kind of frequency divider. This technique uses the least amount of hardware because the program itself contains segments that keep account of the BUSY/DONE state of the device. The program can be changed easily without upsetting the synchronization because program segments execute in the same time regardless of the instructions in the segment. Computer scientists, for no good reason, abhor real-time synchronization, so it should never be used, even on a microcontroller dedicated to a single control function. Real-time synchronization cannot be used to synchronize error conditions because we cannot predict the time of the next error. Gadfly synchronization uses hardware to track the device state, and the program watches the outputs from this hardware. Therefore feedback from the device controls I/O operation so it can be completed as soon as possible. Nevertheless, real-time synchronization is always faster than gadfly synchronization, because the former is always timed for the minimum time to complete an action in the device.

11. Write a C program that punches paper tape using interrupt synchronization. Analogous to the first procedure `void main()` in §5.2.2, data are output through `PORTA` from `buffer[0x80]` each time an interrupt is generated by an edge on bit 7 of `PORTB` causing *flag* to be set in the interrupt handler: `interrupt 23 handler() { KWIFJ = 0x80; flag++ }`.

12. Write a C program that punches paper tape using interrupt synchronization. Analogous to the second procedure `main()` in §5.2.2, data are output through `PORTA` from `buffer[0x80]` each time an interrupt is generated by an edge on bit 7 of `PORTB`.

13. Design a hardware breakpoint device. When the expanded mode 'A4's address and RW signals are determinate, this device compares them against a breakpoint address *ba* written in `PORTD` (high byte) and `PORTA` (low byte) and a *brw* bit written in `PORTB` bit 0. The address is compared with *ba,* and the RW signal with *brw,* by open collector quad 2-input exclusive NOR gates (74HC266s; see Figure 5.25) whose output is low if the inputs differ. The 74HC266's outputs are connected in a wire-AND bus with a 4.7-KΩ pull-up resistor. If each pair of inputs are equal, the outputs of the gates will be high, otherwise the outputs will be low. They connect to `PORTB` bit 1 to generate an interrupt that stops the program when the address matching the number in the output ports is generated.

a. Show a complete logic diagram of the system. Show all connections and pins to 74HC gates. However, don't show pin numbers on the 'A4 itself, or +5, or Gnd.

b. Show a self-initializing procedure `setBreakpoint(a,rw)` to generate a key wakeup `PORTB` bit 1 interrupt when the address `a` is written into if `rw` is 0, or read from if `rw` is 1.

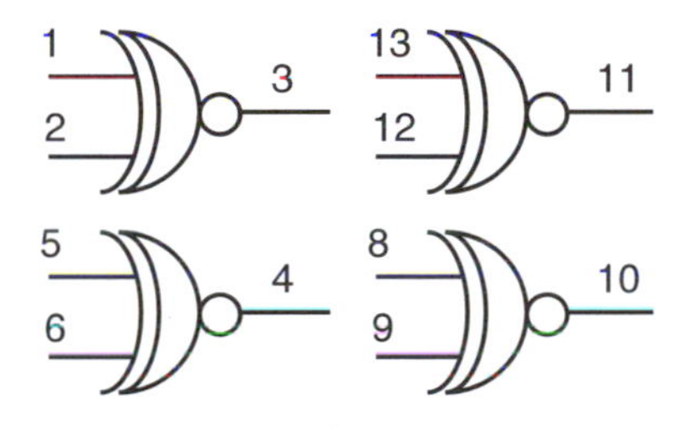

Figure 5.25. 74HC266

14. Assume that IRQ is a falling edge-sensitive interrupt handler, because bit 7 of the port at 0x1E was set. A polling routine like `interrupt 23 handler(){ if(srvc1() || srvc2() || srvc3()) ; }` needs to check the `IRQ` pin, which is `PORTE` bit 6, after any service procedure such as `srvc1()` removes its source of the interrupt to repoll the devices if `IRQ` remains asserted low. Rewrite `handler()` to handle such a falling edge-sensitive `IRQ` interrupt.

15. Write a round-robin IRQ handler `interrupt 23 handler()` (§5.2.3) that checks only `PORTJ`'s key wakeup interrupt bits. If `KWIEJ` bit 0 is set so that `PORTB` bit 0 sees a rising or falling edge, and `KWIFJ` bit 0 therefore becomes set (as determined by bit 0 of `KPOLJ`), the service procedure `h0()` is executed. The analogous actions occur for the other bits of `PORTB`. If `PORTB` bit 0 did not see an edge, `PORTB` bit 1 is checked similarly, and so on. If `PORTB` bit *n* had an edge and procedure `hn()` was executed, $0 \leq n < 8$, then upon the next `PORTB` key wakeup interrupt, bit $n + 1$ is tested first, then bit $n + 2$, and so on, where bit 0 is tested after bit 7. In this version of handling key wakeup interrupts, the service procedures do not return a value and do not clear the interrupt, but `interrupt 23 handler()` does clear the interrupt.

16. Compute the difference in worst-case latency between polled interrupts and vectored interrupts for key wakeup interrupts. Compare the first `interrupt 23 handler()` in §5.2.3 to `interrupt 23 handler()` and `interrupt 24 handler()` in §5.2.4. Use §5.2.3's assembly language for the first `interrupt 23 handler()` handler.

17.* Interrupts permit the computer to perform some useful function while waiting for a device to become BUSY or for an error condition to arise. Interrupts are always faster than a simple gadfly loop because they save the state of the machine and restore it, while the gadfly technique has to loop a long time. When a key wakeup J device requests an interrupt, if the device is enabled by clearing the corresponding bit of `KWIEJ`, the flag flip-flop, which is the corresponding bit of `KWIFJ`, is set. When this flag flip-flop is set, the 6812 immediately honors the interrupt, saving the values in all output registers on the stack and jumping directly to a handler routine. The handler may have just an RTS instruction to return to the program that was originally running. Vectored interrupts use external hardware to eliminate the polling routine in the device handler, so the interrupt handler can be executed immediately. Interrupts, and vectored interrupts in particular, should be used whenever the latency-time requirement is critically small or something useful can be done while waiting for an interrupt; otherwise, real-time or gadfly synchronization should be used.

18.* Key bounce is a problem when the user bangs on a key repeatedly to get something done. It can be eliminated by a sample-and-hold circuit that takes a "snapshot" of the signal just once. Software debouncing is rarely done in microcomputers because it takes too much of the microcomputer's valuable time to monitor the key. Keyboards are often used on microcomputers because the user may want to enter different commands, and this is most easily done with a set of keys. Linear selection is often used for a keyboard that has a lot of keys, like a typewriter. *N*-key rollover is a property of keyboards whereby the microcomputer can correctly determine what keys were pressed and in what order they were pressed, provided that no more than *N* keys are pressed at one time. Two-key rollover is commonly used in microcomputer systems but is rather inadequate for most adept users, who often hold several keys down at once. The LED seven-segment display is used on many inexpensive microcomputer systems because it uses very little power. LCDs have to be multiplexed carefully, because they require AC (square wave) signals, and the difference in RMS voltage between a clear and an opaque segment is rather small. Therefore we do not multiplex more than two or three LCDs on the same drive circuitry. A typical keyboard system is often integrated with a multiplexed display system because the hardware to scan the keyboard can also be used to scan the display.

19. The coincident select keyboard shown in Figure 5.15c is modified to have 16 columns. The left 8 columns of switches are connected to `PORTA` as shown, and the right 8 columns are similarly connected to `PORTD`. Rewite the matrix or coincident select handler `interrupt 23 handler()` in the middle of §5.2.5 to push a low-order 3-bit row concatenated with a higher-order 4-bit column number into the queue, where `PORTH`'s bits correspond to columns 15 to 8 and `PORTD`'s bits correspond to columns 7 to 0.

20. The coincident select keyboard shown in Figure 5.15c is modified to have 16 rows. The top 8 rows of switches are connected to `PORTB` as shown, and the bottom 8 rows are similarly connected to `PORTD`. Rewrite the matrix or coincident select handler `interrupt 23 handler()` in the middle of §5.2.5 as handlers `interrupt 23 handler()` and `interrupt 6 handler()` to push a low-order 4-bit row concatenated with a higher-order 3-bit column number into the queue, where `PORTJ`'s bits correspond to rows 15 to 8 and `PORTD`'s bits correspond to rows 7 to 0.

21. Write a self-initializing procedure `barCode(char *s, char *t, int n)` that gives in `s` a series of `n` pulse widths, such as the vector `pulses` generated by the interrupt handler `interrupt 23 handler`, and produces an ASCII string of characters in *t.* Assume the bar-code reader is uniformly swiped over the whole code, but in either forward or reverse motion. See Figure 5.17.

22. Write an interrupt synchronized class `tapePunch` that is a derived class of `IntSyncPort`, but its constructor blesses an object of class `BitPort` (Chapter 4, problem 15) and its constructor's arguments are the parallel port for data, the port and bit for key wakeup interrupt and the port and bit for handshaking. It uses an output queue of 10 byte. Then write a procedure `main()` that selects `PORTA` as the parallel port for data, `PORTB` bit 0 as the bit for key wakeup interrupt, and `PORTB` bit 6 as the bit for handshaking. Use the `enum` statement of Chapter 4's problem 14 for `BitPort`'s argument, and numeric constants, not symbolic names, for the other arguments.

23. Assume that three threads `thread[1]`, `thread[2]`, and `thread[3]` have priorities 5, 3, and 3. Identify which will be executed during each tick time, from time *t0* when they are all forked by `thread[0]`, for the next 30 ticks. Give a general rule for assigning priorities to threads (i.e., give the same priority to thus and such threads, higher priority to thus and such threads) and give a rough estimate of the amount of CPU time a thread gets as a function of its priority for two cases: (1) when *n* threads all have priority *p,* and (2) when there are just two threads having priorities *p1* and *p2.*

24. Replace procedures `main1()` and `main2()` of §5.3.2 with a single procedure `main1()` so that each thread controls a different traffic light, as is done in §5.3.3, but both threads execute only one procedure `main1()`. The first thread's north and south lights are red, and the east and west lights are green, for 10 seconds; then the north and south lights are red, and the east and west lights are yellow, for 2 seconds; then the north and south lights are green, and the east and west lights are red, for 16 seconds; then the north and south lights are red, and the east and west lights are yellow, for 2 seconds. This pattern is stored in global vector `char tbl1[4][2]`. The second thread's light pattern is the same as the first, except north and south are exchanged with east and west. This pattern is stored in global vector `char tbl2[4][2]`. (Hint: `main()` should check if `thisThread` is `&threads[1]` or `&threads[2]`, to set a local variable pointer `tbl` to global vector `tbl1` or `tbl2`.

25. Replace procedures `main1()` and `main2()` of §5.3.2 so that `main1()` reads tape and `main2()` punches tape. Initially `main1()` goes to sleep; when `PORTB` bit 0's signal falls `main1()` wakes up and reads data from `PORTA`, pushing it into a queue. Then `main1()` goes to sleep again until another byte arrives. If data are available in the queue, `main2()` pulls a byte into `PORTB`, asserts `PORTB` bit 1 for one tick time, and negates `PORTB` bit 1 for two tick times. These threads continue working this way indefinitely. Use §2.2.2's procedures `pstop` and `plbot`, and assume the queue doesn't overflow.

26. Replace procedures `main1()` and `main2()` of §5.3.2 so that `main1()` punches tape and procedure `main2()` sends Morse code for the global null-terminated ASCII string:

```
char message[70] =
"Now is the time for all good men to come to the aid of their country.";
```

The `main()` procedure sets up 8.192 ms ticks (periodic interrupts) and then forks `main1()` and `main2()`. Either `main1()` or `main2()` may execute before the other or might be used without the other, so each should initialize its I/O as if the other doesn't run, and shouldn't interfere with the other thread's I/O. `main1()` causes the characters in `message` to be punched on paper tape. The paper-tape punch will energize solenoids to punch holes if a T (1) bit is in a corresponding `PORTA`, and to punch a sprocket hole if a T is in `PORTB` bit 7; otherwise, if an F (0) is in `PORTB` bit 7, the paper tape is pulled forward by a motor (unless tape runs out). The holes are to be punched for about 16.4 ms and paper tape is to be advanced for about 50 ms After all characters are punched, `main1()` will "kill" itself by sleeping indefinitely as tape runs out. `main2()` causes the characters in `message` to be sent on `PORTB` bit 6 in Morse code. The Morse code for characters is stored in the global vector `int Morse[2][128]` where `Morse[0][]` is the number *n* of dots and dashes, and `Morse[1][]` is the (right-aligned) pattern: a 0 is a dot, a 1 is a dash, and unused bits are 0. Do not write the vector `Morse`. A dot is sent by making `PORTB` bit 6 T (1) for about one second, and a dash sent by making `PORTB` bit 6 T (1) for about 3 seconds; there is a one-second F (0) between dots and dashes in a letter, and a three-second F (0) between letters. After all characters in *message* are sent, `main2()` will "kill" itself by sleeping indefinitely and outputting an F (0).

27. The WAI instruction reduces I/O latency and can be used to provide more accurate synchronization, so that the instructions execute in precise memory cycles after the interrupt edge has occurred. Rewrite the procedure `main()` in §5.2.5 immediately under the declaration `int index, start, pulses[256];` to use the 6812 WAI instruction while awaiting key wakeup interrupts. This improved routine will read the running counter TCNT at the same time after each key wakeup edge occurs, so the difference between these read values is a more accurate measure of pulse width.

28. Rewrite `main()` in §5.3.2 to use the 6812 STOP instruction when all threads other than thread 0 sleep. This reduces the 'A4's power consumption when no threads need to be executed. Note: condition code bit `S` must be cleared to be able to execute `STOP`.

29. Write a handler `handler()`, whose address is put into locations 0xfff8-9, that treats opcodes from 0x1840 as `TRAP` instructions to implement an operating system. The procedures `systemCall0()`, `systemCall1()`, `systemCall2()`, etc. will perform various operating system functions. If the opcode is 0x1840, execute `systemCall0()` and then `RTI`. If the opcode is 0x1841, execute `systemCall1()` and then `RTI`. If the opcode is 0x1842, execute `systemCall2()` and then `RTI`. And so on. Note that you need embedded assembly language to read the program counter from the stack, but this address is the address of the next instruction after the `TRAP` instruction.

30. Rewrite §5.3.5's `Keycheck`, to translate key row-columns to ASCII characters; when control-C (0x4) is received, puts the thread using the device to sleep indefinitely.

31. Repeat problem 25 using a class `pipe` object to hold the data moved from `main1` to `main2`, and use 5.2.6's class `IQFPort` to read data from the paper-tape reader device.

32.* Direct memory access is a synchronization technique that uses an extra processor that is able to move words from a device to memory, or vice versa. With an output device, when the device is able to output another word, it will assert a request to the DMA chip, which checks its BUSY/DONE state, and, if DONE, it requests that the microprocessor stop and release its control of R/W and the address and data buses. The microprocessor will tell the DMA device when it has released control; the DMA device will output on the data bus and will send a signal to the I/O device to put a signal on the R/W line and an address on the address bus. A DMA chip itself is an I/O device, whose BUSY state indicates that a buffer full of data has been moved. The BUSY state then is an interrupt request. Either gadfly or interrupt synchronization can be used to start a program when the buffer has been moved.

33. A pair of indirect memories using an MCM6264D-45 will be implemented, where one is shown in Figure 5.23, so that when one is being accessed by the 'A8, the other is free to be used for I/O. The first memory is connected as shown in that figure. Show the logic diagram, including pin numbers, of the connections between the second memory and the 'A8, which uses `PORTA` (high 5 bits) and `PORTB` (low byte) of the address, and `PORTT` for data. This memory's enable *E1* is on `PORTA` bit 7, its output enable *G* is on `PORTA` bit 6, and its write control *W* is on `PORTA` bit 5.

34. Complete the logical design of an indirect memory using an MCM6264D-45 (Figure 5.23) for input from a fast 8-bit data source `data`. Use 74HC590 counters to supply addresses and a 74HC244 tristate driver to supply `data` to the MCM6264D-45 when the I/O device needs to write into the memory. When it does so, it pulses control signal WRITE first low, then high. The 74HC590 counters are written into by pulsing `PORTF` bit 7, which is normally high, low, and high when the address to be written into the counters is in `PORTA` and `PORTB`.

a. Show the logic diagram for the complete circuit, but excluding connections already shown on Figure 5.23.

b. Show a self-initializing procedure `void setAddress(a) int a;` that writes the address into the counter.

Adapt912 is Technological Arts' version of Motorola's 912EVB evaluation board for the 68HC912B32 microcontroller. Offering the same modular hardware design as other Adapt12 products, Adapt912 includes Motorola's DBug-12 in on-chip Flash. This gives it the versatility to function as a standalone development system, a BDM Pod, or even a finished application (when combined with the user's circuitry on a companion Adapt12 PRO1 Prototyping card).

6

System Control

The 'A4 has a number of ports that control bus signals, memory management, and the E clock rate to simplify the implementation of external memories. The 6812 instruction set also has some instructions related to these functions. While these ports and instructions are not fundamental to I/O interfacing generally, they are important for the intelligent use of the powerful 'A4. These ports and instructions are discussed herein.

Technical summaries, MC68HC812A4TS/D and MC68HC912B32TS/D, provide a very good reference for the control ports discussed in this chapter. Rather than duplicate that document, we will concentrate on some of the more useful features and show examples of their use. The reader should access Motorola's technical summaries for a more complete discussion of the use of these ports and of the ports that we do not cover.

6.1 6812 Chip Modes

The 6812 has a `MODE` port to specify whether single-chip or multiple-chip operation is to be used, and a `PEAR` port that specifies whether certain control signals, useful in multiple-chip mode, are to be made available. These are described in turn.

6.1.1 MODE Control Port

The 6812 has three pairs of modes, single-chip, narrow expanded, and wide expanded, in two sets of modes, normal and special. They are selected by initially asserting or negating three inputs, `BGND`, `MODB`, and `MODA` (see Figure 6.1); these values are loaded into the most significant bits of the `MODE` port when the 'A4 comes out of reset. The user can write `MODE` any time, thereby changing the mode; however other ports initialized out of reset as a function of the mode are not affected when `MODE` is written into. If the most significant bit of the `MODE` port, `SMODN`, is clear, the 'A4 is in a special mode, otherwise it is in a normal mode. If the next two significant bits of the `MODE` port, `MODB` and `MODA`, are 00, the 6812 is in a single-chip mode; if 01, then it is in a narrow expanded mode; and if 11, then it is in a wide expanded mode. The 'B32's expanded modes differ from the 'A4's primarily by using a time-multiplexed bus.

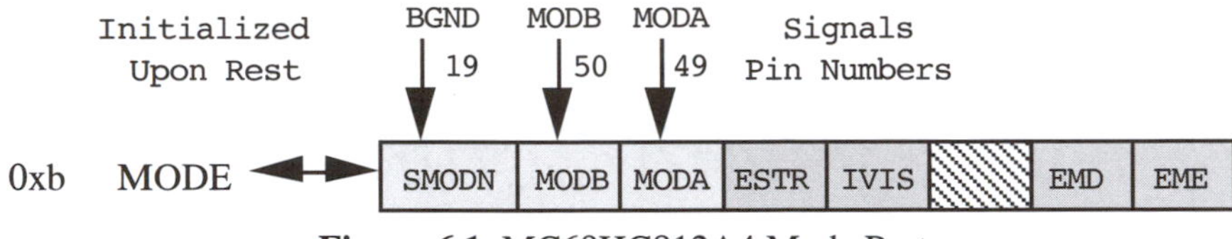

Figure 6.1. MC68HC812A4 Mode Port

For the 'A4, in the single-chip mode, the address and data buses and several other control signals, like the E clock, are externally unavailable, but *PORTA* through *PORTD* are available. In the narrow expanded mode, the 16-bit address bus and an 8-bit data bus are available in place of *PORTA* through *PORTC*, respectively. *PORTD* is available as a parallel port. In the wide expanded mode, the 16-bit address bus and a 16-bit data bus are available in place of *PORTA* through *PORTD*.

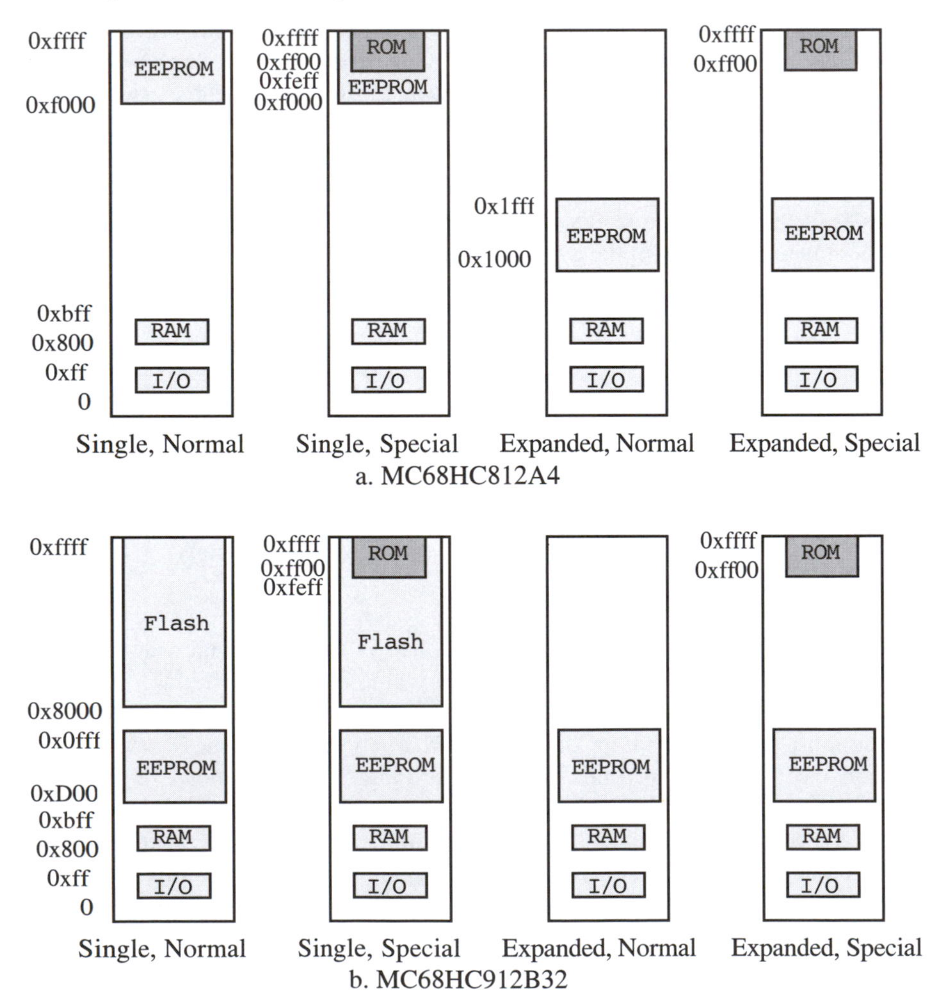

Figure 6.2. Memory Maps

For the 'B32, in the single-chip mode, the address and data buses, and several other control signals, like the E clock, are externally unavailable, but *PORTA* and *PORTB* are available. In the narrow expanded mode, the 16-bit address bus is output on *PORTA* and *PORTB* in the first part of the memory cycle, and the 8-bit data is input or output on *PORTA* in the second part of the cycle. In the wide expanded mode, the 16-bit address bus and 16-bit data bus are time-multiplexed in place of *PORTA* and *PORTB*.

Special modes are used for testing, debugging programs, and for downloading data into EEPROM when the chip is first initialized. Some functions and ports are only available in special modes; these are primarily used for factory testing, but are available to sophisticated users to assist in debugging programs running on the 6812.

Figure 6.2 shows single-chip and expanded-mode memory maps for the 'A4 and the 'B32. In all modes, I/O devices are at locations 0 to 0xff, and RAM is between 0x800 and 0xbff and in special modes, a ROM at 0xff00 to 0xffff is used for debugging. In the 'A4, in single-chip modes, EEPROM is at 0xf000 to 0xffff, and in expanded modes, it is at 0x1000 to 0x1fff. In the 'B32, EEPROM is at $d00 to $fff, and flash memory is at 0x8000 to $ffff in single-chip modes. However, these locations of RAM, EEPROM, flash, and I/O devices can be changed, as we discuss in a later section. If two modules appear to conflict at the same address, the priority is ROM (highest), internal I/O, RAM, EEPROM, flash (in the 'B32), and external memories and I/O devices (lowest).

The single-chip mode obviously should be used whenever the RAM, EEPROM, flash, and I/O devices within the 'A4 or 'B32 are adequate for the application. The wide expanded mode is useful in increasing performance, because each memory cycle can read or write 16 bits of data, while the narrow expanded mode is useful in reducing cost, because less expensive memories generally have 1-bit wide or 8-bit wide data pins.

The MODE port can be initialized for special narrow expanded mode as follows:

```
enum {SnglChip,Narrow,Peripheral,Wide,Normal,Special=0,ModeFld=5};
MODE = ( Special + Narrow ) << ModeFld;
```

6.1.2 Port E Assignment

PEAR determines whether a given pin is a *PORTE* data pin or a control signal pin (see Figure 6.3). Control signals include the extra interrupt request *XIRQ*, shared with *PORTE* bit 0; interrupt request *IRQ*, shared with *PORTE* bit 1; read/write R/W, shared with *PORTE* bit 2; LSTRB, shared with *PORTE* bit 3; and the E clock, shared with *PORTE* bit 4. R/W, LSTRB, and the E clock will be discussed in §6.5. Except for the E clock, a T (1) makes the pin a control function, and an F (0) makes it a *PORTE* parallel I/O bit. However, *PEAR* does not enable or disable *IRQ* or *XIRQ*.

The initial value of the *PEAR* port also depends upon the initial mode that is determined by *SMODN*, *MODB*, and *MODA*. In normal single chip mode, *NECLK* is asserted, making *PORTE* bit 4 available. In all normal modes, *LSTRE* and *RDWE* are negated, making *PORTE* bits 3 and 2 available. For all other modes, *NECLK* is negated, making the E clock available in place of *PORTE* bit 4; and *LSTRE* and *RDWE* are asserted, making *RD* and LSTRB available in place of *PORTE* bits 3 and 2.

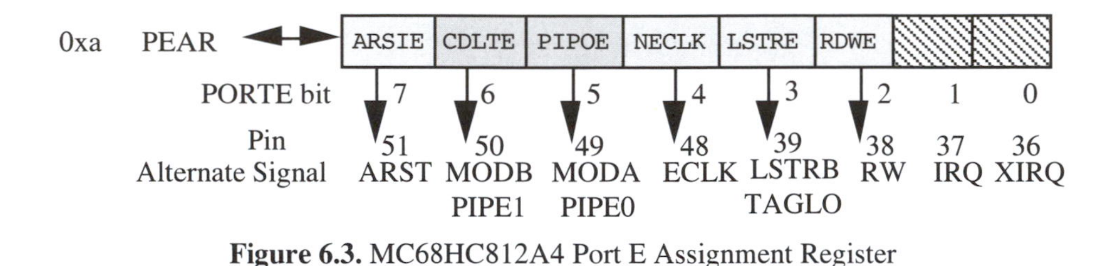

Figure 6.3. MC68HC812A4 Port E Assignment Register

PEAR must be initialized in expanded modes if the E clock, R/W, or LSTRB are used in external decoders, before external devices are used. *PEAR* can be initialized for expanded wide mode by enabling the use of the E clock, R/W, and LSTRB as follows:

```
enum { RDWRE = 4, LSTRE = 8, NECLK = 0x10 };

PEAR = RDWRE + LSTRE;
```

The E clock is enabled by not including the *NECLK* bit when initializing *PEAR*. In the 'B32, the bit *NDBE* replaces the 'A4's bit *ARST*. Assert *NDBE* if you do not want the DBE signal to control bus demultiplexing, or negate it if you want to use *PORTE* bit 7.

6.2 6812 Memory Map Control

The 'A4 provides signals for external memory and I/O devices through pins that can alternatively be used as *PORTE* parallel I/O bits. I/O device, RAM, and EEPROM locations within the 'A4 or 'B32, and of memory and I/O devices outside the 'A4, are controlled by ports, which we discuss in turn.

6.2.1 6812 Internal Memory Map Control

The location of RAM, I/O ports, and EEPROM can be repositioned in the 6812 memory map by writing in the ports *INITRM*, *INITRG*, and *INITEE* (see Figure 6.4). The high-order bits of each register are the high-order bits of the address of the device.

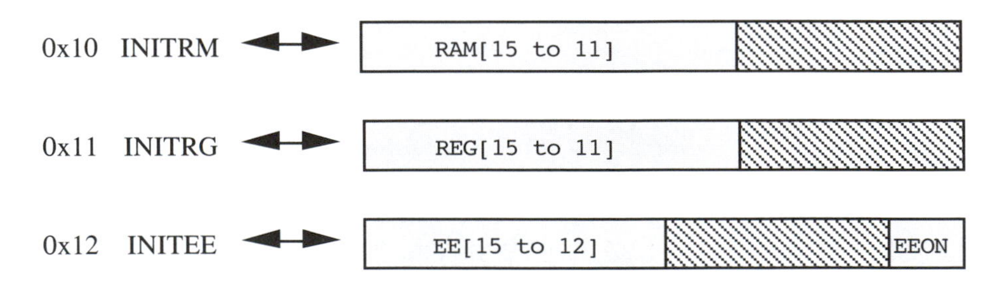

Figure 6.4. MC68HC812A4 Internal Memory Map Control Ports

To initialize RAM to be at 0 to 0x3ff and I/O ports at 0x800 to 0xbff, the following statements can be executed: `INITRM = 0;` and `INITRG = 0x800 >> 8;`. Be careful to use symbolic names after the remapping is done that are different from those used before the remapping is done.

If executing code in EEPROM, to initialize EEPROM to be at 0x1000 to 0x1fff, the following procedure can be written into RAM and then executed:

```
SETEEPROM: LDAA #$11       ; high byte of address + 1
           STAA $12        ; put into INITEE
           JMP  NEWSTART   ; go to entry point at new EEPROM location
```

The least significant bit of `INITEE` has to be set to enable EEPROM; if it were F (0), EEPROM is removed from the memory map. The procedure is loaded into RAM and executed there because the address of EEPROM cannot be changed while executing code that is in EEPROM.

The 'B32's flash memory can be put in the low half (0 to 0x7fff) or high half (0x8000 to 0xffff) depending on whether bit 1 of the `MISC` port at 0x13 is 0 or 1.

6.2.2 MC68HC812A4 Chip Selects

The 'A4 has negative-logic chip selects to enable external memory or I/O devices. Four of them, CS0, CS1, CS2, CS3, are designed to enable I/O devices; they enable small ranges of addresses and appear at locations between internal I/O ports and internal RAM (see Figure 6.5a). Three chip selects, CSD, CSP0, and CSP1, are designed for memories; they enable larger ranges of addresses and appear at higher addresses in the memory map (see Figure 6.5b). There are a fair number of alternative mechanisms that use these chip selects. In this section, we focus on the simpler mechanisms.

Chip selects can be initialized by writing 16-bit constants into `int CSCTL` and `int SCSTR`. These chip select outputs may be used in place of `PORTF` I/O bits, under control of `CSTL` (see Figure 6.6). For instance, if `CS0E`, `CSTL` bit 8 is T (1), chip select 0 will be on pin 68 instead of `PORTF` bit 0.

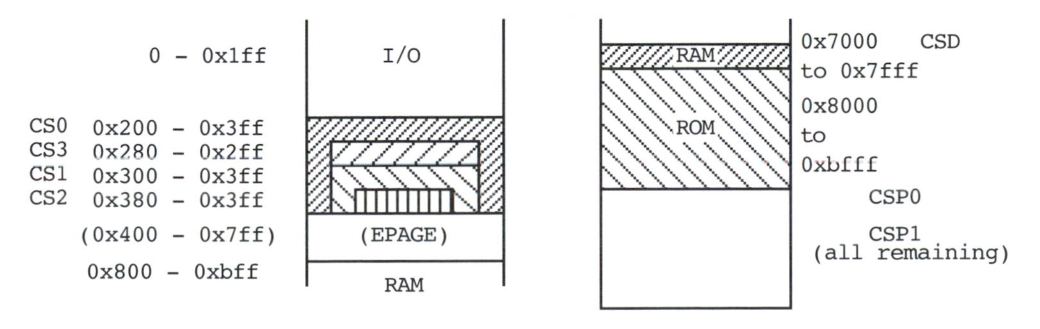

Figure 6.5. MC68HC812A4 Chip Select Memory Map

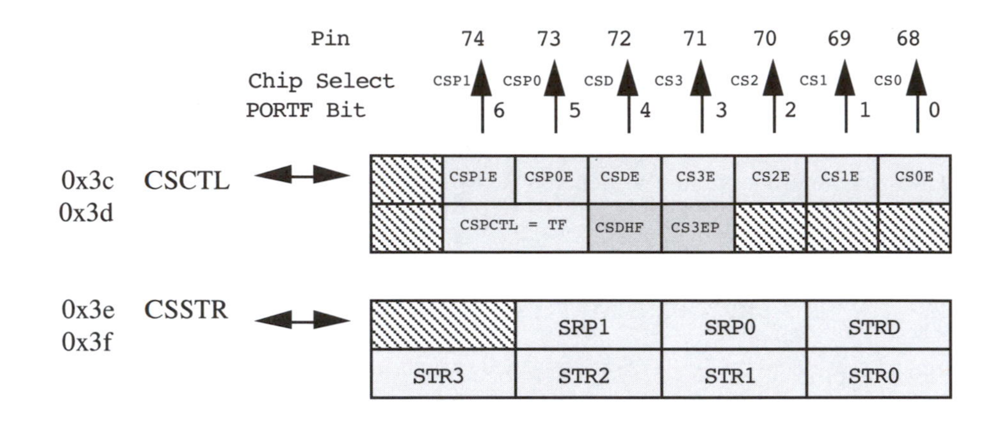

Figure 6.6. MC68HC812A4 Chip Select Registers

The chip selects have a priority order. If two chip selects are enabled and appear to conflict at the same address, the priority is: `CS3` (highest), `CS2, CS1, CS0, CSP0, CSD,` and `CSP1` (lowest). This priority order permits some alternative assignments of addresses. If `CS0` is enabled but `CS1, CS2,` and `CS3` are not (e.g. `CSCTL` is 0x0100), `CS0` is asserted low whenever an address between 0x200 and 0x3ff is generated as an instruction's effective address. However, if `CS0` and `CS2` are enabled but `CS1` and `CS3` are not (e.g., `CSCTL` is 0x0500), `CS0` is asserted low whenever an address between 0x200 and 0x2ff is generated, and `CS2` is asserted low whenever an address between 0x300 and 0x3ff is generated. If `CS3, CS2, CS1,` and `CS0` are enabled (`CSCTL` is 0x0f00), then each enables 128 consecutive locations. If `CSPCTL` in `CSCTL` is TF (10), `CSP1` is enabled whenever no other device is enabled. (We do not further describe the other interesting uses of chip select 3, `CS3`)

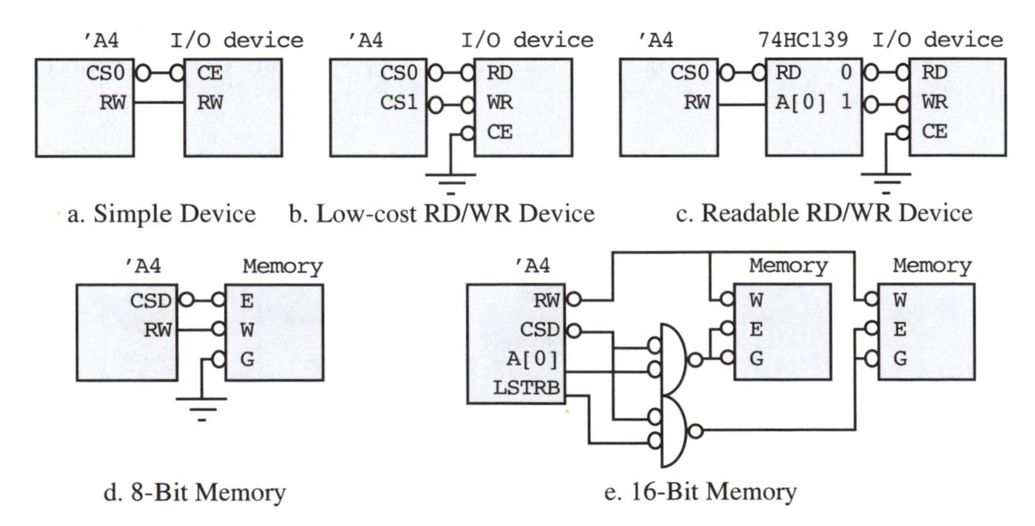

Figure 6.7. Use of Chip Select Lines

Chip selects may be used to enable I/O devices or memories, as shown in Figure 6.7. For 6800-style I/O devices having a chip select and a read/write line R/W, the chip's enable is connected to one of 'A4's chip selects such as `CS0`, and the chip's R/W is connected to 'A4's R/W (see §9.3.3). Intel-style I/O devices have read RD, write WR, and chip select CE pins. The device's RD can be connected to an 'A4 chip select such as CS0, while WR is connected to another 'A4 chip select such as `CS1` (see §10-2.2). Reading and writing the same register must be done at different addresses; for instance, reading a device register might be done using one chip select, `CS0`, by a statement `i = *(char *)0x200;`. Writing would be done using a different chip select, `CS2`, by a statement `*(char *)0x300 = i;`. A statement `*(char *)0x200++;` or `*(char *)0x300++;` wouldn't increment the register, since the port is no longer a readable output port but is a basic output port and an input port at different locations. If the port must be a readable output port, external logic is needed to derive the RD and WR signals from the chip select and R/W signals, such as the decoder in Figure 6.7c. An 8-bit memory's enable can be connected to a chip select such as CSD, and the memory's R/W can be connected to the 'A4's R/W; the output enable *G* can be permanently asserted because the memory's tristate drivers are disabled when R/W is asserted to write in the memory. This configuration is shown in Figures 6.7d and 6.12. A 16-bit memory comprising two 8-bit memories can be connected as shown in Figures 6.7e and 6.14. The 'A4's R/W signal is attached to both memories' W pin. Each memory chip select E and output enable *G* is tied together, and both are asserted as a function of the 'A4 enable CSD, address A0, and low strobe LSTRB. The 'A4 signal A0 is ANDed with CSD to get E and W for the (right) memory holding odd-numbered locations. Similarly, the 'A4 signal LSTRB is ANDed with `CSD` to get E and W for the (left) memory holding even-numbered locations. In each case mentioned, SSI gates, a decoder, or a PAL can be used with each chip select to obtain chip selects for multiple devices.

Each chip enable may be stretched under control of two bits of the `CSSTR` port; for instance, bits 1 and 0 of `CSSTR` stretch CS0. If these two bits are FF (00), CS0 is not stretched, so it is asserted for about 2/3 of an 8-MHz memory clock cycle (80 ns). If they are FT (01), `CS0` is stretched an additional clock cycle, so CS0 is asserted for about 1 2/3 of an 8 MHz memory clock cycle (200 ns). If they are TF or TT, CS0 is stretched two memory clock cycles (325 ns) or three memory clock cycles (450 ns). Chip select signals become asserted low when address and R/W become asserted, 40 ns into the memory cycle, and become negated high within 10 ns after the E clock falls. Note that after reset, `CSSTR` is 0x3fff.

For instance, to use `CS2` at locations 0x380 to 0x3ff with a 325-ns access time (which requires a two-cycle stretch), `CS3` at locations 0x280 to 0x2ff, with a 450-ns access time (which requires a three-cycle stretch), and `CSP0` everywhere no other device is selected (to detect any illegal addresses), we execute the following:

```
enum{ CSP1E = 0x4040, CSP0E = 0x2000, CSDE = 0x1000, CS3E = 0x800,
      CS2E = 0x400, CS1E = 0x200, CS0E = 0x100};

enum{ STR0, STR1=2, STR2=4, STR3=8, STRD=10, STP0=12, STRP1=14},

CSCTL = CSP1E | CS2E | CS3E;

CSSTR = (2 << STR2) | (3 << STR3);
```

6.2.3 MC68HC812A4 Memory Expansion

In many applications, the 64K-byte memory space is inadequate. The 'A4's memory expansion uses `PORTG` pins to output higher-order address bits derived from `DPAGE`, `PPAGE`, and `EPAGE` ports, if `DWEN`, `PWEN`, and `EWEN` have values of T (1) in the `WINDEF` port. See Figure 6.8. The corresponding *expansion windows* described below are designed for data, program, and "extra" memory expansion.

These pages map a 16-bit *internal address*, generated by the programmer, into a 22-bit *external address*, which appears on the pins. The higher-address bits appear in place of `PORTG` bits if corresponding bits of the `MXAR` port are T (1). If `PWEN` is T (1), and a internal address appears in the range 0x8000 to 0xbfff (a 16K-byte range, shown in Figure 6.9), `PPAGE` is appended above the low-order 14 bits of the internal address to obtain the external address. For instance, if the internal address 0x9004 (in binary, 1001 0000 0000 0100) appeared and `PPAGE` had 0x34 (in binary 0011 0100), then the external address would be 0x0d1004 (in binary 00 1101 0001 0000 0000 0100). If `DWEN` is T (1), and a internal address appears in the range 0x7000 to 0x7fff (a 4K-byte range, shown in Figure 6.9), two 1s and `DPAGE` are appended above the low-order 12 bits of the internal address to obtain the external address. For instance, if the internal address 0x7012 (in binary, 0111 0000 0001 0010) appeared and `DPAGE` had 0x56 (in binary 0101 0110), then the external address would be 0x356012 (in binary 11 0101 0110 0000 0001 0010). If `EWEN` is T (1), and a internal address appears in the range 0x400 to 0x7ff (a 4K-byte range, shown in Figures 6.5a and 6.9), four 1s and `EPAGE` are appended above the low-order 10 bits of the internal address to obtain the external address. For instance, if the internal address 0x0789 (in binary, 0000 0111 1000 1001) appeared and `EPAGE` had 0xab (in binary 1010 1011), then the external address would be 0x3eaf89 (in binary 11 1110 1010 1111 1000 1001).

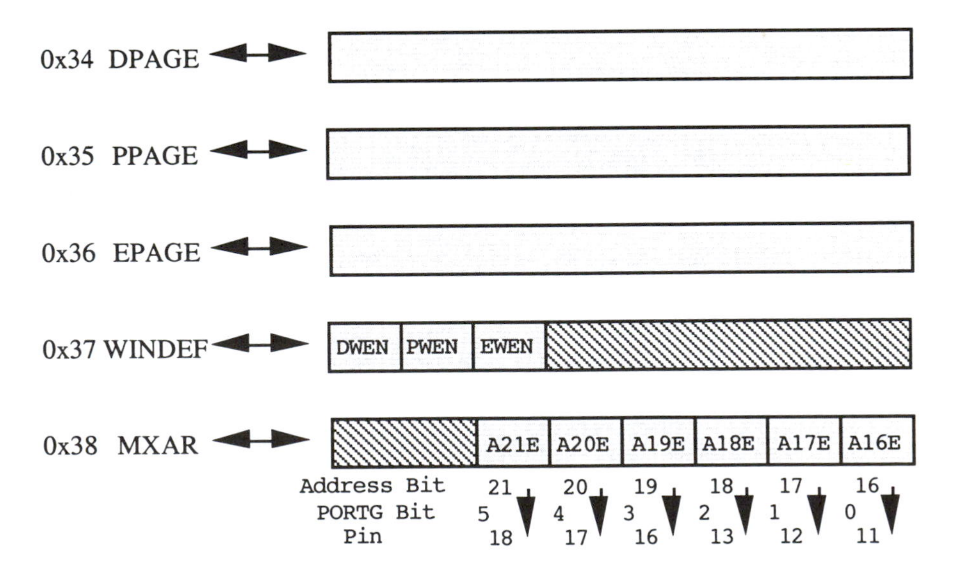

Figure 6.8. MC68HC812A4 Memory Expansion Ports

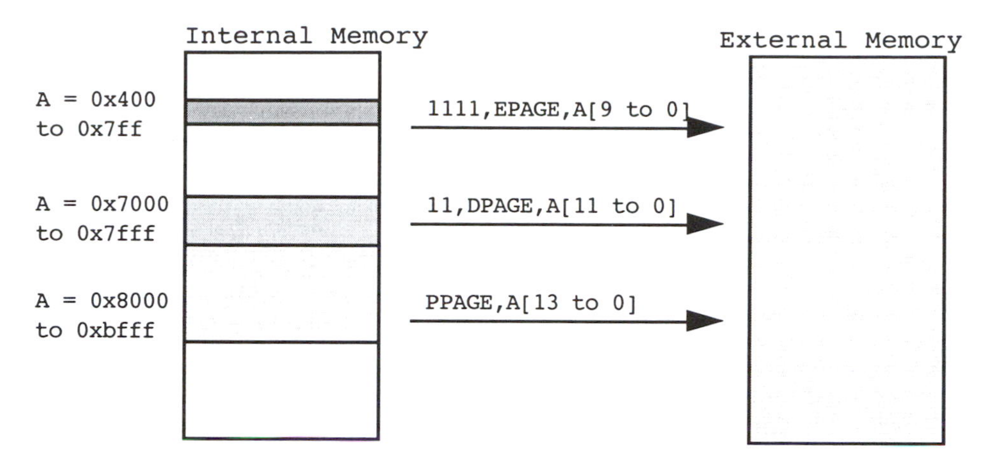

Figure 6.9. MC68HC812A4 Memory Expansion Mapping

Expansion pages can overlap at the top of the external memory map, but chip selects can separate data from program and extra pages. Chip selects enable the memories or devices. `CSD` is only asserted when a data page is accessed, and `CSP0` or `CSP1` when a program page is accessed; `CS3` enables the "extra" page in a more complex manner.

Alternatively, if external memory 0x3c0000 to 0x3fffff is only used for "extra" pages, 0x300000 to 0x3bffff is only used for data pages, and 0 to 0x2fffff is only used for program pages, then there are no potential duplicate uses of memory, and chip selects are not needed to separate them out. In this technique, the external decoder would AND the inverted E clock with high-order address signals rather than use the chip enables.

External memory can be allocated and accessed using these pages. Suppose data is at external address A[21 to 0], and you wish to read it through a data page. Then put A[19 to 12] into DPAGE and read the address 0x7000 + A[11 to 0]. For instance, external byte 0x312345 can be read into *i* by `DPAGE = 0x12; i = (char *)0x7345;`.

Recall that the 6812 instruction CALL has a 3-byte subroutine address in external space. The contents of `PPAGE` are saved on the stack together with the program counter; this is essentially a return address in external space. Then the first two bytes are put into the program counter and the third byte is put in `PPAGE;` this is essentially a subroutine address in external space. The corresponding return instruction, RTC, pops three bytes back into the PPAGE register and program counter. Note that the left two bits of the first address byte of the internal address have to be TF (10) to use the program expansion window; if a program runs in the data expansion window, the left four bits of the internal address need to be FTTT (0111).

6.2.4 Object-oriented Programming of Memory Expansion

Object-oriented programming realizes a clean way to use expansion memory. Overloaded cast, assignment, and index operators make *n*-dimensional arrays stored in expanded memory as easy to use as 'A4 RAM global variables. The class `dataPage` shows a two-dimensional array stored in a data page window to expanded memory. `dataPage`'s constructor just saves the array base and dimensions. Overloaded operator `[]` saves the index and also saves the previously entered index so that two indexes, for row and column, are available. Overloaded operators = and *cast* use the saved indexes in the private function member *ptr,* which computes the external memory address, puts the page number in the *DPAGE* port, and returns the 6812 internal address. The overloaded operators = and *cast* use this internal address to read or write the data in the expansion memory. This example can be easily modified to use the program or "extra" window, and can handle one-dimensional vectors or arrays of more than three dimensions.

```
volatile unsigned char DPAGE@0x34, WINDEF@0x37;

template <class T> class dataPage{int r,c,previousIndex,index;long base;

 public:dataPage(long b,int ra,int ca){WINDEF|=0x80;base=b;r=ra;c=ca;}

   virtual T operator = (T data) { return *ptr() = data; }

   operator T () { return *ptr(); };

   virtual dataPage operator [] (char i)
       { previousIndex=index; index = i; return *this; }

   private : T *ptr(void) { long extAddr;
       extAddr = (index + previousIndex * c) * sizeof(T) + base;
       DPAGE = extAddr >> 12;
       return (T *) ((extAddr & 0xfff) | 0x7000);
   }
};

void main(){char i;dataPage<char>a(0,1000,1000);a[1][5]=3; i=a[1][5];}
```

6.3 EEPROM and Flash Memory Programming

We describe the programming of EEPROM words. Memory from 0xF000 to 0xFFFF are in EEPROM. Though they can be read in one E cycle, they require 10 ms to be written in. They can be written in or erased by writing bit patterns into the *EEPROG* port at 0xF3 (see Figure 6.10). Four modes of erasure or writing into these EEPROM words are (1) full EEPROM erase, (2) 32-byte erase, (3) 8- or 16-bit word erase, and (4) 8- or 16-bit word write. For mode 2, the EEPROM memory is organized into *rows* of 32 bytes, whose addresses have the same high-order 11 bits.

EEPROM programming is protected from accidental erasure. The *EEMCR* port (at 0xf0) has bit 1, *PROTLCK*, which must be F (0) to write in the protection port *EEPROT* port (at 0xf1), and is writable only once after reset. Each bit of *EEPROT* controls a part of EEPROM and must be F (0) to program that part. In the 'A4, bit 6 controls the 2K-byte first half, 0xff00 to 0xf7ff; bit 5 controls the 1K-byte next quarter, 0xf800 to 0xfbff; and bit 4 controls the 512-byte next eighth, 0xfc00 to 0xfdff. In the 'B32, bit 4 controls the low-addressed 256 bytes. In both implementations, if the EEPROM is located at the top of memory, bit 3 controls the 256-byte next sixteenth, 0xfe00 to 0xfeff; bit 2 controls the 128-byte next thirty-second, 0xff00 to 0xff7f; bit 1 controls the 64-byte next sixty-fourth, 0xff80 to 0xffbf; and bit 0 controls the 64 byte last sixty-fourth, 0xffc0 to 0xffff, where the interrupt vectors are located.

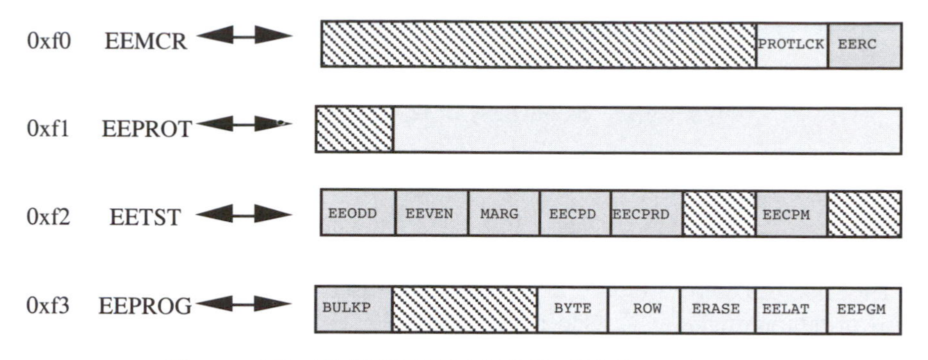

Figure 6.10. MC68HC812A4 EEPROM Control Ports

To program EEPROM, one must have `PROTLCK` F (0) (it is F after reset), and the protection bits in `EEPROT` F (0) (they are T after reset) for the parts to be erased and reprogrammed. The unprogrammed state of EEPROM bits is T (1). Writing can change a T into an F, but to return the bit to T, it must be erased. Erasing is done to all of EEPROM memory (*bulk erase*), all of an 8-bit or 16-bit word, or all of 32 bytes in a row. If any bits are to be changed from F to T, the bytes or rows to be reprogrammmed must be erased before they are written into. Then they can be written; individual bits of an 8- or 16-bit word can be made F.

The bit patterns written into the `EEPROG` register have meaning as follows: `BYTE` means writing or erasing an 8-bit or 16-bit word at a time, `ROW` indicates erasing a row at a time, `ERASE` indicates erasing, `EELAT` latches the next address and data that appear after this bit is set for programming or erasing, and `EEPGM` applies the programming voltage. If `BYTE` is F, `ROW` is F, and `ERASE` is T, the entire EEPROM will be erased (a bulk erase). If `BYTE` is T and `ERASE` is T, an 8-bit or 16-bit word will be erased. If `BYTE` is T and `ERASE` is F, an 8-bit or 16-bit word will be written. If `BYTE` is F, `ROW` is T, and `ERASE` is T, a row of EEPROM will be erased. Erasure or writing takes 10 ms. These modes of erasing or writing into the EEPROM are described as a C program, shown here:

```
enum{T10ms=5000,wordWrite=0x10,wordErase=0x14,rowErase=0xc,bulkErase=4};

void eeProgram(char command,int data,int address){int i; char saveProt;
    saveProt = EEPROT; EEPROT = 0; /* remove protection */
    EEPROG = command; /* put parameters into EEPROM control port */
    EEPROG += 2; /* assert EELAT to latch data */
    *(int *)address = data; /* put data into any word of the EEPROM */
    EEPROG++; /* turn on erasure voltage */
    i = T10ms; do ; while(--i); /* wait 10 msec */ EEPROG--;/* turn off vlt */
    EEPROG = 0; /* return to read mode */ EEPROT = saveProt; /* reapply prot. */
}
```

For erasure, the address must be in the range of the words that will be erased, and data must be written into the 16-bit word. Any address in the range, and any data, can be used. To erase and then write a 16-bit word 0x5678 at 0xff34, we can execute as follows:

Figure 6.11. MC68HC912B32 Flash Control Ports

```
eeProgram(wordErase,0x5678,0xff34);eeProgram(wordWrite,0x5678,0xff34);
```

To erase or write an 8-bit rather than a 16-bit word, the preceding procedure `eeProgram` need only be changed by substituting `char *address` for `int *address`.

The 'B32 has both EEPROM, programmed as discussed above, and flash memory, whose programming is discussed next. The FEECTL port controls flash memory programming as the EEPROG port controls EEPROM programming. See Figure 6.11.

The entire flash memory is erased by calling `eraseFlash`, shown below. Then, a 16-bit word can be programmed by calling `writeFlash` with a 16-bit data word and an address in flash memory. Both operations require 12 V applied to the VFP pin, which should otherwise be at 5 V. Actually, the topmost 2K bytes of flash memory is neither erased nor written into unless special efforts are made to unlock this area; which normally contains a bootstrap program to reload memory. The erasing of all of memory is accomplished by giving up to five "shots" to the memory, each of which takes 111 ms, until the memory is checked and found to be erased. Writing a 16-bit word is accomplished by giving up to 50 "shots" to the memory, each of which takes 25 μs, until the word is checked and found to be written. For both erasure and writing, the "shots" are repeated to ensure compete erasure or writing.

```
#define mc * 8 / 3
enum{ SVFP = 8, ERAS = 4, LAT = 2, ENPE = 1 };
void delay(int count) { while(--count); }
void shot(int d0, int d1, int d2)
    { FEECTL|=ENPE;do delay(d1);while(--d0);FEECTL&=~ENPE;delay(d2);}
char eraseFlashProgram() { char i, j, k;
    if( ! (FEECTL & SVFP)) return 0; FEECTL = ERAS + LAT;
    for(*(int *)0x8000=j=1;(i=cleared())&&(j<=5);j++)
        shot(110,1000 mc,1000 mc );
    if(i) do shot( 110, 1000 mc, 1000 mc ); while(--j);
    i = cleared();   FEECTL &= ~LAT;   return i;
}
unsigned char writeFlashProgram(int data, volatile int *address) {
unsigned char i, j, k;
    FEECTL = LAT; *address = data;
    for(j=1; (i=*address!=data) && (j<=50);j++)shot( 1 ,25 mc,10 mc );
    if( ! i ) do shot( 1, 25 mc, 10 mc ); while(--j);
    i = *address == data;   FEECTL &= ~LAT; return i;
}
char cleared() {for(int *a = (int *)0x8000; a < (int *)0xf800;)
    if(*a++ != 0xffff) return 0; return 1;
}
void main()
    {if(eraseFlashProgram()) writeFlashProgram(0x1234, (int *)0x8000);}
```

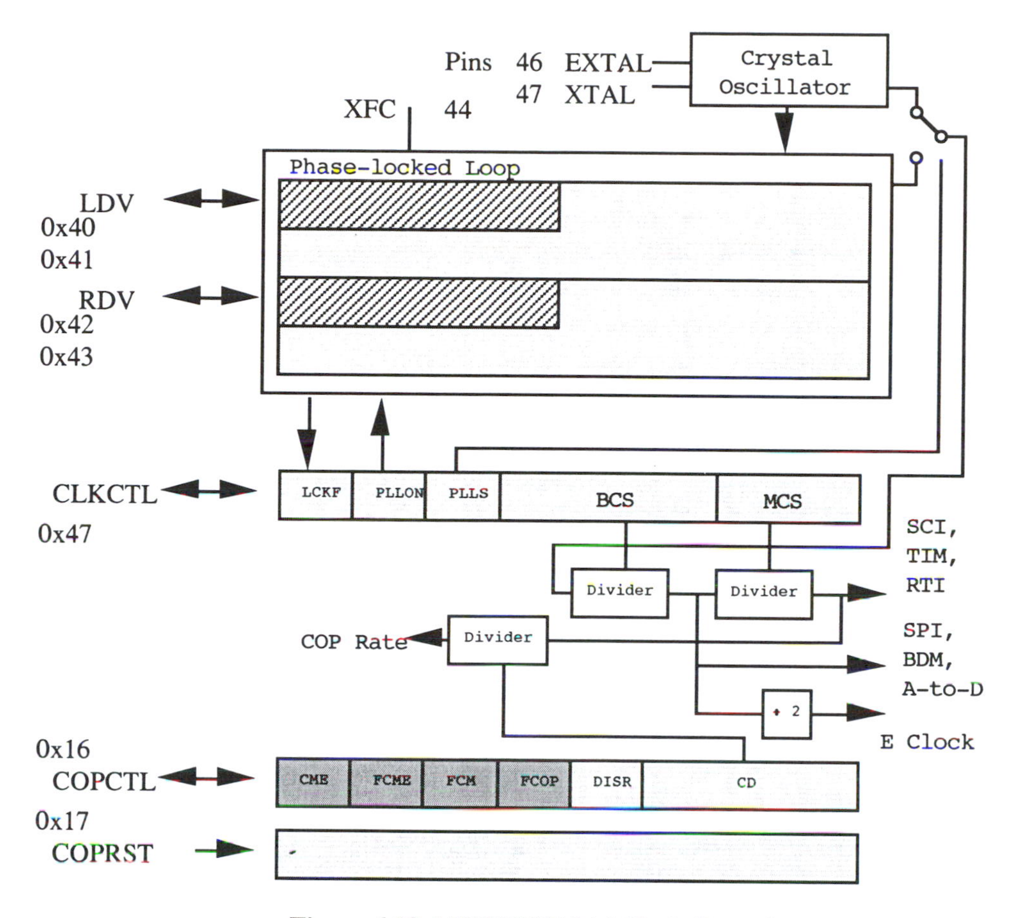

Figure 6.12. MC68HC812A4 Clock Control

6.4 MC68HC812A4 Timing Control

This section covers the main clock and the *computer operating properly* (COP) timing control. We discuss the main clock controls first, and then the COP controls.

The 'A4 clock period is derived from an oscillator that uses a crystal on its XTAL and EXTAL pins, an optional phase-locked loop (PLL, §7.4.3), and dividers (see Figure 6.12). The initial clock source is the crystal if *PLLS* is F (0), otherwise it is the PLL. If the latter, power must be applied to the PLL by asserting *PLLON*, and the PLL is locked if read-only flag *LCKF* is T (1). This initial clock is divided by *BSC* to give the SPI, BDM, and A-to-D clocks, that is in turn divided by two to give the 6812 processor E clock, and is then divided by *MCS* to give the serial communication interface SCI, timer module TIM, real-time interrupt RTI, and COP clocks.

The PLL allows slight adjustments to the clock frequency, in the order of 10-KHz steps, by selecting the 12-bit loop divider *LDV* and reference divider *RDV* values. However, some care is required to implement a loop filter to achieve acceptable stability.

The COP can be used to catch a runaway program that, when operating correctly, will execute in a loop having a known maximum period. The COP mechanism, shown in Figure 6.12, is enabled if *DISR* is F (0) and *CR* is not 0. When it is enabled, the *COPRST* port must be written into with 0x55 and 0xaa within a period specified by CR. If incorrect data is written, or if the pair of correct data are not written in the prescribed period, a COP interrupt occurs using the 16-bit vector at 0xfffa. The handler can restart the program in the loop.

The COP mechanism is enabled after reset to catch a program that runs away right out of reset. In systems that do not use this COP mechanism, it must be defeated by clearing *COPCTL* as soon as possible after reset.

6.5 An External Memory for the MC68HC812A4

In this section we implement an 8K- and a 16K- byte external memory. The address and data bus and control signal connections are discussed first, then timing is analyzed.

As shown in Figure 6.13, the memory has a chip enable E1, read/write W, output enable G, a 13-bit address A[12 to 0], and eight data pins D[7 to 0]. Using the narrow expanded mode, address and data buses are connected to *PORTA* through *PORTC* pins. Data chip select, CSD, enables the memory, and R/W is used to select reading or writing into it. These connections are almost identical to those that we used for indirect memory, in §5.4.3 as shown in Figure 5.15c. Indeed, the reader is invited to first implement the indirect memory, and then implement this expanded memory.

In a write cycle (Figure 5-15b), we assume that the W signal is asserted low before E1 is asserted low and remains low until E1 rises. To prevent writing in the wrong word, the E1 signal should not fall until *A* is determinate and should be low for at least 35 ns, while *A* is determinate. The data input on the data pins should be stable 20 ns before the rising edge of E1 and remain determinate until then. These are the MCM6264D-45's setup and hold times. If these requirements are met, the data input on the data bus pins will be stored in the bit at the address given by *A*.

In a read cycle (see Figure 5.15a), the cycle begins when chip enable E1 falls. It must be low for 45 ns before data is output. Data are available on the data bus pins from the byte addressed by *A*, 45 ns after *A* is determinate, and *A* must be determinate as long as data are being output. Data may become indeterminate 15 ns after E1 rises.

We now consider the timing compatibility of this external memory. These are especially simple if G is connected to ground. To see if the MCM6264D-45 memory is compatible with the 'A4, we match the 'A4's timing against the memory timing for the write cycle and then for the read cycle. The key to timing analysis is that the 'A4 chip select (§6.2.2) will be the memory enable. In both cases, adjust the timing scale so both diagrams have the same dimensions, and slide the 'A4 timing diagram (Figure 3.10) so the chip select lines up with the memory enable (Figure 3.6). See Figure 6.14.

The timing requirements are concisely explained using intervals. An interval $<a, b>$, $a \le b$, is a range of values between a and b. All intervals will be relative to the falling edge of the E clock. The interval of acceptance of a signal is the interval over which the signal must be determinate (stable); it is the setup and hold time. For instance, if a register has a setup time of 10 ns and a hold time of 5 ns, relative to the falling edge of E, the interval is <-10, 5> (with reference to the falling edge of E). The interval of supplying a signal is the interval over which it is guaranteed to be determinate. For example, if a signal will be determinate 5 ns before, and 5 ns after, the falling edge of E, then it is determinate for <-5, 5> with reference to the falling edge of E. For an interval <a, b> to contain an interval <c, d>, every value in <c, d> must be in <a, b>; therefore a ≤ c and b ≥ d. The interval <-5, 5> does not contain <-10, 5> because there is at least one value (-7) that is in <-10, 5> but is not in <-5, 5>; equivalently, -5 is not less than or equal to -10. Generally, for address, data, and control signals, the supply interval must contain the acceptance interval.

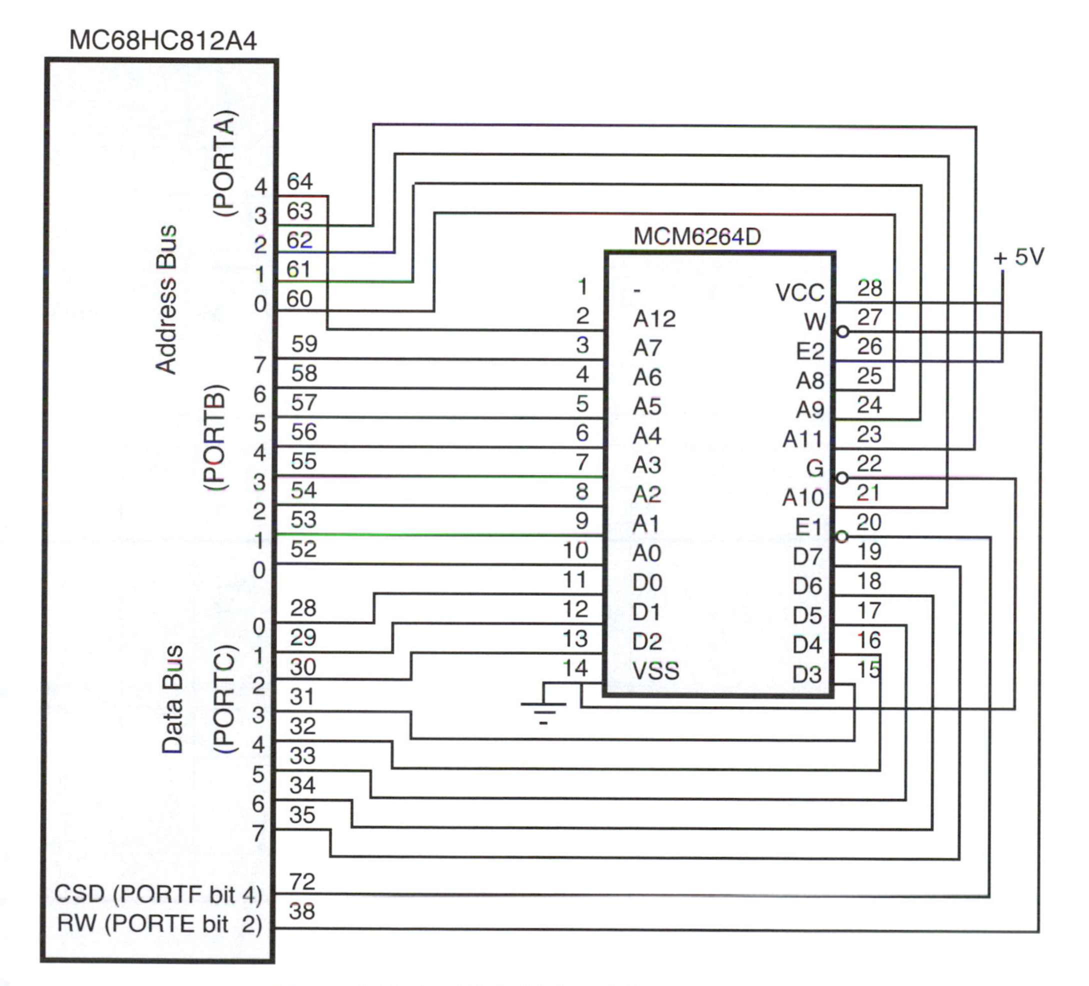

Figure 6.13. An MCM6264D-45 Memory System

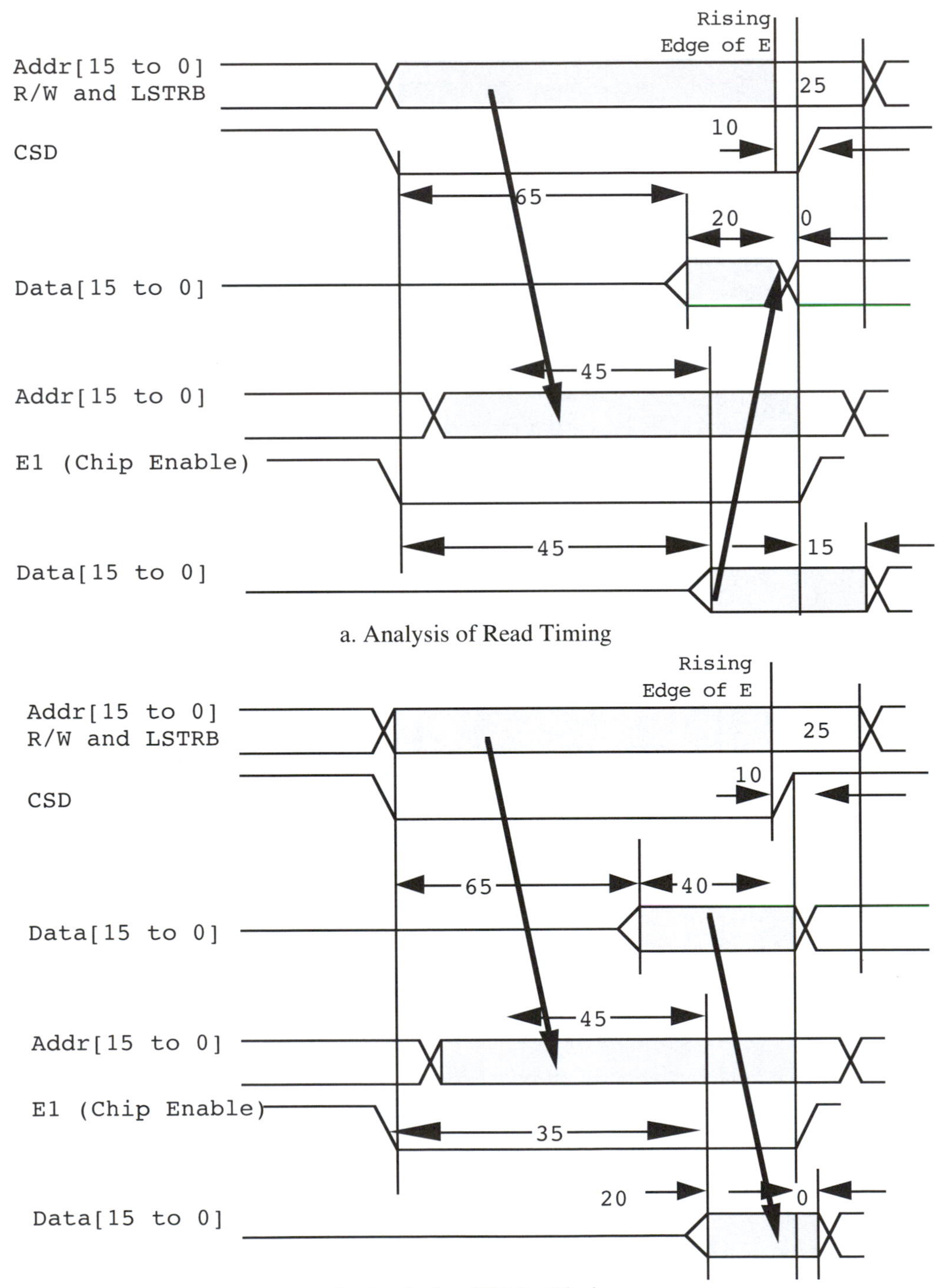

Figure 6.14. Analysis of Memory Timing

We analyze the write cycle first. The lower bound of the address and control signal supply interval (Figure 3.10) is the 125-ns cycle time less the first 40 ns that the signal is indeterminate, and the upper bound is the 25 ns the signal remains determinate into the next cycle: it is <- (125 - 40), 25> = <-85, 25>. This must contain the address acceptance interval. For the write cycle, its lower bound is the 35 ns that these signals must be determinate while the data is written, and its upper bound is 10 because `CSD` can rise as late as 10 ns after E falls, and the address must remain determinate as long as the memory chip's E1 signal is asserted, so it is <- 35, 10>. The address supply interval clearly contains the address acceptance interval during a write cycle. The 'A4 supplies determinate data 40 ns after the E clock rises, which is 60 ns before the end of the memory cycle until 25 ns after it falls; the data supply interval is <-20, 25>. The memory data acceptance interval is <-20, 10>, so the data supply interval clearly contains the data acceptance interval during a write cycle. Write cycle timing is verified.

For the read cycle, the data supply interval (from the memory) begins 45 ns after `CSD` falls, which is 40 ns into the memory clock cycle, and ends 15 ns after the E clock falls, so it is <125 - (40 + 45), 15> = <-30, 25>. The data acceptance interval is <-20, 0>. The address supply interval is <-85, 25>, as it was for the write cycle. The address acceptance interval is <-(45 + 20), 15>. The supply intervals contain the data acceptance interval. Read cycle timing is therefore verified.

An 8K- by 16-bit memory can be implemented in the wide expanded mode as shown in Figure 6.16. The A[0] signal negative-logic ANDed with CSD enables the even-numbered byte locations, and the LSTRB signal negative-logic-ANDed with CSD enables the odd-numbered byte locations, as discussed in §6.2.2. The timing analysis of this circuit will be the same as for the narrow expanded mode (Figure 6.13).

If we get into timing problems, several possible solutions can be tried. The following are the most practical: slow down the processor, lengthen the chip enable, use a faster memory, or modify the memory enable signal with some extra hardware. For the first solution, slow down the 'A4's clock to lengthen the cycle so stable data will reach the 'A4 earlier in its cycle. Use a lower-frequency crystal or use the phase-lock loop (§6.4). To lengthen I/O memory cycles, stretch the chip selects (§6.2.2). You may use faster memory – a faster version of the MCM6264D-45. For instance, the MCM6264D-35 has access times of 35 ns. To utilize external hardware, the MCM6264D-45's E1 signal could be generated with one or more one-shots.

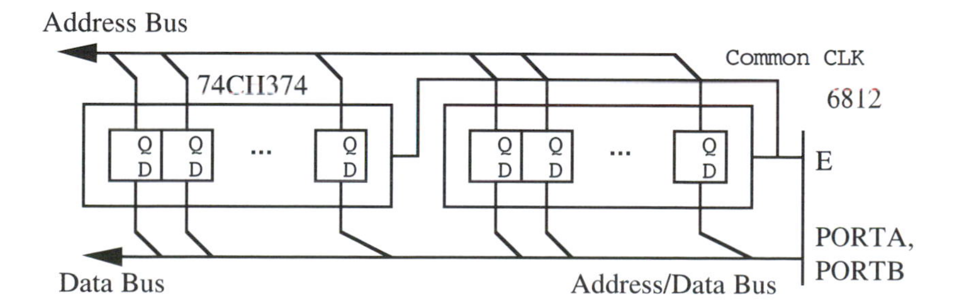

Figure 6.15. MC68HC912B32 Address Demultiplexing

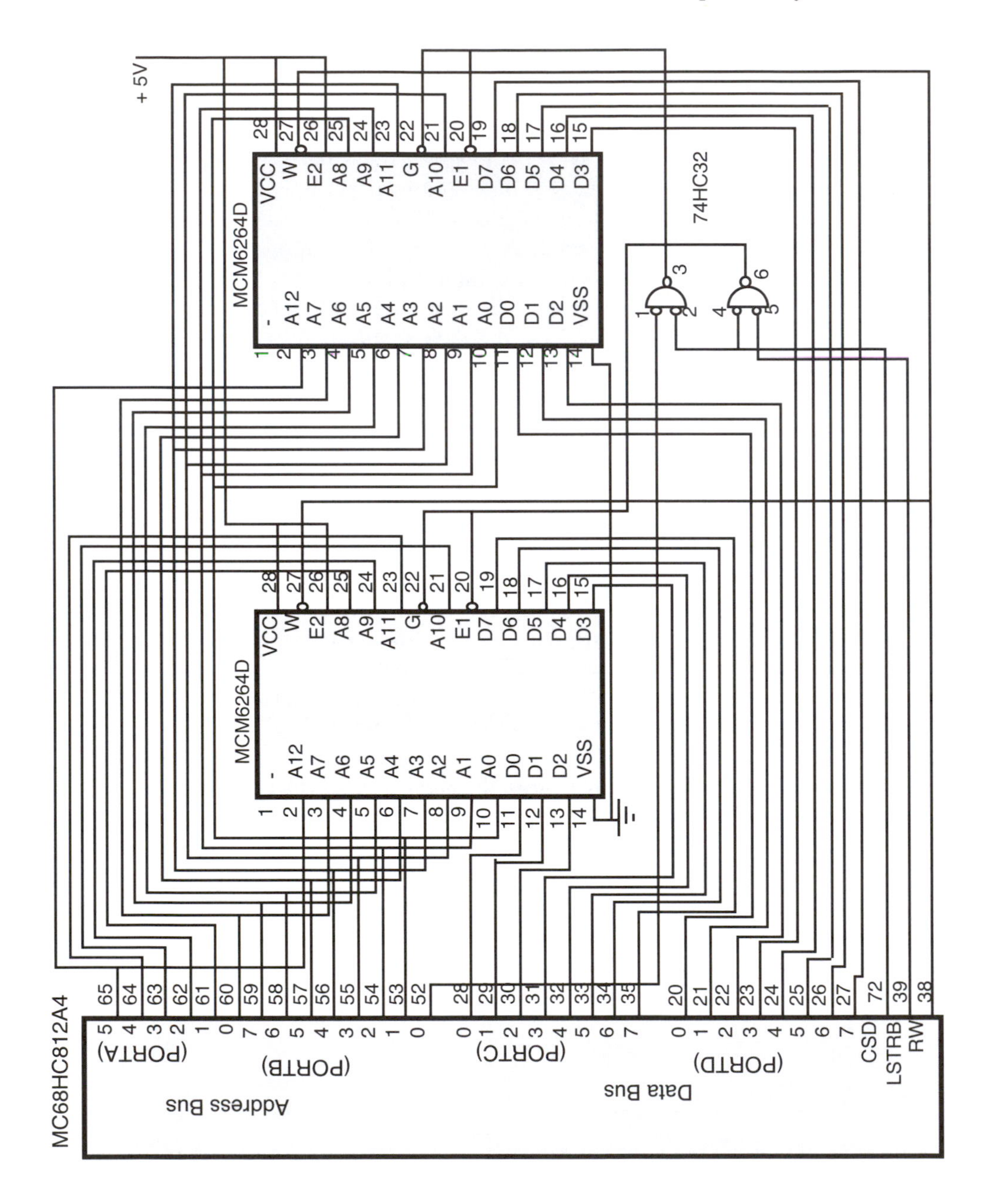

Figure 6.16. 8K-by-16-Bit-Word Wide Expanded-Mode Memory System

The 'B32 has essentially the same mechanism as the 'A4, except that the buses are time-multiplexed. The address can be obtained from the multiplexed address-data bus using a 74HC374, as illustrated in Figure 6.15. Also (negative-logic) DBE should enable memory and device outputs to the data bus so the device doesn't try to mess up the address output in a stretched memory cycle (stretching controlled by `MISC` at 0x13).

This section shows a simple example of a complete interfacing design to connect a memory to a microcomputer. The same principles are used in interfacing I/O registers, discussed in future chapters, and these may be the object of many design travails.

6.6 The 6812 Background Debug Module

The 6812 has a *background debug module* (BDM) that communicates over a single wire (BGND, pin 19). The *host* is generally a microcontroller that can translate commands from the user, via a personal computer, to the BDM and send data to and from the BDM

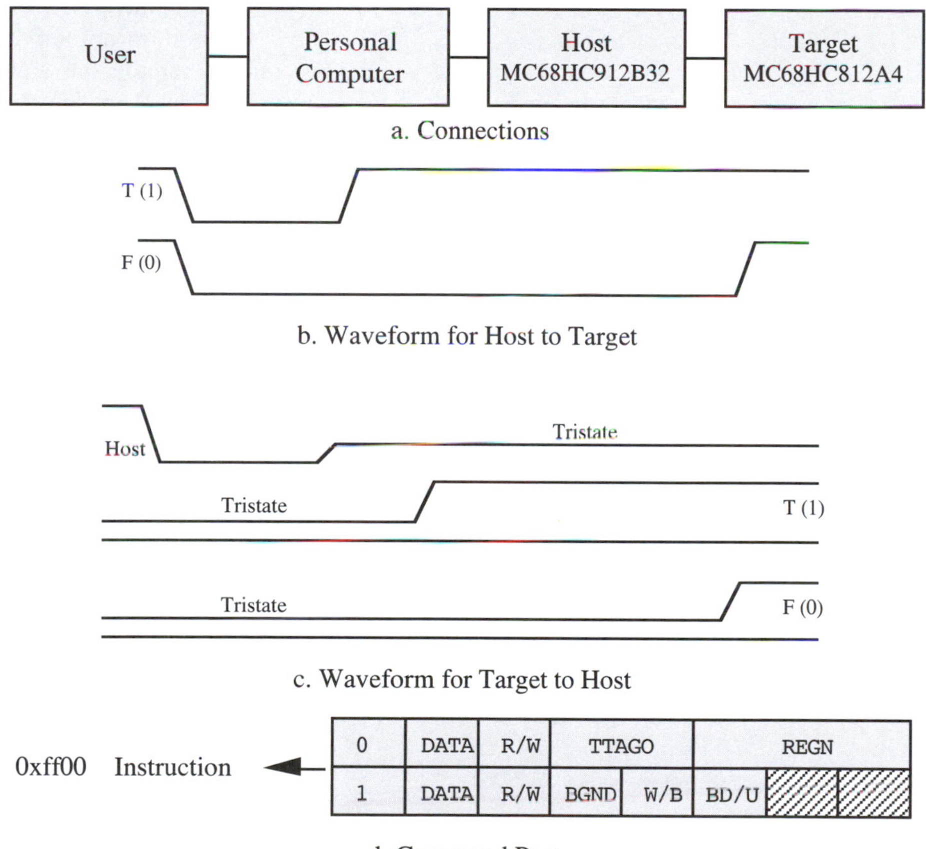

Figure 6.17. Background Debug Module

(see Figure 6.17a). The 'B32 is an excellent microcontroller host. The *target* is the microcontroller that is being debugged. Debugging is simplified because the target being debugged is essentially not used by the debugger; the target can malfunction but the host will work correctly, so the user can determine the malfunction. Also the BDM is used to program the EEPROM of a single-chip microcontroller when its EEPROM is in an unknown state, such as when EEPROM has been accidentally overwritten. In this section, we present the basic ideas of the BDM and show a program that can send data and commnds to, and receive data from, the BDM.

The host sends an 8-bit command, followed by zero, one, or two 16-bit operands, depending on the command. Bits sent most significant bit first are: a T (1), low for a short time, and an F (0), low for a longer time (Figure 6.17b). Essentially, the BDM waits for a falling edge, which signifies the start of a bit, then waits until the middle of the bit time to sample the input. The target's BDM may respond with 16 bits of data, again sent most significant bit first. The host pulses the BGND line low to give a falling edge and then tristates its driver; the target waits for the falling edge and responds with a T (1), pulsing the line high right away, or an F (0), pulsing the line high a little bit later (Figure 6.17c). Capacitance holds the line low when it is not driven. Essentially, the host waits until the middle of the bit time to sample the input. Assembler-language procedures are given below; they will be called from C and therefore their arguments will be on the stack.

```
#pragma NO_ENTRY
#pragma NO_EXIT
#pragma NO_FRAME
void send(int s, int i){
asm{
 ldy 2,sp       ; get number of bits to send
 lsld           ; get msbit
 xgdx           ; shift data in x
 ldd #1         ; acca is 0, accb is 1
 stab 2         ; set DIRA to output on lsb
loop1: staa 0   ; put out low
 nop            ; pad to 16 cycles per loop
 nop            ; pad to 16 cycles per loop
 rol 0          ; insert carry into lsb output
 xgdx           ; shift pattern in d
 lsld           ; get next bit, msb first, into carry
 xgdx           ; shift pattern in x
 stab 0         ; set output bit to 1
 dbne y,loop1   ; count out 16 bits
 rts
}}
```

```
#pragma NO_ENTRY
#pragma NO_EXIT
#pragma NO_FRAME
int receive(){
asm{
```

```
 ldy #16          ; pick up 16 bits
 ldd #1           ; acca is 0, accb is 1
loop: std 0       ; clear data, set DIRH to output on lsb
 nop              ; pad to 16 cycles per loop
 staa 2           ; clear DIRH to input from lsb
 xgdx             ; shifted data in d
 rolb             ; move previous bit into d
 rola             ; msb first
 xgdx             ; shifted data in x
 lsr 0            ; carry is input bit
 dbne y,loop      ; count out 16 bits
 xgdx             ; shifted data in d
 rolb             ; move last bit into d
 rola             ;
 rts
}
  return 0; /* keep C++ happy */
}
#pragma NO_ENTRY
#pragma NO_EXIT
#pragma NO_FRAME
void wait(){
asm{
 ldd #150/3
loop: dbne d,loop
 rts
}}
```

These assembler-language procedures, `out8(char c)`, `out16(int i)`, and `receive()`, can be used to issue commands alone or from C procedures shown below. The first byte sent through the BGND pin is a command and is put into a port at 0xff00 (Figure 6.17d). If the most significant bit is T (1), then the command is executed in hardware and the target continues to run normally. Such hardware commands can be executed at any time, even when the target is executing a user program. The remaining bits indicate if data accompanies the command `(DATA)`, if the command reads or writes `(R/W)`, if the command causes the target to suspend running `(BGND)` if data is 8 bits or 16 bits `(W/B)`, and if memory includes the BDM registers and ROM `(BD/U)`. An `enum` statement is shown, that gives the command byte for operations which can be executed by the BDM. A single-byte command to cause the target to suspend running is executed in C as `out8(BACKGROUND)`. A command to write 16-bit data `d` at address `a` is executed as `write(WRITE_WORD, a, d);`. A command to read 16-bit data at address `a` is executed as `read(READ_WORD, a);`. Commands that read and write in the backround memory map can read or write the memory-mapped registers used by the background debug module.

```
enum {BACKGROUND = 0x90, WRITE_BYTE = 0xc0, WRITE_WORD = 0xc8,
   WRITE_BB = 0xc4, WRITE_BW = 0xcc, READ_BYTE = 0xe0,
   READ_WORD = 0xe8, READ_BB = 0xe4, READ_BW = 0xec,
   GO = 0x08, WRITE_NEXT = 0x42, WRITE_PC = 0x43,
   WRITE_D = 0x44, WRITE_X = 0x45, WRITE_Y = 0x46,
   WRITE_SP = 0x47, READ_NEXT = 0x62, READ_PC = 0x63,
   READ_D = 0x64, READ_X = 0x65, READ_Y = 0x66,
   READ_SP = 0x67};

char read(char c,int a)
   {int i;send(8, c);send(16, a); wait();i=receive(); wait();return i;}
/* use with READ_BYTE, READ_WORD, READ_BB, and READ_BW */

void write(int c,int a,int d){send(8,c);send(16,a);send(16,d);wait();}
/* use with WRITE_BYTE, WRITE_WORD, WRITE_BB, and WRITE_BW */

char readreg(char c)
   {int i;send(8,c);wait();i=receive();wait();return i; }
/* for READ_NEXT, READ_PC, READ_D, READ_X, READ_Y, and READ_SP */

void writereg(char c, int d){ send(8, c); send(16, d); wait(); }
/* for WRITE_NEXT, WRITE_PC, WRITE_D, WRITE_X, WRITE_Y, WRITE_SP */

void main(){ int i; /* example program */
   i = read(READ_BYTE, 0x1000); write(WRITE_BYTE, 0x1000, 0x1234);
   i = readreg(READ_D); writereg(WRITE_PC, 0x1234);
}
```

If the most significant bit of a command is F (0), then the command is executed by target software, and target execution must be stopped before the command can be executed. The software is stored in a ROM in the range 0xff00 to 0xfff, which is only in the memory map when the target is executing background debug commands. Such commands may not be executed when the target is executing a program being debugged. In order to suspend operations later, the BDM must first be given a command `write(WRITE_BB, 0xff01, 0x80);` This sets a bit in a BDM port at location 0xff01 that enables operation of the background debug mode. The target suspends operation when it executes the `bgnd` instruction, or when the host executes a command `out8(BACKGROUND).` Then the target can execute software commands having most significant bit F (0), as shown on the top of Figure 6.17d. In these commands the *DATA* and R/W bits are the same as for hardware BDM commands, but the *TAGGO* field indicates commands to make the target go, to trace an instruction, or to control the instruction decoder so that "tagging" can be used with a logic analyzer or with a hardware breakpoint mechanism. Finally, a field *REGN* permits selection of specific registers, D, X, Y, SP, or PC, to be written into or read from.

One of the cutest experiments for the 'A4 is to have it execute the procedures above, that use only the hardware commands, in its own BDM module. The 'A4 `PORTH` bit 0 is connected to the 'A4 `BGND` pin. It is possible to read or write memory, and even to program EEPROM. However, to execute any software background debug command using the 6812, a separate 6812 must be used for the host and for the target; the 6812 can't execute both programs simultaneously.

Generally, the host can be any machine, such as the MC68HC912B32. When the host and target are different machines, the software-based debug commands can be executed as well as the hardware-based commands.

6.7 The 6812 Reset Handler

In the 6812, RESET must be asserted low after power is first applied and reaches 4.5 V. Also, if the RESET pin is asserted low at any time, processing stops (a kind of ultimate interrupt). The RESET line is generally connected to RESET pins on all I/O devices; it initializes the I/O registers, normally clearing all I/O ports.

When RESET rises again, the 16-bit word at location 0xfffe, 0xffff is put into the program counter. These locations are normally in read-only memory, for if we didn't have these locations and programs there before executing the first instruction, we would have no way to start the machine. (You may laugh at this, but machines proposed even in learned papers have had this problem – that they can't be started.) A nonvolatile memory must be enabled to read out the initial value of the program counter and stack pointer, and must remain available to fetch instructions from it until the bootstrap loads programs into RAM. In normal single-chip mode, 'A4 reads the contents of 0xfffe and 0xffff from EEPROM, and the address stored therein will be an EEPROM location between 0xf000 and 0xffff.

However, it may be desirable to effectively remove that nonvolatile memory from the high end of the memory map after the bootstrap program in it has served its purpose, so that interrupt vectors can be written in RAM that are made accessible in high memory. The concept of including a memory or device in the memory map and then excluding it from the memory map is called *phantoming* the memory.

The program that is entered handles the initialization ritual after a reset, so we call it the *reset handler*. The reset handler configures the 'A4 and many of the I/O devices not already configured as desired by the RESET signal on their RESET pins and not to be configured later as part of the program that uses the device. In earlier discussions, we said that the rituals to be run just after power is applied are all put in this reset handler.

In the 'B32, a reset handler `handler()` should initialize the stack pointer, bus mode, and clock ports for the application. For instance, the mode may be changed to expanded wide mode after being reset and coming up in single-chip mode, and control signals may be assigned to get the DBE, E clock, R/W, and LSTRB signals in place of `PORTE` bits. The COP usually must be disabled. The handler generally initializes global variables and one or more I/O devices, such as the SCI, and calls the C procedure `main()`, which returns to the handler when it is finished. The handler's function for the rest of the time is to reinitialize devices and memory, and then restart `main()`, each time `main()` returns to it. A 'B32 handler is illustrated below.

```
volatile unsigned int PEARMODE@)0x0A, CSCTL@0x3C;
volatile unsigned char COPCTL@0x16;
enum { Special=0,Wide=0x60,RDWRE=0x0400,LSTRE = 0x0800,NDBE = 0x8000};
extern void main(void);
interrupt 0 void handler() {
    do { asm lds #0xc00 /* set the stack pointer */
         PEARMODE=Special + Wide + RDWRE + LSTRE; /* init PEAR, MODE*/
         COPCTL = 0; /* disable the COP monitor; initialize global variables here */
         main(); asm SEI /* disable interrupts after returning from main */
    } while(1);
}
```

In the 'A4, a reset handler `handler()` should initialize the stack pointer and then configure memory map and clock ports for the application. For instance, the mode may be changed to expanded wide mode after being reset and coming up in single-chip mode, and control signals may be assigned to get the E clock, R/W, and LSTRB signals, in place of `PORTE` bits. The chip selects may be assigned in place of `PORTF` bits. If memory expansion is needed, `PORTG` bits may be reassigned as high-order address bits. The COP usually must be disabled. Finally, one or more I/O devices, such as the SCI, must be initialized. Then this reset handler calls up the C procedure `handler()`, which can return to the handler when it is finished. Before it does so, the handler will generally clear and initialize global variables so that each time `handler()` is executed, it has consistent starting data. The handler's function for the rest of the time is to reinitialize devices and memory, and then restart `handler()`, each time `handler()` returns to it.

```
#volatile unsigned char
    MODE@0x0B, PEAR@0x0A, CSCTL@0x3C, CSSTR@0x3E, COPCTL@0x16;
enum { Special = 0, Wide = 0x60, RDWRE = 4, LSTRE = 0x08,
    CSP1E = 0x4000, CS2E = 0x400, CS3E = 0x800, STR2 = 4, STR3 = 6};
extern void main(void);
interrupt 0 void handler() {
    do {      asm lds #0xc00 /* set the stack pointer */
         MODE = Special + Wide; /* switch to special wide bus */
         PEAR = RDWRE + LSTRE; /* R/W, LSTRB and the E clock for decoders */
         CSCTL = CSP1E | CS2E | CS3E; /* enable chip selects */
         CSSTR = (2 << STR2) | (2 << STR3); /* set chip select stretch cycles */
         COPCTL = 0; /* disable the COP monitor; initialize global variables here */
         main(); asm SEI /* disable interrupts after returning from main */
    } while(1);
}
```

The handler may also run diagnostic programs to check the microprocessor, memory, or I/O devices; clear all or part of memory; or initialize some variables. This handler might be a *bootstrap program* whose purpose is to load into memory and then execute the program that follows, or an *operating system* to manage memory and time resources for a microcomputer.

6.8 Conclusions

The 'A4 has considerable flexibility in configuring its memory and timing. We have seen how the 'A4's buses can be expanded, how the internal memories and ports can be remapped, how control signals and chip selects can be used for external devices, and how memory can be expanded. We saw how EEPROM can be programmed and how the E clock can be controlled. Then we saw applications of these concepts, including the design of an external memory and the writing of a reset handler. We also covered the background debug module and showed how to send and receive packets from it.

For more concrete information on the 'A4, consult the MC68HC812A4TS/D. In particular, §7 describes the EEPROM, §8 describes memory mapping and chip selects, §11 describes clock control, and §12 describes the phase-locked loop. As noted earlier, we have not attempted to duplicate the diagrams and discussions in that book because we assume you will refer to it while reading this book; and, since we present an alternative view of the subject, you can use either or both views.

You should now be familiar with memory and clock configuration in general and with the 'A4 memory and clock control ports in particular. Selecting configurations of the 'A4 memory and clock, and writing software to initialize and use them, should be well within your grasp.

Do You Know These Terms?

See page 36 for instructions.

internal address
real address
dataPage
computer operating properly
background debug module
host
phantoming
reset handler
bootstrap program
operating system

Problems

Problem 1 is a paragraph correction problem. See page 38 for guidelines. Guidelines for software problems are given on page 86, and for hardware problems, on page 115.

1. Write a program `main` that sets the 'A4 in the normal wide expanded mode and makes available the E clock and R/W signals but does not make available the LSTRB.

2. Show a program *main* that puts the I/O ports at 0xf000-0xf1ff and RAM at 0xf800-0xf9ff, and disables EEPROM. This might be useful if a 64K-byte memory is externally connected, and EEPROM is phantomed after it is initially used to configure I/O devices and load the external memory. Where can this procedure be run from?

3. Show the logic diagram of a group of eight 8-bit basic output ports at location 0x200 to 0x207, respectively. Use chip select 0 to enable a 74HC138, which in turn enables one of eight 74HC374 registers (show only two of them). Show a statement that initializes the `CSTL` port for this application. Assume that the 'A4 has been configured for normal expanded narrow mode and that there are no other external devices or memories.

4. Show the logic diagram of a group of eight 16-bit input ports at locations 0x200 to 0x20f, contiguously, for the normal expanded wide mode. Use chip select 0 to enable a 74HC138, whose output in turn enables two of the sixteen 74HC244 tristate drivers (show only two pairs of them). Show a statement that initializes `CSTL` for this application. Assume there are no other external devices or memories.

5. Show the logic diagram of a group of four 16-bit basic output ports at locations 0x200-0x201, 0x280-0x281, 0x300-0x301, and 0x380-0x381, respectively. Show only the chip select connections, not the data bus connections. Use chip selects 0 to 3 tc enable two of eight 74HC374 registers. Show a statement that initializes the `CSTL` port for this application. Assume that the 'A4 has been configured for normal expanded wide mode and that there are no other external devices or memories.

6. Show the logic diagram of a group of four 8-bit input ports at location 0x200, 0x280, 0x300, and 0x380, respectively. Show only the chip select connections, not the data bus connections. Use chip selects 0 to 3 to enable four 74HC244 tristate drivers. Show a statement initializing `CSTL` for this application. Assume that the 'A4 has been configured for normal narrow expanded mode.

7. Use an external Dallas Semiconductor DS1650Y/AB 512K SRAM for a DPAGE expansion window for a normal narrow expanded mode 'A4. The SRAM CE is connected to `CSD`, its W pin is connected to the 'A4 R/W pin, and its OE is asserted low. The SRAM data bus is connected to the 'A4 narrow expanded-mode bus, and its address pins are connected to equivalent 'A4 address pins, but 'A4's unused address pins are made available for `PORTG` parallel port bits.

a. Show the logic diagram. The 32-pin DS1250Y/AB has the same "lower" pins as the MCM6264D-45 (Figure 6.13): the first's pins 4 to 27 are the second's pins 2 to 25; the first's pins 1 to 3 are address lines 18, 14, and 16, and its pins 28 to 32 are A13, W, A17, A15, and Vcc.

b. Show a procedure `main()` that configures the `MODE, PEAR, CSCTL, WINDEF,` and `MXAR` ports, and writes 0x1234 into internal memory locations 0x56788 and 0x56789.

8. Use a pair of external MCM6246 512-K-by-8-bit SRAMs for a 1-MB DPAGE expansion window for a normal expanded wide mode 'A4. Both SRAM E pins are connected to CSD, and both SRAM G pins are asserted low. The "left" SRAM's W pin is connected to the 'A4 R/W pin ANDed with A0, and the "right" SRAM's W pin is connected to the 'A4 R/W pin ANDed with LSTRB, where ANDing is in negative logic. The SRAM data bus is connected to the 'A4 wide expanded-mode bus and its address pins are connected to equivalent 'A4 address pins plus one, but 'A4's unused address pins are made available for `PORTG` data bits.

a. Show the logic diagram. Any SRAM's address pins are interchangeable. The 36-pin MCM6246 has address pins 1 to 5, 14 to 18, 20 to 24, and 32 to 35. Any SRAM's data pins are interchangeable. Its data pins are 7, 8, 11, 12, 25, 26, 29, and 30. The MCM6246's chip enable E is pin 6, its W is pin 13, its G is pin 31, its Vcc are pins 9 and 27, and its Vss are pins 10 and 28. Its pins 19 and 36 are not connected. Use vector notation on pins to describe the buses more compactly.

b. Show a procedure `void main()` that configures the `MODE, PEAR, CSCTL, WINDEF`, and `MXAR` ports, and that writes 0x1234 into internal memory locations 0x56788 and 0x56789.

9. Design an external M29F800A3 1-MB-by-8 flash memory to be used for a 1-MB `PPAGE` expansion window for a normal narrow expanded mode 'A4. We assume that the flash memory is written into elsewhere, and here is only read. The flash E pin is connected to `CSP1,` its W is connected to R/W, and its OE pin is asserted low. The flash data bus is connected to the 'A4 narrow expanded-mode bus, and its address pins are connected to similar 'A4 address pins, but 'A4's unused address pins are made available for `PORTG` parallel port bits. Reading the flash requires stretching the E clock for 1 cycle.

a. Show the logic diagram. The 48-pin M29F800A3's address A[19 to 0] are pins 16, 17, 48, 1 to 8, 18 to 25, and 45. Its data D[7 to 0] are pins 29, 31, 33, 35, 38, 40, 42, and 44. Its W is pin 11, its E is pin 26. Pins 46 and 27, write protect 14, BYTE 47, and OE pin 28, are grounded, and pins RP 12 and 37 are connected to +5 V. Data pins 30, 32, 34, 36, 39, 41, and 43, and pins 9, 10, 13, and 15 are not connected.

b. Show a procedure `void main()` that configures the `MODE, PEAR, CSCTL, CSSTR, WINDEF,` and `MXAR` ports and CALLs a subroutine, using embedded assembler language that begins at internal address 0x56788.

10. Design an external M29F800A3 512K-by-16 flash memory for a 1-MB `PPAGE` normal wide expanded-mode expansion window. Assume that the flash memory is written into elsewhere, and here it is only read. The flash's E pin is connected to CSP1, its W is connected to R/W, and its OE pin is asserted low. Its data bus is connected to the 'A4 wide expanded-mode bus and its address pins are connected to similar 'A4 address pins, but 'A4's unused address pins are made available for `PORTG` parallel port bits. Reading the flash requires stretching the E clock for one cycle.

a. Show the logic diagram. The 48-pin M29F800A3's address A[19 to 1] are pins 16, 17, 48, 1 to 8, and 18 to 25. Its D[15 to 0] are pins 45, 43, 41, 39, 36, 34, 32, 30, 29, 31, 33, 35, 38, 40, 42, and 44. Its W is pin 11 and its E is pin 26. Pins 46, 27, write protect 14, and OE pin 28, are grounded, and pins RP 12, BYTE 47, and 37 are connected to +5v. Pins 9, 10, 13, and RY/BY 15 are not connected.

b. Show a procedure `void main()` that configures the `MODE, PEAR, CSCTL, CSSTR, WINDEF,` and `MXAR` ports, and CALLs a subroutine, using embedded assembler language, that begins at internal address 0x56788.

11. The Intel-family bus has negative-logic `RD` and `WR` control variables instead of the 6812's R/W signal and E clock. When the Intel processor asserts `RD` low, it has put a stable address on the address bus; memory should read the word at this address. When the Intel processor asserts `WR` low, it has put stable data on the data bus and a stable address on the address bus; memory should write the data at the address it gave. Show the logic diagram of a decoder, using a minimum number of gates.

12. Rewrite the templated class `dataPage` of §6.2.4 to permit the programmer to treat scalars, vectors, and two- and three-dimensional arrays in the data expansion window like conventional memory variables. Use a procedure `char*vAllocate(int i)` to allocate `i` bytes of internal memory space, which returns a internal memory location to be used for the variable. For instance, if when we declare `dataPage <char> a; vAllocate(1)` returns 0x4ff9c, then *i* = *a* reads internal memory location 0x4ff9c, using `DPAGE`, and *a* = *i;* writes *i* into internal memory location 0x4ff9c, using `DPAGE`. Thus if when we declare `dataPage <int> a(2,3,4); vAllocate(48)` returns 0x4ff00, then `i =` *a[1][2][3];* reads internal memory locations 0x4ff17 and 0x4ff18, using `DPAGE,` and *a[1][2][3]* `= i;` writes `i` into internal memory location 0x4ff17 and 0x4ff18, using `DPAGE.`

13. Write a templated class `extraPage` that accesses data through the extra page window using port `EPAGE,` which

a. permits the programmer to use the constructor and access two-dimensional arrays as in §6.2.4.

b. permits the programmer to call the constructor and to access scalars, vectors, two-dimensional arrays, and three-dimensional arrays, as in problem 12.

14. Write a templated class *EEProm* that accesses EEPROM variables the same as variables in conventional memory. For example, if *char* array *a* is located at 0xff90, `i = a[3][4]` reads EEPROM location 0xff9c, and `a[3][4] = i;` erases location 0xff9c if that location has an F (0) where *i* has a T (1), and then writes *i* into location 0xff9c if *i* is different from the contents of 0xff9c. Inside your class function members, to program EEPROM, call §6.3's procedure *eeProgram*, which is copied into RAM. Write the class and its members to

a. permit the programmer to call the constructor and access two-dimensional arrays as in §6.2.4.

b. permit the programmer to call the constructor and to access scalars, vectors, and two-, and three-dimensional arrays, as in problem 12. Use a procedure `char *eAllocate(int i)` to allocate *i* bytes of EEPROM memory space, which returns an EEPROM location to be used for the variable.

15. Write a C statement to change the 'A4 clock timing.

a. the E clock is 2 MHz, and the clock provided to the counter/timer, RTI, and SCI, before their prescalars divide the clock down, is 2 MHz.

b. the E clock is 8 MHz, and the clock provided to the counter/timer, RTI, and SCI, before their prescalars divide the clock down, is 2 MHz.

16. Write a C procedure *main* that uses a target's background debug module to access its data. Use the procedures written in assembler language and in C, and the I/O ports that are shown in §6.5 to access the data in the target's SRAM.

a. writes 0x12 into location 0x800 in an 'A4

b. reads location 0x800 into local *char* variable *c* in an 'A4

17. Write a procedure `main()` that writes 0x12 into location 0xff80 in a target, using its background debug module to write the data. Assume the 'A4 is in single-chip mode, and use the procedure written in assembler language and in C, and the I/O ports, that are shown in §6.6, to write in the EEPROM programming control ports, erase the EEPROM byte and then write the data in EEPROM. Use symbolic names *#define*d in §6.5 and §6.6 to make your answer more self-documenting.

18. Write a reset handler `void _startup()` for single-chip mode that intitializes the stack pointer to the high end of SRAM, disables the COP monitor, clears the 0x133 bytes of low SRAM, writes 0x11bd in 0x810, and calls `main()`. After returning from `main()`, it disables interrupts and repeats this handler.

19. Write a reset handler `void _startup()` for narrow expanded mode that initializes the stack pointer to the high end of SRAM, disables the COP monitor, enables E and R/W and chip selects *CS0* and *CS1*, outputs all high-order address bits, clears the 0x133 bytes of low SRAM, writes 0x11bd in 0x810, and calls `main()`. After returning from `main()`, it disables interrupts and repeats this handler.

This inexpensive Axiom PB68HC12A4 board is well suited to senior design, and other prototyping projects. Its wire-wrap pins can be reliably connected to external wire-wrap sockets and connectors.

7

Analog Interfacing

Analog circuits are commonly used to interconnect the I/O device and the "outside world." This chapter will focus on such circuits as are commonly used in microcomputer I/O systems. In this chapter, we will assume you have only a basic knowledge of physics, including mechanics and basic electrical properties. While many of you have far more, some, who have been working as programmers, may not. This chapter especially aims to provide an adequate background for studying I/O systems.

Before analog components are discussed, some basic notions of analog signals should be reviewed. In an *analog* signal, voltage or current levels convey information by real number values, like 3.1263 V, rather than by H or L values. A *sinusoidal alternating current* (AC) signal voltage (or current) has the form $v = A \sin(P + 2\pi F t)$ as a function of time t, where the *amplitude* A, the *phase* P, and the *frequency* F can carry information. The *period* is $1/F$. (See Figure 7.1a.) One of the most useful techniques in analog system analysis is to decompose any *periodic* (that is, repetitive) *waveform* into a sum of sinusoidal signals, thus determining how the system transmits each component signal. The *bandwidth* of the system is the range of frequencies that it transmits faithfully (not decreasing the signal by a factor of .707 of what it should be). A square wave, shown in Figure 7.1b, may also be used in analog signals. Amplitude, phase, and frequency have the same meaning as in sinusoidal waveforms.

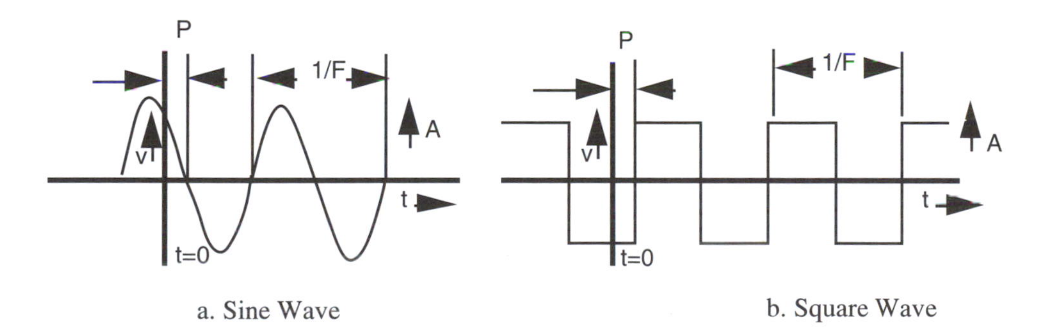

a. Sine Wave b. Square Wave

Figure 7.1. Waveforms of Alternating Voltage Signals

Two kinds of analog signals are important. These correspond to AM and FM radio. In this chapter, we consider analog signals whose amplitude carries the value, whether the signal is direct current or alternating current, as AM radios carry the sound. In the next chapter, we consider analog signals whose frequency or phase carries the value of the signal, as FM radios carry the sound. Amplitude analog signals are more pervasive in interface design. It is hard to find examples of interface hardware that do not have some analog circuitry (we had to search long and hard to find some decent problems for Chapter 3 that did not have analog circuits in them). It is even hard to discuss frequency analog circuits without first discussing amplitude analog circuits. So we study amplitude analog circuits in this chapter and frequency analog circuits in Chapter 8.

Analog signals are converted to digital signals by *analog-to-digital converters* (A-to-D converters), and digital signals are converted to analog by *digital-to-analog converters* (D-to-A converters) such that the digital signal – usually a binary number or a binary-coded decimal number – corresponds in numerical value to the analog signal level. Analog signals are also converted to a single digital bit (H or L) by a *comparator*, and digital signals control analog signals by means of *analog switches*. The frequency of an AC signal can be converted to or from a voltage by *voltage-to-frequency converters* (V-to-F converters) or by *frequency-to-voltage converters* (F-to-V converters). Finally, analog signals are generated by *transducers* that change other measurements into voltages or currents, such as temperature-to-voltage transducers, and are amplified and modified by *operational amplifiers* (OP AMPs).

A basic theme of this chapter is that many functions can be done using digital or analog hardware, or using software. The smart designer determines the best technique from among many alternatives to implement a particular function. Thus, the designer should know a little about analog circuitry. On the one hand, a basic understanding of the operation and use of analog circuits is essential in making intelligent hardware/software trade-offs and is quite useful even for programmers who write code to interface to such devices. So we want to include the required material in this chapter. On the other hand, to use them well, one can devote an entire year's study to these devices. We have to refrain from covering that much detail. Therefore, our aim is to give enough detail so readers can make good hardware/software trade-offs in the design of microprocessor analog systems and to encourage those who seek more detail to read some of the many excellent books devoted to the topic of analog signal processing.

In the following sections, we will discuss conversion of physical quantities to voltages and from voltages, the basics of operational amplifiers and their use in signal conditioning and keyboard/display systems, digital-to-analog conversion, analog-to-digital conversion, and data acquisition systems. Much of the material is hardware oriented and qualitative. However, to make the discussion concrete, we discuss in some detail the use of the popular CA3140 operational amplifier and the 4066 and 4051 analog switches. Some practical construction information will be introduced as well. The reader might wish to try out some of the examples to understand firmly these principles.

This chapter should provide enough background on the analog part of a typical microcomputer I/O system so the reader is aware of the capabilities and limitations of analog components used in I/O and can write programs that can accommodate them.

7.1 Input and Output Transducers

A transducer changes a physical quantity, like temperature, to or from another quantity, often a voltage. Such a transducer enables a microcomputer that can measure or produce a voltage or an AC wave to measure or control other physical quantities. Each physical property – position, radiant energy, temperature, and pressure – will be discussed in turn, and for each we will examine the transducers that change electrical signals into these properties and then those that change the properties into electrical signals.

7.1.1 Positional Transducers

About 90% of the physical quantities measured are positional. The position may be linear (distance) or angular (degrees of a circle or number of rotations of a shaft). Of course, linear position can be converted to angular position by a rack-and-pinion gear arrangement. Also, recall that position, speed, and acceleration are related by differential equations: if one can be measured at several precise times, the others can be determined.

A microcomputer controls position by means of *solenoids* or *motors*. A solenoid is an electromagnet with an iron plunger. As current through the electromagnet is increased, an increased force pulls the plunger into its middle. The solenoid usually acts against a spring. When current is not applied to the solenoid, the spring pulls the plunger from the middle of the solenoid; and when current is applied, the plunger is pulled into the solenoid. Solenoids are designed to be operated with either direct current or alternating current, and are usually specified for a maximum voltage (which implies a maximum current) that can be applied and for the pulling force that is produced when this maximum voltage (current) is applied. A *direct-current motor* has a pair of input terminals and a rotating shaft. The (angular) speed of the shaft is proportional to the voltage applied to the terminals (when the motor is running without being loaded) and the (angular) force or torque is proportional to the current. A *stepping motor* looks like a motor, but actually works like a collection of solenoids. When one of the solenoids gets current, it pulls the shaft into a given (angular) position. When another solenoid gets current, it pulls the shaft into another position. By spacing these solenoids evenly around the stepping motor and by giving each solenoid its current in order, the shaft can be rotated a precise amount each time the next solenoid is given its current. Hence the term stepping motor. The *universal motor* can be given either direct current or alternating current power. Most home appliances use these inexpensive motors. Their speed, however, is very much dependent on the force required to turn the load. *Shaded pole motors* require alternating current, and the shaft speed is proportional to the frequency of the AC power rather than the voltage. The torque is proportional to the current. These inexpensive motors often appear in electric clocks, timers, and fans. *Induction motors* are usually larger-power AC motors, and their speed is proportional to frequency like the shaded pole motor. Finally, the *hysteresis synchronous motor* is an AC motor whose speed is accurately synchronized to the frequency of the AC power. These are used to control the speed of high-fidelity turntables and tape decks.

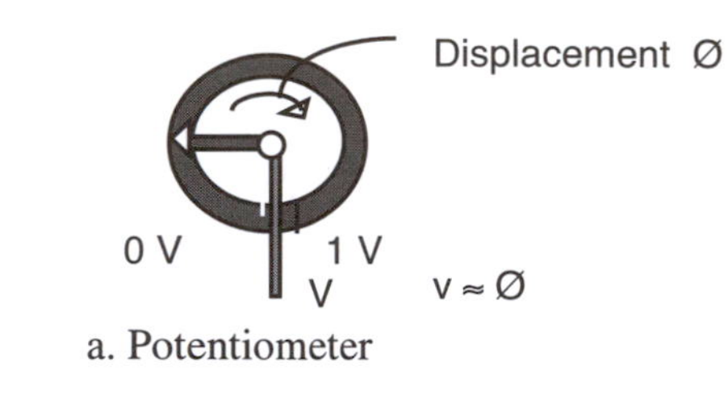

a. Potentiometer

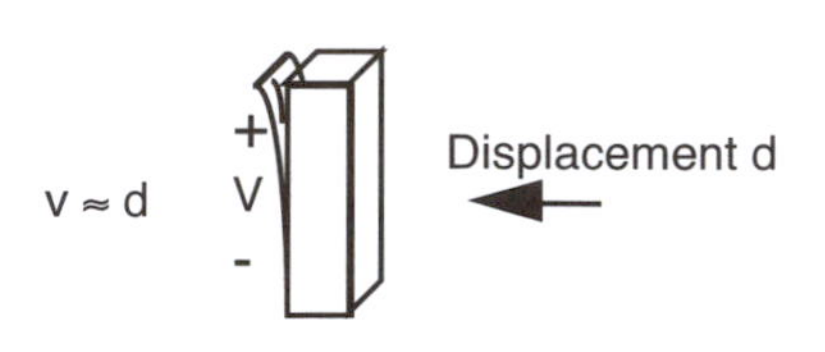

b. Strain Gauge

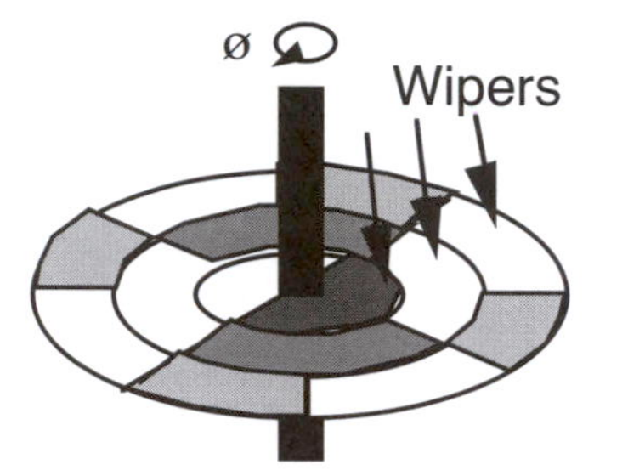

c. Shaft Encoder

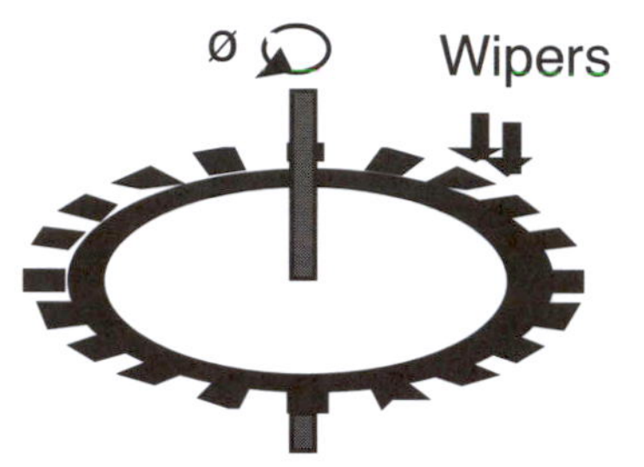

d. Tooth Counter

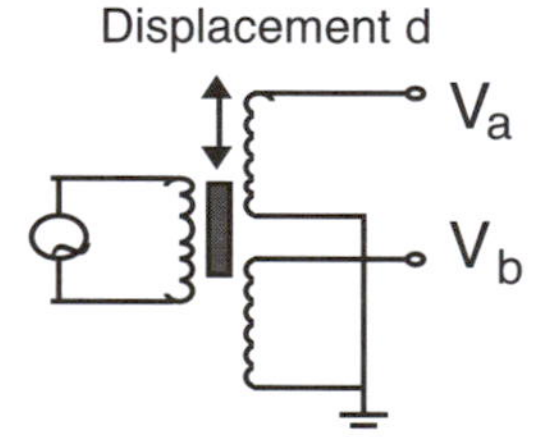

e. Linear Variable Displacement Transformer

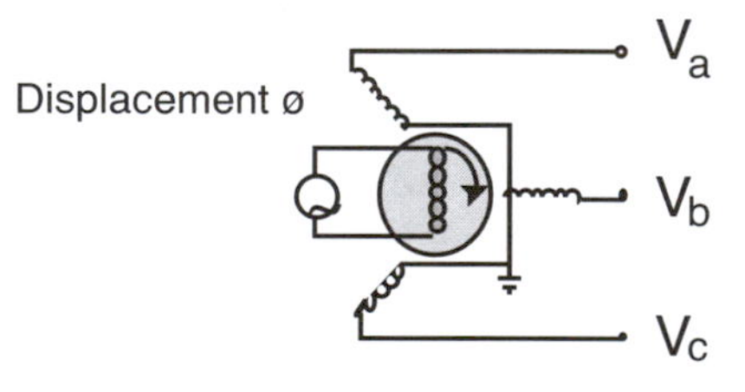

f. Control Transformer

Figure 7.2. Position Transducers

In inexpensive systems, linear position or angular position is usually converted into a resistance, which determines a voltage level in a voltage divider circuit, or a frequency of some kind of RC oscillator. A *potentiometer* converts angular position to resistance. (See Figure 7.2a.) A *slide potentiometer* converts linear position to resistance. Both transducers are inexpensive but are prone to inaccuracy as the wiper arm in the potentiometer wears down the resistor or as a coat of dirt or oil builds upon the resistor. Also, these transducers are sensitive to vibration. Overall accuracy is limited to about 3%. Minute position displacements can be measured by piezoelectric crystal, such as in commercial *strain gauges*. A crystal phonograph cartridge uses the same mechanism. (See Figure 7.2b.) Angular position of a disk can be converted directly into a digital signal by *shaft encoders*, which use mechanical wipers or photodetectors to read a track on the disk, the tracks being laid out so that the wipers or detectors read a digital word corresponding to the angle of rotation ø of the disk. (See Figure 7.2c.) Also, a pair of wipers or detectors can sense the teeth of a gear or gearlike disk, so they can count the teeth as the gear turns. (See Figure 7.2d.) Two wipers are needed to determine both the motion and the direction of motion of the teeth. Finally, the most accurate and reliable position transducer is the *linear variable displacement transformer*. (See Figure 7.2e.) This device is a transformer having a primary winding and two secondary windings and a movable slug. As the slug moves, the two secondary windings of the transformer producing V_1 and V_2 get more or less alternating current from the primary winding, and the relative phase of the sine waves output from the windings changes. Either the voltage level of the sine wave or the relative phase difference between the sine waves may be used to sense the position of the slug. The linear variable displacement transformer is the most accurate device for measuring linear distances because it is not affected by dirt, wear, or vibration as are other devices; however, it is the most expensive. Angular position can be measured by a *control transformer* using the same kind of technique. (See Figure 7.2f.) This device's rotor has a primary coil and secondary windings are in the housing, which is held stationary, surrounding the rotor. The angular position of such a device's rotor determines the amount and phase of a sine wave that is picked up by the secondaries of the transformer.

Velocity and acceleration can be determined by measuring position using one of the aforementioned transducers above and then differentiating the values in software or using an electrical circuit that differentiates the voltage. Also, a *direct-current tachometer* is a direct-current generator. Being an inverse of a DC motor, its output voltage is proportional to the rotational speed of the shaft. An AC tachometer is an AC motor run as a generator; its output frequency is proportional to the (angular) speed of its shaft. Finally, acceleration can also be measured by producing a force F as a mass m is accelerated at rate a $(F = ma)$, then letting the force act against a spring and measuring the displacement of the spring. This type of device, an *accelerometer,* can convert the acceleration into a position using mechanical techniques, thereby measuring acceleration at the output of the transducer. This is an alternative to measuring position, then using software to differentiate the values to determine acceleration. Conversely, an accelerometer can be used to measure acceleration, which can be integrated by a software program to derive velocity, or integrated twice to get position. This is the basis of an inertial guidance system. The above examples show that functions can be done by means of mechanical, electrical hardware, or software, or a combination of the three.

7.1.2 Radiant Energy Transducers

Radiant energy – light and infrared – can be produced or controlled by a microprocessor using lamps, *light-emitting diodes* (LEDs), and *liquid crystal displays* (LCDs). The terms used for light and infrared radiant energy are those used for radio waves. In a continuous wave (CW) or pulse-coded mode (PCM), the radiant energy is either on (high) or off (low). In an amplitude-modulated mode (AM), the amplitude of the radiation varies with a signal that carries analog information. In frequency-modulated mode (FM), the frequency varies with an analog signal. The common incandescent lamp is lit by applying a voltage across its terminals. The radiant energy is mostly uniformly distributed over the light spectrum and includes infrared energy. Gas-discharge lamps and fluorescent lamps work in a similar fashion but require current-limiting resistors in series with the lamp and usually need higher voltages. Their radiant energy is confined to specific wavelengths that are determined by the material in the lamp. While these are sometimes used with microprocessors, their relatively high voltage and current requirements, and the electrical noise generated by gas discharge lamps and fluorescent lamps, limit their usefulness. More popular are the LEDs and LCDs. An LED is basically a diode that will emit light if about 10 mA are sent through it. The light is generated in specific wavelengths: red and infrared are the easiest to generate; but green, yellow, and orange are also widely available. Current passing through an LED drops about 1.7 to 2.3 V, depending on the diode material. LEDs are often used in displays to indicate some output from a microcomputer and are also used in communications systems to carry information. Inexpensive LEDs can be pulse modulated at better than 10 KHz, and special ones can work at around 1 GHz. An LCD is electrically a capacitor that is clear if the RMS voltage across it is less than about a volt and opaque if more than about two volts; it consumes very little power. The voltage across an LCD must be AC, however, because DC will polarize and destroy the material in the LCD. Usually, one terminal has a square-wave signal. If the other terminal has a square-wave signal in phase with the first, the display is clear, and if it has a square-wave signal out of phase with the first, the display is opaque.

Radiant energy is often measured in industrial control systems. A *photodetector* converts the amplitude to a voltage or resistance for a given bandwidth of the very high frequency sine wave carrier. Often this bandwidth covers part of the visible spectrum and/or part of the infrared spectrum. The *photomultiplier* can measure energy down to the photon – the smallest unit of radiation – and has an amplification of about one million. However, it requires a regulated high voltage power supply. The *photodiode* is a semiconductor photodetector able to handle signals carried on the amplitude of the radiant energy around 10 MHz. The current through the diode is linearly proportional to the radiation if the voltage drop across it is kept small. This is done by external circuitry. However, it is inefficient because a unit of radiant energy produces only 0.001 units of electrical energy. A photodiode might be used in a communication linkage to carry a signal on a light beam because of its high bandwidth and ease of use with integrated circuits. If the diode is built into a transistor, a *phototransistor* is made that relates about one unit of electrical energy to one unit of radiant energy, but the signal carried on the amplitude is reproduced up to about 100 KHz. Finally, a *photoresistor* is a device whose resistance varies with the intensity of the light shone upon it. While this device is also temperature sensitive, has poor frequency response, and is quite nonlinear, it can be used to isolate a triac, as we discuss later.

Photodiodes, phototransistors, photoresistors, and other detectors are often used with LEDs or lamps to sense the position of objects or to isolate an external system from the microcomputer. Photodetectors are commonly used with an LED light source to detect the presence or absence of an object between the light source and the photodetector. To sense the pattern on the disk under the contacts, a shaft encoder or tooth counter can use this kind of sensor in place of a mechanical contact. Similar techniques place an LED and a phototransistor inside an integrated circuit package, called an opto-isolator, to isolate the circuitry driving the LED from the circuitry connected to the detector so that they can be kilovolts apart and so that electrical noise in the driver circuitry is not transmitted to the detector circuitry.

Temperature is controlled by means of heaters or air conditioners. To control the temperature of a small component, such as a crystal, the component is put in an *oven*, which has a resistive heater and is fairly well insulated. As more current is passed through the heater, it produces more heat; as less current is passed, the natural loss of heat through the insulated walls brings down the temperature. The temperature of a large room or building is controlled by means of a furnace or air conditioner, of course. Since these usually require AC power at high currents and voltages, the microcomputer has to control a large AC current. An interesting problem in controlling air conditioners is due to the back pressure built up in them. If the air conditioner has just been running, is then turned off and is quickly turned on, the motor in it will stall because it cannot overcome the back pressure in it. So in controlling an air conditioner, if it is turned off, it must not be turned on for an interval of time that is long enough for the back pressure to drop off.

Temperature is often sensed in a microprocessor system. Very high temperatures are measured indirectly, by measuring the infrared radiation they emit. Temperatures in the range -250°C to +1000°C can be measured by a *thermocouple*, which is a pair of dissimilar metals (iron and constantan, for instance), where the voltage developed between the metals is around 0.04 mV times the temperature. Note that such a low-level signal requires careful handling and amplification before it can be used in a microprocessor system. The most popular technique for measuring temperatures around room temperature is to put a constant current through a diode (or the diode in the emitter junction of a bipolar transistor) and measure the voltage across it. The output voltage is typically 2.2 mV times the temperature in degrees kelvin (°K). This voltage level requires some amplification before conversion to digital values is possible. Provided the current through the diode is held constant (by a constant current source), the transducer is accurate to within 0.1°K. While a common diode or transistor can be used, a number of integrated circuits have been developed that combine a transistor and constant current source and amplifier. One of these (AD590) has just two pins, and regulates the current through it to be 1 μA times the temperature in kelvin. Converting to and then transmitting a current has the following advantage: the voltage drops in wires whose sensor is a long distance from the microprocessor, or in switches that may have unknown resistance, and thus does not affect the current. The current is converted to a voltage simply by passing it through a resistor. Finally, temperature can be sensed by a temperature sensitive resistor called a *thermistor*. Thermistors are quite nonlinear and have poor frequency responses, but relatively large changes in resistance result from changes in temperature.

7.1.3 Other Transducers

Pressure can be produced as a by-product of an activity controlled by a microcomputer. For instance, if a microcomputer controls the position of a valve, it can also control the flow of liquid into a system, which changes the pressure in the system. Pressure is sometimes measured. Usually, variations in pressure produce changes in the position of a diaphragm, so the position of the diaphragm is measured. While this can be implemented with separate components, a complete system using a Sensym chip in the LX1800 series of chips (formerly a National Semiconductor series) can measure absolute or relative pressure to within 1% accuracy. These marvelous devices contain the diaphragm, strain gauge position sensor, compensation circuits for temperature, and output amplifier on a hybrid integrated circuit. Finally, weight is normally measured by the force that gravity generates. The weighing device, called a *load cell*, is essentially a piston. Objects are weighed by putting them on top of the piston, and the pressure of the fluid inside the piston is measured.

Other properties – including chemical composition and concentration, the pH of liquids, and so on – are sometimes measured by transducers. However, a discussion of these transducers goes beyond the scope of this introductory survey.

7.2 Basic Analog Processing Components

Basic analog devices include power amplifiers, operational amplifiers, analog switches, and the timer module. These will be discussed in this section. The first subsection discusses transistors and SCRs, the next discusses OP AMPs and analog switches in general, and the last discusses practical OP AMPs and analog switches.

7.2.1 Transistors and Silicon Controlled Rectifiers

To convert a voltage or current to some other property like position or temperature, an amplifier is needed to provide enough power to run a motor or a heater. We briefly survey the common power amplifier devices often used with microcomputers. These include power transistors, darlington transistors, and VFETs for control of DC devices; (motors, heaters, and so on) and SCRs and triacs for control of AC devices.

The *(bipolar) transistor* is a device that has terminals called the collector, base, and emitter. (See Figure 7.3a.) The collector current I_c is a constant (called the *beta*) times the base current I_b. The *power transistor* can be obtained in various capacities, able to handle up to 100 A, and up to 1000 V. These are most commonly used for control of DC devices. A *darlington transistor* has a pair of simple transistors connected internally so it appears to be a single transistor with very high beta. (See Figure 7.3b.) Power darlington transistors require less base current I_b to drive a given load, so they are often used with microprocessor I/O chips that have limited current output. *Field-effect transistors* (FETs) can be used in place of the more conventional (bipolar) transistor. In an FET, the current flowing from drain to source is proportional to the voltage from gate to source. Very little current flows into the gate, which is essentially a capacitor with a small leakage current. (See Figure 7.3c.) However, a *vertical field-effect transistor* (VFET) is faster than a standard FET and can withstand larger voltages (about 200 V) between drain and source. The VFET is therefore a superb output amplifier that is most compatible with microcomputers. Suffice to say that for this survey, a power transistor, a darlington, or a VFET is usually required to drive a DC device like a motor, heater, or lamp.

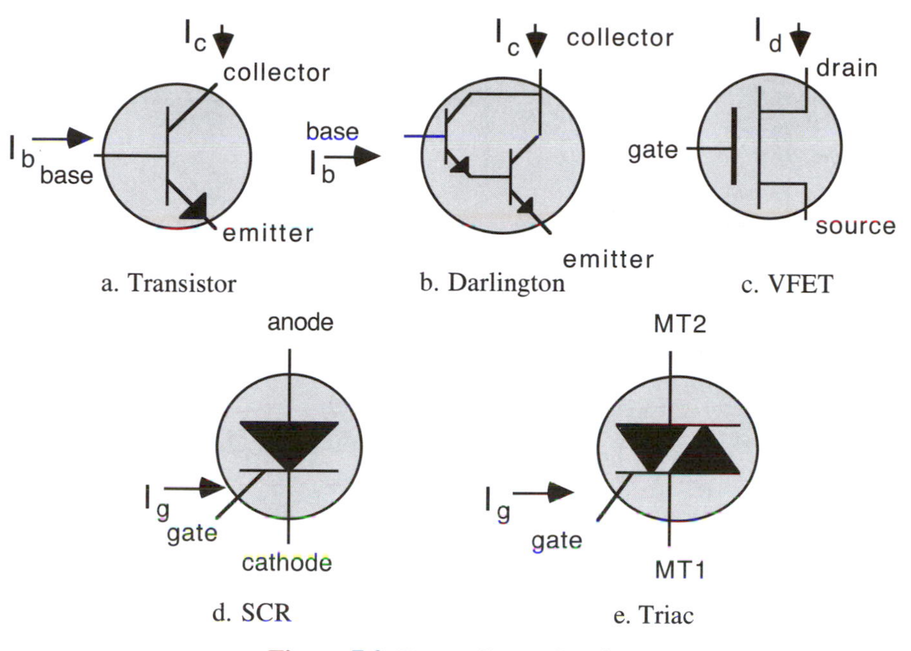

Figure 7.3. Power Output Devices

An AC device like an AC motor uses a *silicon controlled rectifier* (SCR) or a *triac* to amplify the voltage or control signal output from a microcomputer. The SCR has anode, cathode, and gate terminals, as in Figure 7.3d. When sufficient current I_g (about 50 mA) flows into the gate through the anode, the device looks like a diode, passing positive current from anode to cathode but inhibiting flow from cathode to anode. That is why it is called a controlled rectifier, since a rectifier is an older name for a diode. Moreover, the SCR has memory; once turned on, it remains on, regardless of the current through the gate, until the current through the anode tries to reverse itself and is thus turned off. The gate controls only half a cycle, since it is always turned off for the half cycle when current tries to but cannot go from cathode to anode; and it is turned on only when the gate is given enough current and the current will then flow from anode to cathode. To correct this deficiency, a pair of SCRs are effectively connected "back to back" to form a triac. (See Figure 7.3e.) The power current flows through main terminal 1 (MT1) and main terminal 2 (MT2) under the control of the current I_g through the gate and MT1. If the gate current is higher than about 50 mA either into or out of the gate, MT1 appears shorted to MT2 and continues appearing as such regardless of the current through the gate until the current through MT1 and MT2 passes through 0. Otherwise, MT1 and MT2 appear disconnected. SCRs and triacs handle currents from 0.5 A up to 1000 A and can control voltages beyond 800 V.

SCRs and triacs control motors, heaters, and the like by controlling the percentage of a cycle or the number of cycles in which full power is applied to them. The types of control are discussed next in terms of triacs, but they also apply to SCRs.

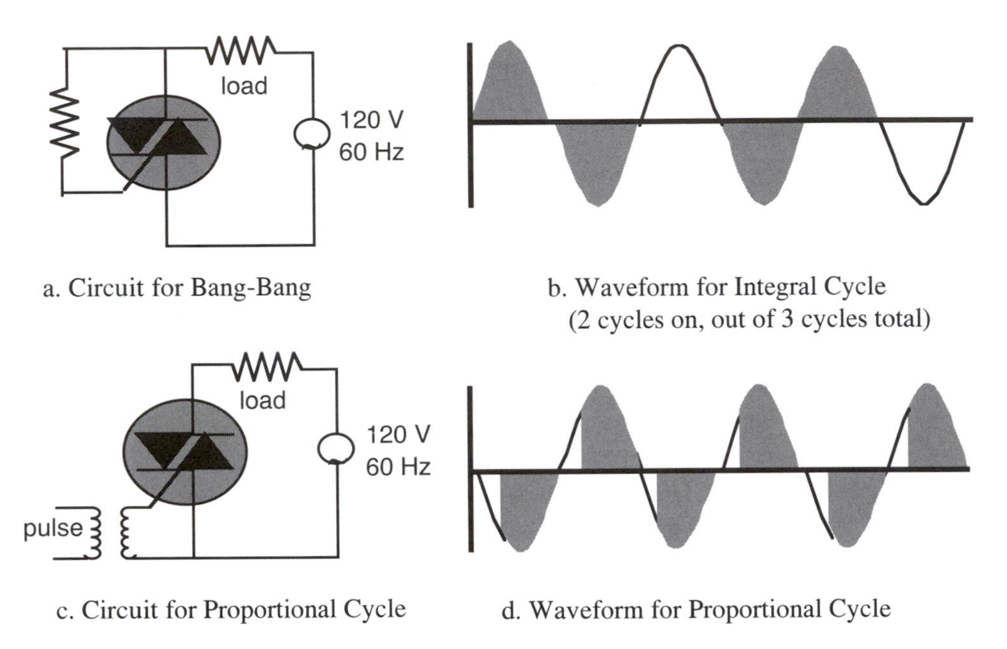

a. Circuit for Bang-Bang

b. Waveform for Integral Cycle (2 cycles on, out of 3 cycles total)

c. Circuit for Proportional Cycle

d. Waveform for Proportional Cycle

Figure 7.4. Triac Control Techniques

In *on/off control,* also called *bang-bang control,* the triac applies either full power or no power to the motor. To do this, either full current or no current is applied to the gate. A simple variation of this technique applies gate current from MT2 through a resistor, so that if the resistance is low, when voltage on MT2 builds up, current flows through the resistor to turn on the triac. (See Figure 7.4a.) As soon as it is turned on, however, the voltage on MT2 disappears. Thus, the current through the resistor stops as soon as it has done its work. This reduces the power dissipated in the resistor. If the resistance is large, no current flows through the gate, so the triac is off. The resistor can be a photoresistor, coupled to an incandescent lamp or an LED in an optocoupler. When the LED or lamp is lit, the photoresistor has a low resistance, which turns the triac on. Otherwise, the resistance is high and the triac is off. This configuration is particularly suited to on/off control of large AC loads by means of triacs.

A simple variation is called *integral cycle control.* Here, the triac is turned on for `n` out of every `m` half-cycles. Figure 7.4b shows `n` = 2, `m` = 3. The gate current is turned on at the beginning of each half-cycle when the triac is on. A final variation is called *proportional cycle control.* A pulse generator of some kind is commonly used to send a current pulse through the gate at a precise time in each half-cycle. (See Figure 7.4c.) For a fraction `F` of each half-cycle, the triac is turned on. (See Figure 7.4d.) Full power is applied to the device for the last `F`th of the cycle. Roughly speaking, the device gets power proportional to the fraction `F`. A pulse transformer is often used to isolate the controller from the high voltages in the triac circuitry. (See Figure 7.4c.) The controller provides a short (5-μs) voltage pulse across the primary winding (shown on the left) of the transformer, which provides a current pulse to the triac to turn it on. The controller has to provide this pulse at the same time in each half-cycle. If the pulse is earlier, more current flows through the triac and more power goes to the load.

On/off control is used where the microprocessor simply turns a device on or off. A traffic light would be controlled like this. Bang-bang control is commonly used in heating systems. You set your thermostat to 70°F. If the temperature is below 70°F, the heater is turned on fully; and, if above 70°F, the heater is completely off. Integral cycle control is useful in electric ranges, for instance, to provide some control over the amount of heating. Finally, variable-duty cycle control is common in controlling lighting and power tools, since the other types of control would cause the light to flicker perceptibly or the tool to chatter. However, this type of control generates a lot of electrical noise whenever the triac is turned on fully in the middle of a half-cycle. This kind of noise interferes with the microcomputer and any communications linkages to and from it. So variable-duty cycle control is normally used when the other forms generate too much flicker or chatter.

7.2.2 Basic Linear Integrated Circuits

The basic module used to process analog signals is the *operational amplifier,* or *OP AMP*. It is used in several important configurations, which we will discuss here. We will then discuss the analog switch, which allows convenient microprocessor control of analog signals, and consider several important applications of this switch.

The OP AMP has two inputs, labled + and -, and an output. (See Figure 7.5.) The output voltage signal `Vout` is related to the signals `V+` on the + input and `V-` on the - input by the expression

$$Vout = A\ (V+ - V-)$$

where `A` is a rather large number, such as 100,000. The OP AMP is in the *linear mode* if the output voltage is within the range of the positive and negative supply voltages, otherwise it is in the *saturated mode* of operation. Clearly, to be in the linear mode, `V+` has to be quite near `V-`.

The first use of the OP AMP is the *inverting amplifier*. Here, the + input is essentially connected to ground, so `V+` is 0, and *feedback* is used to force `V-` to 0 V, so the OP AMP is in the linear mode. In Figure 7.5a, if `Vin` increases by one volt, then `V-` will increase by a small amount, so the output `Vout` will decrease 100,000 times this amount, large enough to force `V-` back to 0. In fact, `Vout` will have to be

$$Vout = -\ (Rf/Rin)Vin$$

in order to force `V-` to 0. The *amplification* of this circuit, the ratio `Vout/Vin,` is exactly `Rf/Rin` and can be selected by the designer as needed. In a slight modification of this circuit, one or more inputs having signals `Vin1, Vin2,...` can be connected by means of resistors `Rin1, Rin2, ...` (as in Figure 7.5b) and the output voltage is then

$$Vout = -\ \{\ [\ (\ Rf/Rin1\ )\ Vin1\ ]\ +\ [\ (\ Rf/Rin2\ )\ Vin2\ ]\ +\ .\ .\ .\ \}$$

in a circuit called a *summing amplifier*.

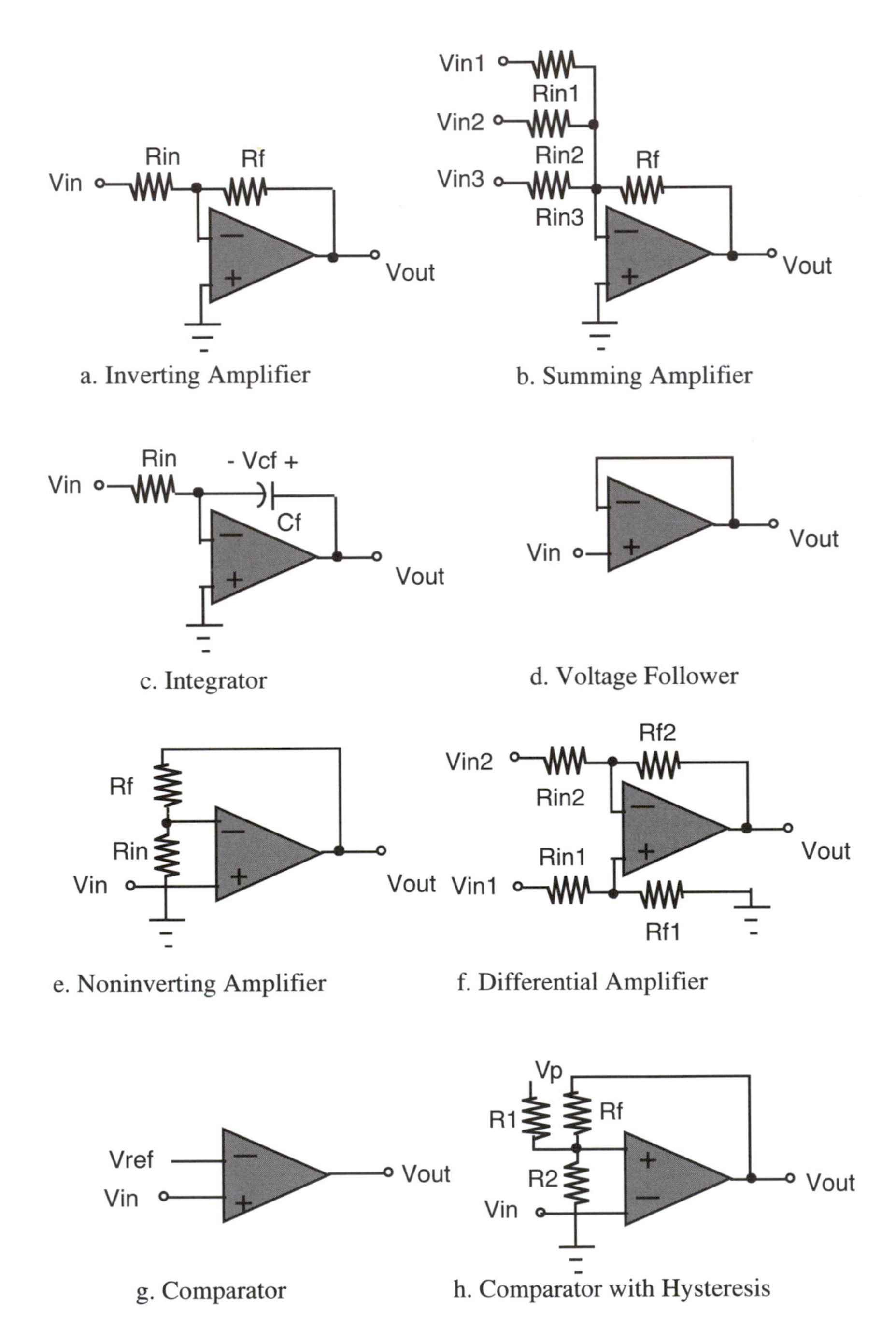

Figure 7.5. Operational Amplifier Circuits

Another classical use of an OP AMP is integration of a signal. A capacitor has the relation of current through it, i, to voltage across it, v, as follows:

$$i = Cf\ dv/dt$$

where Cf is the capacitance. In Figure 7.5c, if Vin increases by one volt, then $Vout$ will have to change by one volt per second so the current through the capacitor can offset the current through Ri to force $V-$ to 0. Generally, the relationship is

$$Vout = VCfi - [1 / (Ri \times Cf)] \int_{t_0}^{t} Vin\ dt$$

where $VCfi$ is the voltage across the capacitor at the time we began integrating the input signal.

In these three techniques, the voltage $V-$ is forced to 0. $V-$ is called a *virtual ground*. Of course, it cannot really be connected to ground or no current would be available for the OP AMP $V-$ input. However, complex circuits, such as amplifiers, integrators, differentiators, and active filters, are analyzed using circuit analysis techniques, assuming that $V-$ is effectively grounded.

A different use of the OP AMP puts the incoming signal on the + input and uses feedback to try to force $V-$ to the same voltage as $V+$. The *voltage follower* (shown in Figure 7.5d) does this by connecting $V-$ to $Vout$. The *noninverting amplifier* uses the same principle (as shown in Figure 7.5e) and satisfies the relationship

$$Vout = [1 + (Rf / Rin)]Vin$$

and the output voltage has the same polarity as the input voltage. Combining the ideas underlying the summing amplifier with those of the noninverting amplifier, we have the *differential amplifier* (shown in Figure 7.5f). One or more inputs such as $Vin1$ are connected via resistors like $Rin1$ to the + input of the OP AMP, and one or more inputs such as $Vin2$ are connected via resistors such as $Rin2$ to the - input of the OP AMP. The output is then

$$Vout = K1\ Vin1 - K2\ Vin2$$

where $$K1 = \frac{Rf1}{Rin1 + Rf1}\left(1 + \frac{Rf2}{Rin2}\right) \quad \text{and} \quad K2 = \frac{Rf2}{Rin2}$$

In this circuit, if more than one input Vin is connected to the + OP AMP input via a resistor Rin, it appears like the term for $Vin1$ adding its contribution to $Vout$; and if connected to the - input, it subtracts its contribution to $Vout$ like the term for $Vin2$.

A final saturation mode technique used in the OP AMP depends on the finite output range it has. The comparator has the connections shown in Figure 7.5g. Here, the output is a high logical signal, H, if $Vin > Vref$, else it is L. The comparator can be reversed, so that $Vref$ is on the + input and Vin is on the - input.

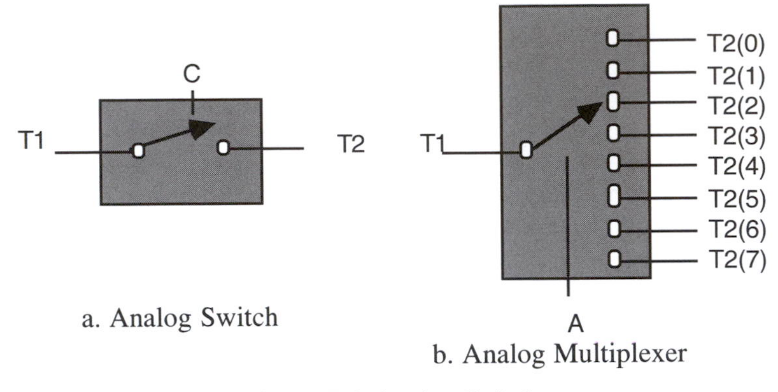

Figure 7.6. Analog Switches

Then `Vout` is high if `Vin` < `Vref`. Also, `Vref` can be derived from a voltage divider:

```
Vref = Vp R2 / (R1 + R2)
```

where `Vp` is an accurate voltage. Finally, feedback can be made to change the effective `Vref` a little. Using this variation, the comparator can be insensitive to small changes in `Vin` due to noise. By connecting the output to `V+` (as shown in Figure 7.5h) the effective `Vref` can be changed so it is higher when the output is H than when the output is L. Then the output remains H or L even when the input varies a bit. Suppose, for instance, that the input is low and the output is high. When the input exceeds the higher reference, the output goes low. The output remains low until the input drops below the lower reference. When the input drops below the lower reference and the output goes high, the input has to exceed the higher reference again before the output can go low, and so on. This stubborn-like mechanism is called *hysteresis* and is the basis of the *Schmitt trigger* gate used to ignore noise in digital systems.

The analog switch is implemented with field-effect transistors. It has a control C and two terminals T1 and T2. (See Figure 7.6a.) If C is high, T1 is connected to T2 and the switch is said to be *on*, else they are disconnected and the switch is *off*. Some number, say, eight, of such switches can be connected at the T1 terminal, and a decoder with 3-bit address A can be on a chip. The Ath switch can be turned on by the decoder, the others being turned off. This chip then behaves like an eight-position rotary switch that is controlled by the address input. (See Figure 7.6b.) This kind of chip is an *analog multiplexer*. Single analog switches and analog multiplexers are valuable as ways to control analog signals from a microcomputer.

The final important device is the *timer*. A timer outputs a periodic signal whose period is proportional to the value of a resistor and a capacitor connected to it. (Often, the period can be adjusted by a control voltage, but the voltage-to-frequency converter is generally better.) The timer allows resistor-based transducers to generate AC signals, where the information is carried by the frequency (period). Such signals are easy to handle and measure, as we will see in the next chapter.

Table 7.1. Characteristics of the CA3140

Characteristic	Value
Maximum (Vs+ - Vs-)	36 V
Minimum (Vs+ - Vs-)	4 V
Maximum (V+ or V-)	Vs+ + 8
Minimum (V+ or V-)	Vs- - .5
Max. common-mode input	Vs+ - 2.5
Min. common-mode input	Vs- -.5
Input resistance	1 TΩ
Input capacitance	4 pF
Input current	2 pA
Input offset voltage	5 mV
Input offset current	.1 pA
Output resistance	60Ω
Maximum output voltage	Vs+ - 3 v
Minimum output voltage	Vs- +.13 v
Maximum sourcing current	10 mA
Maximum sinking current	1 mA
Amplification	100,000
Slew rate	7 V/μs
Gain bandwidth product	3.7 MHz
Transient response	80 ns
Supply current	1.6 mA
Device dissipation	8 mW

7.2.3 Practical Linear Integrated Circuits

We now consider an operational amplifier, the CA3140, which is particularly suitable for microprocessors. It has low-current CMOS inputs and bipolar transistor outputs able to supply significant current. Its characteristics are listed in Table 7.1.

In this book, `Vs+` is the positive supply voltage and `Vs-` is the negative supply voltage. The first two table entries indicate that the total supply voltage may not be greater than 36 V nor less than 4 V. Two very common connections are the *dual supply,* where `Vs+` is exactly the negative of `Vs-` (for example, `Vs+` = +15 V, `Vs-` = -15 V), and the *single supply,* where either `Vs+` or `Vs-` is 0 (for example, `Vs+` = 5 V, `Vs-` = 0 V). A ±15-V dual supply is useful when almost maximum output voltage range is needed, and a single +5-V supply is useful when the only power supply is the one supplying 5 V to the logic part of the system. A good OP AMP for microcomputer applications should be capable of operating in either of the preceding cases. Clearly, both connections are within the specifications listed in the first two rows of Table 7.1 for the CA3140. To make the information in the table more concrete, we will consider its significance for a single-ended +5-V supply application. The reader is invited to consider the significance of these parameters for a ±15-V dual-supply application.

The next four entries indicate the range of input voltages. The maximum and minimum values of the positive signal input `V+` and the negative signal input `V-` should not be exceeded or the OP AMP may be destroyed. For example, if `Vs+` is 5 V and `Vs-` is 0 V, then neither `V+` nor `V-` should have a voltage higher than 13 V, nor lower than -0.5 V. The full range of voltages can be used in the saturated mode of operation. This OP AMP has adequate capabilities using a +5-V single supply for comparator applications. However, if the linear mode of operation is used, `V+` and `V-` should be kept within the maximum and minimum *common-mode voltages*. For our previous example, using the same supply voltages, `V+` and `V-` should be kept within 2.5 V and -0.5 V for operation in the linear mode. Note that inputs above 2.5 V will pull the OP AMP out of the linear mode, and this can be a problem for voltage-follower, noninverting amplifiers or for differential amplifiers. However, since the common-mode voltage range includes both positive and negative voltages around 0 V, inverting amplifiers, summers, and integrators can be built using a single +5-V supply. This is a very attractive feature of an OP AMP like the CA3140.

The next five lines of Table 7.1 show the input characteristics that cause errors. The `V+` and `V-` inputs appear to be a resistor, a capacitor, and a current source, all in parallel. The equivalent input resistance, 1 TΩ, is very high. This high input resistance means that a voltage follower or noninverting amplifier can have such high input resistance, which is especially important for measuring the minute current output from some transducers like pH probes and photodetectors. Moreover, it means that quite large resistors (100 KΩ) and quite small capacitors (0.01 μF) can be used in the circuits discussed earlier, without the OP AMP's loading down the circuit. Especially when the rest of the system is so miniaturized, larger capacitors are ungainly and costly. The input capacitance, 4 pF, is very low but can become significant at high frequencies. The current source can cause some error but is quite high in this OP AMP, and the error can often be ignored. `V+` and `V-` have some current flowing from them, which is less than 2 pA according to Table 7.1. If the `V+` input is just grounded but the `V-` input is connected by a 1-MΩ resistor to ground, this input current causes 2 μV extra, which is multiplied by the amplification (100,000) to produce an error of 0.2 mV in `Vout`. The error due to input current can be minimized by making equal the resistances that connect the `V+` and `V-` inputs to ground. In Figure 7.5a, a resistance equal to `Rin` in parallel with `Rf` can be connected between the + input and ground, just to cancel the effect of the input current on the output voltage. However, this particular OP AMP has such low input current that the error is usually not significant, and the `V+` input is grounded.

The offset voltage is the net voltage that might be effectively applied to either `V+` or `V-`, even when they are grounded. The offset current is the current that can be effectively applied to either input, even when they are disconnected. These offsets have to be counterbalanced to get zero-output voltage when the input is zero. An *offset adjustment* is available on OP AMPs like the 3140 to cancel the offset voltage and current.

The next five entries describe the output of the OP AMP. The output resistance is the effective resistance in series with the output of the amplifier considered as a perfect voltage source. In this case, it is 60 Ω. A high output resistance limits the OP AMP's ability to apply full power to low resistance loads, such as speakers. However, the effective output resistance of an amplifier is substantially decreased by feedback. The output voltage can swing over a range of from 2 to 0.13 V, if the power supply $Vs+$ is 5 V and $Vs-$ is 0 V. This means that for linear operation the amplifier can support about a 1.8-V peak-to-peak output signal, but this signal has to be centered around 1.07 V. Note that the output range is a serious limitation for a comparator whose output drives a digital input, because a high signal is usually any voltage above 2.7 V. An external (10-KΩ) pull-up resistor, from the output to +5 V, can be used such that whenever the OP AMP is not pulling the output low, the output is pulled up to nearly 5 V. The output can source (supply) 10 mA and can sink (absorb) 1 mA to the next stage. It can supply quite a bit of current to a transistor or a sensitive gate triac because these devices require current from the output of the OP AMP. However, this OP AMP's ability to sink only 1 mA restricts its use to low-power (CMOS, LSTTL, microprocessor HCMOS) digital inputs; and it cannot sink 1.6 mA reliably as is required to input signals to conventional TTL gates.

Recall that the bandwidth of an amplifier is the range of frequencies over which the gain is at least $1/\sqrt{2}$ times the maximum gain. If the bandwidth of an amplifier is 100,000 Hz, then any small-signal sine wave whose frequency is between direct current and 100,000 Hz will be correctly amplified. Moreover, any complex periodic waveform can be decomposed into a sum of sine waves. To correctly amplify the waveform, all the component sine waves must be amplified correctly. (The phase delays also must be matched for all components.) Generally, a square wave of frequency F will be reproduced fairly accurately if the amplifier bandwidth is at least 10 F.

For most OP AMPs, the bandwidth decreases as the gain increases, so the product is constant. In the 3140, this constant is 3.7 MHz. That means that if the circuit amplification is 1, the bandwidth is 3.7 MHz. For an OP AMP (shown in Figure 7.5a) with an amplification of 10, the bandwidth is 370 KHz. The bandwidth is an important limitation on the OP AMP's ability to amplify small high-frequency signals. The slew rate is the maximum rate at which the output can change (due to a sudden change on the input). The slew rate usually limits the effective bandwidth of large signals less than the available bandwidth of small signals, because the output cannot change fast enough. This OP AMP has a very good slew rate; the output can change at a rate of 7 V in 1 μs. The transient response is the time delay between a sudden change in the input and the corresponding change in the output. A related parameter, the *settling time,* is the time it takes for the output to reach the desired voltage. It is not specified in Table 7.1 because it depends on the external component configuration and on what we mean by reaching the desired voltage. The transient response and settling time can be of concern to a programmer who must compensate for such delays. In circuits where a digital device interfaces with an OP AMP, the slew rate and transient response may be the limiting factor on the use of an OP AMP.

Finally, the power requirements of the device are given. It dissipates about 8 mW when operated using a single 5-V supply, taking 1.6 mA from the power supply under normal conditions. It takes about 6 mA and dissipates about 180 mW when operated from dual ±15-V supplies. This parameter determines how big the power supply has to be to supply this device and can be significant when little power is available.

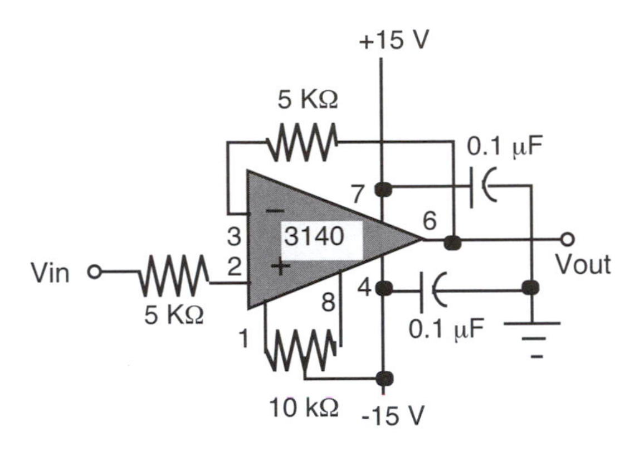

Figure 7.7. A Practical Voltage Follower

Figure 7.7 shows the pin connections for a CA3140 and some practical considerations in using it for a dual-supply voltage follower. To avoid noise input and unwanted oscillation, 0.1-μF capacitors, called *bypass capacitors,* are connected between the `Vs+` pin and ground and between the `Vs-` pin and ground. The connection should be made as close to the pin as possible. Wherever practical, every OP AMP should be bypassed in this manner. The 10-KΩ potentiometer between pins 1 and 8 is used to counterbalance the voltage offset. The inputs (to the whole circuit, not the OP AMP) are connected momentarily to ground, and this potentiometer is adjusted to output 0 V. Although the voltage follower needs no resistors (as in Figure 7.5d), resistors are put in the feedback loop and the input to prevent excessive currents from flowing when the OP AMP is driven out of its linear mode of operation. Since the inputs have very high resistance in normal operation, these resistors have no effect in that mode. However, they should be put in if the OP AMP can enter a saturation mode of operation. Note that if the power to this OP AMP is off and a signal is applied to the input, excessive current can flow unless these resistances are put in because that operation will be in the saturated mode.

Some other considerations are offered. When handling devices with such high input resistances, tools, soldering irons, and hands should be connected via a large (15-MΩ) resistance to ground. Such a device should never be inserted or removed from a socket when power is on, and signals should not be applied to inputs (unless a series resistor is used, as in the voltage follower recently described) when power is off. Especially if high (1-MΩ) resistances are used, keep them clean, keep the leads short, and separate the components on the input of an OP AMP as far as possible from the output circuitry. A sheet of metal connected to ground provides some isolation from electrical noise, and all components and wires should be close to this *ground plane*. However, the ground reference points for such high-gain OP AMPs should be connected at one single point, running separate wires from this point to each ground point, to avoid so-called *ground loops*. If this advice is ignored, the OP AMP may become an oscillator because the minute voltages developed across the small but finite resistance of a ground wire could be fed back into an input of the OP AMP.

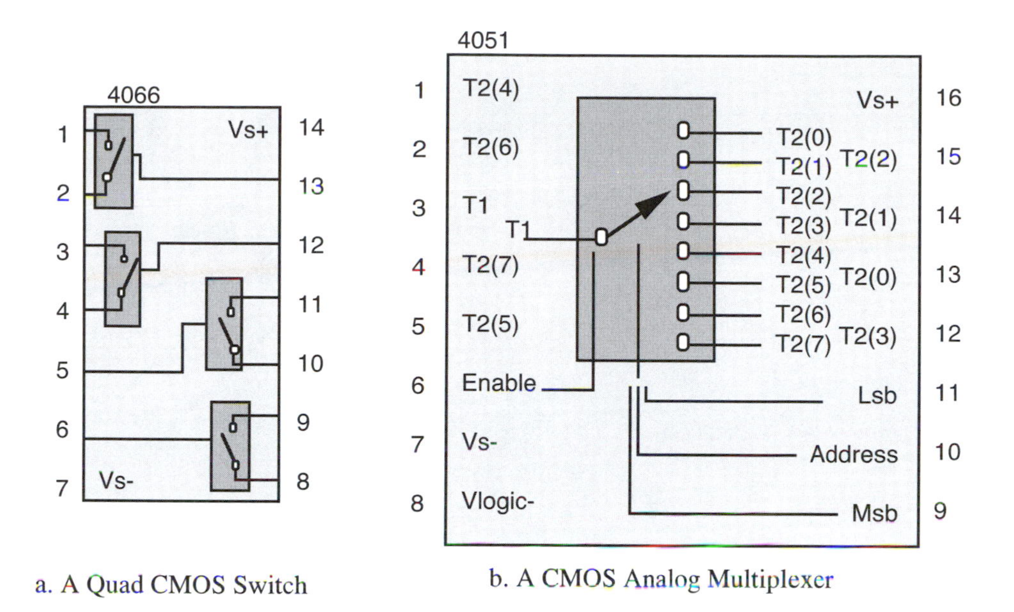

a. A Quad CMOS Switch

b. A CMOS Analog Multiplexer

Figure 7.8. Practical Analog Switches

We now turn to some practical aspects of using CMOS analog switches. The analog switch is almost perfect: its bandwidth is about 40 MHz; when closed it is almost a short circuit, and when open it is almost an open circuit. We now focus on the meaning of "almost." Look at Figure 7.8, which shows the 4066 and the 4051.

We consider the problem of supplying control signals that are compatible with the voltage levels on the terminals of the switch. The maximum `Vs+` minus `Vs-` voltage across the 4066 is 15 V. Sometimes, a dual ±7.5-V supply is used. If so, the control signals on pins 5, 6, 12, and 13 must be around -7.5 V to be considered low enough to open the corresponding switch, and around +7.5 V to be considered high enough to close the switch. Control signals from a microcomputer are normally in the range of 0 to 5 V and must be translated to control the switch. However, the 4051 has some level translation ability. The logic signals to address and enable the switches are referenced to `Vs+` and pin 8, so a high signal is close to `Vs+` and a low signal is close to the voltage level on pin 8. However, the analog levels on the switch's terminals can be between `Vs+` and `Vs-`, which is on pin 7. Commonly, `Vs+` is +5 V, pin 8 is grounded, and `Vs-` is -5 V, to directly use control signals from a microcomputer yet provide some range of analog voltages on the terminals.

When a switch is closed, it appears as a small resistance, about 80 Ω for the 4066 or about 120 Ω for the 4051. This resistance is not exactly linear, varying over a range of 2 to 1. The resistance is more linear if (`Vs+` minus `Vs-`) is as large as possible. However, if used with external resistances around 10 KΩ in series with the switch, less than 0.5 percent distortion is introduced by the nonlinear resistance of the 4066, even for (`Vs+` minus `Vs-`) = 5 V.

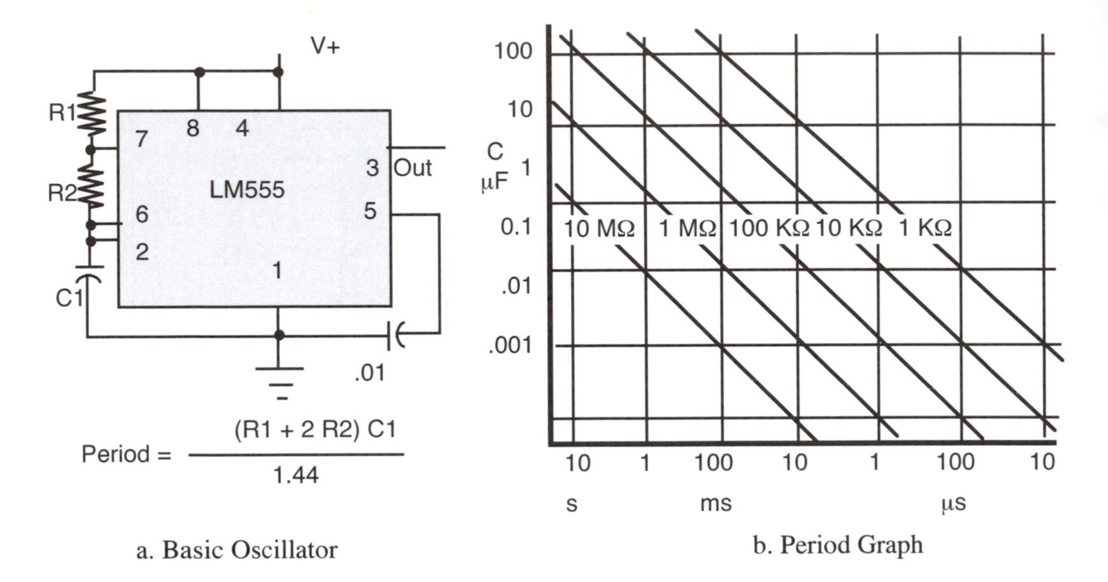

Figure 7.9. 555 Timer

When the switch is off, each terminal appears to be a small current source, about 100 nA for the 4066 and about 500 nA per analog switch in the 4051. This small current increases substantially with temperature and can be a serious source of error if the associated circuitry has very high resistance. To minimize it, we sometimes see a *heat sink* (a metal attachment to a transistor or integrated circuit to dissipate the heat) on an analog switch, and it is sometimes placed away from heat-producing components. Finally, a substantial amount of unwanted current flows from the power supply to the terminals if the voltage from a terminal to `Vs-` is greater than 0.6 V and positive current flows from pin 3 of the 4051, or from pins 2, 3, 9, or 10 in the 4066. One should ensure that positive current flows into these pins or that the voltage drop across the switch is never more than 0.6 V. In summary, the 4066 has a bit better performance, lower "on" resistance and lower "off" current, and may be used individually; but the 4051 incorporates eight switches into one chip and translates the control signal level from 0 to 5 V to control signals between ±5 V.

Finally, we discuss the timer module. The ubiquitous 555 is the most popular and least expensive timer. (See Figure 7.9a for the circuit that generates repetitive signals.) In Figure 7.9b we see a graph that gives the period of the signal as a function of the resistance, which is the value `R1 + 2R2` in Figure 7.9a, and the capacitance, which is the value of C1.

7.3 OP AMP and Analog Switch Signal Conditioning

OP AMPs and analog switches are often used with microcomputers to condition analog signals before converting them to digital signals, to control analog signals used for other purposes, or to clean up or modify analog signals generated by D/A converters. The four main aspects of conditioning a signal are the filtering of frequency components, the selection of inputs, the amplification or scaling of input levels, and the nonlinear modification of signal voltages. These are now considered in turn.

7.3.1 Filters

Recall that any periodic waveform can be considered a sum of sine waves. Frequency filtering is commonly done when the signal of interest is accompanied by unwanted noise, and most of the noise is at frequencies other than those of the signal's sine wave components. If the signal frequencies are low and the noise frequencies are high, a *low-pass filter* is used. (See the amplitude-versus-frequency characteristic of a low-pass filter in Figure 7.10a and the circuit diagram in Figure 7.10b.) Intuitively, capacitor C1 tends to integrate the signal, smoothing out the high-frequency components, and capacitor C2 further shorts out the high-frequency components to ground. Some D/A conversion techniques generate high-frequency noise, so a low-pass filter is commonly used to remove the noise from the signal. If the signal frequencies are higher than the noise frequencies, a *high-pass filter* is used to reject the noise and pass the signal. (See Figure 7.10c for the amplification characteristics and Figure 7.10d for a high-pass filter circuit.) Intuitively, we also see that the capacitors pass the high-frequency components, bypassing the low-frequency components through the resistors. A signal from a light pen on a CRT gets a short pulse every time the electron beam inside the CRT writes over the dot in front of the light pen. The signal has high-frequency components, while the noise – mostly a steady level due to ambient light – is lower in frequency. A high-pass filter passes the signal and rejects the noise. Finally, a *bandpass filter* can reject both higher- and lower-frequency components, passing only components whose frequencies are between the lower and upper limits of the band, and a *notch filter* can reject frequencies within the upper and lower limits of a band. (See Figures 7.10e through 7.10h for the amplification characteristics and circuit diagrams of these filters.)

Compound filters can be used to reject various frequencies and emphasize other frequency components. Two techniques can be used: in one, the output from one filter feeds the input to the next filter to *cascade* them in a chain configuration; and in the other, the signal is fed to both filters and the outputs are added by a summing amplifier in a *parallel* configuration. For instance, a bandpass filter can be made from a low-pass filter that rejects components whose frequency is above the band, cascaded into a high-pass filter that rejects components whose frequency is below the band. A notch filter can be made by summing the outputs of parallel high-pass and low-pass filters. Compound filters can be used to more sharply attenuate the signals whose frequencies are above the low-pass band or below the high-pass band. The best way to cascade n low-pass filters to more sharply attenuate high-frequency components and thus get a *2nth-order filter* is a nice mathematical study, and three types of filters have been shown mathematically optimal in one sense or another. The *butterworth filter* has the flattest amplification-versus-frequency curve in the low-frequency band where we pass the signal in a low-pass filter. However, the phase delays are quite different for different components. A square wave comes out with a few notches and humps. The *bessel filter* has the most linear relationship between frequency and phase delay and is especially useful for processing signals whose information is carried, in part, by the phase of its components and its pulse edges and shapes. The *chebyshev filter* is characterized by a designer-specified irregularity in the amplification-versus-frequency curve in the low-frequency band and maximum rejection just outside this band in a low-pass filter. All these filters look alike, but differ in the precise values of the components. These precise values can be obtained from tables, using simple transformations on the values in the tables, or by means of commonly available computer programs. Finally, while the preceding discussion concentrated on low-pass filters, the same terms and concepts apply to high-pass filters. And high-pass filters can be cascaded with low-pass filters to get bandpass filters or paralleled to get notch filters.

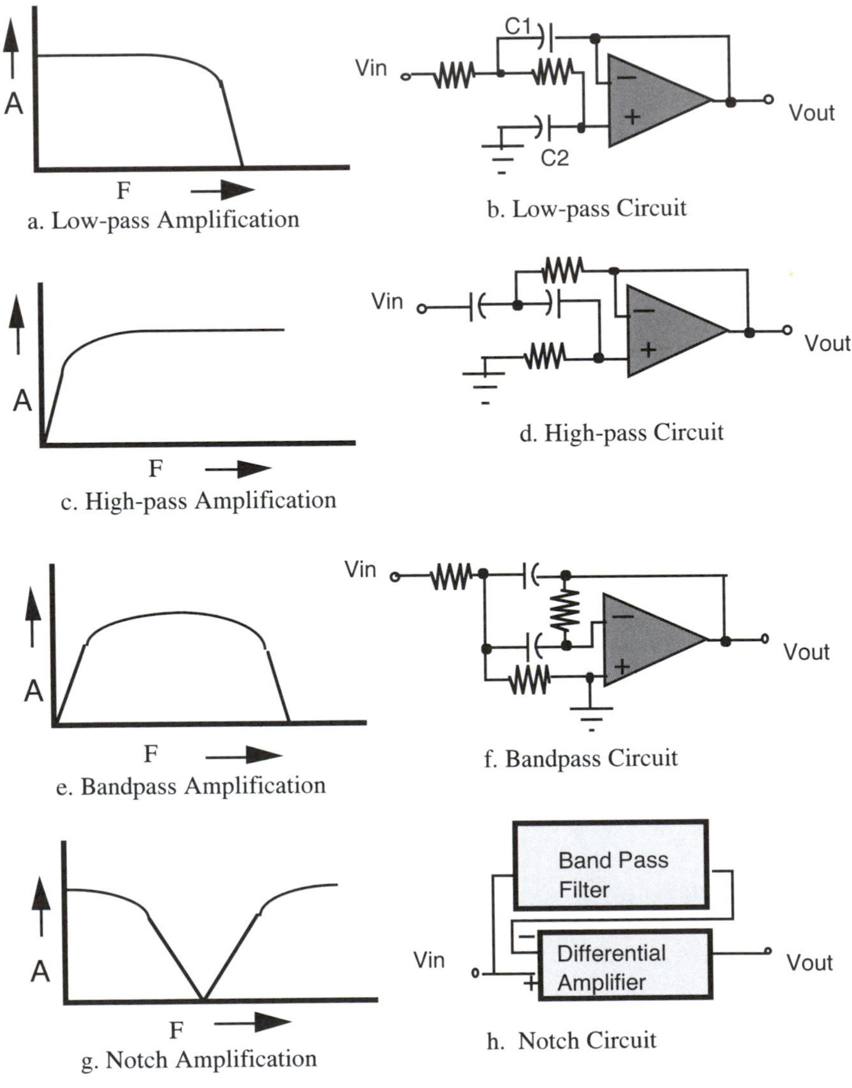

Figure 7.10. Some Filters

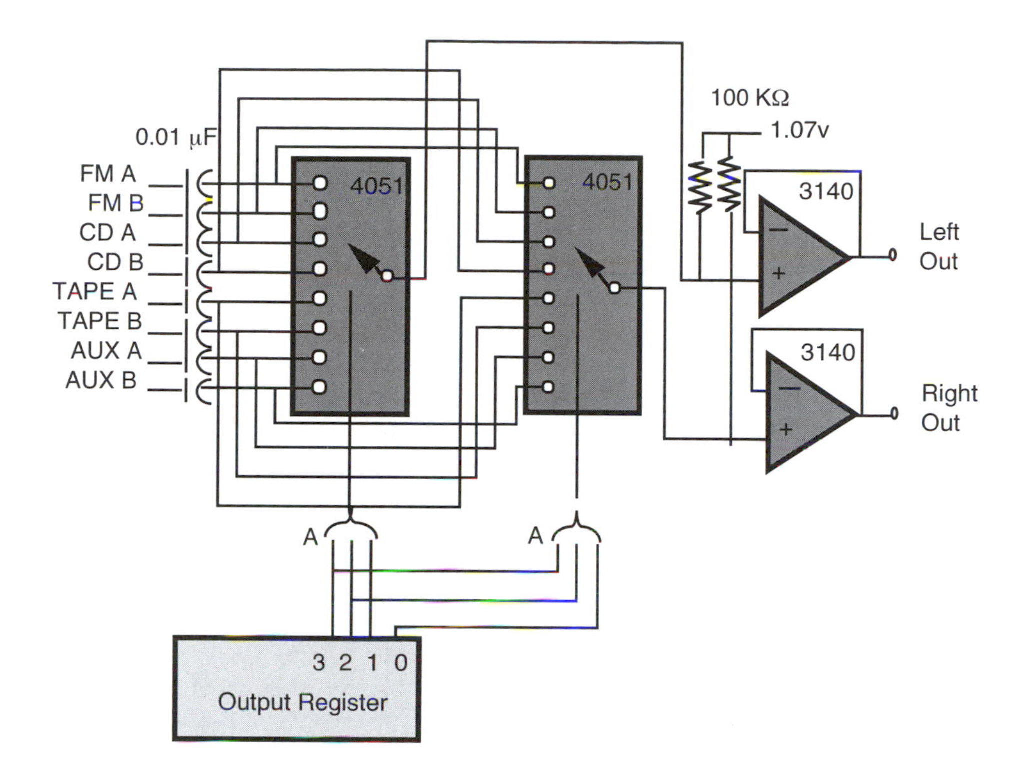

Figure 7.11. Selecting Inputs for a Stereo Preamplifier

7.3.2 Selection of Inputs and Control of Gain

The selection of inputs and distribution of outputs is normally accomplished by means of analog switches under the control of a microcomputer parallel output port. The individual analog switches of a 4066 can be controlled, each by a bit from an output port, to obtain maximum flexibility throughout the system being controlled. Alternatively, the 4051 can select from one of eight inputs, or distribute to one of eight outputs, using 3 bits from a parallel output port.

Microcomputers are appearing in almost every electronic product. They are useful in a stereo system, for example, because the listener can program a selection of music for a day or more. The microcomputer acts as a very flexible "alarm clock." Analog switches can be used to control the selection and conditioning of analog signals in the preamplifier. We now discuss an example of the use of 4051 switches for selection of inputs to a stereo preamplifier. (See Figure 7.11.)

This preamplifier has four sources (FM radio, CD, cassette tape, and auxilliary), and each source has two channels (for example, CD A and CD B). All signals are no larger than 1.5 V peak-to-peak and are as close to that range as possible. The 4 bits from the output port control the two switches such that the high-order 2 bits are the high-order bits of the addresses of both switches, but the lowest-order bit is the low bit of the address of one of the switches, and the next lowest-order bit is the low address bit of the other switch. The 2 high-order bits select the source: FF selects the tuner, FT selects the CD, TF selects the tape input, and TT selects the auxilliary input. The 2 low-order bits select the mode: FF puts the A channel into both speakers, FT puts the A input channel into the A speaker and B input channel into the B speaker (stereo), TF puts the A input into the B speaker and the B input into the A speaker (reverse stereo), and TT puts the B input into both speakers. To select the CD inputs in the stereo mode, the program would put the value 0x5 into the output register.

We note some fine points of the hardware circuit in Figure 7.11. Using a single +5-V supply both for the analog switches and the OP AMP makes level conversion of the control signals unnecessary. To achieve this, the direct current component of the analog signal must be *biased* by adding a constant to it. The OP AMP has its + input connected to a voltage midway between the limits of the input and output voltage of the CA3140 to keep it in its linear mode of operation. The inputs are connected through capacitors to shift the input signal so it is between 0.2 V and 2.5 V.

The analog signal often has to be conditioned either by amplifying or scaling down its magnitude. This is often required because to get maximum accuracy, A-to-D converters require a voltage range as wide as possible without exceeding the range of the converter; and D-to-A converters produce an output in a fixed range that may have to be amplified or reduced before it is sent out of the system. Two techniques for scaling down a signal are discussed first, then a technique for amplifying a weak signal with a computer-selected amplification is discussed. The first technique for scaling down a signal is not unlike the selection of inputs discussed earlier; the scale factor is selected by a switch. The second technique uses a fast switch to sample the input at a given duty cycle. We will discuss examples of these techniques now. Then we explain how the amplification of a weak signal can be controlled by a computer.

Consider a mechanism for reducing an analog signal by a factor controlled by an output port of a microcomputer. This mechanism might be used on a microcomputer-controlled digital meter to select the range of the voltmeter. Suppose an input voltage in the range of 0 to 500 V is to be reduced to a voltage in the range of 0 to 0.5 V for use in the next stage of the meter. (See Figure 7.12a.)

The 4051 selects one of the resistors, connecting it to ground. That resistor becomes part of the voltage divider that reduces the input voltage to within the range needed by the next stage. The other resistors not selected by the 4051 are effectively connected to very large resistors (turned-off analog switches), so they disappear from the circuit. The voltages across all the switches are kept within 0.6 V because the computer will select the appropriate resistor to divide the input voltage so that the next stage gets a voltage within its range. Thus, the analog switch is not corrupted by unwanted current flow, as we worried about in the last section. This technique can be used to reduce the magnitude of incoming analog signals under the control of a microcomputer.

Another very useful technique is to open and close a switch at a very fast rate, about ten times the maximum frequency of the analog signal being processed. (See Figure 7.12b.) If the analog switch is closed, the amplification is unity. If open, the amplification is 0; if open 50% of the time, the amplification is one-half. The microcomputer can control the duty cycle of the switch (the percentage of the time the

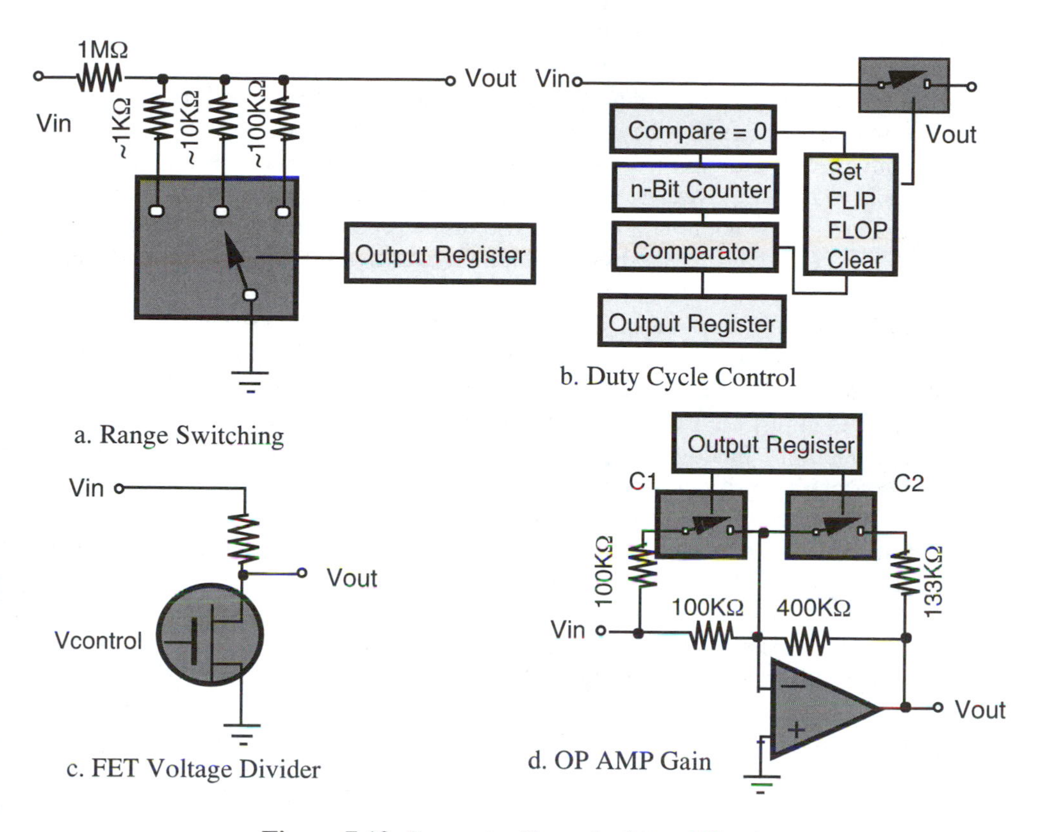

Figure 7.12. Computer Control of Amplification

switch is closed) to control the scaling of the analog signal. The output of this switch has a fair amount of high-frequency noise, which can be eliminated by passing it through a low-pass filter. Since an analog switch can operate well at 10 MHz, making the control signal frequency as high as possible eases the requirements on the low-pass filter. A simple way to control the duty cycle is to use an n-bit binary counter, a comparator fed from an output port, and a set/clear flip-flop. The counter should be clocked fast enough so that it completes its 2**n-count cycle in about ten times the maximum frequency of the analog signal, because that will determine the switch control frequency. When the counter passes 0, the flip-flop is set. When the value in the counter is equal to the value in the output register, as determined by the comparator, the flip-flop is cleared. Its output controls the switch, so the duty cycle of the switch is proportional to the number in the output register. A single counter can be used with a number of comparators, flip-flops, and switches to control several analog signals. For instance, an octave filter used in sophisticated stereo systems. Each octave has a band-pass amplifier so that the listener can compensate for excessive or deficient amplification in its reproduction. Ten comparators, flip-flops, and switches can control the amplification of each octave from a microcomputer. This would enable a microcomputer to automatically calibrate a stereo system by adjusting the amplification of each octave as tones are generated and responses are measured under its control.

Audio integrated circuits generally use the two techniques just discussed. You may not realize that they are being used. You can build these circuits using analog switches to understand how they work, and to experiment with different values of resistance, or chopping frequency, to see how they affect distortion or frequency response.

Two other techniques useful for scaling an analog signal deserve mention. A field-effect transistor (FET) behaves like a fairly linear resistor, provided the voltage across it, from drain to source, is not too high. The resistance is proportional to the voltage from gate to drain. Alternatively, the resistance of a light-sensitive FET is proportional to the light shone on it. Used in an opto-isolator, a light-sensitive FET can be used as any resistor in a voltage divider or an operational amplifier circuit. (See Figure 7.12c.) Finally, some operational amplifiers (like the CA3180) have a pin whose voltage controls the amplification. These devices can be controlled by a microcomputer sending out a voltage to adjust the light of the opto-isolator FET or by the gain of a suitable operational amplifier. Finally, signal level can be determined and used to adjust the amplification of these devices automatically, in an *automatic gain control* (AGC) circuit. An AGC circuit sometimes adjusts a filter's input voltage to prevent saturation.

Amplification (greater than 1) must be done with an OP AMP but can be controlled with analog switches. By effectively connecting or disconnecting a resistor `R1` in parallel with another resistor `R2`, the resistance can be changed from `R2` to `(R1 x R2)/(R1 + R2)`. The two resistors in an inverting amplifier can be switched by this method to alter the gain. (Consider Figure 7.12d.) If control signals C1 and C2 are HL, the amplification is 1; if LL, the amplification is 2; if HH, 4; and if LH, 8. A second stage, cascaded onto this one, could be built to have amplification 1, 16, 256, or (a rather high) 4096, and so on. The computer can select the amplification by setting these control signals to the analog switches. Amplification ratios lower than 2 provide closer control and can be obtained by appropriate resistors in the circuit.

7.3.3 Nonlinear Amplification

Nonlinear modification of analog signals is the final type of signal conditioning. A number of fascinating circuits have been advanced to multiply or divide one analog signal by another, or to output the square root, log, or sine of an analog signal. But unless the signal is too fast, hardware/software trade-offs usually favor microcomputer processing of the signal. Three special cases often favor analog signal conditioning: absolute value, logarithmic function, and sample and hold. (See Figure 7.13.)

A diode is capable of extracting the absolute value of a waveform, and this is the basis of the AM radio detector. An accurate absolute value function is sometimes very useful if, for example, an input voltage whose range is over ± 1V is to be measured by an analog-to-digital converter that can only measure positive signals and perhaps has only a single-ended 5-V supply. Figure 7.13a puts the diode into the feedback loop of an OP AMP to increase the linearity of the absolute value function. For positive inputs, the diode disconnects the OP AMP so the output is connected to the input via the feedback resistor R2. For negative inputs, the OP AMP simply inverts the input to get the output. Using a CA3140, this circuit can derive the absolute value of sine waves even beyond 100 KHz.

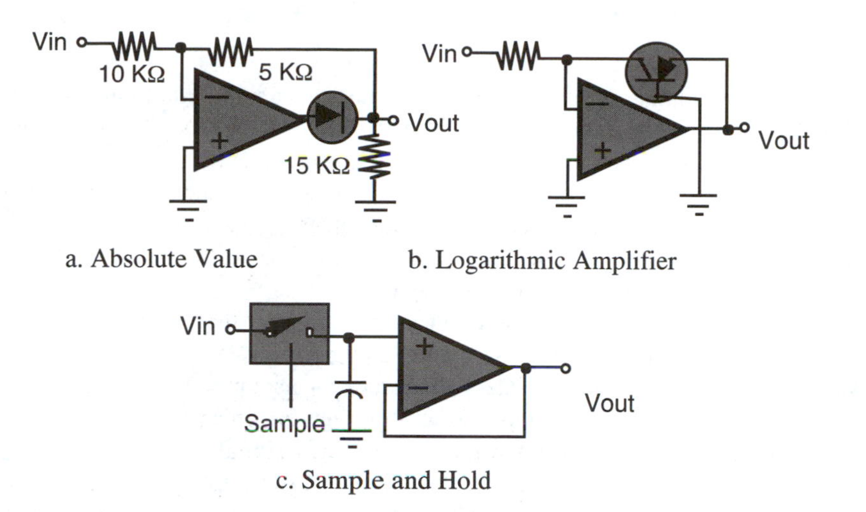

Figure 7.13. Nonlinear Signal Conditioning

The logarithm of an input voltage is sometimes obtained using analog signal conditioners, because audio levels, light levels, and so on are logarithmically related to voltages measured by transducers. Conditioning the analog signal by a logarithmic function drastically compresses the range of signal that must be converted by an analog-to-digital converter. The transistor's emitter current i is related to its emitter voltage V by the exponential law

$$i = (e^{V/a} - 1)$$

where a is a constant. It can be put into a feedback circuit of an OP AMP to derive a logarithmic function signal conditioner. (See Figure 7.13b.) The output $Vout$ is related to the input Vin by

$$Vout = A \log(Vin/B)$$

where A and B are constants that depend on the resistor in the circuit and on the transistor and its temperature.

The sample-and-hold circuit is the last of the nonlinear signal conditioners of particular use in microcomputer systems. Sometimes used by itself to sample an input signal at a precise time, it is also an important building block in digital-to-analog converters, covered in §7.4.1, and in multiple-output data acquisition systems (Figure 7.13c). The input signal passes through an analog switch; when it is on, the voltage on the capacitor quickly approaches the input voltage. When off, the capacitor voltage remains steady. The voltage follower makes the output voltage equal to the voltage across the capacitor without leaking its voltage, even though the output may have to supply considerable current to the device it feeds. Turning the switch on causes this circuit to *sample* the input. A microcomputer output register can control the switch to sample an incoming waveform at a precise time so the output voltage from the sample-and-hold circuit can be converted to a digital value.

7.4 Converters

We often convert analog signals into digital signals, or analog amplitude signals into analog frequency signals, and vice versa. The first subsection describes the digital-to-analog converters that are commonly available for microcomputer I/O systems. The next subsection describes analog-to-digital converters. Though they seem to be more common than digital-to-analog converters, we discuss them later because some analog-to-digital converters use digital-to-analog converters inside them. Finally, the frequency-to-voltage and voltage-to-frequency converters are discussed.

The following are some important concepts that cover the various converters. In general, the analog input is either sampled, using a sample-and-hold circuit or its equivalent, or integrated, using an integrator circuit or its equivalent. Analog output is either produced in samples or is output from an integrator. Integration smooths out the signal, reducing noise, but limits the upper-frequency signal components. Sampling provides a "snapshot" of the data and also of the noise. In sampling converters, another problem is caused by high frequencies. The *sampling rate* is obviously the rate at which the data samples are taken. The *Nyquist rate* is one-half the sampling rate. Components of the signal that have a higher frequency than the Nyquist rate "beat against" the frequency of the sampling rate in the same manner as radio frequency signals are "beat against" the frequency of a local oscillator in a radio, generating *alias* frequency components. For example, if a component has a frequency equal to the sampling rate, it will appear as a direct-current component. To eliminate the generation of these alias components, a low-pass filter is used to eliminate all frequencies above the Nyquist rate.

7.4.1 Digital-to-Analog Converters

Three basic digital-to-analog converters (D-to-As) are introduced now: the summing amplifier, the ladder, and the exponential superposition D-to-As. The summing amplifier converter most readily shows the basic principle behind all D-to-A converters, which is that each digital bit contributes a weighted portion of the output voltage if the bit is true, and the output is the sum of the portions. The ladder converters are easier to build because the resistors in a ladder network can be trimmed precisely without much effort. Ladder networks for these D-to-A converters are readily available, quite fast, and inexpensive. The exponential superposition converter is quite a bit slower and less accurate but doesn't need precision components, so it would be very useful in microcomputer-based toys or appliance controllers. A convenient package of 6-bit D-to-A converters, the MC144110, is considered at the subsection's end.

The summing amplifier can be used in a D-to-A converter, as in Figure 7.14a. Keeping in mind that the output voltage is

```
Vout = - Rf ( V1/R1 + V2/R2 + ... )
```

if `V1` = `V2` = . . . = 1V, and `Ri` is either infinity (an open switch) or a power of 2 times `Rf` (if the corresponding switch is closed), then the output voltage is

```
Vout = C1/2 + C2/4 + C3/8 + ...
```

where `Ci` is 1 if the switch in series with the *i*th resistor is closed; otherwise it is 0. An output device can be used to control the switches, so the *i*th most significant bit controls the *ith* switch. Then the binary number in the output register, considered as a fraction, is converted into a voltage at the output of the summing amplifier. Moreover, if the reference input voltage is made `n` V rather than 1 V, the output is the fraction specified by the output register times `n` V. `n` can be fixed at a convenient value, like 10 V, to *scale* the converter. Usually, a D-to-A converter is scaled to a level, so for the largest output value the summing amplifier is nearly, but not quite, saturated, to minimize errors due to noise and to offset voltages and currents. Alternatively, if `n` is itself an analog signal, it is multiplied by the digital value in the output register. This D-to-A converter is thus a *multiplying D-to-A converter*, and can be used as a digitally controlled voltage divider – an alternative to the range switch and duty-cycle control techniques for amplification control.

Although conceptually neat, the above converter requires using from 8 to 12 precision resistors of different values, which can be difficult to match in the 2-to-1 ratios needed. An alternative circuit, an *R-2R ladder* network, can be used in a D-to-A converter that uses precision resistors, all of which have values `R` or `2R` Ω. This network can be used as a voltage divider or a current divider; the former is conceptually simpler but the latter is more commonly used. (See Figure 7.14b for a diagram of a current ladder D-to-A converter.) A pair of analog switches for each "2R" resistor connect the resistor either into the negative input to the OP AMP or to ground, depending on whether the control variable is high or low, respectively. The current through these switches, from left to right, varies in proportion to 1/2, 1/4, 1/8, . . ., as can be verified by simple circuit analysis. If the *i*th control variable is true, a current proportional to `2**-i` is introduced into the negative input of the OP AMP, which must be counterbalanced by a negative current through Rf to keep the negative input at virtual ground, so the output voltage proportional to 2**-*i* is generated. The components for each input *i* are added, so the output is proportional to the value of the binary number whose bits control the switches. Like the previous D-to-A converter, this can be scaled by appropriately selecting the voltage `Vin` and can be used as a digitally controlled amplification device. It, too, is a multiplying D-to-A converter.

A program for outputting a voltage by means of either a summing or an R-2R D-to-A converter is very simple. One merely stores the number to be converted onto an output register that is connected to the converter.

A ladder network for a converter can be obtained as an integrated circuit for 6 to 12 bits of accuracy. The chip contains the switches and the resistors for the circuit. The output settles to the desired voltage level in less than one microsecond in a typical converter, so the programmer usually does not have to worry about settling time.

The last converter in Figure 7.14 uses a sample-and-hold circuit to sample a voltage that is the sum of exponential voltages corresponding to bits of the digital word being converted. The circuit, in Figure 7.14c, is simplicity itself. We first offer some observations on an exponential waveform and the superposition principle. Consider an exponential waveform as shown in Figure 7.14d. Note that for such a signal there is a time T (not the time constant of the network, though) at which the signal is 1/2 the initial value of the signal. And at times 2T, 3T, 4T, and so on, the signal level is 1/4,

1/8, 1/16 of the initial value, and so on. Furthermore, in a linear circuit, the actual voltage can be computed from the sum of the voltages of each waveform. This is called superposition. Now if a sample-and-hold circuit samples a voltage that is a sum of exponential voltages, an exponential waveform that was started T time units before will contribute 1/2 its initial value; one that was started 2T time units before will contribute 1/4 its initial value; one started 3T units before will contribute 1/8 its initial value; and so on. These waveforms are generated by asserting control variable *P* with the shifted bit, as the least significant bits are shifted out each T time units. Thus, the exponential waveforms from left to right in Figure 7.14d are generated or not; the sampled voltage will or will not have a component of 1/8, 1/4, 1/2, etc. The control variable *S* is asserted to sample the waveform after all bits have been shifted out. The sampled voltage is the desired output of the D-to-A converter. The output can be scaled by selecting an appropriate current source. Some care must be taken to make this A-to-D converter a good multiplying converter because dynamically changing the input current level will alter its accuracy. Nevertheless, this mechanism is essentially the mechanism used in the Crystal Semiconductor CS4330, having 18 bits of accuracy, and the Crystal Semiconductor CS4332, having 24 bits of accuracy.

The Crystal Semiconductor CS4330 is an inexpensive 50-KHz (max), 18-bit D-to-A converter for stereo CD players (see Figure 7.15b). Its serial interface requires but two pins, which is easier to opto-isolate if the analog voltage must be on a different ground system than the microcomputer. A MC74HC4040 counter chip derives a 4-MHz MCLK and a 16-KHz LRCK from the 'A4 expanded bus mode E clock (or an external 8-MHz

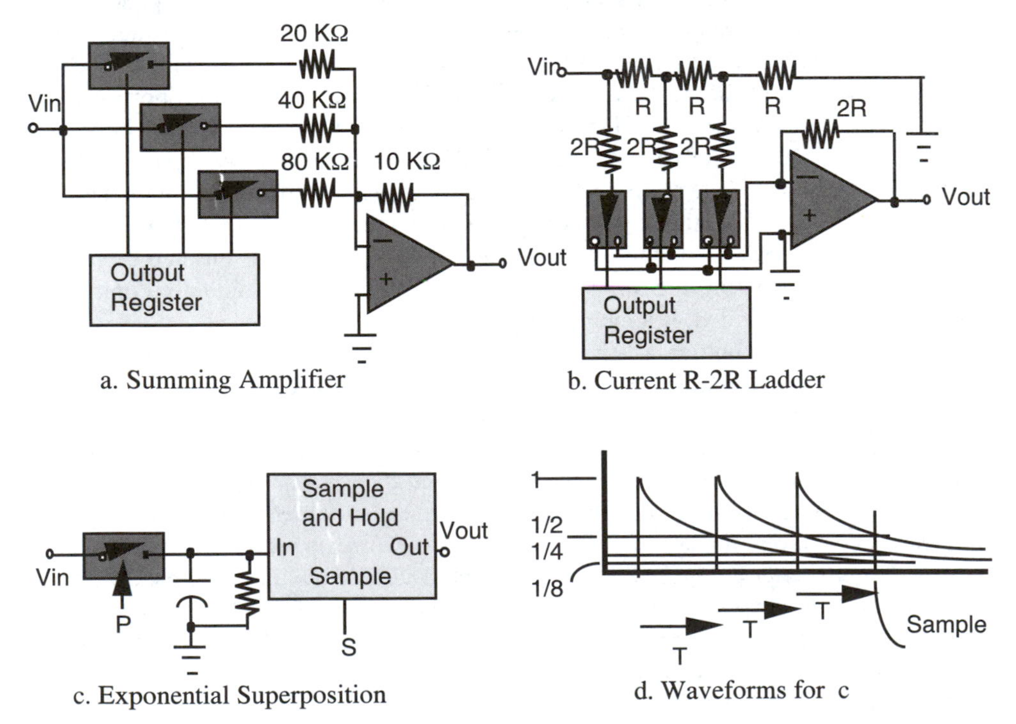

a. Summing Amplifier

b. Current R-2R Ladder

c. Exponential Superposition

d. Waveforms for c

Figure 7.14. D-to-A Converters

clock). A reference 1-MHz SCLK is derived inside the CS4330 from the MCLK (see Figure 7.15a), but this SCLK reference is not externally available. A signal with the same period is on the MC74HC4040's pin 6. Some 18-bit 2's complement data are shifted from SDATA, msb first, onto each falling edge of SCLK. The last 18 bits are applied, when LRCK rises, to the left D-to-A converter, and when it falls, to the right D-to-A converter. The C procedure `DtoA( char hi; int low; unsigned char count)` sends to the CS4330 the two least significant bits of `hi`, and all 16 bits in `low`, repetitively for *count* conversion periods. Note that the 18-bit data in `PORTJ` and accumulator D are shifted out 1 bit per μs. Outputs Right and Left can be OP AMP or analog comparator inputs. The CS4330 has excellent linearity; however its zero-value offset and maximum voltage are somewhat variable, and it requires the 'A4's full attention. But these limitations can be removed (see problems at the end of the chapter).

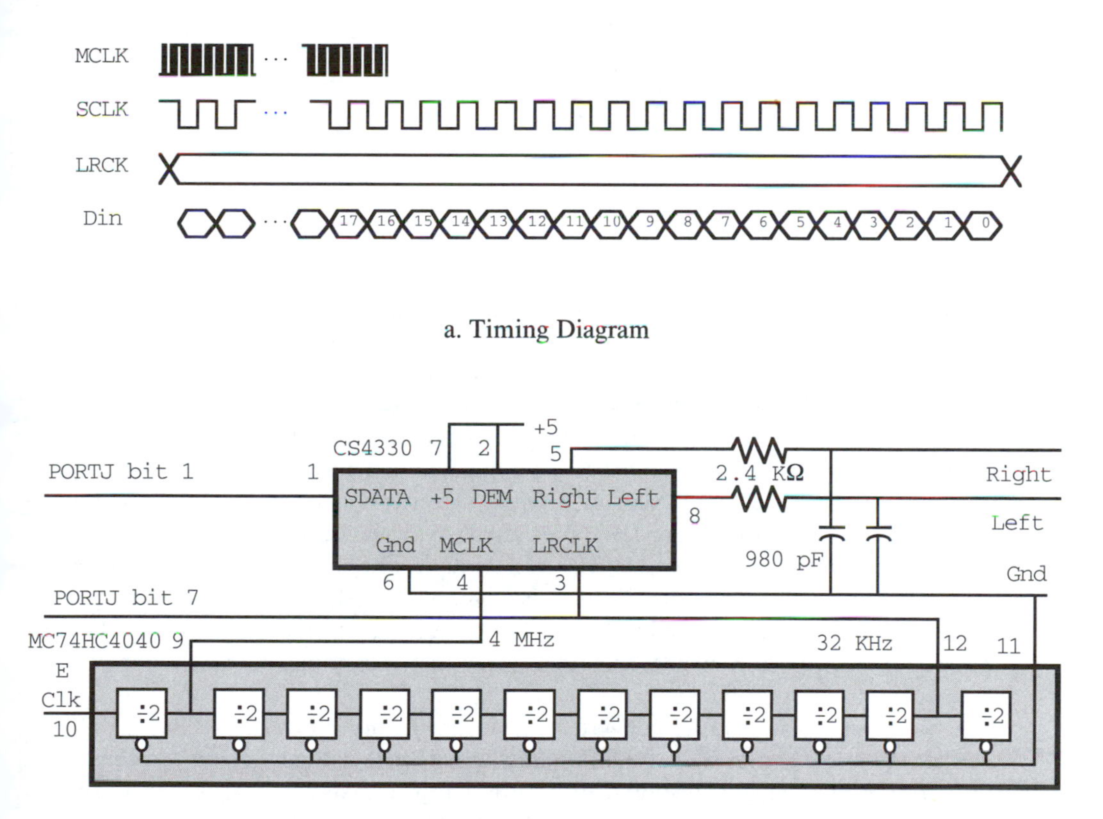

Figure 7.15. The Crystal Semiconductor CS4330

```
DtoA( char hi, int low, unsigned char count) { char compare;
    do { DIRJ = 3; KWIFJ = 0x80; PORTJ = hi;

asm{
    ldx #$17    ; $1b waste first 14 SCLK clock cycles
12  dbne x,12
    nop         ; padd
    ldd 5,sp    ; operand "low" gives low 16 bits of voltage
    ldx #17     ; output 18 bits, but last is left not shifted
13  lsld        ; move up low-order bits
    rol PORTJ   ; move up high-order 2 bits and output
    dbne x,13   ; count down to shift all 18 bits
14  brclr KWIFJ,0x80,14 ; wait for flag on LRCK edge
    ldab PORTJ ; get input in bit 6 from analog comparator
    stab 0,sp   ; return value (for A-to-D converters)
}
  KPOLJ ^= 0x80; /* look for other edge */ }
  while(--count);/* for each requested conversion */ return (compare>>6) & 1;
}
```

7.4.2 Analog-to-Digital Converters

Six analog-to-digital converters (A-to-Ds) are introduced herein. These have different costs, accuracies, and speeds. We discuss them in approximate order of decreasing speed and cost. The parallel and pipeline converters are fastest, followed by the delta and successive-approximation converters and the ramp converters.

The *parallel A-to-D converter* uses comparators to determine the input voltage and can be made to operate almost as fast as the comparators. One avoids using too many comparators because they are expensive, so this converter's accuracy is limited by the number of its comparators. Figure 7.16a illustrates a typical 3-bit converter that has, for ease of discussion, a range of 0 to 7 V. The resistor divider network establishes reference voltages for each comparator, from top to bottom, of 0, 1, 2, . . . , 7 V. If the input voltage `Vin` is between i - 1 and i V, the i bottom comparators output a true value. A priority encoder module encodes this set of values to a binary number that is the most prior true input address, which is i.

A variation of the parallel converter, an `n`-bit *pipeline converter*, consists of `n` identical stages of a comparator and differential amplifier (see Figure 7.16b and Figure 7.16c). In a typical stage illustrated in Figure 7.16b, the signal `Vin` is sent to the input, and the output `Vout` of the differential amplifier on the right of the stage is then sent to the input of the next stage to the right. Suppose the voltage range is `Vmax`. The output of the comparator on the left of the stage is either `Vmax` if the `V+` input is higher than the `V-` input of the comparator or it is 0 V. If the input is above half `Vmax`, then half `Vmax` is subtracted from the input and then doubled – otherwise the input is just doubled – in the differential amplifier that feeds the output `Vout`. If a steady signal is fed into the input, `Vin`, then as the signal flows through the stages,

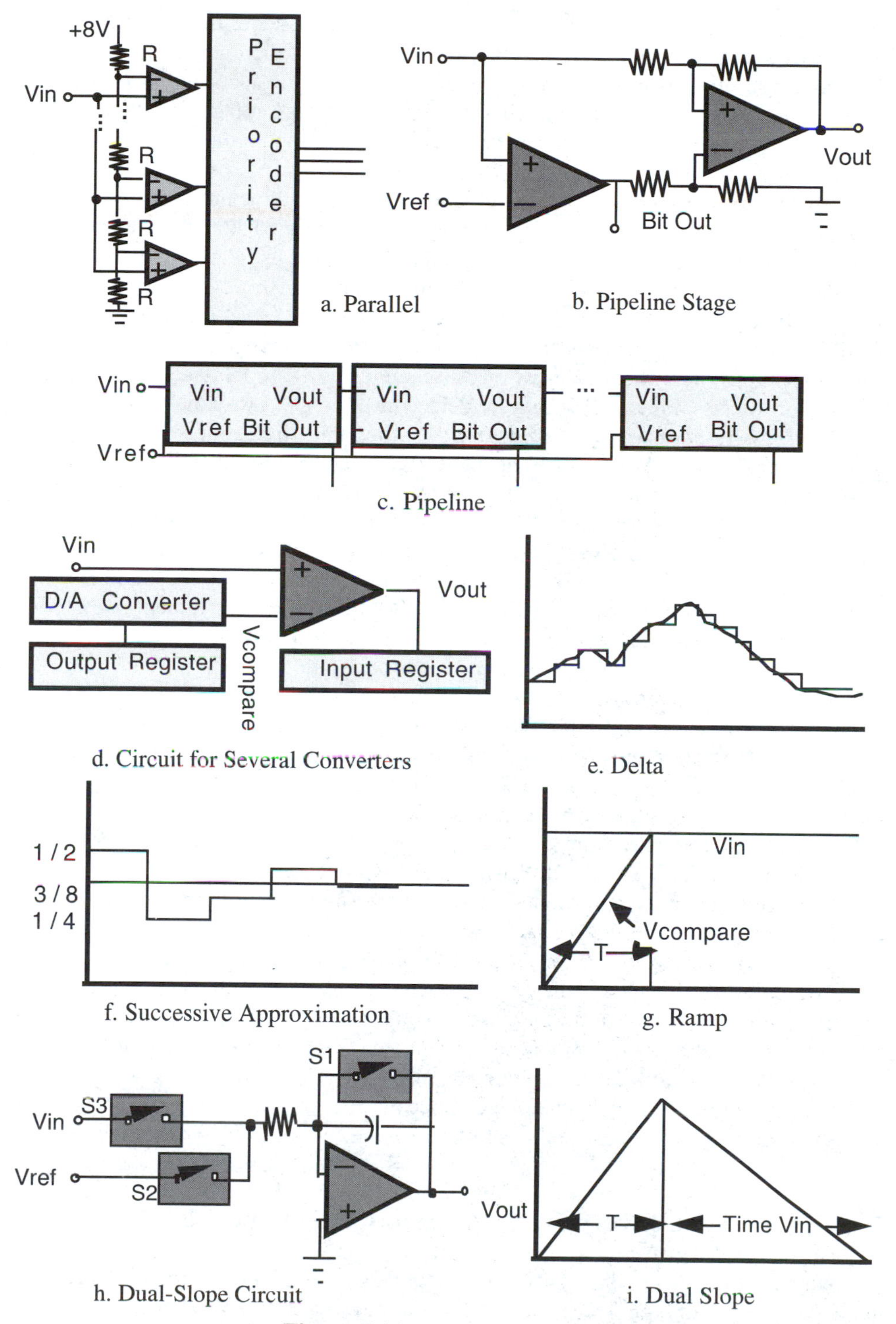

Figure 7.16. A-to-D Converters

bits from the most significant bit are obtained from each stage; they are true if half `Vmax` was subtracted, otherwise they are false. Moreover, the conversion is actually done as the leading edge of the analog signal flows through each stage. It is possible, then, to begin the next conversion when the first stage has settled, even though later stages may yet be settling. It is rather like oil flowing through a pipeline – it can have an incredibly fast conversion rate.

Successive-approximation, delta, and ramp converters can be implemented with the hardware illustrated in Figure 7.16d. The programs differ for each method. For *delta* or *servo conversion,* a D-to-A converter outputs a voltage `Vcomp` that is compared to `Vin`. If `Vcomp > Vin`, then `Vcomp` diminishes; otherwise `Vcomp` increases by a small amount. Assuming `Vin` changes more slowly than `Vcomp` can change, `Vcomp` should "track" `Vin` in the manner of a feedback control or servo system. By analogy to communications systems, the digital output changes by delta increments, as in delta modulation systems. Figure 7.16e shows a delta converter's `Vin` tracking `Vcomp`.

A *successive-approximation converter* uses the same circuit but requires a divide-and-conquer program. We observe the same technique in long division. Suppose the input is in the range 0 to `Vmax`. The D-to-A converter is so loaded as to output `Vmax/2`. If the comparator senses `Vin > Vmax/2`, then the D-to-A converter is set to output `Vmax x 3/4`, otherwise it is set to output `Vmax/4`. Note that this is done by either adding or subtracting `Vmax/4` from the current output, depending on the result of comparing this output with `Vin`. In successive trials, `Vmax/8`, then `Vmax/16`, `Vmax/32`, ... are added or subtracted from the value output to the D-to-A converter. The comparison voltage `Vref` approaches `Vin`, as shown in Figure 7.16f.

A *ramp A-to-D converter* can use the same circuit as in Figure 7.16d or a simpler circuit as in Figure 7.16h. Simply, in Figure 7.16d, the comparator voltage `Vcomp` is initialized to 0 by clearing location 0x8000, then is gradually increased by incrementing 0x8000 until `Vcomp > Vin` is sensed by the comparator. (See Figure 7.16g.) The circuit illustrated in Figure 7.16h uses a *dual-slope converter* that is shown in Figure 7.16i. The output voltage from the integrator, sensed by the comparator, is initially cleared by closing only switch S1. Then, by closing only S2 for a specific time T, the reference voltage `Vref` is integrated, charging the capacitor in the integrator. Lastly only S3 is closed, so that the voltage `Vin` is integrated to discharge the capacitor. The time to discharge the capacitor is proportional to `Vin`. Moreover, the time is proportional to the average value of `Vin` over the time it is integrated, which nicely reduces noise we don't want to measure; and the accuracy of the converter does not depend on the values of the components (except `Vref`), so this converter is inexpensive. However, it is the slowest converter. It finds great use, nevertheless, in digital voltmeters, multimeters, and panel meters, because it can achieve about 12 bits (3 1/2 digits) of accuracy at low cost, and it is faster than the eye watching the display.

7.4.3 Voltage Conversion to or from Frequency

A frequency-to-voltage converter (FVC) outputs a voltage that is proportional to the input frequency. For high frequencies, an FM detector is a good FVC. Several integrated circuits are available for detecting FM signals. For a broad range of frequencies, a phase-locked loop can be used. The error voltage used to lock the oscillator in it to the frequency of the incoming signal is proportional to the difference between the frequency to which the oscillator is tuned and the frequency of the incoming signal. For audio frequencies, a common technique is to trigger a one-shot with the leading edge of the input signal. The output is a constant-width and constant-height pulse train, where pulses occur with the same frequency as the input signal. (See Figure 7.17a.) A low-pass filter outputs a signal proportional to the area under the pulses, which is in turn proportional to the frequency. The LM3905, a one-shot (monostable) with built-in voltage reference and an output transistor that is capable of producing output pulses of precise height, is especially suited to FVC. (See Figure 7.17b.) Another way to convert frequency to voltage is to use an automobile tachometer integrated circuit; it senses spark pulses whose frequency is proportional to engine speed, and it outputs an analog level to a meter to display the engine speed. This technique can be used with subaudio frequencies since it is designed to measure low-rate spark pulse trains. Frequency-to-voltage converters have the advantage that information is carried by the frequency of a signal on a single wire, which can be easily opto-isolated, and is immune to noise and degradation due to losses in long wires from microcomputer to output. However, the signal they carry has to pass through a low-pass filter, so its maximum frequency must be much lower than that of the carrier which is being converted to the voltage.

The final converter of interest is the voltage-to-frequency converter. It generates a square wave whose frequency or period is proportional to the input voltage `Vin`. (See Figure 7.18a.) Internally, `Vin` is integrated, until the integrated voltage reaches a reference voltage `Vref`, when the integrated voltage is cleared. An output pulse, occurring as the integrator is cleared, has a frequency that is proportional to the input `Vin`. If desired, this can be fed to a toggle flip-flop to square the signal as its period is doubled. By reversing the role of the reference and input voltage so the reference voltage is integrated and compared to the input voltage, the period of the output is proportional to the voltage `Vin`. So this makes a voltage-to-period converter. But noise on `Vin` is not averaged out in this technique. Other circuits are used for VFCs, but the principles are similar to those discussed here. VFCs can be quite accurate and reasonably fast; the Teledyne 9400 (see Figure 7.18b) accurately converts voltage-to-frequency to about 13 bits of accuracy and remains equally accurate after two cycles have occurred on the output wave. That means the converter is faster for higher voltages, since they result in higher frequencies, than for lower voltages. Used in an integrating mode, moreover, the VFC can reduce noise the way the dual-ramp converter does. The VFC is of particular value where the microprocessor has a built-in counter to measure frequency, especially since the frequency carrying signal is easy to handle, being carried on only one wire.

The phase-locked loop (PLL) is used to generate higher frequencies than can be achieved by a microcomputer using a counter/timer (see Figure 7.19). They are fundamentally a voltage-to-frequency converter (voltage-controlled oscillator) and a frequency-to-voltage converter (phase comparator). In addition there are usually frequency dividers (counters). A PLL can be used in the 6812 to generate the E clock (§6.4). Figure 7.19a shows the digital part and Figure 7.19b shows the analog part of a PLL that can generate almost any frequency within a wide range. PLLs are used, with a prescaler, for FM radios and television sets.

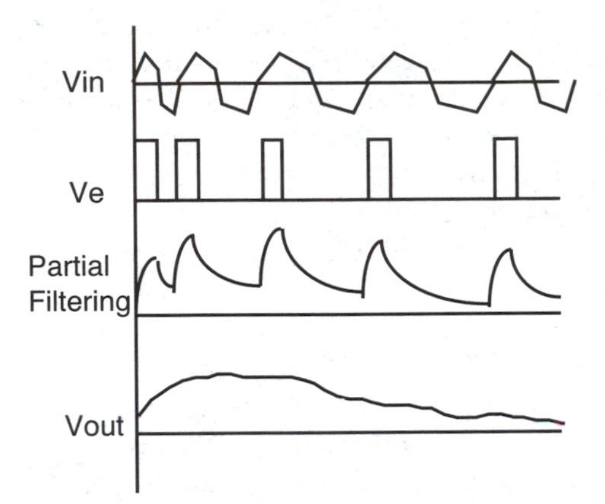

a. Frequency-to-Pulse-Train-to-Voltage Conversion

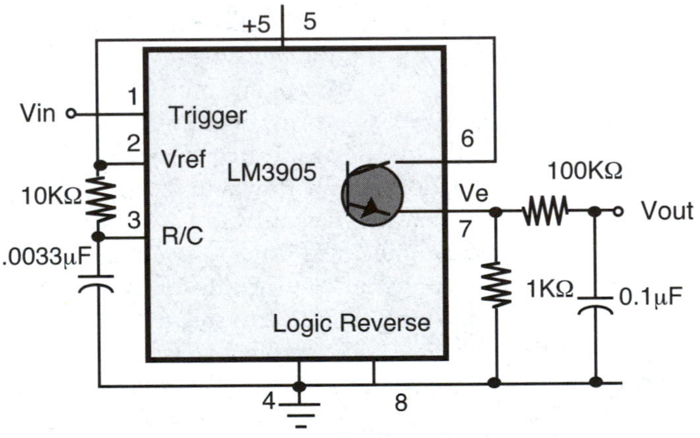

b. Frequency-to-Voltage Converter

Figure 7.17. Frequency-to-Voltage Conversion

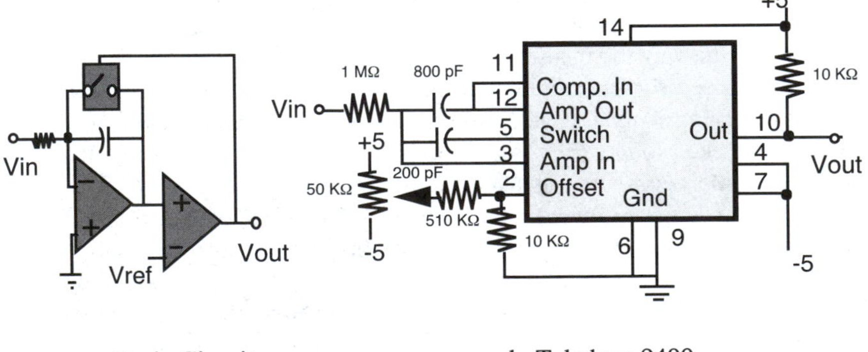

a. Basic Circuit

b. Teledyne 9400

Figure 7.18. Voltage-to-Frequency Conversion

Figure 7.19a diagrams one of several chips that use a serial input for the digital logic of a PLL. Serial input is desirable to save pins because the frequency of the oscillator is not changed that often and does not have to change promptly. The chip contains two down-counters: one divides the variable frequency of a signal `Fin` on pin 9 by `N`, where `N` is a 14-bit number sent from the computer; while the other divides a reference frequency of a signal generated by an oscillator on pin 17 by `R`, where `R` is chosen by the levels on pins 2, 1, and 18. The number `N` is shifted in, using the techniques discussed in §4.4. Actually, the low-order 14 bits of the 16 bits shifted in (most significant bit first) are `N`; the high-order 2 bits are output as signals on pins 13 and 14, which are two open collector output bits for external use. The average voltage on the phase detector output PD (pin 6) is raised if the variable frequency divided by `N` is less than the reference frequency divided by `R`, and it is lowered if the variable frequency divided by `N` is greater than the reference frequency divided by `R`. Other outputs ør, øv, and lock-detect LD can be used in more advanced PLLs.

Figure 7.19b is the analog part of the PLL. The chip contains two variable-frequency oscillators (VCOs). We describe the upper one here. The capacitor Cext between pins 12 and 13 sets the nominal frequency of the oscillator – about (500/Cext) MHz, where Cext is in pF. The frequency of the output on pin 10 is proportional to the voltage on pin 1. The range control Rng on pin 14 sets the sensitivity; the larger Rng voltage makes oscillator frequency changes more sensitive to input voltage changes. The enable E must be low for the output to oscillate.

Figure 7.19c shows a simple application. Some new chips require unusual clock frequencies. For example, we found a very interesting voice-output chip needing a 3.123-MHz crystal, but could not find the crystal. The PLL just mentioned could be used to generate frequencies in the range of 1 to 8 MHz in place of a crystal. It is a feedback control system of the kind described in §6.6.3 that uses voltage analog and frequency analog signals. The 145155 compares a frequency on input Fin to one generated by the OscIn input, and outputs a voltage on PD that is proportional to the difference. This voltage, passed through a low-pass filter, is put into FCnt of the 74LS124 VCO to generate a frequency Fin. When stable, the voltage PD is just the right voltage to generate the frequency Fin that is `R/N` times the frequency of the signal OscIn. In this example, OscIn is generated from the 2-MHz E clock of the MC68HC11A8, and `R` is set to 8192 = 2^{13}. To set up the oscillator at frequency F, the connection from pin 6 of the 145155 to pin 1 of the 74LS124 is broken, a voltage of 2.5 V is put on pin 1, the VCO capacitor Cext is set to cause oscillations at about the desired frequency F, and then a number `N = (F/2MHz) * 8192` is shifted into the 145155 so that the output PD is about 2.5 V. Then the broken connection is put back. The output signal should have frequency `F`. The design of the low-pass filter is quite involved: whole books are available on this topic. In the preceding circuit, the resistor connected to output PD determines the frequency of oscillation of the locking error, and the resistor connected to the 10-μF capacitor determines its damping; these can be twiddled to get acceptable results. This technique is useful in generating high frequencies up to about 15-MHz. Note that the frequency can be changed (over a 2:1 range) by changing the number in the shift register, so the pitch of the voice-output chip can be altered to get a more natural speech.

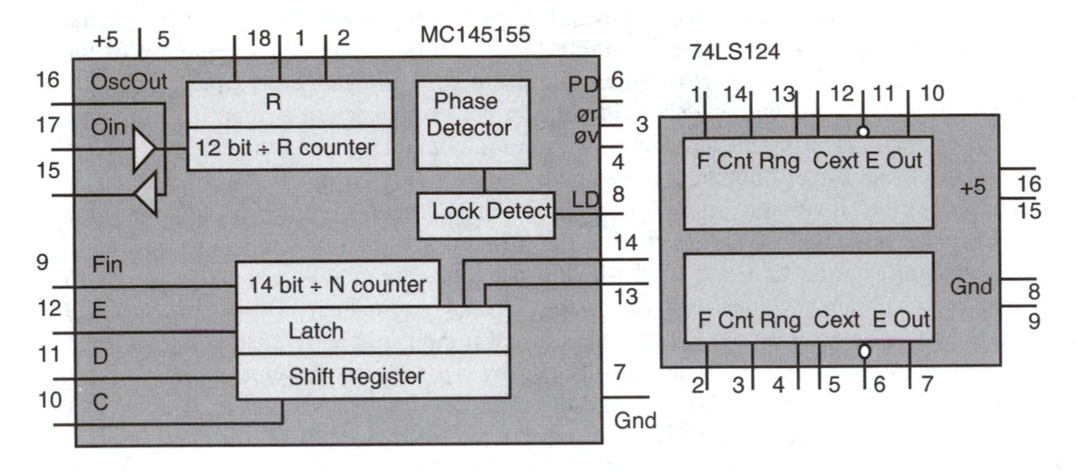

a. The MC145155

b. A Dual Voltage-controlled Oscillator

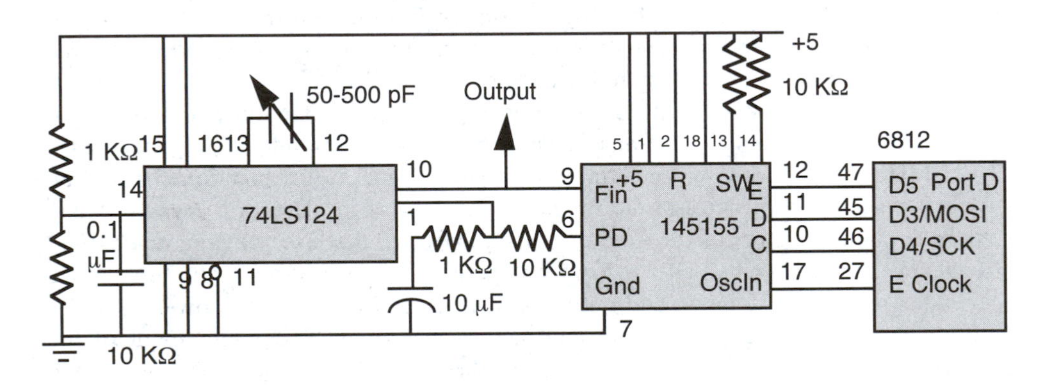

c. A Simple Circuit

Figure 7.19. Phase-locked Loop

7.5 Data Acquisition Systems

A *data acquisition system* (DAS) consists of switches, a D-to-A converter, and signal conditioning circuits so that several inputs can be measured and several outputs can be supplied under the control of a microcomputer. In the first subsection, we consider the basic idea of a data acquisition subsystem; then, we consider the MC145040 A-to-D converter that is the input part of a data acquisition system, and the A-to-D port on the 'A4. The final subsection considers how these data acquisition systems can be used in control systems.

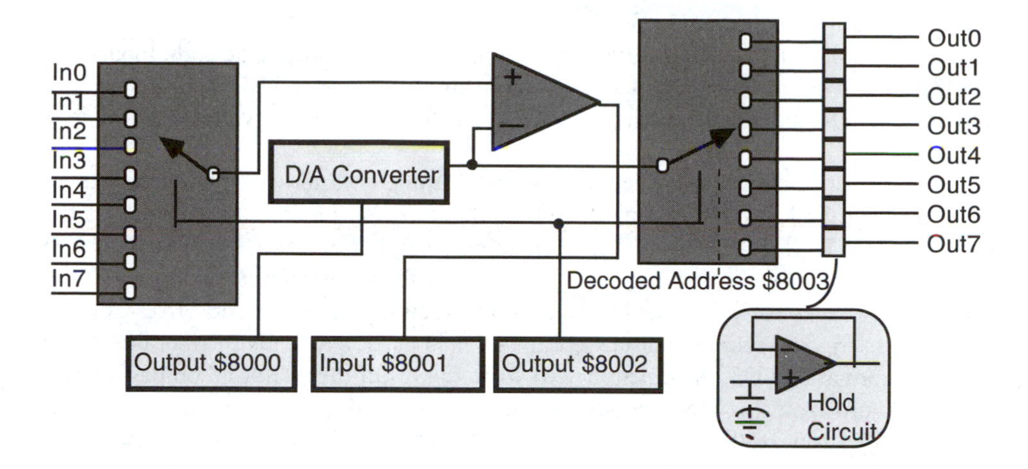

Figure 7.20. Data Acquisition System

7.5.1 Basic Operation of a Data Acquisition System

A DAS can be purchased as a printed circuit board, or even as a hybrid integrated circuit module. Such a DAS would be better to use than the system we discuss, but we introduce the latter to show how such a system works and to bring together the various concepts from previous sections. Finally, in this section we will show how a DAS can be used to implement a digital filter or feedback control system.

The DAS described in this section is diagramed in Figure 7.20. From left to right, an analog switch selects from among eight inputs an input for the comparator. This is used to measure the input voltages. The D-to-A converter is used, with the comparator, for prompting the A-to-D converter to measure the inputs and for supplying the output voltages. The analog switch and voltage followers on the right impel sample-and-hold circuits, which act like analog flip-flops, to store the output voltages after they have been set by the microcomputer. In the following discussion, location 0x8000 will be an output register to the D-to-A converter, and location 0x8001 will be an input whose sign bit is true if `Vin` (selected by the analog switch) is greater than `Vcomp`. Location 0x8002 will be a readable output port that addresses both the analog switches. If, for instance, 3 is put in this register, then input 3 is sent to the comparator (as `Vin`), and the output of the D-to-A converter is made available to the sample-and-hold circuit that supplies output 3. Finally, the analog switch is enabled by an address trigger. If the microcomputer addresses location `0x8003`, as it does in two memory cycles when executing an instruction like `INC 0x8003`, then the address decoder will provide a 250-ns pulse which will enable the analog switch for that time. Recall that when enabled, the addressed input is connected to the output of the switch, but when not enabled, all inputs and the output are not connected.

The DAS is controlled by a program, shown soon, that will be called as a subroutine whenever the inputs are to be measured and the outputs are to be adjusted. Eight output values are stored in a table TBL so that TBL[0] is converted to a voltage on output 0; TBL[1] on output 1; and so on. TBL is loaded with the desired values just before this subroutine is called. After returning from the subroutine, the eight inputs are, to keep things simple, converted and stored in the same table. TBL[0] will store the number equal to the voltage on input 0; TBL[1], that equal to the voltage on input 1; and so on.

The output register that selects inputs and outputs is initialized by the first three lines to select the last input or output, and the index register is initialized to access the last row of the table TBL. Thereafter, in the loop, a number is read from the table to the D-to-A converter via output register 0x8000, and then the output analog switch is enabled for two consecutive memory cycles by executing the instruction INC 0x8003. The address trigger technique discussed in §4.1.1 is used here. This instruction should read the word at 0x8003, increment it, and write the result there. But no output register or RAM word is at this location, so nothing is actually done by the microprocessor. However, location 0x8003 is accessed twice. This enables the analog switch twice. At that time, the output of the D-to-A converter is sampled by the sample-and-hold circuit that feeds output 7, since the analog switch addresses the bottom position. The voltage output from the D-to-A converter is now sampled, and will remain on this output until it is changed when this subroutine is called again. Thus, the sample and hold behaves rather like an analog storage register. Next, a successive approximation subroutine like that discussed in the previous section is called. The subroutine converts the bottom input, since output register 0x8002 is 7, to a digital value that is left in accumulator A. This value is then stored in the bottom row of the table TBL. The index register and the contents of output register 0x8002 are decremented to output row 6 of TBL to the sixth output, and they put the value of the sixth input into row 6 of TBL in the next iteration of the loop. When all rows are output and input, this subroutine is left. The above routine can be made more useful by using two tables, one for outputs and one for inputs; however, we do not show this here because more code would be needed, but no new important ideas would be demonstrated.

7.5.2 The MC145040 A-to-D Converter

Two very convenient ways to measure analog voltages in the 6812 microcomputer are with the 145040 chip or with the A-to-D converter in the 6812 itself. Both of these are similar to a data acquisition system in that they use an analog multiplexor (mux) to select a number of inputs for the converter.

A serial interface is desirable for an A-to-D converter because it uses fewer pins than a parallel interface and thus can be easily isolated using opto-isolators. The MC145040 is one of the better serial interface A-to-D converters. See Figure 7.21.

Data are shifted into and out of the 145040 at the same time, using the exchange technique discussed in §4.4.3. During bit movement, CS must be low and should rise after all bits are moved because that edge transfers the data to the mux address register and begins the conversion. An input address is sent first (to select input *i*, *i = 0 to 11*, send *i* << *4*, msb first), and each bit is clocked in on the rising edge of SClk. Conversion is done with the A-to-D Clk using the successive approximation technique, with `Vref` as the maximum and `VAG` as the minimim reference voltage. If the address is 0xB, then a voltage (Vref + VAG)/2 is input, and should convert to a value of 0x80; this can be used for a check. Wait 32 A-to-D Clk pulse cycles, and then input the 8-bit digital value, msb first, that was converted from the *i*th analog input. Data is sent from the chip on the falling edge of SClk. The "address" for the next conversion can be sent out while the data from the previous conversion are being read in, in an exchange operation.

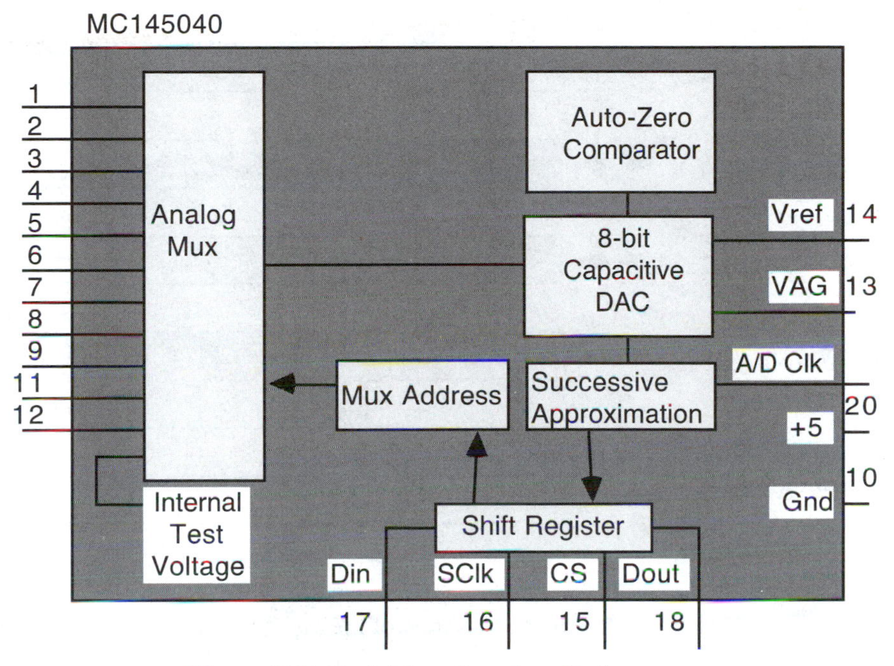

Figure 7.21. Serial Interface A-to-D Converter

7.5.3 The MC68HC812A4 A-to-D Converter

The 'A4 has an onboard A-to-D converter, which, unless PORTAD pins are needed for some other application, can be used for measurement of analog voltages. Figure 7.22 shows the block diagram of this subsystem. Using VRH (pin 85) as high-reference voltage and VRL (pin 86) as low-reference voltage, the analog voltages on inputs PAD0 (pin 87) to PAD7 (pin 94) can be converted to 8-bit digital values and put into registers ADR0 to ADR7. Initially, ADPU must be set to apply power to this subsystem (100 μs are needed for the voltages to become stable). The conversion is begun when ATDCTL5 is written into, and the mode of conversion is dictated by the value put into it. ATDCTL5's low-order four bits, C. If C is less than 8, these bits determine which input pin voltages are converted. Essentially, if C is 5, then pin PORTAD bit 5's voltage is put into port ADR5. Values of C > 8 can be used for testing: if C = 0xC, VRH is converted; if C = 0xD, VRL is converted; and if C = 0xE, VRH/2 is converted. If S8CM is T (1), eight conversions are done; otherwise, four are done. If SCAN is T (1), the inputs are sampled continuously; otherwise they are sampled just once after conversions have begun. We discuss the effect of `MULT` when `S8CM` is T (1) and C < 8. If `MULT` is 0, the voltage on `PORTAD` bit i is sampled eight times and put in each `ADR0 ... ADR7`, but if ***MULT*** is 1, the voltage `PORTAD` bit 0 to `PORTAD` bit 7 is put into `ADR0 ... ADR7`. If `CC` is 5, then `PORTAD` bit 5 is being converted and is about to be loaded into port `ADR5`. Each time `ADR`j is reloaded, `CCFj` is set, and `CCFj` is cleared when this port is read.

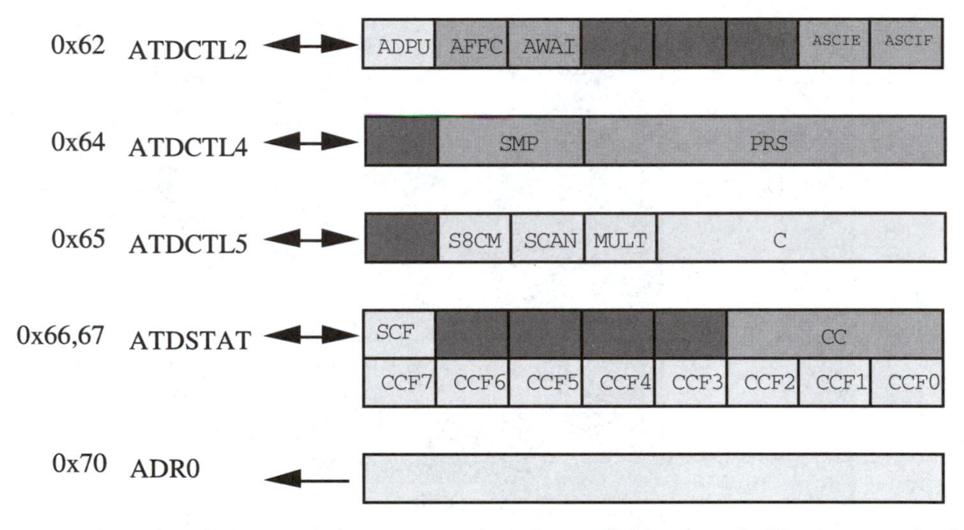

Figure 7.22. A-to-D Subsystem of the MC68HC812A4

Assuming VRH is 5 V and VRL is ground, the following procedure continuously converts the voltage (times 256/5) on `PORTAD` bit 0 into i; it obtains the average of eight samples. Conversion of eight separate pin voltages is left as an exercise.

```
enum { ADPU = 0x80, S8CM = 0x40, SCAN = 0x20, MULT = 0x10 } ;
void main() { int i;
    ATDCTL2 = ADPU; ATDCTL5 = S8CM + SCAN; i = 1; while(--i);
    do {
        do ; while(!(ATDSTAT & 0x8000)) ;
        i = (ADR0+ADR1+ADR2+ADR3+ADR4+ADR5+ADR6+ADR7) >> 3;
    } while(1);
}
```

7.5.4 Object-oriented Programming of Converters

Object-oriented programming is well-suited to analog I/O. In this section we consider a class for the 6812 A-to-D device. If control ports `S8CM`, `SCAN`, and `MULT` are T, then voltages on `PORTAD` bits 0 to 7 are continuously converted and put into ports at 0x70 to 0x7e. All that is needed to read these ports is the class `Port`. For instance, object `AtoD0` can be declared as in `Port<char> AtoD0(0x70);` to read the voltage on `PORTAD` bit 0. However, the `get` function member reads the voltage, which may be a voltage that has already been read by `get`, rather than a new voltage after conversion has been done. To synchronize the A-to-D device, to wait for conversion to take place, a derived class should incorporate gadfly or interrupt synchronization on a conversion complete flag. The class `AtoD12` shown below gadflies on the flags in `ATDSTAT`.

```
enum { ADPU = 0x80, S8CM = 0x40, SCAN = 0x20, MULT = 0x10 } ;
class AtoD : public Port<unsigned char> { unsigned char mask;

    public : AtoD (char id) : Port (0x71 + (id << 1)) {  int i;
        mask = 1 << id;
        if( ! (ATDCTL2 & ADPU)) { /* if A-to-D is off */
            ATDCTL2 = ADPU; /* turn on A-to-D */
            ATDCTL5 = S8CM + SCAN + MULT; /* initialize control */
            for(i=0; i<2000; i++);/* wait for A-to-D to turn on fully */
        }
    }

    virtual unsigned char get(void)
        {while(!(ATDSTAT & mask)); return Port<unsigned char>::get(); }

    virtual int option (int c = 0, int mask = 0) { // safe way to access regs
        if(c == 0) return Port::option(0, m); if(!(c & 0x20)) return 0;
        else if(c & 8){((char *)0x60)[c & 7] = d; return 0; }
        else return ((char *)0x60)[c & 7];
    }
};

void main() { AtoD device(0); char i; i = device.get(); }
```

The constructor's argument `id` is a device number; 0 indicates that `PORTAD` bit 0 is being measured. The constructor calls `Port`'s constructor with port address `0x70 + (id << 1)` so that for `id` = 0, `inherited::get()` reads `ADR0`. The constructor also makes data member `mask` have a T bit in the `id` bit position. The constructor then checks if the A-to-D converter is on because an earlier invocation of this constructor turned it on; if it is off the constructor turns it on and waits for its power to stabilize. The `get` function member gadflies on the `id` bit of `ATDSTAT`, a conversion-complete flag such as `CCF0`, which becomes set when `PORTAD` bit `id` is converted. Reading `ATDSTAT` followed by reading `ADR0` clears `CCF0`, `ATDSTAT` bit 0; reading `ATDSTAT` followed by reading `ADR1` clears `CCF1`, `ATDSTAT` bit 1; etc. Thus we can be sure that `get` will not return the same reading twice, but will wait for a conversion to give a new value each time it is called.

7.5.5 Applications in Control Systems

The DAS in §7.5.1 or the A-to-D converters in §7.5.2 and the D-to-A MC144110 converter in §7.4.1 can be used in control systems. The three main applications are the collection and generation of analog data, and feedback control.

The microcomputer is admirably suited for collecting analog data. The DAS and subroutine recently discussed can collect a sample of up to eight analog inputs. The collected data could be stored in a table, transmitted across a data link, or operated on. The programs for these operations should be simple enough, thus they are not spelled out here. However, it should be stressed that data collection using microcomputers has a unique advantage over simpler techniques: its software can execute functions on the incoming data. In particular, functions can, as we discuss, correct errors in the measurement apparatus.

Suppose the incoming data actually has value `x`, but the measurement apparatus reports the value as `y = F(x)`. The function `F` can be empirically obtained by inputting known values of `x`, then reading the values of `y`. Suppose `F` is an invertible function and the inverse function is `G`; then `x = G(y)`. Software can read `y` from the measurement apparatus, then compute `G(y)` to get the accurate value of `x`.

A number of techniques can be used to evaluate some arbitrary function `G(y)`, such as might be obtained for correcting errors. The well-known Taylor series expansion is sometimes useful; but to evaluate such a polynomial may take a long time and accumulate a lot of rounding error. A better technique is to evaluate `G(y)` as a continued fraction `G(y) = A / B + G'(y)`, where `G'(y)` is either `y` or a continued fraction. The most suitable for microcomputers, however, is the *spline* technique. Just as a complex curve is often drafted by drawing sections of it with a French curve, the complex function `G(y)` is approximated by sections of simpler functions (called splines) like parabolas. (See Figure 7.23.) Given a value `y`, we determine which section of `G(y)` it is in, to choose which spline to evaluate. We do this by comparing `y` against the values `yi` that separate the splines. A fast way is to test `y` against the middle `yi`; then, if `y < yi`, check `y` against the `yi` one-quarter of the way across the scale. Otherwise, check against the `yi` three-quarters of the way; and so on in the same manner as the successive approximation technique for A-to-D conversion. Once the section is determined, evaluate the function by evaluating the spline. If the spline is a parabola, then `X = A y**2 + B y + C` for some constants `A`, `B`, and `C`. The values yi for the boundaries and the constants `A`, `B`, and `C` can be stored in a table. Software for searching this table to select the correct spline and for evaluating the spline is quite simple and fast on a microcomputer.

Analog signals can be converted to digital values, then filtered using digital techniques, rather than filtered using OP AMPs, as discussed earlier in this chapter. The following is a discussion of digital filtering as a feedback control technique.

In a manner similar to that just discussed, if analog values are to be output from a microcomputer, errors in the output apparatus may be corrected in software. If the true output value is `y` but `x` is sent to the output, the output is actually `y = F(x)`; then if `F` is invertible and `G` is the inverse of `x [ x = G(y) ]`, the microcomputer can evaluate `G(y)` and send this value to the output system. The program that evaluates `G(y)` compensates ahead of time for the error to be made in the output apparatus.

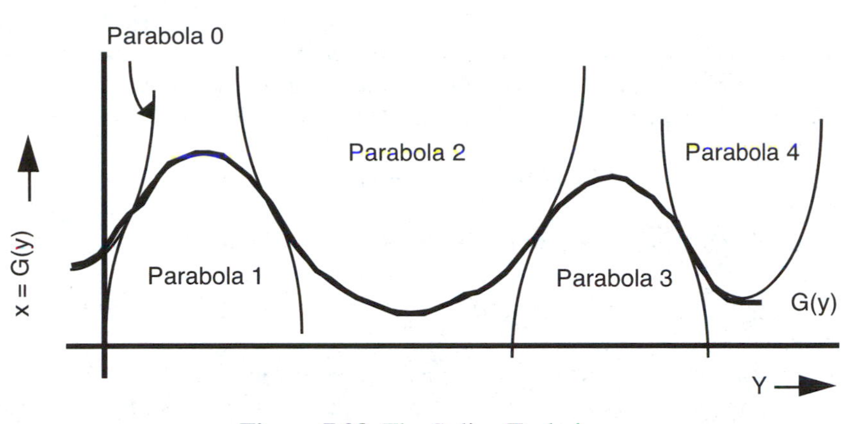

Figure 7.23. The Spline Technique

A test system might be designed using the preceding techniques to output some analog voltages to the object being tested and then measure the voltages it returns. While these systems are important, the feedback control system is even more important and interesting. Figure 7.24 shows the classic model of the feedback control system. The entire system has a stimulus x (or a set of stimuli considered as a vector) as input, and an output z (or a set of outputs, a vector z). The system that outputs z is called the *plant*. The plant usually has some deficiencies. To correct these, a *feedback system* is implemented (as diagramed in Figure 7.24), which may be built around a microcomputer and DAS. The output of this system, an error signal, is added to the stimulus signal x, and the sum of these signals is applied to the plant. Feedback-control systems like this have been successfully used to correct for deficiencies in the plant, thus providing stable control over the output z.

Three techniques have been widely used for feedback control systems; the proportional integral differential, the linear filter, and the multi-input-output controllers.

The simplest and most popular controller is called the *proportional integral differential* (*PID*) *controller*. Its form is easy to implement on a microcomputer. The output of the feedback system U is a weighted sum of the current input to the feedback E, the integrated value of E, and the differential value of E:

$$U = A\ E + B \int E(t)\ dt + C\ d(E(t))/dt$$

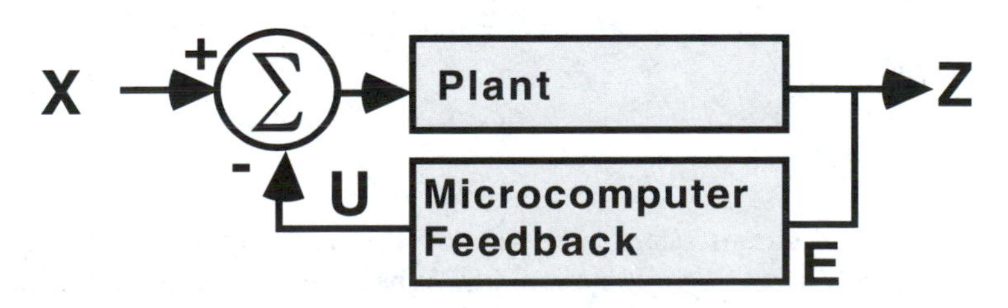

Figure 7.24. Feedback Control

Integration is nicely approximated in a microcomputer by adding each input value to a number in memory each time the inputs are sampled. If the feedback control system is working correctly, the inputs will be positive and negative, so this running sum will tend to 0. The differential is simply approximated by subtracting the current value of the input from its last value.

A more general kind of controller can be implemented as a digital version of a filter. (As a filter, it can be used to correct errors in analog measurement and output systems, as we previously discussed.) A *digital filter* is defined by a *Z-transform.*

```
D(z) = U(z)  = A0 + A1 Z**-1 + A2 Z**-2 + ... + An Z**-n
       E(z)     1 + B1 Z**-1 + B2 Z**-2 + ... + Bn Z**-n
```

This expression is evaluated in a microcomputer as follows. Call the input at time `k` `Ek` and the output at time `k` `Uk`. Then the output `Uk` at any given time is just the weighted sum of the inputs and outputs of the `n` prior times:

```
Uk = A0 Ek + A1 Ek-1 + ... + An Ek-n - B1 Uk-1 - ... - Bn Uk-n
```

The program should keep the vectors `A`, `B`, `E`, and `U`. Each time it updates the most recent output value `Uk`, it can shift all values of `E` and `U` back one place to get the output `Uk` at the next time `k`.

A particularly suitable technique is the *multi-input/multi-output controller,* which has a mathematical definition as follows. Let **E** be an (*i*-variable) input and **U** be a (*j*-variable) output, and **S** be an (*n*-variable) state vector, stored in a table in memory. **A, B, C,** and **D** are matrixes having suitable dimensions. Then the controller is defined by matrix multiplication equations:

$$\mathbf{S} = \mathbf{A}\,\mathbf{S} + \mathbf{B}\,\mathbf{E}$$

$$\mathbf{U} = \mathbf{C}\,\mathbf{S} + \mathbf{D}\,\mathbf{E}$$

These equations can be implemented by subroutines that perform matrix multiplication and vector addition, together with the subroutine that exercises the DAS to get the input vector **X** and to output the values of **Z**.

These techniques show how simply a microcomputer with a DAS can, with programs to correct for nonlinear errors or to digitally filter the data or with one of several feedback controllers, implement multiple-input analog measurement systems or multiple sources of analog output voltages. All we have to do is determine the coefficients for the aforementioned formulas. That is not a trivial problem, but it is treated in many excellent texts on control theory. Our only intent in this chapter was to show that once a desired control system has been defined, it can be implemented easily in a microcomputer.

7.6 Conclusions

In this chapter, we studied transducers and analog devices, A-to-D and D-to-A converters, and data acquisition systems. A good designer must be aware of the analog devices and circuits and must be aware of the advantages of the different ways to implement some functions in analog or digital hardware, or in software.

You should now be ready to use analog circuits in microcomputers. We now turn our attention to frequency analog signals, and then to communications systems and storage and display systems that use frequency analog signals.

Do You Know These Terms?

See page 36 for instructions.

sinusoidal alternating current
amplitude
phase
frequency
period
periodic waveform
bandwidth
analog-to-digital converter
digital-to-analog converter
comparator
analog switch
voltage-to-frequency converter
frequency-to-voltage converter
transducer
operational amplifier
solenoid
motor
direct-current motor
stepping motor
universal motor
shaded-pole motor
induction motor
hysteresis synchronous motor
potentiometer
slide potentiometer
strain gauge
shaft encoder
linear variable displacement transformer
control transformer
direct-current tachometer
accelerometer
light-emitting diode (LED)
liquid crystal display (LCD)
photodetector
photomultiplier
photodiode
phototransistor
photoresistor
opto-isolator
oven
thermocouple
thermistor
load cell
bipolar transistor
transistor
beta
power transistor
darlington transistor
field-effect transistor
vertical field-effect transistor
silicon control rectifier (SCR)
on-off control
bang-bang control
integral cycle control
proportional cycle control
linear mode
saturated mode
inverting amplifier
feedback
amplification
summing amplifier
virtual ground
voltage follower
noninverting amplifier
differential amplifier
hysteresis
Schmitt trigger
analog multiplexer
timer
dual supply
single supply
common-mode voltage
offset adjustment
settling time
bypass capacitor
ground plane
ground loop
heat sink
low-pass filter
high-pass filter
bandpass filter
notch filter
cascade
parallel
*n*th order filter
butterworth filter
bessel filter
chebyshev filter
biased
automatic gain control (AGC)
sample
sampling rate
Nyquist rate
alias
multiplying D-to-A converter
R-2R ladder
parallel A-to-D converter
pipeline converter
delta converter
servo converter
successive approximation converter
ramp A-to-D converter
dual-slope converter
data acquisition system
spline
plant
feedback system
proportional integral differential (PID) controller
digital filter
Z-transform
multi-input/multi-output controller

Problems

Problems 1, 5, 13, and 22 are paragraph correction problems. See page 38 for guidelines. Guidelines for software problems are given on page 86, and for hardware problems, on page 115.

1.* A transducer changes physical properties such as distance to or from voltages or frequencies. The most accurate position measurements can be made using a potentiometer. Acceleration can be measured by measuring distance at several specific times and integrating the results in software. A light-emitting diode is clear when no voltage appears across it and is opaque when a voltage appears across it. The photoresistor generates a current proportional to the light that shines on it, but it is nonlinear. Very high temperatures are best measured with a diode, whose output voltage is 2.2 V/^{O}K. Pressure is generally measured by converting pressure to heat and then measuring the temperature.

2. A stepping motor has three windings and the power signals for them are called A, B, and C. These are singulary (only one is asserted at a time) and the stepping motor is made to go clockwise by making `PORTA` bits 2 to 0 run in the sequence TFF FTF FFT TFF ... and counterclockwise by making them run in the sequence FFT FTF TFF FFT .... When the next pattern is output in either sequence, the stepping motor rotates 7 $^{1}/_{2}{}^{O}$ in the direction indicated by the sequence.

a. Write a C procedure `main()` to initialize the ports, and a procedure `movecw()` that will cause the stepping motor to rotate 7 $^{1}/_{2}{}^{O}$ clockwise, and `moveccw()` that will cause the stepping motor to rotate 7 $^{1}/_{2}{}^{O}$ counterclockwise.

b. Write a C procedure `rotate()` to rotate the motor at one rotation per second clockwise. Use the procedure in part a to move 7 $^{1}/_{2}{}^{O}$ clockwise, and use real-time synchronization.

c. Assume the most significant bit of `PORTA` inputs true when the motor has moved an object to a terminal (most counterclockwise) position; otherwise the bit is false. Write C procedures `main()` which moves the motor to its terminal position, and `move(int p)` to position the object `p` positions clockwise from the terminal position. Position `p` can be more than 48 to give the motor a full revolution. The motor is given each pattern, such as TFF, FTF, or FFT, etc. for 10 ms, using real-time synchronization. (This is basically the mechanism used to position the head on a floppy disk.)

3. A gear has a tooth every 10° of rotation, the teeth are 5° wide, and two wiper contacts are 2^{O} apart. The contacts output a low signal when they touch a tooth.

a. Write a (Mealy) sequential machine description of the state transitions, where the internal states are the signal pair on the contacts, LL, LH, HL, and HH, and the input states are CW for moving clockwise, CCW for moving counterclockwise; and NULL for no movement. The leftmost contact value of the pair corresponds to the more counterclockwise of the two wiper contacts as they appear close together.

b. Declare global variables, and write an initialization in C procedure `main()` and interrupt handler *cckj();* that will keep track of the position of the gear. The more counterclockwise of the two contacts is connected to `PORTJ` bit 0 and is to cause a key wakeup interrupt each time the signal falls, and the other contact can be read as `PORTJ` bit 1. The position of the gear, in 36ths of a revolution, will be kept as an `int position,` positive numbers indicating clockwise rotation. (When the machine is turned on, we will just assume that the position of the gear is defined as 0.)

4. A home thermostat will control a furnace and an air conditioner, based on the home temperature. The temperature in ^{o}C can be read from an A/D input at 0x70. The furnace and air conditioner are controlled by `PORTA` bits 0 and 1, such that writing 0 in bit 0 turns off the air conditioner, writing 1 there turns it on, writing 0 in bit 1 turns off the furnace, and writing 1 there turns it on. Suppose the desired temperature is stored at global variable `setTemp`. Write a C procedure `main()` to control the furnace and air conditioner to adjust the home temperature to this value. If the temperature is 3° cooler than it, turn on the furnace (fully), but turn it off when the temperature is 1° lower than it (because residual heat will cause the temperature to rise after the furnace is off); if the temperature is 3° higher, turn on the air conditioner (fully), but if the air conditioner was on within the last two minutes, do not turn it on (because the back pressure in the compressor has to dissipate or the motor will stall when it is turned on). Turn off the air conditioner when the desired temperature is reached. Use real-time synchronization, using an empirically defined constant *N* to effect the delay.

5.* Bipolar transistors have very high input impedances and are commonly used to measure bipolar (AC) voltages. The VFET or darlington transistor is a good output amplifier for a microcomputer for controlling mechanical motion. SCRs can be used to control AC. A single SCR alone can control both positive and negative cycles of an AC power signal. Proportional cycle control is most attractive for microcomputer systems because it provides the most precise control over the amount of power applied to the load. Proportional cycle control is commonly used to control the heat of an electric range. An operational amplifier has two inputs, and it outputs a voltage that is a large number times the difference between the voltages on the inputs, as long as the OP AMP is in the saturated mode. To sample the incoming voltage, an integrator uses a capacitor on the input to an OP AMP. A comparator is an OP AMP used in the saturated mode. A modern OP AMP such as the CA3140 has very high input impedance, which is useful in microcomputer I/O systems because it allows the use of small capacitors and permits the sensing of minute currents from transducers such as pH measurement devices. The analog switch allows one to switch a digital signal after it has been compared to an analog signal. The timer is a useful device that is most commonly used to convert a voltage into a frequency.

6. For each of the circuits in Figure 7.25, show the output voltage `Vo` as a function of `V1`, `V2`, ... (and the initial value of `Vo` for circuit c):

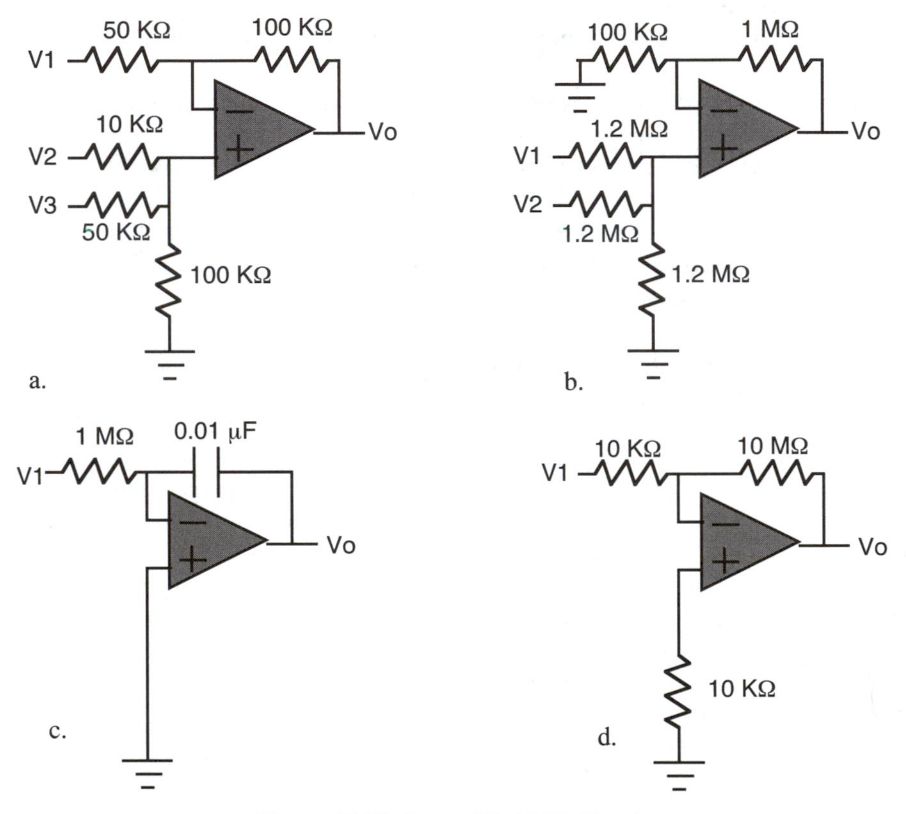

Figure 7.25. Some OP AMP Circuits

7. §7.2.3 discussed in detail the limits of input voltages and other parameters for +5-V single-supply use of the CA3140. Discuss all the limits and parameters that are different and give their values for ±15-V dual-supply use of this same OP AMP.

8. Show all connections in the logic diagram for a 555 timer and eight resistors that can be put into the timing-resistor position (R2 in Figure 7.9a) using a 4051 analog multiplexor controlled by `PORTA`. Do not show pin numbers.

9. Show all connections and values of all components in a diagram of an amplifier whose gain is controlled by an output port. The (4-bit) unsigned binary number *N* in the register sets the gain to -*N*. Use a CA3140 OP AMP, 4066 analog switch, and resistors in the range of 100 KΩ to 10 MΩ. Do not show pin numbers.

10. Show all connections and values of all components in a diagram for a circuit that uses two CA3140 OP AMPs to output the absolute value on `Vabs` and the (logical) sign on SGN of an input voltage `Vin`. Do not show pin numbers.

11. A 555 timer and a binary counter are to be used to generate a square wave as determined by 6812 `PORTA`. The low-order three bits `L` of `PORTA` generate a square wave from the 555 with period `P = 2** (L / 8)` by selecting different capacitors, and the next more significant 3 bits `N` select a tap from the binary counter so the output period is 2^N x `P`. Show a logic diagram using some 74HC161, 4051, and 555 chips. Do not show values of resistors or capacitors. Show pin numbers on the 555 only.

12. A 555 timer and a binary counter are to be used to generate a pulse as determined by the 6812's `PORTA`. The low-order four bits *L* of `PORTA` select different capacitors, and the next-more-significant 3 bits `N` select different resistors so that the the output period is 2^N x $2^{L/8}$. Show a logic diagram using some 4051 and 555 chips. Do not show values of resistors or capacitors. Show pin numbers on the 555 only.

13.* Filters are used to remove noise from the signal being measured or output. A high-pass filter has capacitors connected to ground and to the output to short out the low-frequency components and integrate them out of the signal. A light pen develops a signal that corresponds to the average brightness of the CRT screen, so a high-pass filter is commonly used to eliminate the noise, which is a higher-frequency "signal." A butterworth filter is often used in digital systems because it delays all frequency components about the same, so that square waves do not develop humps or grooves. Analog switches are useful for selecting inputs and for nonlinear signal conditioning. One of the ways a microcomputer can control the loss of a signal is to chop it at a fixed frequency so that it is chopped off for a proportion of each cycle that corresponds to the loss. Three nonlinear functions that are often performed in analog hardware are absolute value, logarithm, and sample and hold. The logarithm function is often used before an A-to-D converter because microcomputers have difficulty multiplying, and multiplication can be performed by adding the logarithm.

14. Show a diagram of a summing D-to-A converter (like that in Figure 7.14a) that outputs the value of a 4-bit 2's complement number, stored in `PORTA` bits 3 to 0, in `V` (for example, 1100 puts out -4 V). Use a voltage reference of 1 V, a CA3140 OP AMP, and 4066 analog switches; and use resistors in the range 100-KΩ to 10-MΩ. Show all pins and component values, and show also the bypass capacitors so that the circuit can be built from your diagram.

15. A D-to-A converter uses the Crystal Semiconductor CS 4330 and 74HC4040 (see §7.4.1), but also allows the 'A4 to do some other work while an interrupt handler feeds data to be output to the CS 4330. It uses all of `PORTA`, bits 1 and 0 of `PORTJ`, one 74HC589 shift register, and some other logic. A byte that is to be shifted is put in `PORTA` and another byte has been put in the 74HC589's input register. A 74HC138 decoder attached to the 74HC4040 causes these bytes to be shifted forward bytewise into the 74HC589's shift register, and it can cause a key wakeup interrupt through `PORTJ` bit 0. The 'A4 E clock causes the shift register's bits to be shifted out, msb first, at 8 MHz. Pulsing `PORTJ` bit 1 will also cause data in `PORTA` to be loaded into the 74HC589's input register (a diode and resistor can be used to provide this alternative way to load the 74HC589's input register).

a. Show the logic diagram of the entire converter device, except what is already shown in Figure 7.15b, showing pin numbers, resistor values, and other component names.

b. Show an interrupt handler `cckj`, written entirely in assembler language, that will output 3 consecutive bytes pointed to by global `char *ptr;` to the CS 4330, moving this pointer ahead 3 bytes. Write comments indicating when each byte is loaded into either the 74HC589's input register or its shift register.

c. Write a C procedure `main()` that plays a tune, and whose digital samples are stored in a global vector `char tune[6000];`, by setting up pointer `ptr` of part b, starting the converter device, and stopping it when the tune has been played.

16. Write a C procedure `main()` that uses the D-to-A converter of problem 15 to convert from A to D. For each case below assume the 18-bit converted left-channel value is put in global `char value[3];` with lsb in `value[2],` and an input on `PORTA` bit 2 is 1 if the external input voltage being converted is higher than the left D-to-A converter output.

a. Use delta conversion.

b. Use successive approximation conversion.

17. Write a C procedure `main()` that configures the MC145155 (§7.4.3) to divide the input frequency by 0x1234. Its E (pin 12) is connected to `PORTS` bit 7, its D (pin 11) is connected to `PORTS` bit 5, and its C pin(10) is connected to `PORTS` bit 6.

18. Write a C procedure `main()` to read all 11 inputs of the MC145040 into global vector `char data[12]`. The serial interface has Din connected to `PORTS` bit 4, SClk to `PORTS` bit 6, CS to `PORTS` bit 7, and Dout to `PORTS` bit 5. (See §7.5.2.)

19. Write a C procedure `main()` to input the voltage on pin PAD bit 2 for eight consecutive samples, putting the sampled values in `ADR0` to `ADR7`.

20. Problem 4 of this chapter assumed the temperature was available at 0x70 (assuming I/O is from 0 to 0x1FF), where it is applied as a voltage on `PORTAD` pin 4. Write an initialization ritual to implement this in the 'A4.

21. Write a C procedure `spline(int d)` to evaluate a spline as specified by an `n` row table `struct{ int yi, A, B, C} t[n]`. The table has a row for each parabola, a column for the lower limit `yi` on the interval where that parabola is to be used, and three columns for the constants `A, B,` and `C,` from left to right in the table, for the coefficients of the parabola (`G(y) = A y 2 + B y + C`). The parabolas are in successive rows of the table in order of increasing values of `yi,` so the `i`th parabola should be evaluated by your routine if the input `y` is between `yi` in that row and `yi` in the row above it. Assume the last parabola covers the highest value of `d` to be evaluated.

22. Write a digital filter `z()` equivalent to the Z-transform $(3 - 2\ Z^{-1}) / (1 + 4\ Z^{-1})$. Evaluate the filter function repetitively on the input, read at 0x200, so that the filtered output is fed out the output at location 0x300. Use 8-bit signed numbers throughout.

23.* The sampling converters have a Nyquist rate, which is the rate at which they sample the analog signal. A high-pass filter is commonly used to remove frequencies below this Nyquist rate to prevent alias signals from appearing. D-to-A converters include the ladder networks, commonly available on integrated circuits, and exponential superposition converters, which are exceptionally accurate yet very cheap. D-to-A converters, such as the successive-approximation converter, are able to sample the input signal quite rapidly; but parallel and pipeline converters are the fastest, using more hardware to achieve the greater speed. A frequency-to-voltage converter is based on a very accurate one-shot that is triggered at the rate of the input frequency, and whose output is filtered through a low-pass filter to recover the voltage. A tachometer is a good voltage-to-frequency converter, but is limited to low frequencies. A data acquisition system uses sample-and-hold circuits to sample the input signals and uses a D-to-A converter to develop the output voltages and a reference voltage for an A-to-D converter. The A-to-D converter can use delta, ramp, or successive-approximation programs to measure input voltages.

The Motorola M68HC12A4EVB board can implement all the experiments including those of Chapter 10. The wire-wrap area shows a shift-register that implements the device diagrammed in Figure 10.6.

8

Counters and Timers

The *counter/timer* is one of the most flexible parts of a single-chip microcomputer. It can generate a square wave. The square wave can be used to generate sine waves, or any periodic wave. Sine waves can be used in telephone systems (touch-tone), and signals to the user (bleeps). The counter/timer can be used to generate single-shot pulses. These can control motors, solenoids, or lights to give precisely timed pulses that are independent of the timing errors to which the real-time programmed microprocessor is susceptible, such as dynamic memory and DMA cycle steals, and interrupts. The counter/timer can itself provide interrupts to coordinate a program, to effect an instruction step, or to effect a real-time clock. To effect an instruction step, the timer is set up as the monitor is left so that it allows one instruction to be executed in the user program before the interrupt returns control to the monitor. The monitor is used to examine or modify memory or registers, then the monitor is left and the next instruction in the user program is executed, and so on. Or a real-time clock can be effected if the timer interrupts every millisecond. The device can be used to count the number of events (falling edges of a signal input to the device), and thus the number of events in a fixed interval of time (the frequency). It is also capable of measuring pulse width and period. Several things can be converted to the period of a signal: voltage can be converted using the voltage-to-frequency converter's integrated circuits, and resistance or capacitance can be converted to the period of a waveform using a linear timer integrated circuit like the ubiquitous 555. We also observe that a single signal can be easily isolated using optical isolators, so the voltage of the system being measured can be kept from the microcomputer and the user. The 'A4 counter/timer was designed for these purposes.

The counter/timer is the principal component, then, in interfacing to frequency analog signals. These signals, like FM radio signals, are easier to handle than amplitude analog signals and are comparatively free of noise. We observe that, at the time of writing, amplitude analog signals are pervasively used in interface circuits; but we believe that frequency (or phase) analog signals will become equally important.

The primary objective of this chapter is to explain the principles of using the counter/timer device. To make these principles concrete, the 'A4 counter/timer system is introduced. A further objective of this chapter is to emphasize a fundamental principle of top-down design. A counter/timer in a microcomputer is so fascinating that the designer may decide to use it before examining alternative hardware and software techniques. This is bottom-up design: I've got this marvelous counter in my microcomputer, now where can I use it? As we pointed out in §4.3.5, this is rather like the once-popular TV character, Carnac the Magnificent, who answers a question sealed in an envelope before he knows the question. Bottom-up design is especially evident whenever a new and powerful integrated circuit, like the 'A4, appears on the market. This design approach generally leads to bad designs. So we emphasize the need to examine alternatives, and we discuss some of the alternatives to using this counter/timer subsystem. This chapter should acquaint you with the hardware and software of the counter/timer in the 'A4 and with alternative techniques using a simple parallel I/O port and more hardware, or a parallel I/O port and more software. Upon finishing the chapter, you should be able to connect a counter/timer in a microcomputer like the 'A4 and write software to generate square waves or pulses, or to measure the frequency or period of a periodic wave or the pulse width of a pulse. With these techniques, you should be able to interface to I/O systems that generate or use periodic signals or pulses, or to interface through voltage-to-frequency or frequency-to-voltage converters to analog I/O systems.

8.1 The MC68HC812A4 Counter/Timer Subsystem

We introduce the block diagram of the 'A4 basic counter/timer subsystem in this section for further reference in this chapter. (See Figure 8.1.) The subsystem has the main counter and the control ports for enabling the counter/timer components. We also discuss `PORTT` parallel port bits and their controls.

Counter/timer control ports are shown in Figure 8.1. The counter `TCNT` is enabled when `TEN,` bit 7 of `TSCR,` is T (1) and can be read at any time. It can be incremented at a rate determined by §6.4 (Figure 6.10), which can be the E clock rate, or at that frequency divided by 2^{PR}, where `PR` are the 3 least significant bits of `TMSK2`.

`TOF,` which is bit 7 of `TFLG2,` is set when the 16-bit counter `TCNT` has an overflow. This bit can be tested in a gadfly loop, or if `TOI`, bit 7 of `TMSK2,` is also set, it causes an interrupt vectored through 0xFFDE, DF. The gadfly loop exit routine or the interrupt handler can increment a software counter to extend the number of bits beyond the 16 bits of the hardware counter. A `TFLG2` bit *must* be cleared before it can be sensed again; it is cleared by writing a T (1) into it (it is a clear port).

For each bit position, `PORTT` can be a parallel I/O port, whose direction is specified by the corresponding bit in `DDRT;` however, if a bit is assigned to be an output compare bit (§8.2.1), the corresponding direction port bit is ignored. The `PORTT` bits can be given pull-ups if `TPU,` which is `TMSK2` bit 5, is T (1), and their output power can be reduced if `TDRB,` which is `TMSK2` bit 4, is T (1).

Under control of each bit of `TIOS,` each `PORTT` bit may be assigned to be an output compare, discussed in §8.2.1, or an input capture, discussed in §8.3.1. In addition, bit 7 has extra capabilities including those of a vector output compare, discussed in §8.2.7, and a pulse accumulator, discussed in §8.3.4.

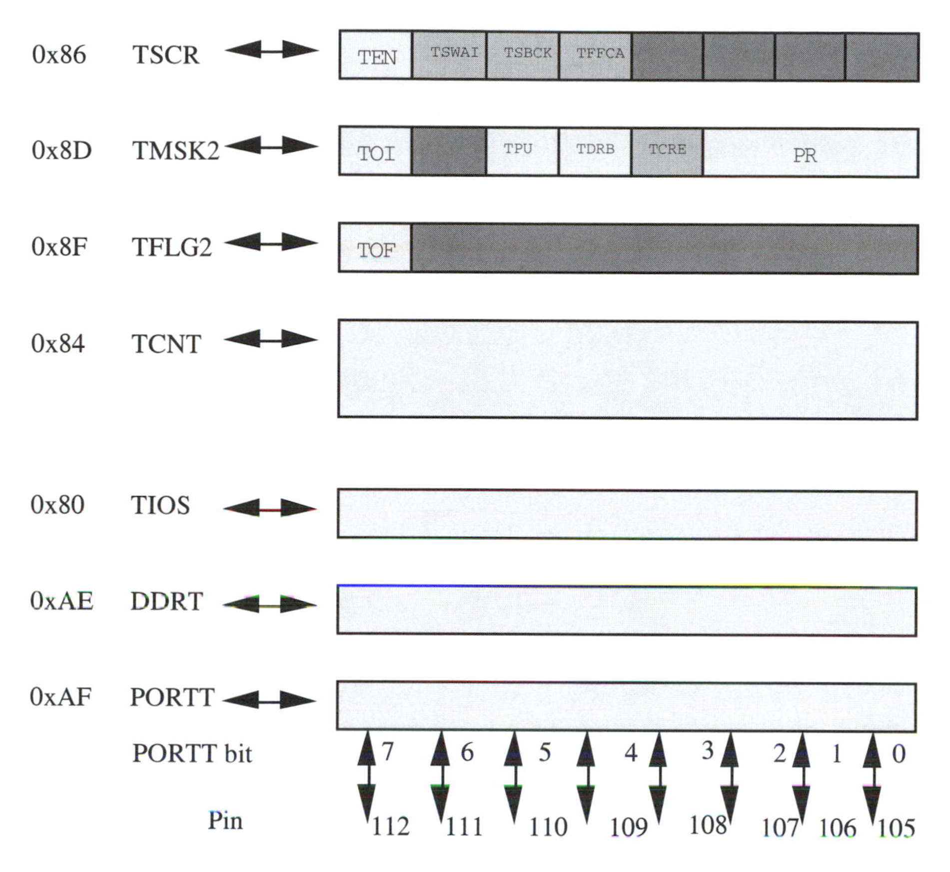

Figure 8.1. The Counter/Timer Subsystem

8.2 Signal Generation

We want to cover the generation of square waves and pulses for external hardware first because you can implement these examples as experiments and see results on the output pins. Later, we look at frequency-measurement techniques, which can be studied using a microcomputer to generate the signals using techniques introduced in this section. We can also generate interrupts for the microcomputer that can be used to time operations, which include the timing of output signals. This section covers the generation of signals with the 'A4 counter/timer and 'B32 pulse-width modulator devices. We describe the hardware used in generating either square waves or pulses. The generation of square waves and subsequent generation of arbitrary repetitive waveforms in the 'A4 will be considered in the next subsection. The next subsection covers 'B32 techniques for pulse-width modulation. We then illustrate the generation of touch-tone signals. The next subsection covers the techniques for pulse generation. Another subsection shows how to generate interrupts, which are used to implement real-time clocks. Finally, we discuss special pulse-generation capabilities of bit 7's device and object-oriented programming.

8.2.1 Output Compare Logic

Signal generation uses eight identical output compare devices, one of which (bit 7) is further modified as described in §8.2.7. The term *output compare* comes from the notion that a fixed number in a port is compared to the running counter; when a compare is sensed, an output operation is done. Figure 8.2a shows output compare ports. Figure 8.2b shows the hardware associated with bit 0, which we use in most of our examples.

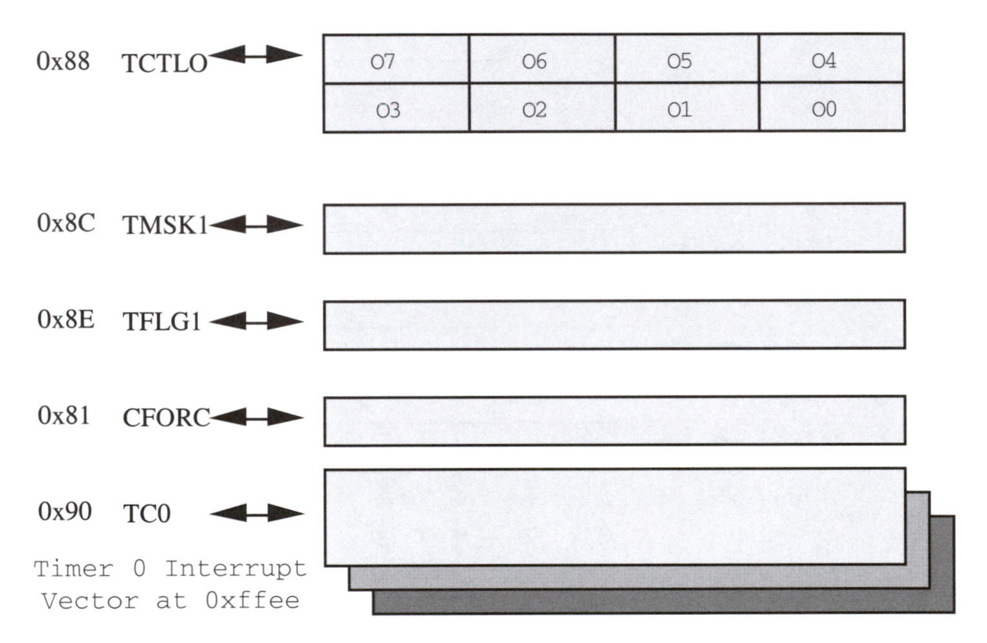

a. Ports

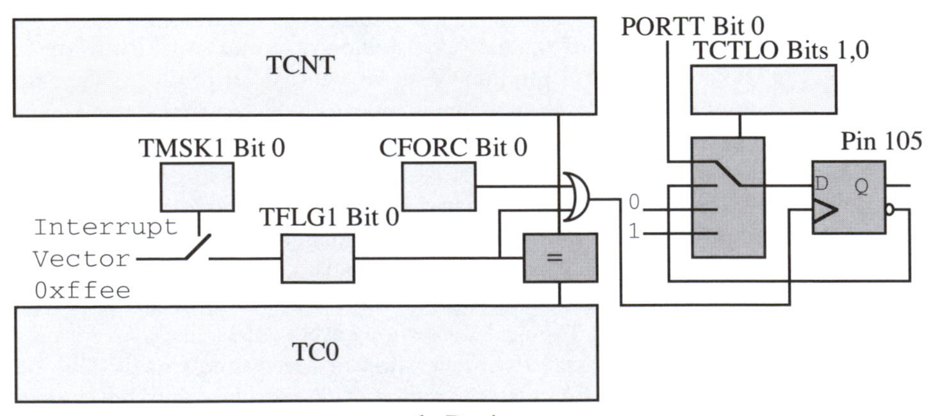

b. Device

Figure 8.2. Output Compare Logic

A programmer uses the output compare 0 device by loading the `TC0` port with a "time" at which an output operation is to occur. When `TCNT`'s "time" equals the "time" in the `TC0`, the output operation will take place. The operation is specified by 2 least significant bits in the 16-bit `TCTLO` port. If FF (00), no change takes place; if FT (01), the output is toggled; if TF (10), the output is cleared; if TT (11), the output is set to 1. The comparison match also sets `TFLG1` bit 0, which can be tested by a gadfly loop, and if `TMSK1` bit 0 is set, an interrupt is vectored through 0xFFEE, EF. The output operation also takes place if a 1 (T) is written into `CFORC` bit 0, as if `TCNT` matched `TC0`, but `TFLG1` is not changed and an interrupt is not generated.

8.2.2 The Counter/Timer Square-Wave Generator

A very economical way to generate a square wave is by means of a software loop that outputs alternate ones and zeros to an output port. The only hardware needed is a single 1-bit output port, say, `PORTT` bit 0. The following program shows how simple this operation is:

```
void main() { int i;
    DDRT = 1; /* set direction to output */
    do{
        PORTT ^= 1; /* flip output data */
        i = N; while(i--) ; /* delay loop */
    }while(1) ;
}
```

The outer loop complements `PORTT` bit 0 each time it is executed. A delay loop is executed a number of times empirically determined to obtain the desired frequency.

This simple approach is often indicated because of its low cost. However, it has some basic limitations. The minimum period (maximum frequency) is limited to the microcomputer clock period divided by two times the time to execute the outer loop. This time depends on the compiler. However, the program can be written in assembly language to get more precise timing. You can control the period by supplying the appropriate value of `N`, which is determined empirically. Other values can be selected by putting instructions with the appropriate execution time inside the loop to stretch it out. However, a different routine is needed for each desired value of the period. An alternative routine to handle arbitrary period values would be rather hard to implement. Moreover, if the microcomputer is interrupted while in this loop, or direct memory access or dynamic memory refresh occurs, the timing will be upset. Finally, the microcomputer is unable to do any other useful work while executing the routine to output a square wave. Nevertheless, this approach is recommended wherever a square wave with fixed or a small number of period times is needed, and no interrupts, direct memory access, dynamic memory refresh, or running of other programs are done while generating the square wave.

The 'A4 counter/timer system can generate a square wave. The square wave period `P` (divided by 2 because there are 2 half-cycles in a period, then multiplied by 8 because the clock is 8 MHz) is the number `N` in the program. Gadfly synchronization is used. Its minimum period is longer than for the single parallel output port example discussed before it. We do not recommend it, but it leads to the interrupt approach discussed next.

```
void main() {
    TSCR = 0x80; TIOS = 1; /* bit 0 is out. cmp */ TCTLO = 1; /* tggle */
    TCO = N + TCNT; /* half-period */
    do{
        while( ! (TFLG1 & 1 )) ; /* wait until flag is set */
        TFLG1 = 1; /* clear flag */ TCO += N; /* set for next half-period */
    }while(1) ;
}
```

The interrupt-based routine is shown below. The value N is the same as it was for gadfly synchronization. While the interrupt minimum period is longer than the previous approaches, the microcomputer is free to do other work when it uses this approach.

```
interrupt 8 void handler(){ TFLG1 = 1; /* clear flag */ TCO += N; }

void main() {  asm sei
    TSCR=0x80; TIOS=1; TCTLO=1; TCO = N + TCNT; TMSK1= 1;  asm cli
    do ; while(1);
}
```

The analysis of these methods shows that the 'A4 counter/timer system using interrupts may be best for generating square waves with periods above 6.75 μs ($N = 27$), since the processor is free to do other things while waiting for an interrupt. A simple parallel output port may be used for assembly-language-coded generation of square waves (periods ≥ 2 μs) if the processor is doing nothing else, but digital hardware oscillators and counters may be needed for square waves with shorter periods.

8.2.3 The MC68HC912B32 PWM Signal Generator

The 'B32 has a *pulse width modulation* (PWM) device, which is well-suited to generate repetitive digital signals, such as square waves, whose periods can be from 250 ns to over a minute. Unlike the output compare device, once PWM ports are loaded, the wave form is produced without the need for software to reload registers or exercise control. The 'B32's PWM Ports can be declared as follows:

```
volatile char PWCLK @0x40, PWPOL @0x41, PWEN @0x42, PWPRES@0x43,
    PWSCAL0@0x44, PWSCNT0@0x45, PWSCAL1@0x46, PWSCNT1@0x47,
    PWCNT0@0x48, PWCNT1@0x49, PWCNT2@0x4a, PWCNT3@0x4b,
    PWPER0@0x4c, PWPER1@0x4d,PWPER2@0x4e, PWPER3@0x4f,
    PWDTY0@0x50, PWDTY1@0x51, PWDTY2@0x52, PWDTY3@0x53,
    PWCTL @0x54, PWTST @0x55, PORTP @0x56, DDRP  @0x57;

volatile int PWPER01@0x4c, PWPER23@0x4e, PWDTY01@0x50, PWDTY23@0x52;
```

The PWM has four channels and many modes of operation; herein we first describe left-aligned channel 0 pulses that use 8-bit counters. Then we describe left-aligned channel 0 and 1 pulses that use 16-bit counters. The device's 8-bit ports are shown in Figure 8.3a. Pairs of these ports, used for 16-bit counters, are shown in Figure 8.3b.

Left-aligned channel 0 pulses are low for `PWDTY0+1` clock periods and repeat every `PWPER0+1` clock periods. If `k` is 1, channel 0's PWM clock is the E clock divided by $2^i * 2 *$ `j`, otherwise the clock is the E clock divided by 2^i, where `i` is bits 5 to 3 of `PWCLK`, and `j` is `PWSCAL0`, and `k` is `PWPOL` bit 4.

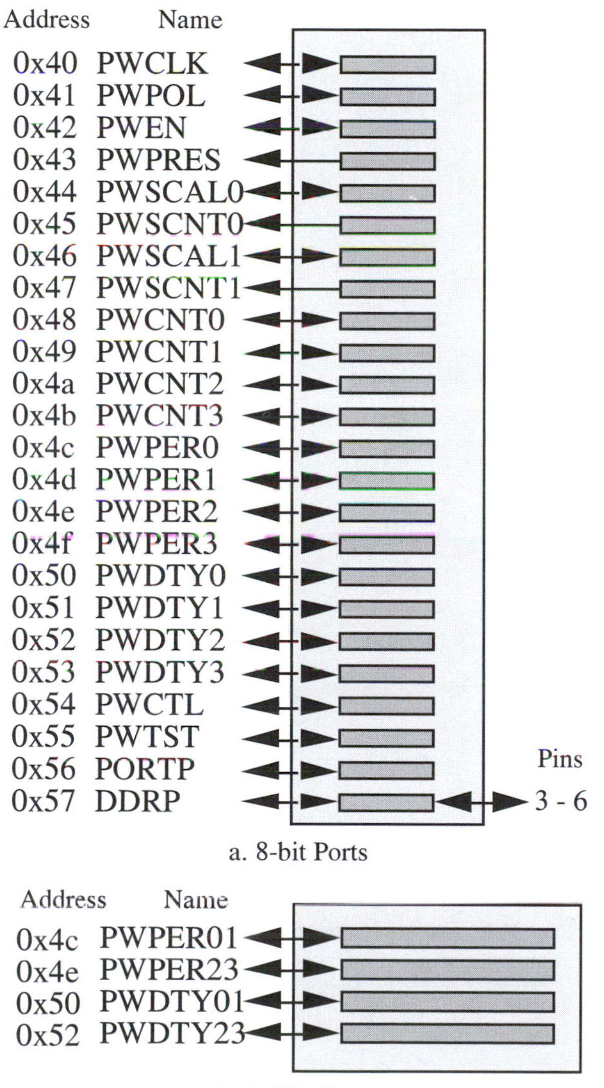

Figure 8.3. Pulse-Width Module Ports

The following *main* procedure shows the use of channel 0 to generate a rather high frequency square wave. It is low for three E clock cycles, and high for one clock cycle.

```
void main(){
    PWDTY0  = 2;  /* duty cycle: low for three E clock cycle */
     PWPER0  = 3; /* period: four E clock cycles */
     PWEN    = 1; /* enable channel 0 */
     do ; while(1);   /* pulses high 125 ns */
}
```

The prescaler, contolled by *PWCLK* bits 5 to 3, can divide down the E clock. The resulting pulse is high for 128 E clock cycles and low for 384 E clock cycles.

```
void main(){
     PWCLK   = 0x38;  /* use clock which is E / 128 = 16 µs */
     PWDTY0 = 2; PWPER0 = 3; PWEN = 1; do ; while(1); /* high 16 µs */
}
```

If *PWPOL* bit 4 is 1, the prescaler, contolled by *PWSCAL0* can also be used to divide down the E clock. The resulting pulse is high for 256 x 128 E clock cycles, and low for three times this time.

```
void main(){
     PWCLK   = 0x38; PWPOL   = 0x10; PWSCAL0 = 0x80; /* E / (256*128) */
     PWDTY0 = 2; PWPER0 = 3; PWEN = 1; do ; while(1); /* high 4.096 ms */
}
```

If *PWCLK* bit 5 is 1, *PWPER0* and *PWPER1* are combined into *PWPER01* and *PWDTY0* and *PWDTY1* are combined into *PWDTY01* to provide 16-bit period and pulse width control. The following program produces a pulse that is high for 65 ms, and low for three times this time.

```
void main(){
     PWCLK   = 0x78;  /* clock with 16 µs, join channel 0 and 1 */
     PWDTY01  = 0x2FFF; PWPER01  = 0x3FFF;  /* uses channel 1 clock */
     PWEN = 3; do ; while(1);  /* pulse high 65,536 ms on channel 0 */
}
```

Such 16-bit period and pulse-width ports can be scaled, as shown below. This program produces a pulse that is high for 16.77 s, and low for three times this time.

```
void main(){
     PWCLK   = 0x78; PWPOL   = 0x20;  /* ch1 clock controls prescaler */
     PWSCAL0 = 0x80;   PWDTY01 = 0x2FFF; PWPER01 = 0x3FFF;
     PWEN = 3; do ; while(1);  /* pulse high 16.77 s high on channel 0 */
}
```

Table 8.1. PWM Channel Ports

Channel	0	1	2	3
enable: *PWEN* bit	0	1	2	3
i = *PWCLK* bits	5 to 3	5 to 3	2 to 0	2 to 0
j = contents of:	*PWSCAL0*	*PWSCAL0*	*PWSCAL1*	*PWSCAL1*
k = *PWPOL* bit	4	5	6	7

If k is 1, PWM clock is $E/(2^i \times 2 \times j)$, otherwise it is $E/(2^i)$

Table 8.1 shows contol ports for other PWM channels. Note that channels 0 and 1 share prescaler ports and can be coupled to implement 16 bit `PWDTY01` and `PWPER01` ports. Similarly channels 2 and 3 share prescaler ports and can be coupled to implement 16-bit duty-cycle and period ports.

The period and pulse width can be independently changed, provided the period is greater than the pulse width. For instance, in controlling an automobile fuel injector, the period can be adjusted to the engine speed, and the duty cycle to the amount of fuel injected. In another example, for pulse-width modulated communication channels, the period can be fixed to the time to send a bit, and the duty cycle can be loaded with different values, sending short or long pulses, to communicate a 0 (F) or 1 (T) signal.

8.2.4 A Touch-tone Signal Generator

We now consider an application of this oscillator with respect to the production of touch-tone signals. A *touch-tone signal* is a pair of sine waves having frequencies that represent the digits used in dialing a telephone. A touch-tone generator can generate these signals so that the microcomputer can dial up on the telephone, and it can generate such tones to be sent via radio remote control or to be stored on cassette tape. In a top-down design, one must consider all the relevant alternatives. One possibility is integrated circuits that output touch-tone signals in response to keyboard contact closings, but these would require the microcomputer to act like the keyboard to such a chip. Another is analog switches in place of the keys that can be controlled by an output port. (A few other alternatives are also possible.) However, the number of chips needed in any alternative would be at least two, if not more. So we next consider generating a square wave with $2N$ times the frequency of the sine waves that make up the touch-tone signal, using an N-stage Johnson counter to generate the sine wave.

A touch-tone signal consists of two sine waves transmitted simultaneously over the phone. The following table shows the tones required for each digit that can be sent. Table 8.2a shows the mapping of digits to frequencies shown in Table 8.2b, and Table 8.2b shows the frequencies in Hz and the corresponding values to be put in 'B32's PWM ports, or to be added to the 'A4's output compare device to generate the desired frequencies, as they will be used later in the example. Thus, to send the digit 5, send two superimposed sine waves with frequencies 770 Hz and 1336 Hz.

Table 8.2. Touch-tone Codes

Digit	Coding
0	R4,C2
1	R1,C1
2	R1,C2
3	R1,C3
4	R2,C1
5	R2,C2
6	R2,C3
7	R3,C1
8	R3,C2
9	R3,C3

a. Codes for Digits

Code	Hertz	Counter
R1	697	717
R2	770	1649
R3	852	587
R4	941	531
C1	1209	414
C2	1336	374
C3	1477	339
C4	1633	306

b. Frequencies for Codes

A Johnson counter is almost ideal for generating a sine wave. It is actually just a shift register whose output is inverted, then shifted back into it. (See Figure 8.4.) A 4-bit Johnson counter would have the following sequence of values in the flip-flops:

```
L L L L
H L L L
H H L L
H H H L
H H H H
L H H H
L L H H
L L L H
```

As described in Don Lancaster's marvelous little book, *The CMOS Cookbook,* these counters can be used to generate sine waves simply by connecting resistors of value 22 KΩ to the first and third stages, and of 30 KΩ to the second stage, of the shift register. (See Figure 8.4.) Although the wave will look like a stair-step approximation to a sine, it is free of the lower harmonics and can be filtered if necessary. Moreover, using more-accurate values of resistors and a longer shift register, it is possible to eliminate as much of the low-order harmonics as desired before filtering. In this case, if we want a sine wave with frequency `F`, we clock the shift register with a "Johnson counter clock" whose period should be `1/(8 x F)`.

Now we consider ways to implement Johnson counter clocks whose frequencies are each one of four values. A real-time-programmed software solution would require two microcomputers since a whole microcomputer is dedicated to outputting just one square wave. Instead, a rather complex program just might be possible; or a table could be read out to supply the right sequences, but a different table is needed for each frequency, so each table would be quite long. The digital and analog hardware solutions require more than two chips, so we consider the 'B32's PWM device, and the 6812 counter/timer device, using interrupts. The Johnson counter clocks have periods of 5.5 to 13 KHz, well within the range of the 16-bit PWM and the counter/timer device using interrupts.

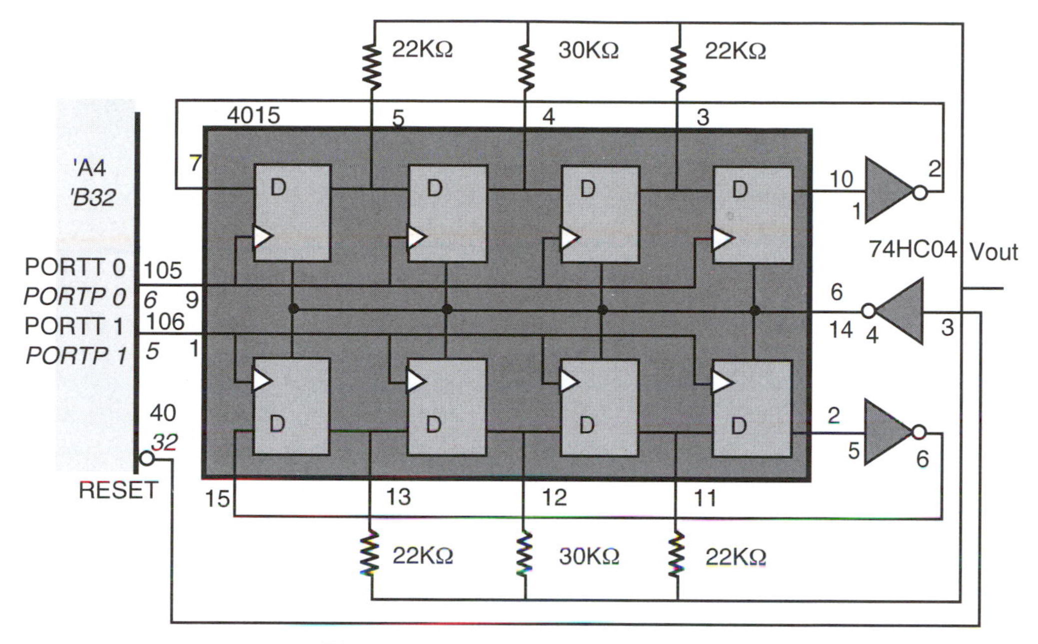

Figure 8.4. A Touch-tone Generator

In the 'B32, the PWM device is best suited to generate two Johnson counter clocks. In the 'A4, two Johnson counter clocks can be implemented using two of the counter/timer devices, which use interrupts to restart the output compare hardware.

In the 'B32, because the periods are longer than 256 E clock cycles, channels 0 and 1 are coupled to produce one Johnson counter clock, and channels 2 and 3 are coupled to produce the other clock. The program below emits the code for the number `d = 5`. A vector `tbl0` converts 5 into the indexes of the second table to generate R2 and C2. This vector implements Table 8.2a. The vector `tbl1` generates the half-period values to be put into `PWDTY01` and `PWDTY23`, whose doubled values are put in `PWPER01` and `PWPER01`. A high-level language like C++ easily translates values using a vector.

```
const unsigned char tbl0[10] =
     {0x35,4,5,6,0x14,0x15,0x16,0x24,0x25,0x26};
const int tbl1[8] = {717,649,587,531,414,374,339,306};

char d = 5;

void main() { char i = tbl0[d];
     PWCLK   = 0xc0; /* join channel 0 and 1, and join channel 2 and 3 */
     PWPER01 = (PWDTY01  = tbl1[i & 0xf]) << 1; /* generate Johnson ctr clk */
     PWPER23 = (PWDTY23  = tbl1[(i >> 4) & 0xf]) << 1; /* generate 2nd clk */
     PWEN = 0xf; /* enable all PWM channels */ do ; while(1);
}
```

The 'A4 counter/timer system generates a Johnson clock in the same way. The values, stored in global vector `tbl1`, are added to the compare registers each time there is an interrupt. `main` and two interrupt handlers, `handler1` and `handler2`, send a touch-tone code as shown below. Initialization is used to begin generating a pattern for a number in global `char d`, and the two handlers are used to respond to interrupts for the two square-wave generators. Timer devices 0 and 1 are used. Device 0 is set up exactly as in the previous example and device 1 is set up as is device 0.

```
const unsigned char tbl0[10] =
    {0x35,4,5,6,0x14,0x15,0x16,0x24,0x25,0x26};

const int tbl1[8] = {717,649,587,531,414,374,339,306};

int T0, T1; char d = 5;

void main() { char i;
    i=tbl0[d];T0=tbl1[i&0xf]<<2;T1=tbl1[(i>>4)&0xf] << 2;  asm sei
    TSCR = 0x80; TIOS = 3; TCTLO = 1; TCO = T0 + TCNT;
    TC1 = T1 + TCNT; TMSK1 = 3; /* enable int */ asm cli
    do ; while(1);
}

interrupt 8 void handler1(){ TFLG1 = 1;  TCO += N; }

interrupt 9 void handler2(){ TFLG1 = 2;  TC1 += N; }
```

Two devices are used to generate the two sine waves required by the touch-tone signal. The external connections are shown in Figure 8.4. On the left, the italic signal names and pin numbers correspond to the 'B32; the other names and pin numbers correspond to the 'A4. Note that the CLEAR control on the shift register is connected to the microcomputer RESET bus line to ensure that the shift register does not have some unusual pattern in it after power is applied. The output signal is simply the sum of the sine waves produced by two shift registers, obtained from the resistors' common points.

A slight variation of this technique can be used to generate any periodic waveform. The square wave can be used to increment a counter that supplies an address to a read-only memory. It can output words to a D-to-A converter and can store the desired pattern to be developed. Thus, generating a square wave can generate other periodic waves. Finally, as discussed in §7.4.2, a square wave can be integrated to get a ramp signal, and this can be shaped by nonlinear analog techniques. Also, as in music generation, a periodic signal can be shaped by attenuating it under control of an output port to apply attack and decay characteristics in a music synthesizer.

The 'B32 PWM and the 'A4 counter/timer are valuable tools. The PWM generates square waves easier than the 'A4 counter/timer does, as can be seen from the preceding examples. Either can generate other periodic waves using Johnson counters or read-only memories and D-to-A converters. However, the designer must not assume that this chip, or any counter/timer chip, is so much better than any other generator. He or she must consider all approaches and pick the best one for a specific application.

8.2.5 The Pulse Generator

Like the square-wave generator, a pulse generator has many uses. The device normally outputs a false value, but outputs a true value for a specified time after it actually is triggered. It can be triggered by software or by an external signal. And, as with the square-wave generator, there are software and hardware techniques to implement it. The software technique to supply a pulse triggered by software merely outputs a true, waits the required time, and then outputs a false value. To react to an external signal, the external signal can be sensed in a gadfly loop or can generate an interrupt so that the pulse is generated when the signal arrives. A program to generate a pulse follows.

```
void main() { int i;
    DDRT = 1; /* set direction to output */
    PORTT |= 1; /* set output data */
    for(i = 0; i < N; i++) ; /* delay loop */
    PORTT &= ~1; /* clear output data */
}
```

But like the software square-wave generator, the software pulse generator is susceptible to timing errors due to interrupts, direct memory access, and dynamic memory refresh cycles. A 555 timer can act like a pulse generator, triggered by a microcomputer or an external signal; and the length of the pulse, determined by the value of a resistor and a capacitor, can be controlled by selecting the resistor by means of an analog switch controlled from an output port. One-shot integrated circuits can be controlled in a like manner. Finally, the 'A4 counter/timer system can be used to generate pulses when the computer starts them (or they can be started by an external signal and then sensed by the computer, as discussed later), and the pulse length can be computer controlled, as we now discuss. First, gadfly synchronization is used to generate a pulse, then interrupts are used. In both programs, care is exercised lest the timer inadvertently cause an input compare, `CFORC` cause a rising edge, and the falling edge occur after device 0 times out.

```
void main() {
    TSCR = 0x80; TIOS = 1;    /* bit 0 is output compare */
    TC0 = TCNT;               /* prevent timer from interfering */
    TCTL0 = 2;                /* set device 0 to clear */
    CFORC = 1;                /* cause it to clear now */
     /* do the following when pulse is to start */
    TCTL0 = 1;                /* set device 0 to toggle */
    CFORC = 1;                /* cause it to set now */
    TCTL0 = 2;                /* set device 0 to clear */
    TC0 = N + TCNT;           /* pulse width */
    TFLG1 = 1;                /* clear flag */
    while( ! (TFLG1 & 1 )) ; /* wait until flag is set */
}
```

The interrupt-based routine is shown below. While this minimum pulse width is longer than that for previous approaches, the 6812 is free to do other work.

```
interrupt 8 void handler(){ TMSK1 = 0; /* disable interrupt */ }

void main() { int i;  asm sei
    TSCR = 0x80; TIOS = 1; TC0 = TCNT; TCTLO = 2; CFORC = 1;
    /* do the following when pulse is to start */
    TCTLO = 1; CFORC = 1; TCTLO = 2; TC0 = N + TCNT;
    TMSK1 = TFLG1 = 1; /* enable interrupt, clear flag */ asm cli
    for(i = 0; i < 0xffff; i++) ; asm sei
}
```

The 'A4 counter/timer system provides significant advantages. The pulse width is precisely timed to within a memory cycle time, which is usually 125 ns, and this time is not affected by processor interrupts, dynamic memory refresh requests, or other subtle problems that affect the timing of real-time programs. Moreover, the program has to put a compare value into the output compare register sometime after the last time the output changed and before the time the next output changes, rather than precisely at the time the output must change. This is quite useful in automobile engine control, where the pulse width controls the amount of gasoline injected into the engine, the spark timing, and other key factors in running the engine. A counter/timer system, with up to eight output compare devices, is sufficient to control an engine so the microcomputer can measure input signals and compute the values of the pulse widths for the timers. In fact, the 'A4 is especially suited for the vast automobile industry.

8.2.6 A Rotary Dialer

Pulses are used for many things, such as a telephone that uses a rotary dialer. A relay connected in series with the dialer contacts will be pulsed to dial the number. The telephone standards require the relay to be closed for at least 40 ms and then opened for at least 60 ms for each pulse, and the number of pulses corresponds to the number being dialed; 600 ms are needed between each number being dialed. We consider a software approach using a single-bit output (`PORTT` bit 0) to control the relay, using the counter/timer system as a pulse generator, and using additional digital or analog hardware. Each designer has individual preferences. In fact, we really wanted to use the counter/timer system. But unless the pulse generation and timing are done by the counter/timer system so that the microcomputer can do something else, the program is actually less efficient than a simpler approach using real-time synchronization. Therefore, we swallow our pride and implement the dialer using real-time programming. We will show an 'A4 counter/timer-based pulse generator in a later example, where it is coupled to an edge detector. The program below outputs a pulse signal for the digit in global variable `d`. `N` is the loop counter for a 1-ms delay.

```
void main() { int i;
  for(DDRT=1,i=0;i<d;i++) {PORTT|=1;del(40);PORTT &= ~1;del(60); }
  del(600);
}
void del(int t) { while(--t) { int i = N; while(--i) ; } }
```

`main` dials a number by outputting a T (1) value to `PORTT` bit 0 for 20 ms, then an F (0) for 20 ms, and so on for each pulse. `del` is then called for a 600-ms delay to provide the spacing between digits, as required by the telephone company.

This example gives us an opportunity to relate one of the truly great stories in electronics – the invention of the dial telephone. It seems that in the 1880s Almond B. Strowger, one of two undertakers in a very small town, couldn't get much business. His competitor's wife was the telephone operator for the town. When someone suffered a death in the family, they called up to get an undertaker. The wife naturally recommended her husband, diverting callers from our poor friend Almond. Necessity is the mother of invention. With a celluloid shirt collar and some pins, he contrived a mechanism that could be operated by the caller, using a stepping relay mechanism that would connect the caller to the desired telephone, so that calls for his business would not go through his competitor's wife. It worked so well that it became the standard mechanism for making calls all over the world. Even today, about a quarter of all telephones use this "step-by-step" or Strowger system.

8.2.7 Real-time Clock and Trace Mechanism

Using a configuration almost identical to that used for a pulse generator, a device can be used as an accurate delay, as a real-time clock, or as a trace-mode interrupt. In this section we will cover these applications, as well as an example using a real-time alarm clock that illustrates how precise timing can be achieved.

Gadfly synchronization can be used to wait a precise amount of time. There is no effect on the output port signal because device 0's `TCTLO` bits are FF (00). The program below will do the "next operation" after a time determined by `N` has elapsed.

```
void main() {
    TSCR = 0x80; TIOS = 1; TCTLO = 0; TFLG1 = 1; TC0 = N + TCNT;
    while( ! (TFLG1 & 1 )) ; /* gadfly until flag is set */ /* do "next operation" */
}
```

When generating pulses using interrupts, `TFLG1` is set when an output compare occurs. This can be used to set an interrupt if the corresponding `TMSK1` bit is also set. A global variable can be set in the handler, and `main`, gadflying on this variable, can do the "next operation" after the global variable changes.

```
char flag;

void main() {  asm sei
    TSCR = 0x80; TIOS = 1; TCTLO = 0;
    TC0 = N + TCNT; /* delay */ TMSK1=TFLG1=1; /* rmv int. flag */ asm cli
    do ; while(! flag) ; /* wait for interrupt */  /* do "next operation" */
}

interrupt 8 void handler(){flag=1;TMSK1=0;/* set global flag, disable device */}
```

This interrupt mechanism can be used for executing a tracing instruction. In §1.2.3 we explained that a monitor is a program used to help you debug programs. The monitor, in the handler, is left by an RTI instruction. To "trace" a user program, we can leave the monitor to execute just one instruction, then reenter the monitor to see what happened. Just before we leave the monitor, we add *TCNT* to a number *N* and write the result *TC0*, after which we execute RTI. *N* will be chosen to give the user program just enough time to execute just one instruction; then an interrupt will occur and the monitor, in the *handler*, will be reentered to display the changes wrought by the executed instruction.

```
interrupt 8 void handler(){ /* first display 6812 registers, memory etc. */
    TFLG1 = 1; /* clear device's flag port */
    TC0 = N + TCNT; /* delay just long enough to execute 1 instruction */
}

void main() {
    asm sei
    TSCR = 0x80; TIOS = TMSK1 = TFLG1 = 1; TCTLO = 0;
    TC0 = N + TCNT; asm cli /* now execute program being debugged */
}
```

The counter/timer real-time interrupt mechanism can be used, in place of real-time interrupts, to implement time-sharing as discussed in §5.3. Code in §5.3.1's *handler* would be put into the timer's *handler*. Instead of being restricted to the real-time interrupt's fixed and awkward periods, such as 1.024 ms, counter/timer real-time interrupts can select any period to within 125 ns, such as 10.000 ms.

The counter/timer chip is seen to be a valuable component for generation of pulses. But don't forget to consider the alternatives to this chip. Another could be better.

8.2.8 Output Compare 7

The special output compare 7 device can affect any or all the output pins at the same time. Bit 7 of *CFORC*, *TFLG1*, and *TMSK1* work the same as the other output compare devices, but when an output comparison match is detected, the bits of data *OC7D* are forced into the output flip-flops wherever mask *OC7M* is T (1). If a change results from another output compare module when a change results from this module, this module takes priority and the other change is ignored. (Note that this means device 7 cannot be used for output compare operations described in §8.2.1 and §8.2.2.) Also, *PORTT* bit 7 (pin 112) is an input/output bit also used by the pulse accumulator, discussed in §8.3.4.

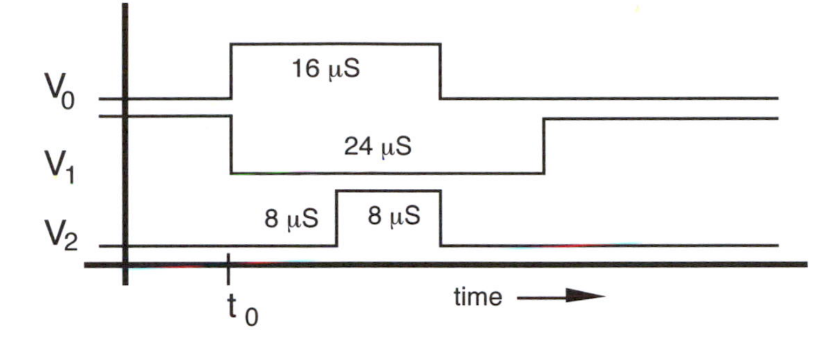

Figure 8.5. Timing of Some Pulses

We wish to generate three waveforms (see Figure 8.5) to control something, which is very nicely handled by the 'A4 counter/timer system. This technique is limited to 8 ms between transitions; longer times can be obtained using the timer prescaler. The program achieves this pulse sequence using gadfly synchronization. Output `Vi` is to be generated by output compare device `i`. Note that devices 0, 1, 2, and 7 are configured for output compare. Figure 8.6 shows output compare 7 ports. The devices' outputs are initialized by writing 0x80 in `CFORC`. We first set up output compare 7's data register, `OC7D`, to give the signals levels at time t_0, which occurs when `CFORC` is written into. The numbers in the compare registers are set up to output the signals at the time the changes are required. Constant `N` is 64000, set for a delay of 8 ms.

```
void main() {
   TSCR = 0x80; TIOS = 0x87; TCTLO = 0; TFLG1 = 0x80; OC7M = 7;
   OC7D = 2; CFORC = 0x80; /* set up initial conditions on ports 0, 1, 2 */
  /* execute the following when the pulse sequence is to be output */
  OC7D = 1; TC7 = N + TCNT; CFORC = 0x80; /* time t0 */
  TFLG1 = 0x80; while(!(TFLG1 & 0x80)) ; OC7D = 5; TC7 += N;
   TFLG1 = 0x80; while(!(TFLG1 & 0x80)) ; OC7D = 0; TC7 += N;
   TFLG1 = 0x80; while(!(TFLG1 & 0x80)) ; OC7D = 2; TC7 += N;
}
```

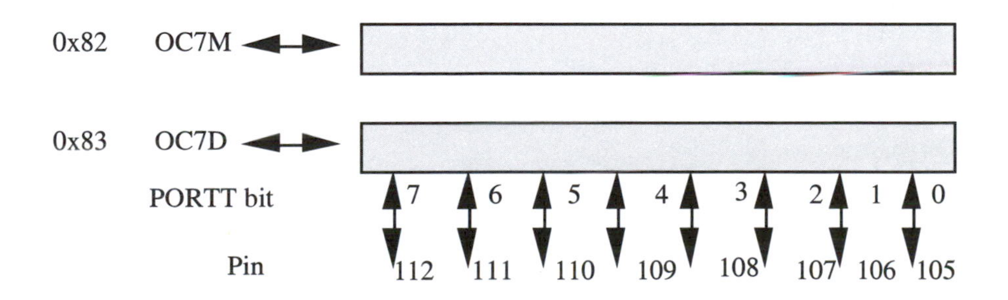

Figure 8.6. Output Compare 7 Ports

Some remarks on this program are offered. Compared to using a parallel port and real-time programming, this approach is much more precise. The real-time programming technique transfers data to the output pins when the program executes the statement *PORTT = d;* which is subject to any number of timing errors, such as interrupts, DMA, dynamic memory refreshing, and so on. The data in *OC7D* are transferred to the output pins on a precise count of *TCNT,* while the program just puts data into *OC7D* sometime before it is needed.

8.2.9 Object-oriented Square-wave and Pulse Generation

Object-oriented programming simplifies generation of square waves and pulses. We demonstrate a class *squareWave* below. Note the use of a destructor to stop output. We also illustrate a technique to use objects in a manner that allows them to be assigned to counter/timer devices when the object is declared or blessed. This is relatively easy when all devices are polled after an interrupt. It is complicated by the use of separate vectors that allows vectoring of different devices to different handlers. To get around this problem, the devices are made elements of a vector of objects. The interrupt handler is largely written in assembly language so as to convert the interrupt vector into a value of local variable *i* that is used as an index to select the device objects. The byte 0x8f is the opcode for the CPS # instruction; it is used to skip two bytes, which another interrupt vector accesses to put a different value in accumulator B, which becomes the variable *i*.

```
interrupt 8 void handler(){ int i;
asm{
    ldab #0
    dc.b 0x8f
    ldab #1
    dc.b 0x8f
    ldab #2
    dc.b 0x8f
    ldab #3
    dc.b 0x8f
    ldab #4
    dc.b 0x8f
    ldab #5
    dc.b 0x8f
    ldab #6
    dc.b 0x8f
    ldab #7
    clra
    pshd
}
    devices[i]->service();
}
```

The constructor is passed a channel number called *id,* from 0 to 7, and a frequency. The channel number selects the port and installs the interrupt vector so as to enter `handler` above at a location that will load accumulator B with the channel number.

```
class squareWave : public Port<int> { char id, mask; int halfPeriod;
    public:squareWave(char id,int frequency):Port (0x90 + (id << 1)) {
        asm sei
        TSCR = 0x80; TIOS |= mask = 1 << (this->id = id);
        TCTLO=(TCTLO& ~(3<<(id <<1))) | (1 << (id << 1)); /* toggle */
        TMSK1=mask;/* enable interrupt for device id */ put(frequency); asm cli
    }

    virtual void put(int data){halfPeriod=data;TC[id]=TCNT+halfPeriod;
}

    virtual ~squareWave(void){TCTLO &= ~(3<<(id<<1));TMSK1 &= ~ mask; }

    void service(void) { TFLG1 = mask; TC[id] += halfPeriod; }
};

squareWave *devices[8];

void main() {
    devices[0]=new squareWave(0,100);devices[0]->put(5);
    devices[0]->~squareWave();
}
```

A class `Pulse` is shown below. The `put` function member causes an output transition to high, and the service routine, entered after the output goes low, disables its interrupt. In this class, each handler is vectored to, and calls the service routine with, the appropriate object. The service routine uses the object data members `mask` and `id2` to clear the interrupt.

```
interrupt 8  void handler0(){ devices[0]->service(); }
interrupt 9  void handler1(){ devices[1]->service(); }
interrupt 10 void handler2(){ devices[2]->service(); }
interrupt 11 void handler3(){ devices[3]->service(); }
interrupt 12 void handler4(){ devices[4]->service(); }
interrupt 13 void handler5(){ devices[5]->service(); }
interrupt 14 void handler6(){ devices[6]->service(); }
interrupt 15 void handler7(){ devices[7]->service(); }

class Pulse : public Port<int> { char mask, id, id2;
    public: Pulse (unsigned char id) : Port (0x90 + (id << 1))  {
        asm sei
        TSCR = 0x80; TIOS |= mask = 1 << (this->id = id);
        TCTLO=TCTLO & ~(1 << (id2 = id << 1))|(2 << id2);CFORC = mask;
        TMSK1 = mask; /* enable interrupt for device id */
        *(int *)(0xb2e+(id << 1))=((int)handler0)+7 * id; asm cli
    }
```

```
    virtual void put(int data) {
        asm sei
        TCTLO |= 3 << id2; CFORC = mask; /* set bit id */
        TC[id] = data + TCNT; TCTLO &= ~(1 << id2);
        TMSK1 = TFLG1 = mask; /* clear and enable device id interrupt */ asm cli
    }

    void service(void) {TMSK1 &= ~mask; TCTLO &= ~(3 << id2);  }
};

Pulse *devices[8];

void main()
    {devices[0]=new Pulse(0);devices[0]->put(5);devices[0]->~Pulse();}
```

The reader is invited to write a class `negPulse` for negative pulse generation. We finally demonstrate a class `pSq` below. The constructor's *mask* argument determines which `PORTT` bits the output sequence is put on. These bits, together with bit 7, are assigned to an output compare function. `put` passes length *n* and vectors *bits[n]* and *time[n]*. Elements *time[i]* are the time in E clock cycles from the *i*-first interrupt to the *i*th interrupt. Device 7's interrupt occurs *time[i]* clock cycles after the last interrupt, and outputs bit pattern *bits[i]* when the period *time[i]* has elapsed.

```
class pSq : public Port<int> { int *time; char *bits, size;
    public : pSq(char mask) : Port(0x9e){
        TSCR = 0x80; TIOS |= (OC7M = mask) | 0x80;
        *(int *)0xb20 = ((int)handler); /* for DBug12 */ asm cli
    }

    virtual void put(int *time, char *bits, char size) {  asm sei
        this->time = time; this->bits=bits;this->size = size;
        OC7D = *bits++; TC7 = TCNT + *time++;
        if(--size)TMSK1|=TFLG1=0x80; /* clear, en. device 7 int. */ asm cli
    }

    void service(void) {
        TFLG1 = 0x80; OC7D = *bits++; TC7 += *time++;
        if((--size)<= 0) TMSK1&=0x7F; /* if no more sequence, disable int. */
    }
};

interrupt 15 void handler(){ device.service(); }

pSq device(7);
#define SIZE 5
int times[SIZE]={ 10, 20, 30, 40, 50 }; char bits[SIZE]={ 1,2,4,3,6 };

void main() { device.put(times, bits, SIZE);  }
```

8.3 Frequency and Period Measurement

Converse to generating square waves or pulses, one may need to measure the frequency or period of a square wave or repetitive signal, or the width of a pulse. Many important outputs carry information by their frequency: tachometers, photodetectors, piezoelectric pressure transducers. The voltage-to-frequency converter's integrated circuit can change voltages to frequencies economically and with high accuracy. The period of output from a timer chip like the 555 timer is proportional to the resistance and capacitance used in its timing circuit. Therefore, a by-product of measuring period is that one can easily obtain measurements of resistance or capacitance. But, for high frequencies, frequency is easier and faster to measure, while for low frequencies, period is easier and faster. The 6812 is quite capable, if necessary, of inverting the value using its IDIV and FDIV instructions. For nonrepetitive waveforms, pulse-width measurement is very useful. Also, the time between two events can be measured by using the events to set, then clear, a flip-flop. The pulse width of the output of this flip-flop can be measured. Sometimes the microcomputer has to keep track of the total number of events of some kind. The event is translated into the rising or falling edge of a signal, and the number of edges is then recorded. Note that the number of events per second is the frequency. Thus, events are counted in the same way as frequency is measured, but the time is not restricted to any specific value.

In this section, we first study the measurement of period and pulse width. We also consider the recording of timing of edges, and gadfly and interrupt sensing of edges. Later we discuss the counting of events and the measurement of frequency.

8.3.1 The Input Capture Mechanism and Period Measurement

The software approach to period measurement is simpler than the software approach to frequency measurement. A counter is cleared, then the input signal is sensed in a gadfly loop until a falling edge occurs. Then the main loop is entered. Each time the loop is executed, the counter is incremented and the input value is saved. The loop is left when the input value from the last execution of the loop is true and the input value in this execution of the loop is false – that is, on the next falling edge. The counter then contains a number *N*, and the period is *N* times the time taken to execute the loop. The analog hardware approach uses a voltage-to-frequency converter, in which the input voltage and the reference voltage are interchanged (see §7.4.3 for details). The digital hardware approach uses a reference clock to increment a counter. The counter is cleared on the falling edge of the input signal and stops counting (or is examined) on the next falling edge of the input signal. Some reflection on these techniques shows that in each case the frequency measurement technique is used, but the roles of the reference frequency and the input frequency are interchanged.

Recall that the popular 555 timer integrated circuit can generate a digital signal with period $P = A + B (R_a + 2 R_b) C$, where A and B are constants, R_a and R_b are the resistors and C is the capacitor (as diagrammed in Figure 6.9.) The resistor R_b can be a volume control or joystick with which the computer may want to sense the wiper's position, or it may be a photoresistor or thermistor. In fact, under the control of a parallel output port and an analog multiplexor, the computer can insert any of a number

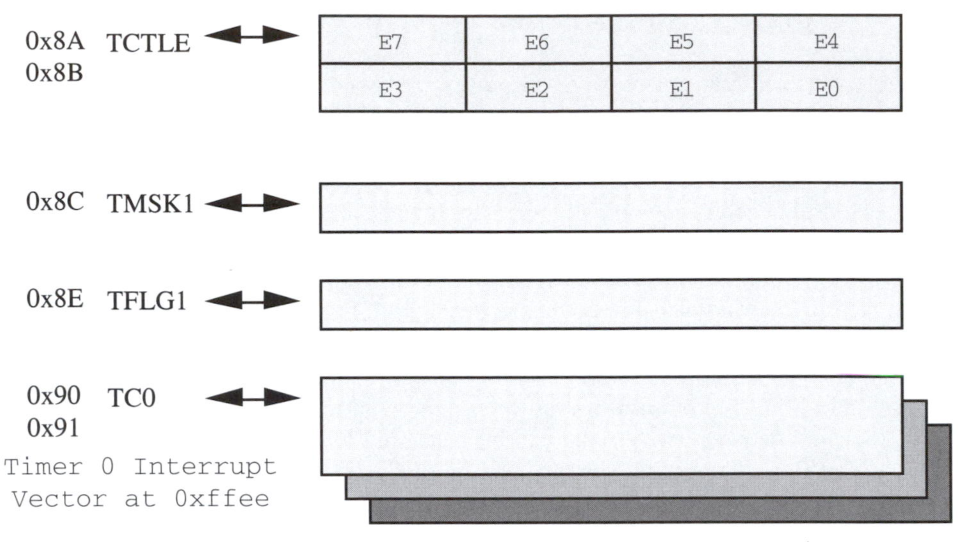

a. Ports

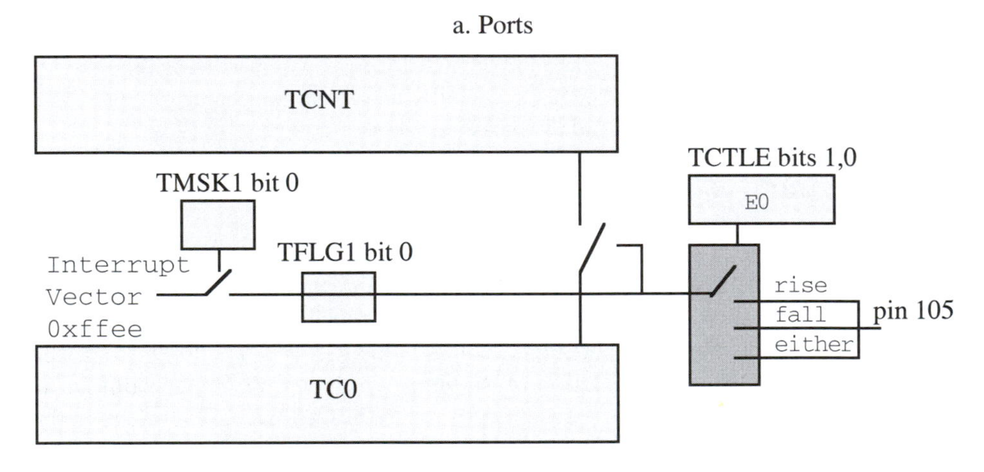

b. Device

Figure 8.7. Input Capture

of resistors in place of R_b, so that the selected resistor's value can determine the signal's period and the period can be measured by a counter/timer. This is a good way to "read" into the microcomputer the potentiometers on a stereo console, a game, or a toy.

The term *input capture* denotes that when an edge occurs, the running counter is "captured" into a register. Figure 8.7 shows these input capture ports and devices.

Eight input capture devices are provided, so eight waveforms can be measured at the same time (see Figure 8.7a). Input capture device 0 is diagrammed in Figure 8.7b. An edge is selected by the two least significant bits of the 16-bit port `TCTLE`. If they are FF (00), no capture occurs; if FT (01), capture occurs on a rising edge; if TF (10), capture occurs on a falling edge; and if TT (11), capture occurs on both edges. A transfer causes the current value of *TCNT* to be copied into *TC0*, and sets *TFLG1* bit 0. This bit can be sampled in a gadfly loop, or if *TMSK1* bit 0 is set, setting *TFLG1* bit 0 causes an interrupt vectored through 0xFFEE and 0xFFEF.

The period of a waveform can very simply be measured by reading the input capture register after two consecutive transfers. The following program *main* shows a gadfly technique for doing this (assuming the period is less than 2^{16} clock cycles).

```
void main() { int period;
    TSCR = 0x80; /* enable counter/timer */
    TCTLE = 0x4000; /* device 7 input capture: capt rise edge */
    TFLG1=0x80; while(!(TFLG1&0x80)); period=TC7; /* get time of 1st */
    TFLG1=0x80; while(!(TFLG1 & 0x80)) ;/* wait for another rising edge */
    period = TC7-period; /* subtract this time from time of 1st transition */
}
```

8.3.2 Pulse-width Measurement

The pulse width of a signal is the time from a rising edge to the next falling edge (that is, as the width of a positive pulse). A negative pulse width is the time from a falling edge to the next rising edge. You may question why we might want to measure pulse width when we can already measure period, or vice versa. Usually the signal being measured is an analog signal, and this is converted to a digital signal by a comparator. Observe that the analog signal's slanting edges generally result in different digital signals depending on the comparator's threshold. Normally however, the period is independent of the comparator's threshold, so it should be measured. For nonrepetitive wave forms where the period cannot be measured, the pulse width of the waveform is measured.

As usual, there are software, analog and digital hardware, and counter/timer techniques for pulse width measurement. Except for the analog hardware approach, they are all similar to the techniques for period measurement. Generally, in the software approach the time is measured when the input signal falls and is again measured when that signal rises. After the pulse is over, the pulse width is the difference in times.

Pulse width is measured by a counter/timer in a manner similar to period measurement. (Again we assume the counter doesn't overflow.) The only difference is that we must change the edge of the second reading of the input capture register to determine the time from a leading edge to the next falling edge.

```
void main() { int pulseWidth;
    TSCR = 0x80; TCTLE = 0x4000; /* device 7 input capture: capture rise edge */
    TFLG1=0x80;while(!(TFLG1 & 0x80)); pulseWidth=TC7;/* get rising edge */
    TCTLE = 0x8000;
    TFLG1 = 0x80; while(!(TFLG1 & 0x80)) ; pulseWidth = TC7 - pulseWidth;
}
```

In §9.3.1, we will study the universal asynchronous receiver-transmitter (UART), which is used to communicate over a serial link. The UART clock signal is 16 times this baud rate; a counter/timer device, configured as a square-wave generator, can generate the UART clock (although special chips are also available for it). An input capture device is also suited to automatic determination of the bit rate and setting of the clock rate, using pulse-width measurement. The sender should send the capital U character – ASCII code 0x55 – repetitively. Each time it is sent, it generates five pulses (or six pulses if the parity is set even). The pulse width can be measured using the counter/timer, then multiplied by 16 to establish the UART clock, to be used by the device that generates a square wave.

Pulse-width measurement provides a key to determine the codes used by a remote control for a TV or stereo. An infrared detector outputs pulses to `PORTT` bit 0. The command may be encoded in the positive pulse widths, negative pulse widths, periods, or counts of pulses. To establish the nature of the coding, we collect some data called a *histogram,* a vector of counts. The element `histogram[i]` is the number of times the pulse width was between `i * range` and `(i + 1) * range`. In the following program `main`, `range` is set to 30. If the period is 0 to 29, `histogram[0]` is incremented; if the period is 30 to 59, `histogram[1]` is incremented; and so on. Examining the histogram, we get an idea of what is a T (1) and what is an F (0).

```
int i, j, histogram[10];
void main(){TSCR = 0x80; TIOS &= ~1; TCTLE = 1; /* bit 0 input capture */
  for( i = 0; i < 100; i++) { /* get 100 samples */
    TCTLE = 1; TFLG1 = 1; while(!(TFLG1 & 1)) ; j = TC0;
    TCTLE = 2; TFLG1 = 1; while(!(TFLG1 & 1)) ; j = TC0 - j;
    if(j < 300) histogram[j / 30]++;
  }
}
```

Suppose many pulses are around either 50 or 120 clock cycles. We then define a T (1) as a period of 120 and an F (0) as a period of 50 (a midpoint threshold is 85); the program `main` below will collect Ts and Fs into vector `history` to determine patterns that are emitted by the remote control. Pressing the key "0" and obtaining its `history,` we can examine `history` to determine how to recognize the code for the key "0" is pressed. Pressing the key "1" and obtaining its `history,` we can examine `history` to determine how to recognize when the key "1" is pressed, and so on. This way, we can build an infrared remote receiver or duplicate an infrared remote transmitter.

```
int i, j, history [10];

void main() {
  TSCR = 0x80; TIOS &= ~1; TCTLE = 1; /* make bit 0 input capture */
  for( i = 0; i < 160; i++) { /* get 160 samples */
    TCTLE = 1; TFLG1 = 1; while(!(TFLG1 & 1)) ; j = TC0;
    TCTLE = 2; TFLG1 = 1; while(!(TFLG1 & 1)) ; j = TC0 - j;
    history [i >> 4] <<= 1; if(j > 85) history [i >> 4]++;
  };
}
```

8.3.3 Triac Control

The counter/timer edge interrupt can be used, like the key wakeup interrupt, to initiate an action when an edge of a signal arrives. Moreover the input capture port provides a precise reference when the edge occurred. This can be used to proportionately control a triac (see Figure 7.4). Device 0 detects either the rising or falling edge of a squared-up 60-Hz power line signal. Device 1 produces a rising edge `offtime` clock cycles after each such edge; its interrupt handler negates this signal after a few microseconds. This pulses the triac gate to turn it on `offtime` clock cycles into each 60-Hz half-cycle.

```
int offtime = 100;

interrupt 8 void handler1(){
      TC1 = TC0 + offtime; /* set device 1 time to cause rising edge and int. */
      TMSK1 |= 2; TFLG1 = 1; /* enable device 1 interrupt, clear its request */
}

interrupt 9 void handler2(){
      TMSK1 &= ~2; /* disable device 1 interrupt */
      CFORC = 2; /* output was just asserted; now negate output */
}

void main() {
      TSCR = 0x80; TFLG1 = TMSK1 = 1; TIOS = 2; /* device 1 is output comp */
      TCTLO = 4; /* timer device 1 toggles each interrupt or CFORC application */
      TCTLE = 3; /* timer device 0 captures on both edges */ asm cli
      do ; while(1) ;
}
```

Up to eight triacs can be individually controlled using similar techniques on all the timer devices, if a pair of key wakeup J devices are used: one set to detect rising and the other falling edges of the 60-Hz square wave.

8.3.4 Pulse Accumulation and Frequency Measurement

As with the square-wave generator and pulse generator, frequency measurement can be done in the 6812 by software using a simple input port, by digital hardware read by a parallel input port, by an analog technique, and by using a counter/timer system. The designer should consider all techniques for frequency and period measurement. We'll mainly discuss how the counter/timer device is used to measure frequency or period, before which we'll consider the other approaches. However, there is no need to study the software approach because the 'A4's counter/timer devices are so easy to use and available that they make this awkward approach merely an academic exercise.

A digital hardware approach would use one or more counter ICs that can be cascaded to make up a counter of sufficient width. A parallel I/O register can be used to clear the counter and then read the counter output word, say, one second later. The counter can count input transitions from low to high between these two times. A note of caution: if the width is greater than 16 bits, then two or more words will be read at different times and the counter could be incremented between successive reads. The count should be examined with this in mind. For instance, a 24-bit counter should be read and read again. If the readings of the upper bytes differ, the reading should be considered erroneous and should be tried again. An analog approach would be to convert the frequency to a voltage, then measure the voltage by an analog-to-digital converter. An FM demodulator, a frequency-to-voltage converter, or a tachometer can convert high frequencies, audio frequencies, or subaudio frequencies to a voltage.

Finally, we focus on the use of a counter/timer system for frequency measurement or event counting. The input capture mechanism can be used; see Figure 8.7. Consider interrupts; one edge-detect interrupt handler can count the number of an input signal's high-to-low transitions. That is, each edge causes an interrupt, and the handler increments a global variable *count*. The fixed interval of time can be measured out by another counter in the chip or by real-time programming. After this is over, if the interval of time is a second, *count* is the frequency. If the interval of time is a fraction of a second, *count* is a submultiple of the frequency.

The counting can be done most effectively by the pulse accumulator in the 6812 (see Figure 8.8). This subsystem uses a 16-bit pulse accumulator counter *PACNT*, pulse accumulator control *PACTL*, and pulse accumulator flag *PAFLG*.

For the pulse counter to change, `PAEN`, bit 6 of `PACTL` must be set. `PAMODE`, bits 5 and 4 of `PAEN`, determine what is counted. If they are FF (00), then falling edges on the signal on the input `PORTT` bit 7 cause the counter to increment; if FT (01), then rising edges of that input cause the counter to increment. If those control bits are TF (10), then as long as the `PORTT` bit 7 input is high, the counter is incremented each 8 μs; and if those bits are TT (11), then as long as the `PORTT` bit 7 input is low, the counter is incremented each 8 μs (assuming a 2-MHz E clock). `PAOVF`, `PAFLG` bit 1, indicates that the counter has overflowed and `PAIF`, bit 0, that the counter has incremented; they are cleared by writing a 1 into them. If the corresponding bits are set in `PACTL`, interrupts will occur via 0xFFDC,DD (for overflow) and 0xFFDA,DB (for incrementing). This subsystem is very flexible and can be used to count events and measure frequency or pulse width, as shown in problems at this chapter's end. Here we concentrate on event counting and frequency measurement.

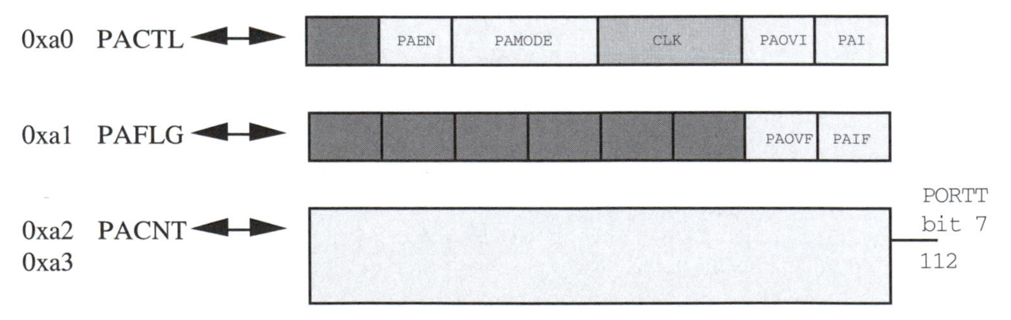

Figure 8.8. The Pulse Accumulator

To count events that correspond to falling edges, set bits 5 and 4 of *PACTL* to FF (00). Then, each time an event occurs, the pulse counter is incremented. When it overflows, an interrupt can be used to count overflows, thus extending the range of the counter to 24 or more bits. To measure frequency, we count events in a fixed time interval, such as 1 s. Alternatively, we count the number of events in a fraction of a second. Device 6 is used to measure the time delay. The resulting number in the pulse accumulator, *PACNT*, is a submultiple of the frequency.

```
void main() { int frequency;
    TSCR = 0x80; TIOS = 0x40;
    TCTLO = 0x1000; /* enable timer 7 to wait 1/128 seconds */
    PACNT = 0; /* clear counter */ TFLG1 = 0x40; /* clear timer flag */
    PACTL = 0x40; /* enable pulse accumulator, count falling edges */
    TC6 = TCNT + N; /* let pulse accumulator measure 1/128 second */
    while( ! (TFLG1 & 0x40 )) ; /* gadfly until flag is set */
    PACTL = 0; /* disable pulse accumulator */
    frequency = PACNT << 7; /* read count (frequency) */
}
```

8.3.5 Object-oriented Period, Pulse-width, and Frequency Measurements

Object-oriented programming can be used to measure period, pulse width, or frequency. We first present the class to measure positive or negative pulse width, or period. This class uses gadfly synchronization; the reader is invited to write a class using interrupts.

```
enum{ PULSE, NEGPULSE, PERIOD };

class width:public Port<long>{char id,mode, mask;unsigned long result;
    void setMode(unsigned char m){
        {TCTLE = (TCTLE & ~ (3 << (id << 1)))|( m << (id << 1));}
    public : width (char id, char mode) : Port (0x90 + (id << 1))
        {TSCR=0x80;TIOS &= ~(mask=1<<(this->id=id));this->mode=mode;}

    virtual long get(void){unsigned int lastTime;int hiByte;long result;
        if( ( mode == PULSE) || ( mode == PERIOD) ) setMode(1);
        else setMode(2);
        TFLG1 = mask = 1 << (this->id = id); /* clear flag */
        while(! (TFLG1 & mask)) ; /* wait for edge */
        lastTime = *(unsigned int *)port; /* get time of edge */
        if(mode==NEGPULSE) setMode(1); else if(mode==PULSE) setMode(2);
        TFLG1 = mask; /* clear flag */
        hiByte=0; while(!(TFLG1 & mask)) if(TFLG2){TFLG2=0x80;hiByte++;}
        result=*(unsigned int *)port+(hiByte<<16)-lastTime;return result;
    }
};
```

```
long i;

void main() { width device(0, PULSE); i = device.get(); }
```

The last gadfly loop in *width*'s function member `get` keeps track of timer overflow. If `TOF` is asserted (Figure 8.1) so that `TFLG2` is nonzero, then `hiByte` is incremented. This value `hiByte` is shifted 16 bits and added to the returned width.

Frequency can be measured continuously with the class `frequency` shown below. Upon each timer 6 interrupt, which occurs every 1/128 second, `PACNT` is saved in `value` and `PACNT` is cleared. `value` is the last pulse count obtained by this handler. The function member `get` gadflies on `flag` to be sure a new measurement has been made, and then returns `value` multiplied by 128.

```
#define N 62500
class frequency : public Port<long> { int value; char flag;

    public : frequency() : Port(0xa2) { asm sei
        TSCR = 0x80; TC6 = TCNT; TIOS |= 0x40; PACTL = 0x40;
        TFLG1|=0x40; TMSK1 = 0x40; TCTLO = (TCTLO & ~0xcff) | 0x1000;
        TC6 += N; flag = 0; *(int *)0xb22 = ((int)handler); asm cli
    }

    virtual long get(void){while(!flag);flag=0; return (long)(value<<7);}

    void service(void)
        { TFLG1 = 0x40; value = PACNT; PACNT = 0; TC6 += N; flag++;}
};

frequency device;

void main() { int i; i = device.get(); }

interrupt 14 void handler(){ device.service(); }
```

8.4 Conclusions

Frequency- or phase-analog signals are often generated naturally, by an AC tachometer, for instance. They may be used directly, for example, in firing a triac. Even when the signal is first an amplitude analog signal, conversion to frequency-analog or phase-analog signals simplifies the noise isolation and voltage-level isolation between transducer and microcomputer. Moreover, several hardware and software techniques, including those that use a counter/timer like the 'A4 counter/timer system, can be used to measure or generate frequency-analog signals.

The counter/timer is a very useful and attractive I/O device for measuring or generating frequency- or phase- analog signals. It is useful in generating square waves, pulses, and timing interrupts; it can measure events, frequency, period, and pulse width; and it can monitor period or pulse width to interrupt the computer if too long a value is noticed. It is very attractive because a single wire sends the signal to or from the counter/timer. To the chip designer, it means that an I/O device can be put on a chip or inside a microcomputer chip without using a lot of pins. While counters take up a nontrivial amount of area on the chip, that area is comparatively cheap, while pins are in much shorter supply. Moreover, to the system designer, a single wire is easy to isolate with an optical isolator, to prevent the voltages of the system under test from getting to the microcomputer and the user, as well as to isolate noise generated in that system and prevent it from getting into the microcomputer.

The 'A4 counter/timer system was introduced to illustrate the discussion of the counter/timer device. This chapter prepares us for similar techniques in communication devices in Chapter 9, and for secondary storage chips, covered in Chapter 10.

For more concrete information on the 'A4, consult the MC68HC812A4TS/D. In particular, §13 describes the programmable timer and pulse accumulator. As noted earlier, we have not attempted to duplicate the diagrams and discussions in that book because we assume you will refer to it, while reading this book; and, since we present an alternative view of the subject, you can use either or both views.

You should now be familiar with the counter/timer in general and with the 'A4 counter/timer system in particular. Connecting the pins of the 'A4 counter/timer system, and writing software to initialize and use it, should be well within your grasp. Moreover, you now have enough information to consider alternatives to this system and chip and to recognize whenever alternatives are superior to the counter/timer.

Do You Know These Terms?

See page 36 for instructions.

counter/timer	touch-tone signal	input capture	pulse width
output compare	pulse	id	histogram
pulse-width modulator (PWM)			

Problems

Problem 1 is a paragraph correction problem. See page 38 for guidelines. Guidelines for software problems are given on page 86, and for hardware problems, on page 115. Problems involving the 6812 counter/timer will all assume that I/O registers are at 0 to 0x1FF, using an 8-MHz E *clock unless otherwise indicated. All global variables needed for the handler should be shown, and* `main()` *procedures should initialize the counter/timer assuming I/O control ports have unknown initial values.*

1.* Communication using the frequency or phase is attractive because it can be accomplished using lower bandwidth channels than amplitude analog communication, and such signals can be generated or measured by microcomputers with counter/timer devices in them or counter/timer integrated circuits. These combination chips appear because the counter/timer does not require much surface area on a chip and the transistors needed to build counter/timers need not be as good as those needed for A-to-D converters, so the inclusion of such a function on another chip does not raise the cost. Moreover, many functions require counting and timing. Pulses from a pulse generator can generate tones, and Johnson counters can generate a staircase approximation of a sine wave that is useful for touch-tone dialing. Pulses can be used to control automobiles, and interrupts from the timer can be used to implement real-time clocks. Frequency measurements can be used with a voltage-to-frequency converter to measure amplitude analog signals, and they can be used to measure intervals of time between two events. Period measurements are an alternative to frequency measurements, and one can always be obtained from the other by division; frequency measurement is preferable if the frequency is low because this will give more accuracy than period measurement. The 6812 counter/timer should be used for these different generation and measurement functions if the software approach is susceptible to errors due to interrupt handling, DMA, or dynamic memory refresh. But digital hardware or analog hardware approaches, such as the phase-locked loop, may be needed if the microcomputer software or counter/timer approach is too slow.

2. Write a self-initializing procedure `square(int i)`. The desired frequency `i`, converted to a period, is less than 2^{16}.

a. Write a shortest C procedure `divide(int *dvdad,int *dvsad,int *qutad, int *remad, int signd)`, using embedded assembly language, that divides a 32-bit number (i.e. a *long*) pointed to by `dvdad` by a `long` pointed to by `dvsad` putting the quotient in a long pointed to by `qutad` if `qutad` is nonzero, and the remainder in a long pointed to by `remad` if `remad` is nonzero. If `signd` is zero, divide unsigned; otherwise divide signed. Use a procedure `negat(l) int *i;` that you write using embedded assembly language, to negate the 32-bit data pointed to by `i`, and use `subq` from Chapter 1, problem 30, part b.

b. Write `square` using gadfly synchronization on device 3. Use part a's procedures.

c. Write an interrupt square-wave generator C procedure `square(int i)` that uses device 3 to generate the square wave on `PORTT` bit 3. Use part a's procedures.

3. Repeat problem 2 for `square(int i,int j)`. The desired frequency is `(i << 16) + j`. Converted to a period, it can be greater than 2^{16}. (Hint: consider toggling, as well as setting and clearing the output, upon output compare, to time out 2^{16} clock cycles.)

4. Write a routine to generate a touch-tone dial sequence, calling up a self-initializing procedure `dial()` that is like `main()` in §8.2.3. The called number is in vector:

```
char numbers[7] = { 5,5,5,1,2,1,2 };
```

Each tone is on for 50 ms, and there is a mute period of 45 ms between tones. (These are the minimum times specified by the telephone industry.) Use devices 0 and 1 to generate sine waves and device 4 to time the delays.

5. A music synthesizer has 16 voices and is to be built with 'A4 counter/timer device 2 and `PORTJ`, a (256,8) PROM, a 74LS161, and an 8-bit D-to-A converter. `PORTT` bit 2's output compare clocks the 74LS161, which provides the low-order 4 bits of the PROM's address so that each voice is generated by 16 samples of the repetitive waveform. The low-order 4 bits of `PORTJ` provide the PROM's high-order address to select a voice. The output from the PROM is an 8-bit 2's complement number, which is converted to a voltage between -128/128 and +127/128 volts by the D-to-A converter.

a. Show a block diagram of the system. Do not show pin numbers, but do identify the pins and lines with meaningful, unambiguous labels.

b. Write a self-initializing procedure `sound(unsigned char octave, unsigned char note, unsigned char voice, unsigned char length)`. The 'A4 counter/timer, with no prescaling, generates a tone clock on `PORTT` bit 2, using interrupts, and times the length of the note using timer device 5, using gadfly synchronization. The first two arguments are *octave* (the lowest octave is represented by zero) and *note* (A is represented by zero, B^{b} is 1, B is 2, . . . , A^{b} is 0xB). Low A is 27.5 Hz and is represented by zero and zero. The frequency of each note is $12\sqrt{2}$ times the frequency of the next lower note. The last two arguments are *voice,* which will be input to the PROM, and *length,* which is the length of the note, in 1/16*ths* (1/4 seconds). (Hint: look up *int* vector *cnvt,* shown in part c, to generate the lowest octave's frequency, then shift it right to derive the required values for the output compare port.)

c. Show the vector, `int cnvt[13]` for the program in part b, using `#define`s.

6. Write a self-initializing procedure `pulse(int w)` to generate a pulse of width `w` for E clock cycles.

a. Generate a positive pulse on `PORTT` bit 2. Assume the bit is initialized low.

b. Generate a negative pulse on `PORTT` bit 2. Assume the bit is initialized high.

7. Write a self-initializing procedure `pulse (iint i, int j)` to generate a positive pulse of width `(i << 16) + j` on `PORTT` bit 2. (Hint: you need to switch between toggling, setting, and clearing the output as you handle the high and low 16-bit parts of the pulse width.) a. Use gadfly synchronization. b. Use interrupt synchronization.

8. The `main()` procedure in §8.2.5 will dial a number on a conventional step-and-repeat telephone, but will tie up the computer while it is dialing the number. Write `main()` procedure and an interrupt handler for the 6812 device 4 that will cause an interrupt every 5 ms. The most significant bit of `PORTJ` is given a value 1 to close the relay in series with the dial contacts. A digit "0" is represented by the number 10. Use global variables to keep track of what part of the sequence of numbers, what part of the number, and what part of the pulse has been output.

a. Write the handler to output just one number, in global variable `int number`.

b. Write the handler to output the seven numbers in the vector `char numbers[7]`, as in problem 4.

9. An "alarm clock" can start a procedure at a specified time, using counter/timer device 2. Write a `main()` procedure and a handler interpreting a table of times that the alarm is supposed to "go off," so that when this happens a program corresponding to the alarm will be called from `main()`. If no procedure is to be executed, `main()` calls a procedure `dummy(){}`. Suppose that a table of 10 "alarms" is stored:

```
struct{ int TH, TL, GO } alarm[10];
```

where *TH* is the 2 high-order bytes, *TL* is the 2 low-order bytes of a time interval in E clock cycles, between the time the "alarm" went off for row `i-1` (or the beginning if `i` is 0) and the time it will go off for row `i`, and `GO` is the address of a subroutine to be started when that interval is over. Each such subroutine ends with an RTS, after it causes `main( )` to call `dummy()` repetitively.

10. Write a gadfly-synchronized, self-initializing, procedure `sequence()` to sequence three output signals using the output compare device 7. Outputs on `PORTT` bits 7, 6, and 5 are initially 0, 1, and 0, respectively. When `PORTA` bit 0 falls, bit 7 rises, and 20 μsec later, bit 6 falls. After another 30 μs, bits 5 and bit 6 rise together, and after yet another 40 μs, bit 6 falls.

11. An X-10 home remote control can use a remote keyboard that communicates to it using ultrasonic signals, which avoids the need to connect the computer to the power line in any way. We would like to control the remote stations by sending to the command device the ultrasonic signals that would have been sent by the remote keyboard. Write a self-initializing procedure `ultra(data, address) char data, address;` that generates ultrasonic signals used for sending commands via `PORTJ` bit 0 to an ultrasonic transmitter, which then transmits to the receiver in the X-10 control device. A true bit is sent for 8 ms – 4 ms of 40-KHz square wave followed by 4-ms mute output – and a false is sent for 4 ms – 1.2 ms of a 40-KHz square wave followed by 2.8-ms mute output. A command to send a data bit `d` which is the least significant bit of `data` to a remote station identified as `a`, the four least significant bits of `address`, is sent as follows: A true bit is sent; then a 4-bit remote station number `a` is sent, most significant bit first; data bit `d` is sent; then the complement of `a` and `d` are sent in the same order; and then 16 ms of a 40-KHz square wave are sent. Use real-time synchronization to time out the sending of bits of the command.

12. Write a class `negPulse` whose `put(int data)` member function generates a negative pulse of width `data` E clock cycles.

13. Frequency can be determined by the pulse accumulator, or by measuring pulse width using an input capture device and getting the inverse using the `divide` procedure of problem 2 part a. The former gives better accuracy for high frequencies, and the latter for low frequencies. Suppose a square wave is simultaneously input to the pulse accumulator and device 7 (on `PORTT` bit 7). Write a self-initializing procedure `freq()` and interrupt handlers for the pulse accumulator overflow and for device 7 and an initialization in `freq()` to wait for the measurement from both the period and frequency handlers, waiting exactly 0.01 seconds using a gadfly loop on device 2; then return the most accurate frequency reading from the procedure `freq()`.

14. Period can be determined using an input capture device, or by measuring frequency using the pulse accumulator and getting the inverse using problem 2 part a's `divide` procedure. Write a self-initializing procedure `period()` and interrupt handlers, as in problem 13, to accumulate pulses, within 1/128 seconds, or capture input edges, so that `period()` returns the most accurate period measurement, in E clock cycles.

15. Write a C self-initializing procedure `pulse()` to measure pulse width using the pulse accumulator device. Assume the procedure is entered while the input is low. Set up the device to count E clock cycles, divided by 64, for an entire time while the input is '1'. Note that you can gadfly on `PORTT` input bit 7. Return `PACNT`, which is the pulse width in 125-μs units.

16. `PORTH` bit 0 is connected to a signal that pulses low at the moment that the 60-Hz power line signal passes through 0. All eight timer devices are connected to pulse transformers that fire eight triacs to implement proportional phase control of eight lamps. Show a C procedure `main()`, interrupt handlers, and a global vector `char d[8]` that set up and maintain these devices so that they output a waveform whose falling edge fires triac `i` at time `d[i]` degrees, `0 ≤ d[i] < 177`, in each half-cycle.

17. Design a voltmeter using an optically isolated voltage-to-frequency converter.

a. Show a diagram of the complete system, giving enough detail that the system could be built from it. Use the Teledyne 9400 (Figure 7.18b), a 4N33 opto-isolator to isolate the voltage sensor from the microcomputer, and the 6812 pulse accumulator to measure the frequency.

b. Write a self-initializing procedure `voltmeter()` to measure the frequency so that the voltage at the input of the hardware in part a, in millivolts, is returned. Measure frequency, using the pulse accumulator, for 1 s using gadfly synchronization on device 5. Assume a frequency of 32768 Hz represents 5 V, and 0 represents 0 V. Use the `divide` procedure of problem 2 part a to scale the result.

18. An AM tuner has a local oscillator which is tuned to a frequency that is 455 KHz higher than the frequency of a station that is tuned in; so that the "beat frequency" is the intermediate frequency amplified by the radio. Write a self-initializing procedure `AM()` to measure the frequency, using the pulse accumulator, for 1 ms utilizing gadfly synchronization on device 6, so that if the pulse accumulator input has the local oscillator frequency, the program will output the frequency of the station being received in kilohertz.

19. Design a capacitance meter. A capacitor of unknown value `C` is put in the timing circuit of a 555 timer, `C` being lower than 0.01 μF.

a. Show the hardware, pin numbers, and component values to build such a meter, using the pulse accumulator device.

b. Write a self-initializing procedure `capac()` that returns the capacitance in pF.

20. Design a phase meter for audio frequencies, using an 'A4. V1 and V2 are periodic waveforms that have the same period, which is less than 8 ms. The phase of V2 with respect to V1 is the number of degrees (of a circle) that periodic wave V2 follows V1. V1 and V2 are converted into square waves having the same phase and period. counter/timer device 0 will obtain the time of a rising edge of the squared-up V1, and device 1 will obtain the time of the rising edge of the squared-up V2. Write a self-initializing procedure `phase()` to output the phase between V1 and V2 in degrees. Use gadfly synchronization, Chapter 2's problem 14, and problem 2a's `divide` procedure.

21. Design a TV sync signal generator using real-time, gadfly, and interrupt synchronization with devices 1 and 2 of an 'A4. Device 1's output is the horizontal sync pulse, which is 4 μs high, 60 μs low, repetitively. Device 2's output is the vertical sync pulse, which is high for 26, and low for 244, horizontal sync pulses. Device 1 should use gadfly synchronization to wait for the 4 μs that vertical sync is high, but interrupt synchronization should be used to time the other signal edges. In order to start both sync pulses together, a small amount of real-time synchronization should be used. Show an initialization procedure `sync()` and interrupt handlers to configure the 'A4 to generate these signals.

22. Write a class `width` that uses interrupt rather than gadfly synchronization. After each first edge occurs in the device's handler, a service routine member of this class saves the time in a data member `int lastTime`. Between edges, timer overflows are counted in the handler by a different service routine member of this class, which increments an *int* value `hiCount` to keep track of the high-order 16 bits of the interval of time. After each second edge occurrence, the device's first service routine member calculates the desired period or positive or negative pulse width, and pushes the result into an input queue, to be pulled by the `get` function member. Explain a timing problem: how can you distinguish if the timer overflow indicated by a `TFLG2` flag occurs just after or just before the input capture, so that you can determine the high-order part of the pulse width in `hiByte`?

23. Design a logic analyzer, using the 'A4, to record and display the bus activity of a target computer. The logic analyzer stores the target computer's address, data, and control signals in a memory, and displays these values that occurred shortly before or after the *n*th time it recognizes a predetermined pattern on the target computer's buses. A logic analyzer has timer/counters to allow `times` occurrences of the pattern `address` recognized by the comparator to occur, and then allow `delay` clock cycles before a time `Tf` occurs. `PORTA` and `PORTB` hold the high byte and low byte of the comparator address, as well as the address to the indirect memory in problem 34 of Chapter 5. If the indirect memory holds `M` words, the `M` patterns, which appeared on the buses before time `Tf,` are displayed by the logic analyzer. The user can select `times` to be 0 and `delay` to be 0, if `Tf` is to be the time the pattern occurred first; but the user can use another number *times* if the first `times` occurrences are to be ignored, as when the pattern appears inside a `for` loop, or another number `delay` if the `delay` words read after the comparator detects a match. Also `(M - delay)` words before the comparator detects a match are to be stored in memory and displayed by the logic analyzer. Indirect memory, of which the `M` word by one byte logical design is shown in problem 5.34, stores each consecutive word read on the target computer's address, control, and data buses, as long as a control signal on `PORT` bit 0 is asserted high. The 'A4 counter/timer can be used to count occurrences and delays. The comparator feeds counter/timer device 7 rather than `PORTJ` bit 0's key wakeup interrupt hardware. The pulse accumulator overflow interrupt sets up output compare device 5, whose interrupt handler clears the indirect memory control signal on `PORTA` bit 0 to stop writing in the indirect memory.

a. Show a block diagram of such a logic analyzer. Include the memory, its address counter, the comparator, the target computer, and port connections to the 'A4.

b. Write a procedure `logicAnalyzer(int address, int count, int delay)` that sets up the logic analyzer, and interrupt handlers. (Some other mechanism reads out the data and displays it later.)

A memory expansion card, **Adapt812 MX1**, plugs onto the rear of Adapt 812, offering the user up to 512K of Flash and 512K of SRAM. A real-time clock/calendar and battery back-up for the SRAM is included, as well as a prototyping area for the user's own application circuitry. A versatile dual-slot backplane/adapter couples the memory card to the micro-controller card so that the entire assembly can be plugged into a solderless breadboard.

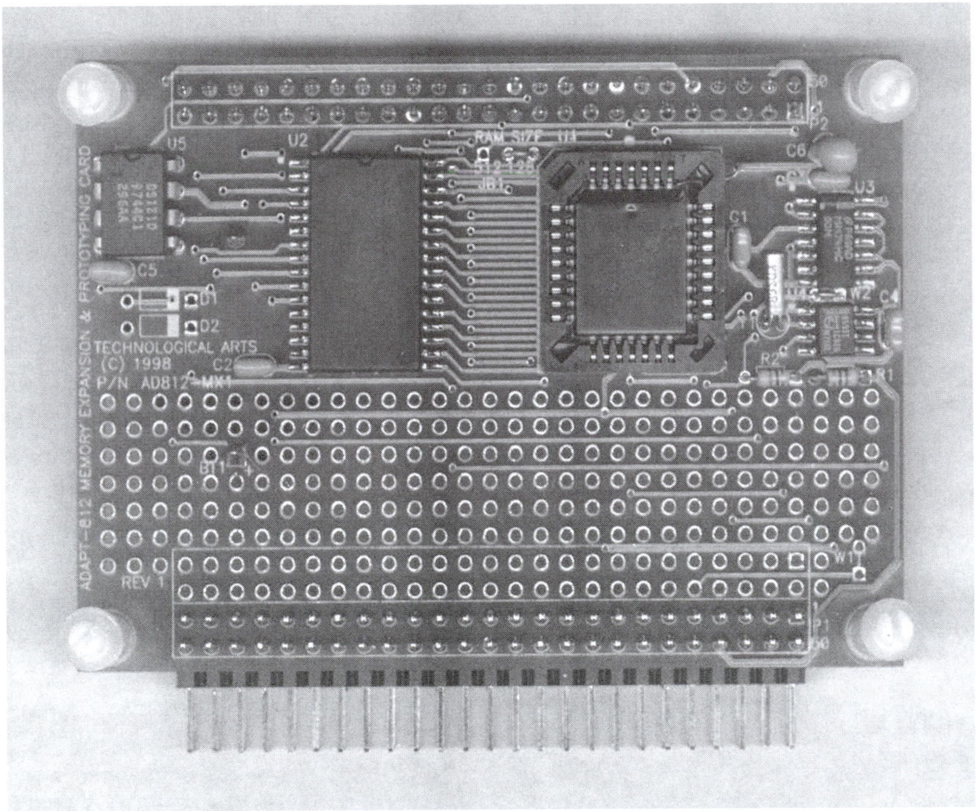

9

Communications Systems

The microcomputer has many uses in communications systems, and a communications system is often a significant part of a microcomputer. This chapter examines techniques for digital communications of computer data.

Attention is focused on a microcomputer's communications subsystem – the part that interfaces slower I/O devices like typewriters and printers to the microcomputer. This is often a universal asynchronous receiver-transmitter (UART). Because of their popularity in this application, UARTs have been used for a variety of communications functions, including remote-control and multiple-computer intercommunications. However, their use is limited to communicating short (1-byte) messages at slow rates (less than 1000 bytes per second). The synchronous data link control (SDLC) is suitable for sending longer messages (about 1000 bytes) at faster rates (about 1,000,000 bits per second), such as for sending data between computers or between computers and fast I/O devices. The IEEE-488 bus, for microcomputer control of instruments like digital voltmeters and frequency generators, and the SCSI bus, for communication to and from intelligent peripherals, send a byte at a time rather than a bit at a time.

The overall principles of communications systems, including the ideas of levels and protocols, are introduced in the first section. The signal transmission medium is discussed next, covering some typical problems and techniques communications engineers encountered in moving data. The UART and related devices that use the same communications mechanisms are fundamental to I/O interface design. So, we spend quite a bit of time on these devices, imparting basic information about their hardware and software. They will probably find use in most of your designs for communicating with teletypes or teletype-like terminals, keyboards, and CRTs, as well as for simple remote control. Finally, we look at the more complex communications interfaces used between large mainframe computers to control test and measurement equipment in the laboratory and to connect intelligent I/O.

Communications terminology is rather involved, with roots in the (comparatively ancient) telephone industry and in the computer industry, and some terms stemming uniquely from digital communications. Communications design is almost a completely different discipline from microcomputer design. Moreover, one kind of system, such as one using UARTs, uses quite different terminology than that used to describe another, similar system, such as one using SDLC links. While it is important to be able to talk to communications system designers and learn their terminology, we are limited in what we can do in one short chapter. We will as much as possible use terminology associated with the so-called X.25 protocol, even for discussing UARTs, because we want to economize on the number of terms that we must introduce, and the X.25 protocol appears to be the most promising protocol likely to be used with minicomputers and microprocessors. However, you should be prepared to do some translating when you converse with a communications engineer.

On completing this chapter, you should have a working knowledge of UART communications links. You should be able to use the 6812 SCI module, connect a UART or an ACIA to a microcomputer, and connect a UART or an M14469 to a remote-control station so it can be controlled through the 6812 SCI module, ACIA, or UART. You should understand the basic general strategies of communications systems, and the UART, SDLC, IEEE-488, and SCSI bus protocols in particular, knowing when and where they should be used.

9.1 Communications Principles

In looking at the overall picture, we will first consider the ideas of *peer-to-peer interfaces,* progressing from the lowest-level to the higher-level interfaces, examining the kinds of problems faced at each level.

Data movement is coordinated in different senses at different *levels of abstraction,* and by different kinds of mechanisms. At each level, the communication appears to take place between *peers,* which are identifiable entities at that level. However, even though the communication is defined between these peers as if they did indeed communicate to each other, they actually communicate indirectly through peers at the next lower level. (See Figure 9.1.)

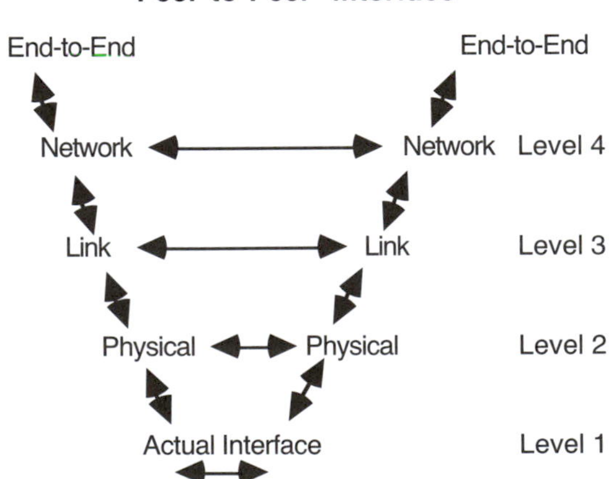

Figure 9.1. Peer-to-Peer Communication in Different Levels

Consider this analogy. The president of company X wants to talk to the president of company Y. This is called *end-to-end communication.* The job is delegated to the president's secretary, who calls up the secretary of the other president. This is referred to as *network control.* The secretary doesn't try to holler to the other secretary but dials the other secretary on the telephone. The telephone is analogous to the *link-control level.* But even the telephone "delegates" the communication process to the electronics and the electrical circuits that make the connection at the telephone exchange. This is the *physical-control level.* End-to-end communication is done between user (high-level) programs. User programs send information to each other like the presidents in the analogy. Network control is done at the *operating system level.* Like the secretary, this software must know where the communications object is and how to reach this object. Link control is done by I/O interface software and is responsible for setting up and disconnecting the link so the message can be sent. Physical control actually moves the data. In the design of I/O systems, we are primarily concerned with link control and secondarily with physical control.

The peer-to-peer interfaces are defined without specifying the interface to the next lower level. This is done for the same reasons that computer architecture is separated from computer organization and realization, as we explained in §1.1.1. It permits the next lower level to be replaced by another version at that level without affecting the higher level. This is like having one of the presidents get a new secretary: the presidents can still talk to each other in the same way, even though communication at the next lower level may be substantially changed.

We now discuss some of the issues at each of the levels. At the lowest level, the main issue is the medium, and a secondary issue is the multiplexing of several channels on one link. The technique used to synchronize the transmission of bits may be partly in the physical-interface level and partly in the link-control level.

The *medium* that carries a bit of information is of great concern to the communications engineer. Most systems would probably use voltage level to distinguish between true and false signals. In other systems, mechanical motion carries information, or radio or light beams in free space or in optical fibers carry information. Even when the carrier is electric, the signal can be carried by current rather than voltage, or by presence or absence of a particular frequency component. The signal can be conveyed on two frequencies: a true is sent as one frequency, while a false is sent as another frequency (*frequency shift keying*). More than one signal can be sent over the same medium. In *frequency multiplexing*, n messages are sent, each by the presence or absence of one of n different frequency components (or keying between n different pairs of frequencies). In *time multiplexing,* n messages can be sent, each one in a time slot every nth time slot. A frequency band or a time slot that carries a complete signal, enabling communication between two entities at the link-control level, is called a *channel.* Each channel, considered by itself, may be *simplex* if data can move in one direction only, *half-duplex* if data can move in either direction but in only one direction at a time, or *full-duplex* if data can move in both directions simultaneously.

Usually, a bit of information is sent on a channel over a time period, the *bit time period*, and this is the same time for each bit. The *baud rate* is the inverse of this bit time period (or the shortest period if different periods are used). The *bit rate*, in contrast, is the rate of transfer of information, as defined in information theory. For simplicity in this discussion, the bit rate is the number of user information bits sent per time unit, while the baud rate is the total number of bits – including user information, synchronization, and error-checking bits – per time unit.

In general, a clock is a regular occurrence of a pulse, or even of a code word, used to control the movement of bits. If such a (regular) clock appears in the channel in some direct way, the system is *synchronous*, otherwise it is *asynchronous*. In a synchronous system, the clock can be sent on a separate line, as the clock is sent inside a computer to synchronize the transmission of data. The clock can also be sent on the same wire as the data – every other bit being a clock pulse and the other bits being data – in the so-called *Manchester code*. Circuitry such as a phase-locked loop detects the clock, and further circuitry uses this reconstructed clock to extract the data. Finally, in an asynchronous link, the clock can be generated by the receiver in hopes that the clock matches the clock used by the sender.

Link control is concerned with how data is moved as bits, as groups of bits, and as complete messages that are sent by the next-higher-level peer-to-peer interface. Link control is usually implemented by I/O device chips.

At the *bit level*, individual bits are transmitted; at the *frame level*, a group of bits called a frame or packet is transmitted; and at the *message level*, sequences of frames, called messages, are exchanged. Generally, at the frame level, means are provided for detection and correction of errors in the data being sent, since the communication channel is often noisy. Also, because the frame is sent as a single entity, it can have means for synchronization. A frame, then, is some data packaged or framed and sent as a unit under control of a communications hardware/software system. The end-to-end user often wishes to send more data – a sequence of frames – as a single unit of data. The user's unit of data is known as the *message*.

At each level, a coordination mechanism, called a *protocol,* is used. A protocol is a set of conventions that coordinate the transmission and receipt of bits, frames, and messages. Its primary functions in the link-control level are the synchronization of messages and the detection or correction of errors. This term protocol suggests a strict code of etiquette and precedence countries agree to follow in diplomatic exchange, so the term aptly describes a communication mechanism whereby sender and receiver operate under some mutually acceptable assumptions but do not need to be managed by some greater authority like a central program. Extra bits are needed to maintain the protocol. Since these bits must be sent for a given data rate in bits per second, the baud rate must increase as more extra bits are sent. The protocol should keep efficiency high by using as few as possible of these extra bits. Note that a clock is a particularly simple protocol: a regularly occurring pulse or code word. An important special case, the *handshake protocol*, is an agreement whereby when information is sent to the receiver, it sends back an acknowledgment that the data is received either in good condition or has some error. Note, however, that a clock or a protocol applies to a level, so a given system can have a bit clock and two different protocols – a frame protocol and a message protocol.

A collection of individual protocols, each at a different level, is called a *stack*. Don't confuse this stack with the stack data structure described in §1.2.1, §1.2.3, and §2.1. This stack defines the overall protocol at all levels of interest to the discussion.

The third level of peer-to-peer interface is the network level. It is concerned about relationships between a large community of computers and the requirements necessary so that they can communicate to each other without getting into trouble.

The *structure* of a communications system includes the physical interconnections among stations, as well as the flow of information. Usually modeled as a graph whose nodes are stations and whose links are communications paths, the structure may be a loop, tree graph, or a rectangular grid (or sophisticated graph like a banyan network).

A path taken by some data through several nodes is called *store and forward* if each node stores the data for a brief time, then transmits it to the next node as new data may be coming into that first node; otherwise, if data pass through intermediate nodes instantaneously (being delayed only by gate and line propagation), the path is called a *circuit,* from telephone terminology. If such a path is half-duplex, it is sometimes called a bus because it looks like a bus in a computer system.

Finally, the communications system is *governed* by different techniques. This aspect relates to the operating system of the system of computers, which indirectly controls the generation and transmission of data much as a government establishes policies that regulate trade between countries. A simple aspect of governance is whether the decision to transmit data is centralized or distributed. A system is *centralized* if a special station makes all decisions about which stations may transmit data; it is *decentralized* or *distributed* if each station determines whether to send data, based on information in its locale. A centralized system is often called a *master/slave system,* with the special station the "master" and the other stations its "slaves." Other aspects of governance concern the degree to which one station knows what another station is doing, or whether and how one station can share the computational load of another.

9.2 Signal Transmission

The signal is transmitted through wires or light pipes at the physical level. This section discusses the characteristics of three of the most important linkages. Voltage or current amplitude logical signals, discussed first, are used to interconnect terminals and computers that are close to each other. The digital signal can be sent by transmitting it at different frequencies for a true and for a false signal (frequency shift keying). This is discussed in the next subsection.

9.2.1 Voltage and Current Linkages

In this section, we discuss the line driver and line receiver pair, the 20-mA current loop, and the RS-232 standard.

Standard high-current TTL or LSTTL drivers can be used over relatively short distances, as the IEEE-488 standard uses them for a bus to instruments located in a laboratory. However, slight changes in the ground voltage reference or a volt or so of noise on the link can cause a lot of noise in such links. A *differential line* is a pair of wires, in which the variable in positive logic is on one wire and in negative logic on the other wire. If one is high, the other is low. The receiver uses an analog comparator to determine which of the two wires has the higher voltage, and outputs a standard TTL signal appropriately. If a noise voltage is induced, both wires should pick up the same

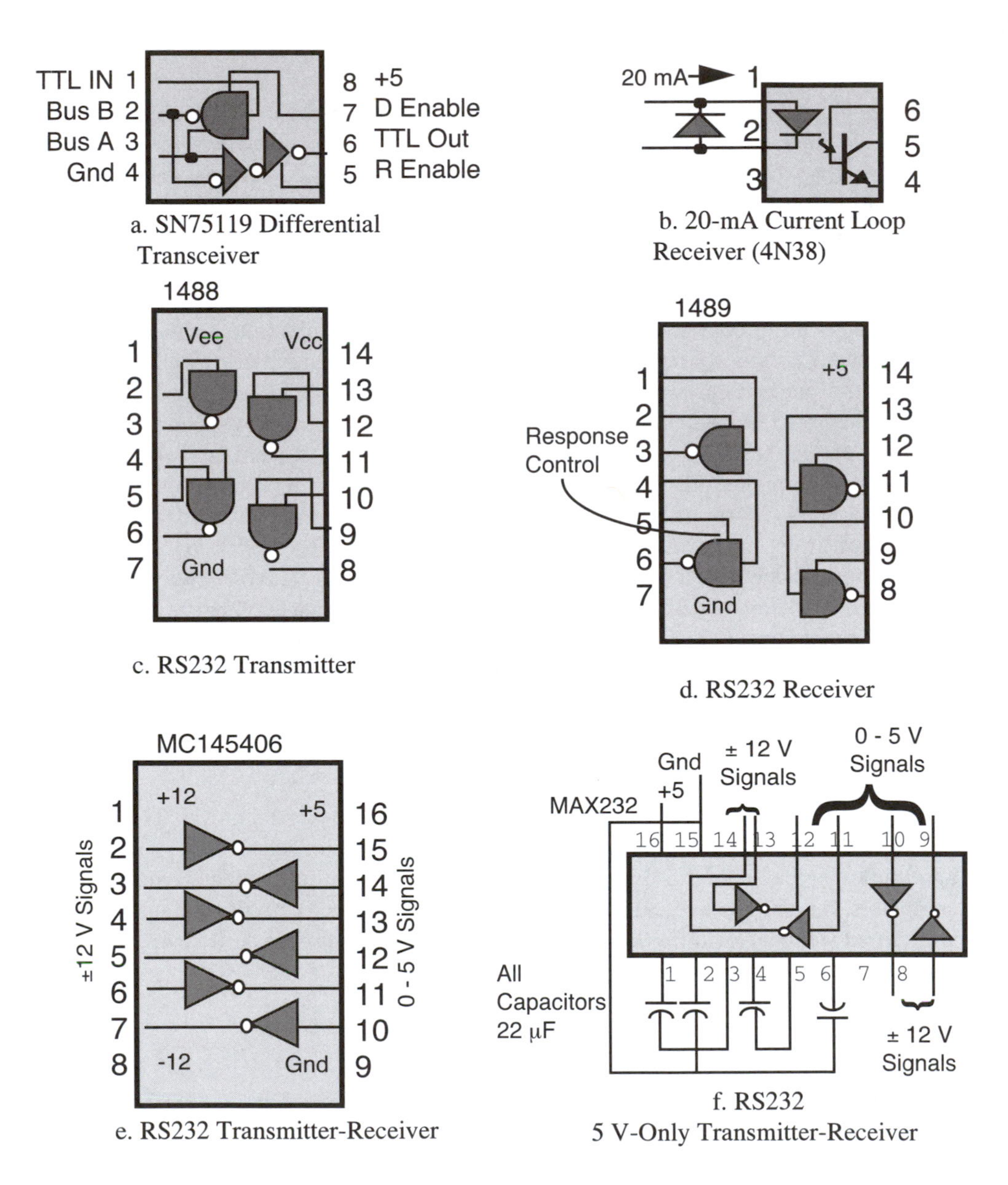

a. SN75119 Differential Transceiver

b. 20-mA Current Loop Receiver (4N38)

c. RS232 Transmitter

d. RS232 Receiver

e. RS232 Transmitter-Receiver

f. RS232 5 V-Only Transmitter-Receiver

Figure 9.2. Drivers and Receivers

noise so the differential is not affected and the receiver gets the correct signal. Similarly, imperfect grounding and signal ringing affect the signal on both wires and their effect is cancelled by the voltage comparator. A number of driver and receiver integrated circuits are designed for differential lines, but some require voltages other than +5, which may not be used elsewhere in the system. An integrated circuit suitable for driving and receiving signals on a half-duplex line, using a single 5-V supply, is the SN75119, shown in Figure 9.2a. If driver enable DE (pin 7) is high, then the signal on IN (pin 1) is put on line LA (pin 3) and its complement is put on line LB (pin 2); otherwise the pins LA and LB appear to be (essentially) open circuits. If receiver enable RE (pin 5) is high, then the output OUT (pin 6) is low if the voltage on LA is less than that on LB, or high if the voltage on LA is greater than that on LB; if RE is low, OUT is (essentially) an open circuit. The *RS-442 standard* (RS means "recommended standard") uses basically this differential line, but a driver such as the Am26LS30 has means to control the slew rate of the output signal.

The 20-mA current loop is often used to interface to teletypes or teletype-like terminals. A pair of wires connect driver and receiver so as to implement an electrical loop through both. A true corresponds to about 20 mA flowing through the loop, and a false corresponds to no current or to 20 mA flowing through the loop in the reverse direction (for "neutral working" or "polar working" loops, respectively). A current, rather than a voltage, is used because it can be interrupted by a switch in a keyboard and can be sensed anywhere in the loop. A current is also used in older equipment because the 20-mA current loop was used to drive a solenoid, and a solenoid is better controlled by a current than a voltage to get faster rise times. The current is set at 20 mA because the arc caused by this current will keep the switch contacts clean.

A 20-mA current loop has some problems. A loop consists of a current source in series with a switch to break the circuit, which in turn is in series with a sensor to sense the current. Whereas the switch and sensor are obviously in two different stations in the circuit, the current source can be in either station. A station with a current source is called *active*, while one without is *passive*. If two passive stations, one with a switch and the other with a sensor, are connected, nothing will be communicated. If two active stations are connected, the current sources might cancel each other or destroy each other. Therefore, one station must be active while the other is passive, and one must be a switch and the other must be a sensor. While this is all very straightforward, it is an invitation to trouble. Also, note that the voltage levels are undefined. Most 20-mA current loops work with voltages like +5 or -12 or both, which are available in most communications systems; but some, designed for long-distance communication, utilize "telegraph hardware" with voltages in excess of 80 V. Therefore, one does not connect two 20-mA current loop stations together without checking the voltage levels and capabilities. Finally, these circuits generate a fair amount of electrical noise, which gets into other signals, especially lower-level signals, and the switch in such a circuit generates noise that is often filtered by the sensor. This noise is at frequencies used by 1200-baud lines, so this filter can't be used in other places in a communications subsystem. The circuitry for a 20-mA current loop can be built with an opto-isolator, as shown in Figure 9.2b. If the current through the LED is about 20 mA, the phototransistor appears to be a short circuit; if the current is about 0 mA, it is an open circuit and the output is high. The diode across the LED is there to prevent an incorrect current from destroying the LED.

An interface standard developed by the Electronic Industries Association (EIA) and other interested parties has evolved into the RS-232C (recommended standard 232 version C). A similar standard is available in Europe, developed by the Comite Consultatif Internationale de Telegraphie et Telephonie (CCITT), and is called the CCITT V.24 standard. These standards are supposed to be simple and effective, so that any driver

Table 9.1. RS-232 Pin Connections for D25P and D25S Connectors

Pin	Name	Function
1	Protective ground	Connects machine or equipment frames together and to "earth"
2	Transmitted data	Data sent from microcomputer to terminal
3	Receive data	Data sent from terminal to microcomputer
4	Request to send	(Full duplex) enables transmission circuits (Half duplex) puts link in transmit mode and disables receive circuitry
5	Clear to send	Responds to Request to send; when high, it indicates the transmission circuitry is working
6	Data set ready	(telephone links) The circuitry is not in test, talk, or dial modes of operation so it can be used to transmit and receive
7	Signal ground	Common reference potential for all lines Should be connected to "earth" at just one point, to be disconnected for testing
8	Data carrier detect	A good signal is being received
9	+P	+12 V (for testing only)
10	-P	-12 V (for testing only)
11 . . . 25		Used for more elaborate options

conforming to it can be connected to any receiver conforming to it, covering the voltage levels used for the signals as well as the pin assignments and dimensions of the plugs. Basically, a false variable is represented by any voltage from +15 to +5 V, and a true by any voltage from -5 to -15 V (negative logic is used). A number of specifications concerning driver and receiver currents and impedances can be met by simply using integrated circuit drivers and receivers that are designed for this interface – RS-232 drivers and RS-232 receivers. The MC1488 is a popular quad RS-232 line driver, and the MC1489 is a popular receiver. (See Figure 9.2c.) The driver requires +12 V on pin 14 and -12 V on pin 1. Otherwise, it looks like a standard quad TTL NAND gate whose outputs are RS-232 levels. The four receiver gates have a pin called response control (pins 2, 5, 9, and 12). Consider one of the gates, where pin 1 is the input and pin 3 is the output. Pin 2 can be left unconnected. It can be connected through a 33-KΩ resistor to the negative supply voltage (pin 1) to raise the threshold voltage a bit. Or it can be connected through a capacitor to ground, thus filtering the incoming signal. This controls the behavior of that gate. The other gates can be similarly controlled. The MC145406 is a chip that combines three transmitter and three receiver gates in one chip (Figure 9.2e); and the MAX232 (Figure 9.2f), made by MAXIM, has two transmitters and two receivers, and a charge-pump circuit that generates ±10 V needed for the transmitter from the 5-V supply used by the microcomputer. (This marvelous circuit is just what is needed in many applications, but some MAX232 chips have a small problem: if the 5-V supply turns on too fast, the charge pump fails to start; put a small

– 10 Ω – resistor in series with the 5-V pin and put a large – 100 μF – capacitor from that pin to ground.)

The RS-232 interface standard also specifies the sockets and pin assignments. The DB25P is a 25-pin subminiature plug, and the DB25S is the corresponding socket; both conform to the standard. The pin assignments are shown in Table 9.1. For simple applications, only pins 2 (transmit data), 3 (receive data), and 7 (signal ground) need be connected; but a remote station may need to make pins 5 (clear to send), 6 (data set ready), and 8 (data carrier detect) 12 V to indicate that the link is in working order, if these signals are tested by the microcomputer. These can be wired to -12 V in a terminal when they are not carrying status signals back to the microcomputer.

9.2.2 Frequency Shift-Keyed Links Using Modems

To send data over the telephone, a *modem* converts the signals to frequencies that can be transmitted in the audio frequency range. The most common modem, the Bell 103, permits full-duplex transmission at 300 baud. Transmission is originated by one of the modems, referred to as the *originate modem,* and is sent to the other modem, referred to as the *answer modem.* The originate modem sends true (mark) signals as a 1270-Hz sine wave and false (space) signals as a 1070-Hz sine wave. Of course, the answer modem receives a true as a 1270-Hz sine wave and a false as a 1070-Hz sine wave. The answer modem sends a true (mark) as a 2225-Hz sine wave and a false (space) as a 2025-Hz sine wave. Note that the true signal is higher in frequency than the false signal, and the answer modem sends the higher pair of frequencies.

Some modems are originate only. They can only originate a call and can only send 1070- or 1270-Hz signals and receive only 2025- or 2225-Hz signals. Most inexpensive modems intended for use in terminals are originate only. The computer may have an answer-only modem with the opposite characteristics. If you want to be able to send data between two computers, one of them has to be an originate modem. So an answer/originate modem might be used on a computer if it is expected to receive and also send calls. Whether the modem is originate-only, answer-only, or answer/originate, it is fully capable of sending and receiving data simultaneously in full-duplex mode. The originate and answer modes determine only which pair of frequencies can be sent and received, and therefore whether the modem is capable of actually initiating the call.

Modems have filters to reject the signal they are sending and pass the signals they are receiving. Usually, bessel filters are used because the phase shift must be kept uniform for all components or the wave will become distorted. Sixth-order and higher filters are common to pass the received and reject the transmitted signal and the noise, because the transmitted signal is usually quite a bit stronger than the received signal, and because reliability of the channel is greatly enhanced by filtering out most of the noise. The need for two filters substantially increases the cost of answer/originate modems.

The module that connects the telephone line to the computer is called a *data coupler*, and there is one that connects to the originator of a call and another that connects to the answerer. The data coupler isolates the modem from the telephone line to prevent lightning from going to the modem and to control the signal level, using an automatic gain control; but the data coupler does not convert the signal or filter it. The data coupler has three control/status signals. *Answer phone* ANS is a control command that has the same effect on the telephone line as when a person picks up the handset to start a call or answer the phone. *Switch hook* SH is a status signal that indicates that the telephone handset is on a hook, if you will, so it will receive and transmit signals to the modem. Switch hook may also be controlled by the microcomputer. Finally, *ring indicator* RI is a status signal that indicates the phone is ringing.

Aside from the fact that data are sent using frequency analog signals over a telephone, there is not much to say about the channel. However, the way an originate modem establishes a channel to an answer modem and the way the call is terminated is interesting. We now discuss how the Motorola M6860 modem originates a call and answers a call. Calling a modem from another, maintaining the connection, and terminating the connection involve handshaking signals *data terminal ready* DTR and *clear to send* CTS in both originate and answer modems. (See Figure 9.3a for a diagram showing these handshaking signals.) If a modem is connected to an RS-232C line, as it often is, data terminal ready can be connected to request to send (pin 4) and clear to send can be connected to the clear to send (pin 5) or the data set ready (pin 6), whichever is used by the computer. Figure 9.3b shows the sequence of operations in the modems and on the telephone line, showing how a call is originated and answered by the Motorola M6860 modem chip.

The top line of Figure 9.3b shows the handshaking signals seen by the originator, the next line shows signals seen by the originator modem, the center line shows the telephone line signals, the next line shows signals seen by the answer modem, and the bottom line shows the handshaking signals seen by the answerer. As indicated, the originator asserts the switch-hook signal. This might be asserted by putting the telephone handset on the modem hook or by an output device that asserts this signal. This causes the command ANS (answer phone) to become asserted, which normally enables the data coupler electronics to transmit signals. The telephone is now used to dial up the answerer (17 seconds is allowed for dialing up the answerer). The answering modem receives a command RI (ring indicator) from the telephone, indicating that the phone is ringing. It then asserts the ANS signal to answer the phone, enabling the data coupler to amplify the signal. The answerer puts a true signal, 2225 Hz, on the line. The originator watches for that signal. When it is present for 450 ms, the originator will send its true signal, a 1270-Hz sine wave. The answerer is watching for this signal. When it is present for 450 ms, the answerer asserts the CTS command and is able to begin sending data. The originator meanwhile asserts CTS after the 2225-Hz signal has been present for 750 ms. When both modems have asserted CTS, full-duplex communication can be carried out.

Some answer modems will automatically terminate the call. To terminate the call, send more than 300 ms of false (space) 1070 Hz. This is called a *break* and is done by your terminal when you press the “break” key. The answer modem will then hang up the phone (negate ANS) and wait for another call. Other modems do not have this automatic space disconnect; they terminate the call whenever neither a high nor a low frequency is received in 17 s. This occurs when the telephone line goes dead or the other modem stops sending any signal. In such systems, the “break” key and low frequency sent when it is pressed can be used as an attention signal rather than a disconnect signal.

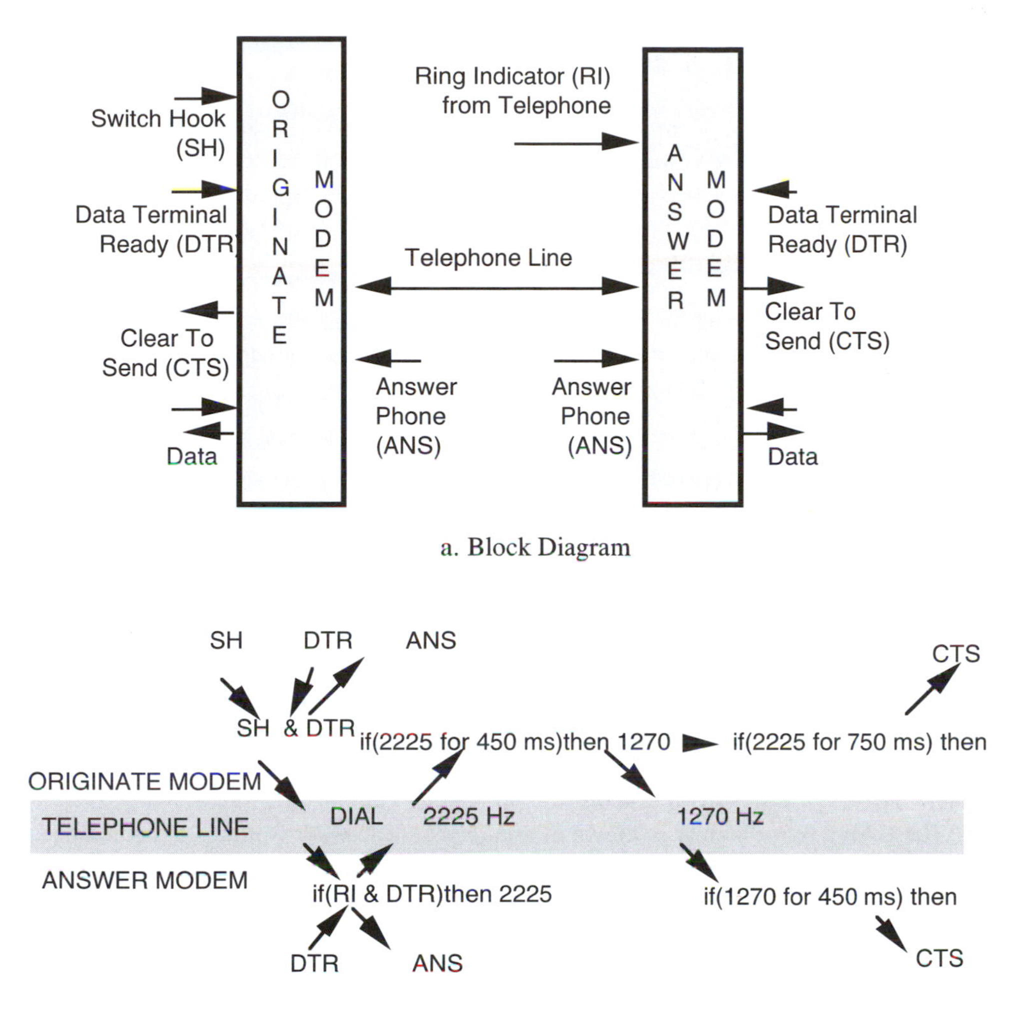

a. Block Diagram

b. Handshake Sequence

Figure 9.3. Originating a Call on a Modem

9.2.3 Infrared Links

Another scheme to transmit serial data is to send it on an infrared carrier, according to the IRDA Serial Infrared Physical Layer Specification. A logic 0 (F) is sent as emission of infrared light for $^{3}/_{16}$ of a bit time, followed by $^{13}/_{16}$ of a bit time of darkness. A logic 1 (T) is sent as a bit time of darkness. Data are sent by driving current through an infrared diode, and data are received by having the infrared light shine on a phototransistor, much as in an opto-isloator. The main advantage of this scheme is that the two communicating devices need not have any electical connections between them, so there is significantly less danger of electrical shock and noise.

9.3 UART Link Protocol

By far the most common technique for transmitting data is that used by the *universal asynchronous receiver-transmitter* (UART). This simple protocol is discussed in this section. Software generation of UART signals, discussed first, is quite simple and helps to show how they are sent. The UART-like chip – the ACIA – designed for the 6812 family is covered in the next subsection. A special remote-control chip that uses the UART protocol is discussed in the next subsection, then the UART chip is discussed. A system inside the 6812 that is capable of UART signal generation and reception, the serial communications interface (SCI), is discussed last because it has several useful but nonstandard extensions to the UART protocol.

9.3.1 UART Transmission and Reception by Software

As noted earlier, the UART is a module (integrated circuit) that supports a frame protocol to send up to eight bit frames (characters). We call this the *UART protocol*. However, the UART protocol can be supported entirely under software control, without the use of a UART chip or its equivalent. A study of this software is not only a good exercise in hardware/software trade-offs, but is an easy way to teach the protocol; the software approach also is a practical way to implement communication in a minimum-cost microcomputer. However, we do warn the reader that most communication is done with UART chips or their equivalent, and low-cost microprocessors such as the 6812 already have a built-in "UART" on the microprocessor chip itself, which is the SCI.

The UART frame format is shown in Figure 9.4. (The UART protocol is contained within the UART frame format.) When a frame is not being sent, the signal is high. When a signal is to be sent, a *start bit*, a low, is sent for one bit time. The frame, from 5 to 8 bits long, is then sent 1 bit per bit time, least significant bit first. A parity bit may then be sent and may be generated so that the parity of the whole frame is always even (or always odd). To generate even parity, if the frame itself had an even number of 1s already, a low parity bit is sent, otherwise a high bit is sent. Finally, one or more *stop bits* are sent. A stop bit is high, and is indistinguishable from the high signal that is sent when no frame is being transmitted. In other words, if the frame has n stop bits (n = 1, 1 1/2, or 2), this means the next frame must wait that long after the last frame bit or parity bit of the previous message has been sent before it can begin sending its start bit. However, it can wait longer than that.

Figure 9.4. Frame Format for UART Signals

In addition to the format above, the protocol has some rules for sampling data and for error correction. A clock, used in the receiver, is 16 times the bit rate, and a counter, incremented each clock time, is used to sample the incoming data. (The same clock is used in the transmitter to generate the outgoing data.) The counter is started when the input signal falls, at the beginning of a frame. After 8 clock periods, presumably in the middle of the start bit, the input is sampled. It should be low. If it is high, the falling edge that started the counter must be due to some noise pulse, so the receiver returns to examine the input for the leading edge of a start bit. If this test passes, the input is sampled after every 16 clock periods, presumably in the middle of each bit time. The data bits sampled are reassembled in parallel. The parity bit, if one is used, is then sampled and checked. Then the stop bits are checked.

The following are definitions of error conditions. If the parity bit is supposed to be even, but a frame with odd parity is received, a *parity error* is indicated. This indicates that one of the frame bits or the parity bit was changed due to noise. Note that two errors will make the parity appear correct – but two wrongs don't make a right. Parity detection cannot detect all errors. Even so, most errors are single-bit errors, so most errors are detetected. If a stop bit is expected, but a low signal is received, the frame has a *framing error*. This usually indicates that the receiver is using the wrong clock rate, either because the user selected the wrong rate or because the receiver oscillator is out of calibration. However, this condition can arise if the transmitter is faulty, sending frames before the stop bits have been timed out, or if more than one transmitter are on a link and one sends before another's stop bits are completely sent. Finally, most UART devices use a buffer to store the incoming word, so the computer can pick up this word at leisure rather than at the precise time that it has been shifted in. This technique is called *double buffering*. But if the buffer is not read before another frame arrives needing to fill the same buffer, the first frame is destroyed. This error condition is called an *overrun error*. It usually indicates that the computer didn't empty the receiver buffer before a subsequent message arrived.

The UART communication technique is based on the following principle. If the frame is short enough, a receiver clock can be quite a bit out of synchronization with the transmitter clock and still sample the data somewhere within the bit time when the data are correct. For example, if a frame has 10 bits and the counter is reset at the leading edge of the frame's start bit, the receiver clock could be 5% faster or 5% slower than the transmitter clock and still, without error, pick up all the bits up to the last bit of the frame. It will sample the first bit 5% early or 5% late, the second 10%, the third 15%, and the last 50%. This means the clock does not have to be sent with the data. The receiver can generate a clock to within 5% of the transmitter clock without much difficulty. However, this technique would not work for long frames, because the accumulated error due to incorrectly matching the clocks of the transmitter and receiver would eventually cause a bit to be missampled. To prevent this, the clocks would have to be matched too precisely. Other techniques become more economical for longer frames.

A C procedure `SUart` to generate a signal compatible with the UART protocol is quite simple. The procedure is shown below and its description follows. It uses the same pins (`PORTS` bits 2 and 3) that will be used later with the SCI1 examples.

```
void SUart(char c) { unsigned char i, parity;
     DDRS = TxD1; parity = PORTS = 0 ; delay(N);
     for(i = 8; i > 0; i-- ) {
          if(c & 1) { PORTS = TxD1; parity++; }
          else PORTS = 0; c >>= 1; delay(N);
     }
     if(parity & 1) PORTS = TxD1; else PORTS = 0;
     delay(N); PORTS = TxD1; delay(N); delay(N);
}

void main(){ char c = RUart(); SUart(c);}
```

In the above procedure we use the following delay procedure, whose argument is the time delay. Let `N` be the parameter that delays for the time to send one bit.

```
void delay(int t){ while(--t); }
```

The start bit is output from the least significant bit of `c`, and a delay subroutine is called to delay one bit time. Then the bits are written to the output port so that the least significant bit is sent out the serial channel, and parity is updated with the exclusive-OR of parity and data so that the least significant bit is the parity of the data sent. This is repeated for 8 data bits. Then the parity bit is output, and the stop bit is output. Appropriate delays are inserted between each bit that is sent serially.

A C procedure `RUart` to receive a UART frame is also quite simple. Again, the subroutine is shown and its description follows it.

```
char RUart(){ unsigned char i, parity, c;
     do {while(PORTS & RxD1) ; delay(N/2); } while (PORTS & RxD1);
     parity = c = 0; delay(N);
     for(i = 8; i > 0; i-- )
          { if(PORTS & RxD1) { c |= 0x80; parity++;} c >>= 1; delay(N);}
     if(PORTS & RxD1) parity++; if (parity & 1) {/* report parity error */;}
     delay(N); if(!(PORTS & RxD1)) { /* report framing error */;}
     delay(N); if(!(PORTS & RxD1)) { /* report framing error */;}
     return c;
}
```

The while loop waits for the input to go low, and the `do while` loop confirms that it is still low after half a bit time (using the procedure `delay` to delay half a bit time). Then, after a delay of a bit time, the least significant bit is picked up, and is exclusive-ORed with the computed parity bit. For eight steps, another bit is picked up, the parity is updated, and a bit delay is wasted. Then the transmitted parity bit is combined with the computed parity bit to determine if a parity error occurred, and the stop bits are checked.

Both of the preceding C procedures are simple enough to follow. They can be done in software without much penalty because the microprocessor is usually doing nothing while frames are being input or output. In an equivalent hardware alternative, essentially the same algorithms are executed inside the UART chip or an equivalent chip like the ACIA. The hardware alternative is especially valuable where the microcomputer can do something else as the hardware tends to transmitting and receiving the frames, or when it might be sending a frame at the same time it might be receiving another frame (in a full-duplex link or in a ring of simplex links). In other cases, the advantages of the hardware and software approaches are about equal: the availability of cheap, simple UART chips favors the hardware approach, while the simplicity of the program favors the software approach. The best design must be picked with care and depends very much on the application's characteristics.

9.3.2 The UART

The UART chip is designed to transmit and/or receive signals that comply with the UART protocol (by definition). This protocol allows several variations (in baud rate, parity bit, and stop bit selection). The particular variation is selected by strapping different pins on the chip to high or to low. The UART can be used inside a microcomputer to communicate with a teletype or a typewriter, which was its original use, or with the typewriter's electronic equivalent, such as a CRT display. It can also be used in other remote stations in security systems, in stereo systems controlled from a microcomputer, and so on. Several integrated circuit companies make UARTs, which are all very similar. We will study one that has a single-supply voltage and a self-contained oscillator to generate the clock for the UART, the Intersil IM6403. See Figure 9.5.

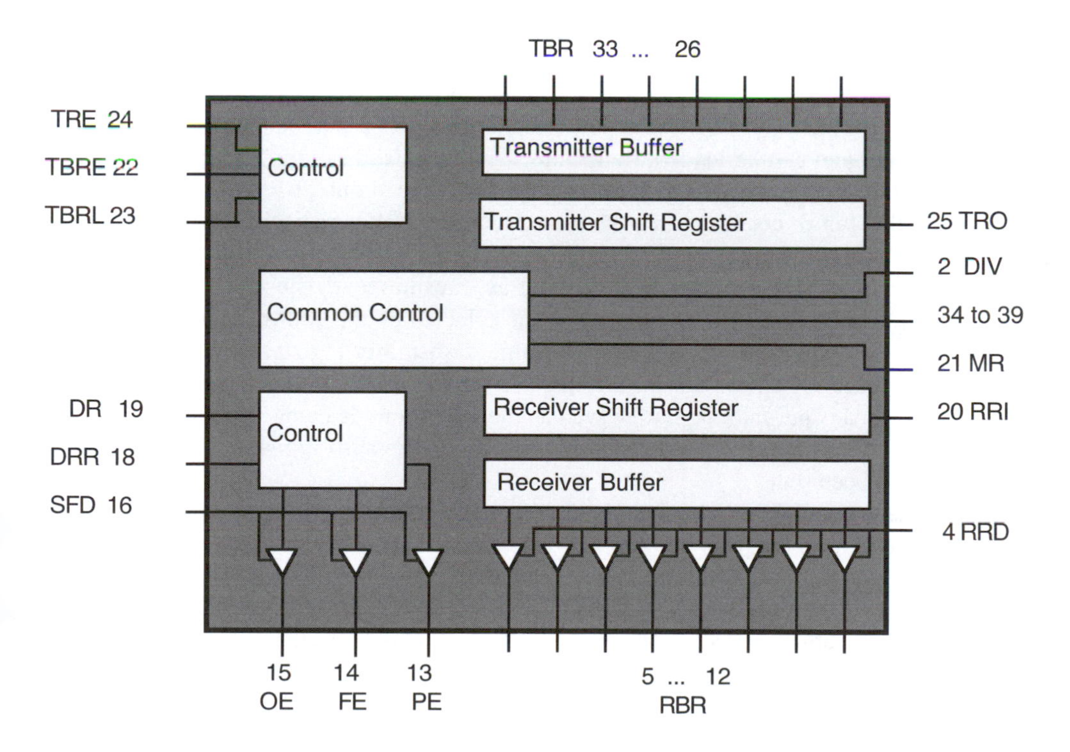

Figure 9.5. Block Diagram of a UART (IM6403)

The UART contains a transmitter and a receiver that run independently, for the most part, but share a common control that selects the baud rate and other variations for both transmitter and receiver. We discuss the common control first, then the transmitter, and then the receiver. The baud rate is selected by the crystal connected to pins 17 and 40 and by the divide control DIV on pin 2. If DIV is high, the oscillator frequency is divided by 16; if low, by 2^{11}. If the crystal is a cheap TV crystal (3.5795 MHz) and DIV is low, the baud rate is close to 110, which is commonly used for teletypes. When master reset MR, on pin 21, is high, it resets the chip; normally, it is grounded. The other control bits are input on pins 39 to 35 and are stored in a latch inside the chip. The latch stores the inputs when pin 34 is high. This pin can be held high to defeat the storage mechanism, so the pin levels control the chip directly. Pin 36 selects the number of stop bits: low selects 1 stop bit, high selects 2 (except for an anomaly of little interest). If pin 35 is high, no parity bit is generated or checked; otherwise, pin 39 selects even parity if high and odd parity if low. Pins 37 and 38 select the number of data bits per frame; the number is 5 plus the binary number on these pins. The user generally determines the values needed on these pins from the protocol he or she is using, and connects them to high or low. However, these inputs can be tied to the data bus of a computer, and pin 34 can be asserted to load the control latch to effect an output register; in the reset handler, the computer can then set the control values under software control.

The operation of the transmitter and receiver is compactly and simply explained in the data sheets of the 6403, which are paraphrased here. The transmitter has a buffer register, which is loaded from the signals on pins 33 (msb) to 26 (lsb) when transmitter buffer register load TBRL (pin 23) rises. If $n < 8$ bits are sent, the rightmost n bits on these pins are sent. Normally, these pins are tied to the data bus to make the buffer look like an output register, and TBRL is asserted when the register is to be loaded. When this buffer is empty and can be loaded, transmitter buffer register empty TBRE (pin 22) is high; when full, it is low. (SFD, pin 16, must be low to read out TBRE.) The computer may check this pin to determine if it is safe to load the buffer register. It behaves as a BUSY bit in the classical I/O mechanism. The data in the buffer are automatically loaded into the transmitter shift register to be sent out as transmitter register output TRO (pin 25) with associated start, parity, and stop bits as selected by the control inputs. As long as the shift register is shifting out part of a frame, transmitter register empty TRE (pin 24) is low. Figure 9.6 shows a typical transmission, in which two frames are sent out. The second word is put into the buffer even as the first frame is being shifted out in this double-buffered system. It is automatically loaded into the shift register as soon as the first frame has been sent.

The receiver shifts data into a receiver shift register. When a frame has been shifted in, the data are put in the receiver buffer. If fewer than 8 bits are transmitted in a frame, the data are right-justified. This data can be read from pins 5 to 12, when receive register disable RRD (pin 4) is asserted low. Normally these pins are attached to a data bus, and RRD is used to enable the tristate drivers when the read buffer register is to be read as an input register. If RRD is strapped low, then the data in the read buffer are continuously available on pins 5 to 12. When the read buffer contains valid data, the data ready DR signal (pin 19) is high, and the error indicators are set. (DR can only be read when SFD on pin 16 is high.) The DR signal is an indication that the receiver is DONE, in the classic I/O mechanism, and requests the program to read the data from the receiver buffer and read the error indicators if appropriate. The error indicators are reloaded after each frame is received, so they always indicate the status of the last frame that was received. The error indicators, TBRE, and DR can be read from pins 15, 14, 13, 22, and 19 when SFD (pin 16) is asserted low, and indicate an overrun error, a framing error, a parity error, an empty transmitter buffer, and a full receiver buffer, respectively, if high. The error and buffer status indicators can be read as another input register by connecting pins 22, 19, and 15 to 13 to the data bus, and asserting SFD when this register is selected; or, if SFD is strapped low, the error and buffer status indicators can be read directly from those pins. When the data are read, the user should reset the DR indicator by asserting data ready reset DRR (pin 18) high. If this is not done, when the next frame arrives and is loaded into the buffer register, an overrun error is indicated.

The UART can be used in a microcomputer system. The control bits (pins 35 to 39) and the transmitter buffer inputs (pins 26 to 33) can be inputs, and the buffer status and error indicators (pins 22 and 19, and 15 to 13) and receive data buffer outputs (pins 5 to 12) can be outputs. All the inputs and outputs can be attached to the data bus. TBRL, SBS, SFD, and RRD (pins 23, 36, 16, and 4) are connected to an address decoder so that the program can write in the control register or transmitter buffer register, or read from the error indicators or the read buffer register. The TBRE signal (pin 22) is used as a BUSY bit for the transmitter; and the DR signal (pin 19) is used as a DONE bit for the receiver. When the UART is used in a gadfly technique,which can be extended to interrupt or even DMA techniques, the program initializes the UART by writing the appropriate control bits into the control register. To send data using the gadfly approach, the program checks to see if TBRE is high and waits for it to go high if it is not. When it is high, the program can load data into the transmitter buffer. Loading data into the buffer will automatically cause the data to be sent out. If the program is expecting data from the receiver in the gadfly technique, it waits for DR to become high. When it is, the program reads data from the receiver buffer register and asserts DRR to tell the UART that the buffer is now empty. This makes DR low until the next frame arrives.

The UART can be used without a computer in a remote station that is implemented with hardware. Control bits can be strapped high or low, and CRL (pin 34) can be strapped high to constantly load these values into the control register. Data to be collected can be put on pins 33 to 27. Whenever the hardware wants to send the data, it asserts TBRL (pin 23) low for a short time, and the data get sent. The hardware can examine TBRE (pin 22) to be sure that the transmitter buffer is empty before loading it, but if the timing works out such that the buffer will always be empty, there is no need to check this value. It is pretty easy to send data in that case. Data, input serially, are made available and are stable on pins 5 to 12. Each time a new frame is completely shifted in, the data are transferred in parallel into the buffer. RRD (pin 4) would be strapped low to constantly output this data in a hardware system. When DR becomes high, new data have arrived, which might signal the hardware to do something with the data. The hardware should then assert DRR high to clear DR. (DR can feed a delay into DRR to reset itself.) The buffer status and error indicators can be constantly output if SFD (pin 16) is strapped low, and the outputs can feed LEDs, for instance, to indicate an error. However, in a simple system when the hardware does not have to do anything special with the data except output them, it can ignore DR and ignore resetting it via asserting DRR. In this case, the receiver is very simple to use in a remote station.

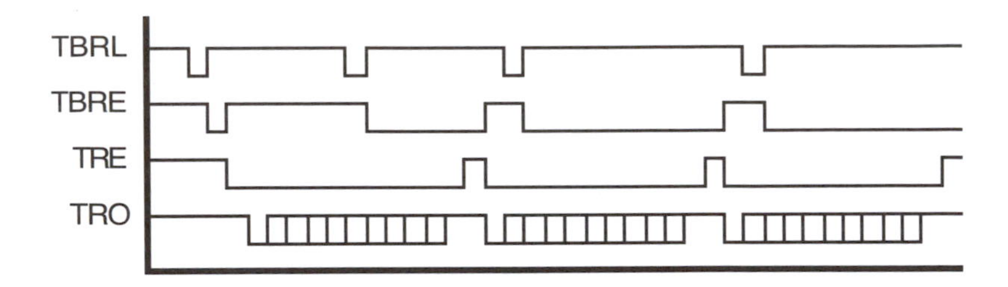

Figure 9.6. Transmitter Signals

9.3.3 The ACIA

The *asynchronous communications interface adapter* (*ACIA*) is a UART that has been specially tailored for use as an external chip for the 6812 microcomputer in normal expanded narrow mode. This section covers the ACIA's highlights. A complete description is available in the ACIA data sheet. The ACIA is designed for the Motorola microcomputer. It can also be used in other microcomputers, and other microcomputer manufacturers have special chips like the ACIA for their systems. The ACIA is different from a UART like the IM6403 in the following ways. To save pins, a bit of the transmitter buffer input, a bit of the receiver buffer output, a bit of the control register, and a bit of the buffer and error status register output are internally connected and then connected to a single pin on this chip. Thus, only 8 pins are used to connect to the data bus. An external clock is needed to set the baud rate, and the transmitter can have a different clock than the receiver. Also, because this chip is designed to connect to a modem, discussed in the next section, it has three pins to control the modem so that the program can control it. Finally, it has a status register with interrupt request logic so that the 6812 can easily examine it in its interrupt handler. (A diagram of the ACIA is shown in Figure 9.7; for simplicity, the system is configured so that this chip is addressed at locations 0x200 and 0x201.)

The transmitter, with its buffer and shift register, and the receiver and its shift register, operate just as in the UART. They are addressed in the same location because the transmitter buffer is write-only, while the receiver buffer is read-only. Once the control register is set up, a word is transmitted simply by writing it in location 0x201, and an incoming word is picked up by reading it from location 0x201.

The 'A4 is configured in narrow extended bus mode to provide data and address buses, E, RW, and CS0. Review §6.4 to determine how to slow the E clock to satisfy the 2-MHz MC68B50's bus timing, and how to provide the data, address, and control signals. As discussed in §8.2.2, timer device 0 provides its transmit and receive clocks for 1200 baud; they are 19.2 KHz (52.08 μs or 52 E clock cycles per interrupt). We want to send 8 data bits, even parity, and 1 stop bit. No interrupts are generated, and RTS is low. Consult Table 9.2; bits 7 to 5 should be F (0) to disable interrupts and set RTS low; bits 4 to 2 should be TTF (110) to select 8 data bits, even parity, and 1 stop bit; and bits 1 and 0 should be FT (01) to divide the clock rate, 153.6 KHz, by 16, to deliver 1200 baud. The control word should be 0x19. The instructive pattern 0xC5 is repetitively output to the data port to immediately refill the buffer every time it is empty. (If the buffer is already full, writing another word into it causes that data to be lost. Normally the program checks a status bit to be sure this buffer is empty before filling it. But in this case, there is no harm in constantly writing the same word into the data port.) The output TxD should appear on an oscilloscope, as in Figure 9.8.

Table 9.2. ACIA Control and Status Bits

Bits	Function
1,0	Clock frequency
0 0	Divide by 1
0 1	Divide by 16
1 0	Divide by 64
1 1	Master reset

Bits	Function
4,3,2	Frame format
0 0 0	7 bits, even parity, 2 stop bits
0 0 1	7 data, odd parity, 2 stop bits
0 1 0	7 data, even parity, 1 stop bit
0 1 1	7 data, odd parity, 1 stop bit
1 0 0	8 data, 2 stop bits
1 0 1	8 data, 1 stop bit
1 1 0	8 data, even parity, 1 stop bit
1 1 1	8 data, odd parity, 1 stop bit

Bits	Function	
6,5	Trans. int.	RTS
0 0	Disable	low
0 1	Enable	low
1 0	Disable	high
1 1	Disable	low *

* Transmit data output is low.

Bits	Function
7	Receiver interrupt
0	Disable
1	Enable

```
enum{SnglChip,Narrow,Peripheral,Wide,Normal,Special=0,ModeFld = 5};

enum { RDWRE = 4, LSTRE = 8, NECLK = 0x10 };

enum{ CSP1E = 0x4040, CSP0E = 0x2000, CSDE = 0x1000, CS3E = 0x800,
    CS2E = 0x400, CS1E = 0x200, CS0E = 0x100};

enum{ STR0, STR1=2, STR2=4, STR3=8, STRD=10, STP0=12, STRP1=14};

interrupt 8 void handler(){ TFLG1 = 1; TC0 += N; }

void main(){ char *acia = (char *)0x200; asm cli
    CLKCTL = 2 << 2; /* E clock is 2 MHz; high or low for 250 ns */
    MODE = ( Special + Narrow ) << ModeFld; /* select 6812 mode */
    PEAR = RDWRE; /* enable port E alternate functions: RW, E */
    CSCTL= CS0E; /* enable chip select 0 */ asm sei
    TSCR = 0x80; TIOS = 1; TCTLO = 1; TC0 = N + TCNT;
    TMSK1 = 1; /* enable interrupt for device 0 */ asm cli
    *acia = 0x03; /*11 to bits 1,0 of ACIA cntrl port to reset it*/
    *acia = 0x19; /* ACIA cntrl.: 8 data, even parity, 1 stop */
    do acia[1]=0xC5; while (1); /*data to ACIA transmitter output buffer port*/
}
```

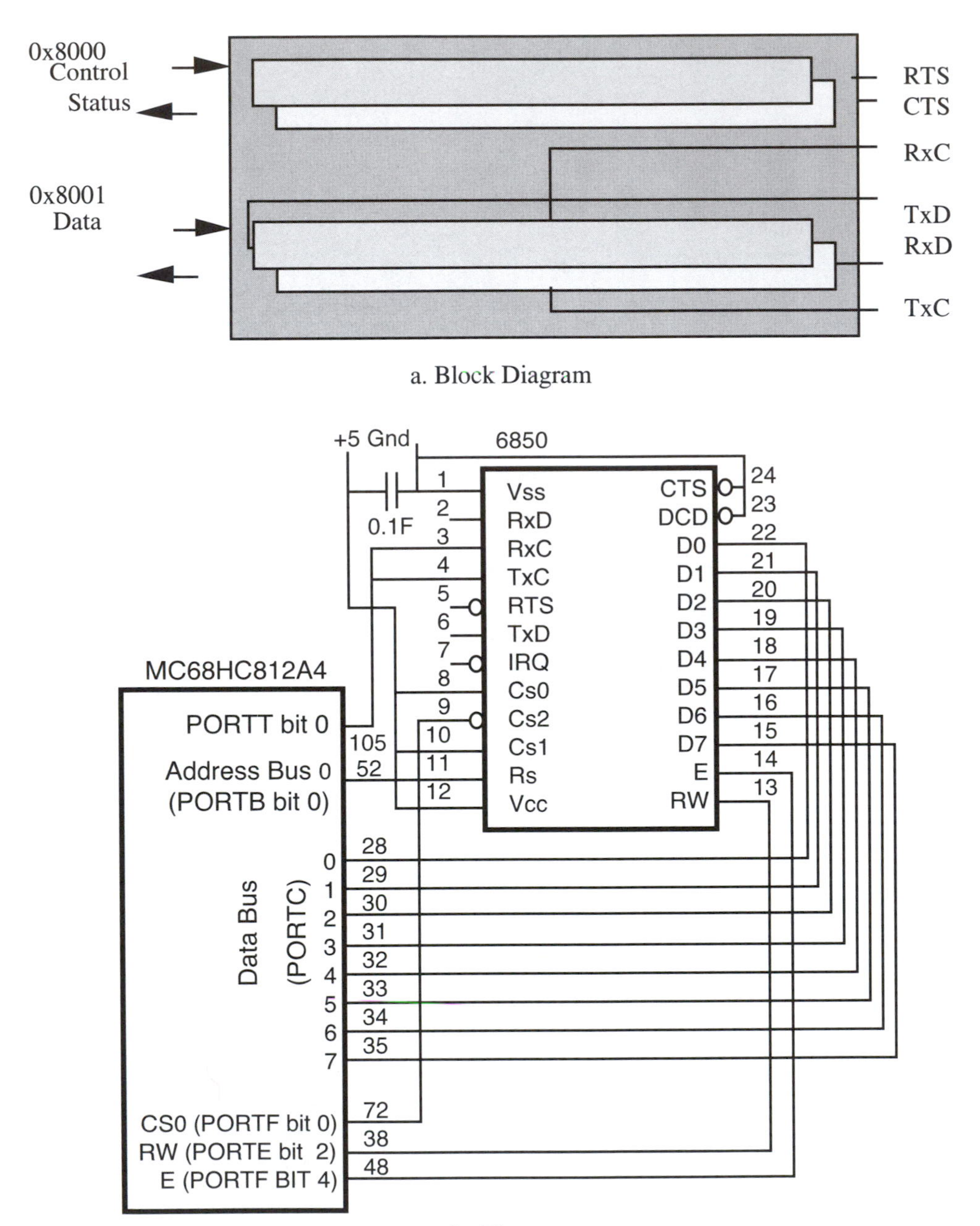

a. Block Diagram

b. Pinouts

IRQ	PE	OVRN	FE	CTS	DCD	TDRE	RDRF

c. Status Register

Figure 9.7. The ACIA

Figure 9.8. Output of a Test Program

The MC68B50's status bits can be read from location 0x200. (See Table 9.2.) RDRF (bit 0) is true if the receiver buffer is full. TDRE (bit 1) is true if the transmitter buffer is empty. Bits 2 and 3 indicate the signals from a modem, DCD and CTS, that normally indicate the data carrier is present and the channel is clear to send. FE, OVRN, and PE, (bits 4, 5, and 6) are the framing, overrun, and parity error indicators. IRQ (bit 7) is a composite interrupt request bit, which is true if the interrupt enable, control bit 7, is true and any one or several of the status bits 0, 2, 4, 5, or 6 are true or if control bits 6 and 5 are FT (01) and status bit 1 is true.

9.3.4 The M14469

The M14469 is a UART specially designed for a remote station. We give a short description here and a full description in its data sheet. A CMOS chip, the M14469 can use an unregulated supply whose voltage can vary between 4.5 and 18 V, and it uses very little current. It features a self-contained oscillator and an address comparator that permits the selection of a station when multiple stations are on the same link. A UART protocol is supported in which the frame has even parity and 1 stop bit. The baud rate is determined by the crystal (or ceramic resonator) connected between pins 1 and 2, or by an external oscillator that can drive pin 1. The oscillator frequency is divided by 64 to set the baud rate. A diagram of the M14469 is shown in Figure 9.9.

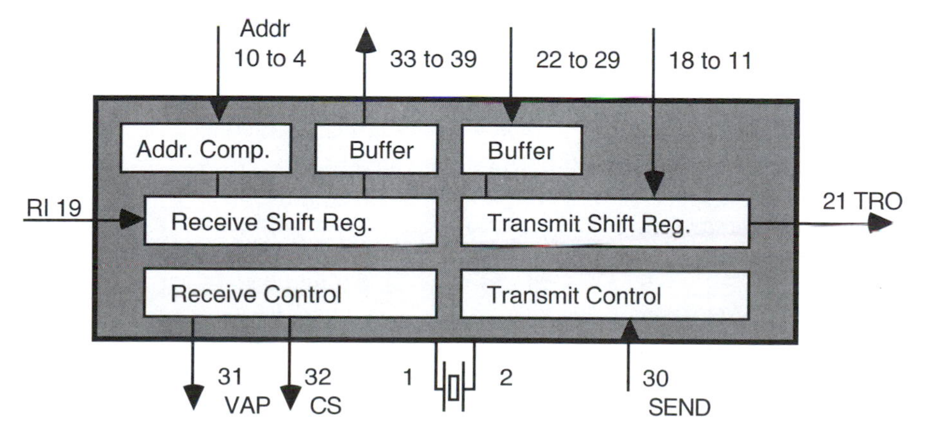

Figure 9.9. The M14469

The receiver is a standard UART receiver with an address comparator. A 7-bit address is sent as the low-order 7 bits of an 8-bit word, the most significant bit being true. The station has a 7-bit address, which is selected by strapping pins 10 to 4 low if a 0 bit is needed, or leaving them open if a 1 bit is needed in the address (these pins have an internal pull-up resistor to make them high if they are not connected). If the incoming address is equal to the station address, the valid address pulse VAP (pin 31) is made high momentarily, and the station is said to be *selected*. A 7-bit data word is sent as the low-order 7 bits of a word, the most significant bit being false. A station that has been selected will put any data word into its receiver buffer when the word is completely shifted in, and make a command strobe CS (pin 32) high momentarily just after this happens. Error status is not available on a pin, but if a parity or framing error is detected, an address will not select a station, data will not be transferred to the receiver buffer, and VAP or CS will not be pulsed.

Note that a typical message will consist of a frame with an address (most significant bit [MSB] true) followed by zero or more frames with data, MSB false. A single address frame can be used to trigger a remote station to do something, by asserting VAP in it when the address is recognized; or a message with an address frame followed by a number of data frames will cause the data to be stored in the receiver buffer each time a data frame arrives, and will pulse CS to command something to be done with the data.

The transmitter is a conventional UART transmitter modified to send out 16 bits of data in two consecutive frames if SEND is made high when VAL or CS is asserted (or within 8 data bit time units after that) and if it is not currently transmitting a pair of frames. Sixteen bits are sent from the signals on pins 11 to 18 and 29 to 22 by transferring the data on pins 11 to 18 directly into the transmitter shift register, and simultaneously transferring the data on pins 29 to 22 into the transmitter buffer. The data in the shift register are sent out (pin 11 data first) in the first UART frame, and the data in the buffer (pin 29 data first) are sent out immediately after that in the next frame. The data appear on the transmitter output TRO (pin 21) in negative logic. This output is in negative logic so that it can drive a transistor, which inverts the signal to power the link out of the station.

The chip is designed for both full and half-duplex, with some special provisions for the latter application. In full-duplex applications, a master (likely an ACIA in a microcomputer) sends to all the slave stations (several M14469s) on one line (ACIA TxD output to RI input of each slave), while all the slave stations send to the master on another line (slave TRO output into transistor base; transistor collectors in each slave tied together, in a wire AND bus line, to RxD input of ACIA), so that the master can be sending to a slave at the same time that a slave is sending to the master. In this case, VAP can be connected to SEND to send back the two frames as quickly as possible after a station is selected. The master should take care it does not send two address frames, one right after another, so that two slaves will send overlapping frames back. In the half-duplex mode, a single bus line is used between master and all slaves so that the master can send data to the slaves or the slaves can send data to the master, but not at the same time. TxD and RxD in the master, and RI and the transistor collector in each slave, would be connected to this single line. In this application, SEND should be connected to CS so the slave that is selected will wait for an address frame and a data frame to be sent over the line from the master, before the slave returns its two frames. The master should wait for both frames to be returned before it sends more data on the same line.

To ensure the data have been received, handshaking is often used; and to permit handshaking, the M14469 is designed to prevent difficulties in the half-duplex mode. The slave can be implemented so that the first frame it returns has its own station address. When the master sends a message, it can wait for the slave to respond with a frame having the address of the slave. If that frame is returned, the message must have been received without error and the slave must be active (as opposed to being shut off). This is a simple handshake protocol. However, if it is used in the half-duplex mode, we don't want the return frame to be received by the same slave and for it to recognize its own address again to trigger itself, nor do we want the return message stored in the receiver buffer. Therefore, this chip is designed so that it deselects the receiver as soon as it begins transmitting a frame. And the frame being transmitted should be a data frame (most significant bit false) to prevent the address decoder from matching it, even though the frame really contains an address. This provision makes handshaking in a half-duplex mode possible. Note that the chip is designed that way; these peculiarities are also apparent in the full-duplex mode.

Before the end of this section, we present a short program that shows how the ACIA can communicate to several M14469s over a full-duplex line. The object of the program is to select station 3, send a word of data to it, and receive a word of data from it. An M14469 is configured as station 3 by wiring pins 10 to 4 and pins 17 to 11 to represent the number 3. The data to be sent back from this station is connected to pins 23 to 29. Handshaking is used, so the transmission on the link will look as follows: the master will send the slave's address and 7 bits of data to the slave on the line from master to slave; then the slave will return its address and 7 bits of data on the other line.

The following program sets up an ACIA to send 8 bits of data, even parity, and 1 stop bit per frame, and to divide the clock by 64. The gadfly technique uses a subroutine WTBRE, shown at the bottom of the program, to wait until the transmitter buffer is empty, and then outputs the word in accumulator A to it. Initially, accumulator A has 0x5A, which is some data for station 3. The address is sent first, then the data. Then the receiver is checked for an incoming frame. While checking for the returned frame, the index register is used to keep track of elapsed time. The contents of this frame are compared with the address that was sent out. If too much time elapses before the frame returns, or if it contains the wrong address, the program exits to report the error. Otherwise, the data in the next frame are left in accumulator B, and this routine is left.

The C program `Remote` sets up an ACIA to send 8 bits of data, even parity, and one stop bit per frame, and to divide the clock by 64. This is accomplished by putting 3 and then 0x1A into the control port. The address (0x83) is sent first, then the data. Then the receiver is checked for an incoming frame. Meanwhile, a counter, initialized high enough to wait for any reasonable response, is decremented to check if there is no response. The contents of this frame are compared with the address that was sent out. If too much time elapses before the frame returns, or if it contains the wrong address, the program exits to report the error. Otherwise, the data picked up from the ACIA that was sent in the next frame after the address are returned.

```
#define N 52 /* for 1200 baud */

interrupt 8 void handler(){ TFLG1 = 1; TC0 += N; }
```

```
int Remote(char c, int *acia) { int i, parity;
    CLKCTL = 2 << 2; MODE = ( Special + Narrow ) << ModeFld;
    PEAR = RDWRE; CSCTL = CS0E; asm sei
    TSCR=0x80; TIOS=1; TCTLO=1; TMSK1=1; TC0 = N + TCNT; asm cli
    *acia = 3; *acia = 0x1a; acia[1] = 0x83;
    while((*acia & 2) == 0); acia[1] = c; i = 33000;
    while((*acia & 1) == 0) if((i--) == 0) {/* report error */};
    if(acia[1] != 0x83) {/* report error */};
    while((*acia & 1)==0); return(acia[1]);
}
```

9.3.5 The Serial Communication Interface System in the 6812

The 'A12 has a pair of UART-like systems in it called the serial communication interfaces (SCI0 and SCI1). We describe the SCI0 device shown in Figure 9.10; the SCI0 device is used by the debugger DBug-12. We describe its data, baud-rate generator, control and status ports. Then we will show how the SCI can be used in a gadfly-, and an interrupt-synchronization interface.

As with the ACIA, the SCI has, at the same port address, a pair of data registers that are connected to shift registers. Eight bits of the data written at `SC0DRL` (0xc7) are put into the shift register and shifted out, as in the ACIA, and 8 bits of the data shifted into the receive shift register can be read at `SC0DRL` (0xc7).

The clock rate is established by the 12-bit `SC0BD` port (0xc0). The number put in this port is the clock going to the SCI (Figure 6.10) divided by 16 times the desired baud rate. For example, to get 9600 baud, put 52 into the `SC0BD` port.

The 16-bit control port, `SC0CR`, at 0xc2, has parity enable `PE` and parity type `PT` to establish the parity, transmitter interrupt enable `TIE` and receiver interrupt enable `RIE` to enable interrupts, and transmitter enable `TE` and receiver enable `RE` to enable the device. The 16-bit status port at 0xc4 indicates what is happening in the transmitter and receiver. `TDRE` is T (1) if the transmit data register is empty; it is set when data are moved from the data register to the shift register, and is cleared by a read of the status port followed by a write into the data port. The remaining status bits are for the receiver. `RDRF` is T (1) if the receive data register is full because a frame has been received. Receive-error conditions are indicated by `OR`, set when the receiver overruns (that is, a word has to be moved from the input shift register before the previously input word is read from the data register), `FE`, T (1) if there is a framing error (that is, a stop bit is expected but the line is low; and `PE`, T (1) if there is a parity error.

In `main`, the SCI is intitialize for gadfly synchronization of 9600 baud and 8 data bits without parity. Reading status and data registers twice clears the `RDRF` flag. The `put` procedure gadflies on transmitter data register empty (`TDRE`); when it is empty, `put` outputs its argument. The `get` procedure gadflies on the receive data register full (`RDRF`); when the receive register is full, `get` returns the data in data port `SC0DRL`.

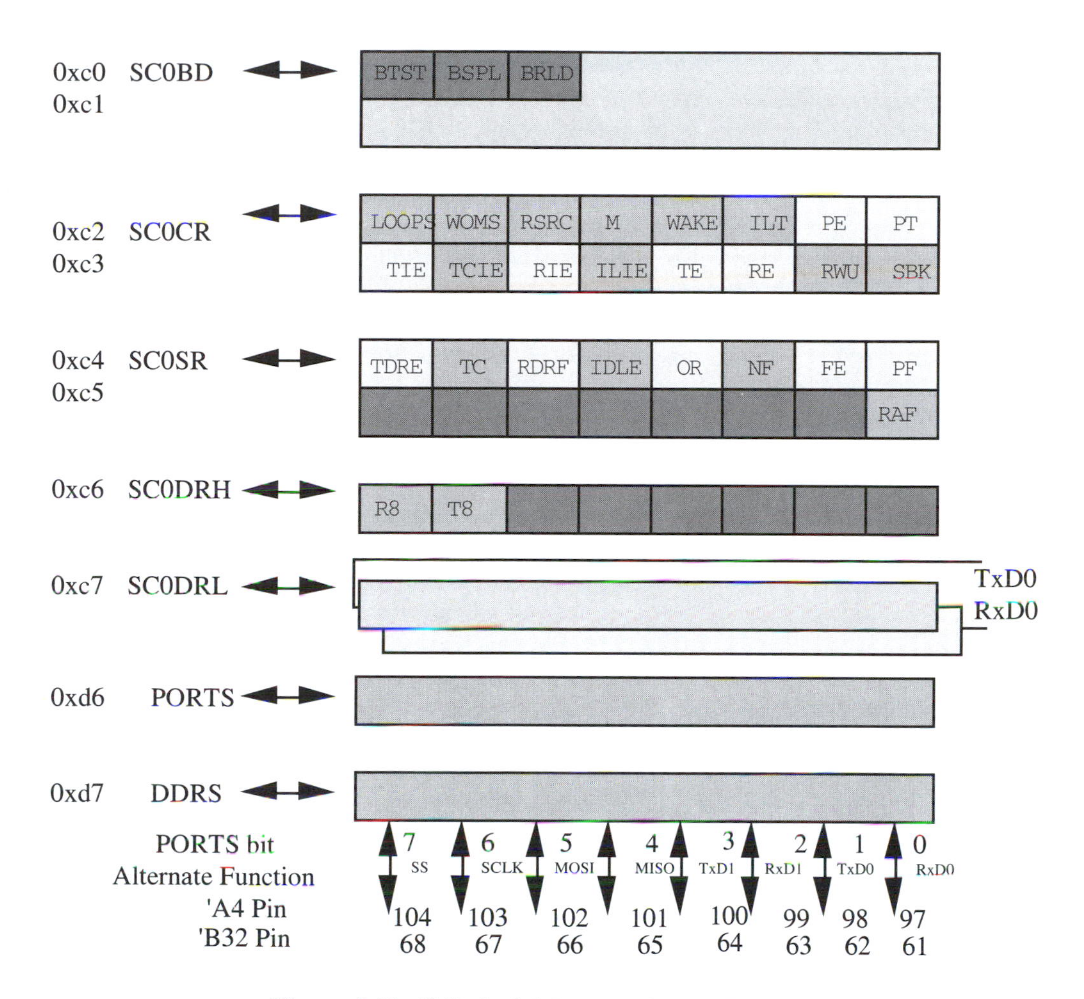

Figure 9.10. 6812 Serial Communication Interface

```
volatile int SC0BD@0xc0, SC0CR@0xc2, SC0SR@0xc4, SC0DR@0xc6,
     SC1BD@0xc8, SC1CR@0xca, SC1SR@0xcc, SC1DR@0xce;

enum{ PEV = 0x200, PT = 0x100, TIE = 0x80, RIE = 0x20, TE = 8, RE = 4 };
enum{ TDRE = -32768, RDRF = 0x2000, OR = 8, FE = 2, PE = 1 };

void put(char d) { while( ( SC0SR & TDRE ) == 0) ; SC0DR = d; }

int get() { while( ( SC0SR & RDRF ) == 0) ; return SC0DR; }

void main() { char i;
     SC0BD = 52; /*9600 baud*/ SC0CR = TE + RE; /* enable Xmt, Rcv devices */
     i=SC0SR; i=SC0DR; i=SC0SR; i=SC0DR;/* clr RDRF */ put(0x55); i = get();
}
```

Interrupt synchronization uses queues (§2.2.2) to permit simultaneous input and output of data, which may be asynchronous with the program. SCI0 initialization includes enabling SCI0 receiver interrupts, which are handled by `handler`. To output data in `put`, gadfly on `oSz` until the output queue has room, then push the data. If the output queue changes from empty to nonempty, enable the transmitter interrupt. To input date from `get`, gadfly on `iSz` until the input queue has data, then pull that data. `handler` checks status port `SC0SR` for receive data or transmitter buffer empty. If the transmitter buffer is empty and the output queue has data, data is pulled from the output queue and is written to the data port. If the output queue becomes empty, disable the transmitter interrupt. If the receive buffer is full, push its data into the input queue.

```
volatile int SC0BD@0xc0, SC0CR@0xc2, SC0SR@0xc4, SC0DR@0xc6,
    SC1BD@0xc8, SC1CR@0xca, SC1SR@0xcc, SC1DR@0xce;

enum{ PEV = 0x200, PT = 0x100, TIE = 0x80, RIE = 0x20, TE = 8, RE = 4};
enum{ TDRE = -32768, RDRF = 0x2000, OR = 8, FE = 2, PE = 1 };

unsigned char oQ[10], iQ[10], oSz, iSz, oTop, oBot, iTop, iBot, error;

void oPush(char i)
    {if((oSz++)>=10) error=1; if (oTop==10) oTop=0; oQ[oTop++]=i;}

char oPull()
    {if((oSz--)<=0)error=1; if(oBot==10)oBot=0; return oQ[oBot++];}

void iPush(char i)
    {if((iSz++)>=10) error=1; if(iTop==10) iTop=0; iQ[iTop++]=i; }

char iPull()
    {if((iSz--)<=0) error=1; if(iBot==10) iBot=0; return iQ[iBot++]; }

void put(char d) {while(oSz>=10) ; oPush(d); if(oSz==1) SC0CR|=TIE; }

char get() { while( ! iSz) ; return iPull(); }

interrupt 20 void handler(){ int status;
    if((status=SC0SR)&TDRE&(SC0CR << 8)) /* if trns int en, TDRE flag set */
        { if( oSz ) SC0DR = oPull(); if( ! oSz ) SC0CR &= ~TIE; }
    if(status&RDRF) { iPush(SC0DR); } /*if RDRF flag set, push received data*/
}

void main() { char i;     asm SEI
    SC0BD = 52; SC0CR = TE + RE + RIE; /* enable Rcv interrupt */
    i = SC0SR + SC0DR; i = SC0SR + SC0DR; asm CLI
    put(0x55); i = get();
}
```

The overwhelming advantage of the SCI system is that it is contained entirely within the 6812 chip. It is therefore free with the computer. Gadfly synchronization is very simple. Interrupt synchronization provides much more power, and is also very easy.

9.3.6 Object-oriented Interfacing to the SCI

We illustrate object-oriented programming of the SCI with a simple gadfly class `SCIg`, an interrupt class `SCIi`, a hardware handshake `SCIh`, and a software handshake `SCIsh` interrupt class. Each example demonstrates increased complexity. The simple `SCIg` class merely implements §9.3.5's gadfly synchronization procedures. We define `baud`, `control`, `status`, and `data` to use `Port`'s vector `port` to access the ports. The function member `option` permits reading and writing the device's control and status ports to gain full use of the device, but if an object is redirected, calling `option` will neither cause the compiler to have an error nor crash the run-time application.

```
#define baud port[0]
#define control port[1]
#define status port[2]
#define theData port[3]

class SCIg : public Port<int> {

    public : SCIg(char id) : Port<int>(0xc0 + (id << 3) )
        {baud = 52; control = TE + RE; }

    virtual int get(void){ while((status&RDRF)==0); return theData&0xff;}

    virtual void put(int d){while((status&TDRE)==0) ; theData = d; }

    virtual int option (int c = 0, int mask = 0) {
        if(c == 0) return Port::option(0, m);
        else if(!(c & 0x10)) return 0;
        else if(c <= 0x13)){((int *)port)[c & 3) = d; return 0; }
        else return ((int *)port)[c & 3];
    }
} *S;

void main()
    {unsigned char i;S=new SCIg(1);S->put(0x55);i=S->get();delete S;}
```

`SCIi`, shown next, implements the procedures in §9.3.5 for interrupt synchronization (using `#define` and `enum` statements from it), but it takes advantage of §2.3.3's `Queue` class. It uses overloaded operators from §4.4.5 and §5.3.5 to implement `IoStreams`, it correctly keeps track of receiver errors, and its destructor properly disables device SCI0's interrupts after its output queue is empty.

```
#define MAXCHARS 80

class SCIi : public Port<int> {

    protected:Queue<char> *InQ,*OutQ;

    public : SCIi(char id):Port(0xc0+(id << 3)) { char dmy;   asm sei
        OutQ=new Queue<char>(10);InQ=new Queue<char>(10);/* create Qs */
        baud=52; control=TE+RE+RIE; /* initialize SCI1's control registers */
        dmy=status; dmy=theData; dmy=status; dmy=theData; asm cli
    }

    virtual void put(int d) {
        while(OutQ->size>=OutQ->maxSize);// wait output queue room
        OutQ->push(d);if(OutQ->size==1)control|=TIE;
    }

    virtual int get(void)
        {char c; while(InQ->size); put(c = InQ->pull()); return c; }

    void service(void){ int i;
        if((i=status)&RDRF){// here if receive register full (save status in i)
            InQ->push(*port);errors|=InQ->error()|(i&0xf);
        }
        if(i&control&TDRE) { // here if transmitter interrupt enbld, TDRE flg set
            if(OutQ->size)*port=OutQ->pull();
            if(OutQ->size==0)control&=~TIE;
        }
    }

    virtual int option (int c = 0, int mask = 0) {
        if(c == 0) return Port::option(0, m);
        else if(!(c & 0x10)) return 0;
        else if(c <= 0x13)){((int *)port)[c&3) = d; return 0; }
        else return ((int *)port)[c & 3);
    }

    ~SCIi(){while(OutQ->size);control=0;/* disable interrupts when empty */ }

}*S;

void main()
    {char i;S = new SCIi(1); S->put(0x55); i = S->get(); delete S; }

interrupt 20 void handler1(){ S->service(); }
```

Class *SCIh*, derived from *SCIi*, uses *hardware handshaking. request to send* (RTS), herein *PORTS* bit 4, is asserted by this microcontroller if its input queue has room, and negated if there isn't enough room. *clear to send* (CTS), herein *PORTS* bit 5, is asserted by another microcontroller and received by this microcontroller if there is room in the other microcontroller's input queue and negated if there isn't enough room.

Because the SCI may have data when CTS becomes negated, RTS is negated when the input queue still has a little room. A constant *THRESHOLD* is defined so as to provide enough room in the input queue to absorb all of the SCI1 device's registers' data.

The input portion of the interrupt handler negates RTS if the input queue is essentially full (actually if the queue has a few bytes available for more data). *get* asserts RTS if the input queue sufficiently empties so there will be room in the input queue to handle an influx of data that might be sent before RTS is negated.

CTS is checked at least once in the *put* function member, and it is checked repetitively while *put* is waiting for the output queue to have some room in it, so that if CTS becomes asserted, the transmitter interrupt is enabled, and if negated, the transmitter interrupt is disabled. Note that the transmitter interrupt can be enabled again, even if it already is enabled, but that won't hurt anything. The output portion of the interrupt handler also disables the transmitter interrupt if CTS is found to be negated.

```
#define THRESHOLD 2
enum{ RTS = 0x10, CTS = 0x20};

class SCI3 : public SCIi { friend void ccsci0(void), ccsci1(void);

    public : SCI3(char id) : SCIi (id) { DDRS = RTS; }

    virtual void put(char d) {// this function overrides SCI2's put function
        do { if(PORTS & CTS) control |= TIE; else control &= ~TIE; }
            while(OutQ->size >= OutQ->maxSize);
        OutQ->push(d);
    }

    virtual int get(void) {
        if(InQ->size == (InQ->maxSize - THRESHOLD)) PORTS |= RTS;
        return SCI2::get(); // use SCI2's get function to pull byte from queue
    }

    virtual void service(void){ short i;
        if((i = status) & RDRF) { // input handler
            InQ->push(SCI2::get());errors|=InQ->error()|(i & 0xf);
            if(InQ->size==(InQ->maxSize - THRESHOLD)) PORTS &= ~RTS;
        }
        if(i & control & TDRE) { // output handler
            if((OutQ->size)&&(PORTS&CTS)) SCI2::put(OutQ->pull());
            if((OutQ->size == 0)||!(PORTS & CTS)) control &= ~TIE;
        }
    }
} *S;

void main(){char i;S=new SCI3(1); S->put(0x55); i=S->get(); delete S; }
```

The class `SCIsh`, derived from `SCIi`, uses *software handshaking*. An ASCII character XON (0x11) will be sent by this microcontroller if there is room in its input queue and another ASCII character XOFF (0x13) will be sent by this microcontroller if there isn't enough room. These characters may be received from another microcontroller.

Because SCI1 registers may have data when XOFF is sent, as in the previous class, a constant *T* is defined so as to provide enough room in the input queue to absorb all of the SCI1 device's outstanding data. XON or XOFF are sent by putting them in a data member `Msg`, which is 0 if no special signal is to be sent. If `Msg` is nonzero, the transmitter interrupt is enabled, and `Msg` is sent in place of any output queue data. This decreases the delay time from when the input queue state change requires sending an XON or XOFF until the sender reacts by enabling or disabling its transmitter interrupt.

The input interrupt handler generates XOFF if the input queue is essentially full (actually if the queue has a few bytes available for more data). `get` generates XON if the input queue empties enough so there will be room in the input queue.

The receipt of XON and XOFF characters indicated the same as the CTS signal. As XON or XOFF arrive, they set or clear a data member `CTS`, which will emulate the CTS signal used in the class `SCIh`; data member `CTS` is checked as CTS was there.

```
#define T 3
enum{ XON = 0x11, XOFF = 0x13};

class SCI4:public SCI2{friend void ccsci0(void),ccsci1(void);char Msg, CTS;
    public : SCI4(char id) :  SCI2(id) { Msg = 0; CTS = 1;}
    virtual void put(char d) { // this function overrides SCI2's put function
        do {if(CTS) control |= TIE; else control &= ~TIE; }
        while(OutQ->size >= OutQ->maxSize);
        OutQ->push(d);
    }
    virtual int get(void) {
        if(InQ->size==(InQ->maxSize-T)) {Msg=XON; control|= TIE; }
        return SCI2::get();// use SCI2's get function to pull byte from queue
    }
    virtual void service(){int i; char c;// this handler replaces SIC2's handler
        if((i = status) & RDRF) { // input handler
            if((c=SCI2::get())==XON) CTS=1; else if(c==XOFF) CTS=0;
            else {InQ->push(c);
            if(InQ->size==(InQ->maxSize-T))Msg=XOFF;control |= TIE;}
            errors |= InQ->error() | (i & 0xf);
        }
    if(i & control & TDRE) { // output handler
            if(Msg) { SCI2::put(Msg); Msg = 0; }
            else if((OutQ->size) && CTS) SCI2::put(OutQ->pull());
            if((OutQ->size == 0) || ! CTS) control &= ~TIE;
        }
    }
} *S;

void main(){char i; S=new SCI3(1); S->put(0x55); i=S->get(); delete S;}
```

9.4 Other Protocols

Besides the UART protocol, the two most important protocols are the synchronous bit-oriented protocols that include the SDLC, HDLC, and ADCCP, the X-25, the IEEE-488 bus protocol, and the smart computer system interface (SCSI) protocol.

These are important protocols. We fully expect that many if not most of your interfaces will be designed around these protocols. If you are designing an I/O device to be used with a large mainframe computer, you will probably have to interface to it using a synchronous bit-oriented protocol. If you are designing a laboratory instrument, you will probably interface to a minicomputer using the IEEE-488 protocol, so that the minicomputer can remotely control your instrument. We will survey the key ideas of these protocols in this section. The first subsection describes bit-oriented protocols. The second subsection discusses the 488 bus. The final subsection covers the SCSI interface.

9.4.1 Synchronous Bit-oriented Protocols

Synchronous protocols are able to move a lot of data at a high rate. They are primarily used to communicate between *remote job entry* (RJE) terminals (which have facilities to handle line printers, card readers, and plotters) and computers, and between computers and computers. The basic idea of a synchronous protocol is that a clock is sent either on a separate wire or along with the data in the Manchester coding scheme. Since a clock is sent with the data, there is little cause to fear that the receiver clock will eventually get out of sync after a lot of bits have been sent, so we are not restricted to short frames as we are in the UART. Once the receiver is synchronized, we will try to keep it in synchronism with the transmitter, and we can send long frames without sending extra control pulses, which are needed to resynchronize the receiver and which reduce the efficiency of the channel.

Asynchronous protocols, like the UART protocol discussed in the last section, are more useful if small amounts of data are generated at random times, such as by a computer terminal. Synchronous protocols would anyway have to get all receivers into synchronism with the transmitter when a new transmitter gets control of the channel, so their efficiency would be poor for short, random messages. Synchronous protocols are more useful when a lot of data is sent at once because they do not require the overhead every few bits, such as start and stop bits, that asynchronous protocols need. Bit-oriented synchronous protocols were developed as a result of weaknesses in byte- or character-oriented synchronous protocols when they were used in sending a lot of data at once.

The precursor to the bit-oriented protocol is the binary synchronous *Bisync* protocol, which is primarily character oriented and is extended to handle arbitrary binary data. This protocol can be used with the ASCII character set. The 32 nonprinting ASCII characters include some that are used with the Bisync protocol to send sequences of characters. SYN (ASCII 0x16) is sent whenever nothing else is to be sent. It is a null character used to keep the receiver(s) synchronized to the transmitter. This character can be used to establish which bit in a stream of bits is the beginning of a character. Two Bisync protocols are used: one for sending character text, and the other for sending binary data such as machine code programs, binary numbers, and bit data.

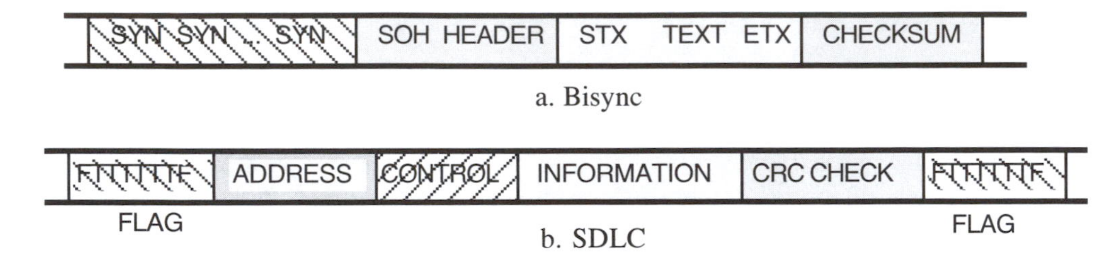

Figure 9.11. Synchronous Formats

Character text is sent as follows (see Figure 9.11a): A header can be sent, beginning with character SOH (ASCII 0x01); its purpose and format are user defined. An arbitrary number of text characters are sent after character STX (ASCII 0x02), and is terminated by character ETX (ASCII 0x03). After the ETX character, a kind of checksum are sent.

To allow any data – such as a machine code program – including characters that happen to be identical to the character ETX, to be sent, a character DLE (ASCII 0x10) is sent before the characters STX and ETX. A byte count is established in some fashion. It may be fixed, so that all frames contain the same number of words; it may be sent in the header; or it may be sent in the first word or first two words of the text itself. Whatever scheme is used to establish this byte count, it serves to disable the recognition of DLE/ETX characters that terminate the frame so that such patterns can be sent without confusing the receiver. This is called the *transparent mode* because the bits sent as text are transparent to the receiver controller and can be any pattern.

Bisync uses error correction or error detection and retry. The end of text is followed by a kind of checksum, which varies in differing Bisync protocols. One good error detection technique is to exclusive-OR the bytes that were sent, byte by byte. If characters have a parity bit, that bit can identify which byte is incorrect. The checksum is a parity byte that is computed "at 90 degrees" from the parity bits and can identify the column that has an error. If you know the column and the row, you know which bit is wrong, so you can correct it. Another Bisync protocol uses a *cyclic redundancy check* (*CRC*) that is based on the mathematical theory of error-correcting codes. The error-detecting "polynomial" X**16 + X**15 + X**2 + 1, called the CRC-16 polynomial, is one of several good polynomials for detecting errors. The CRC check feeds the data sent out of the transmitter through a shift register that shifts bits from the 15th stage towards the 0th stage. The shift register is cleared, and the data bits to be transmitted are exclusive-ORed with the bit being shifted out of the 0*th* stage and then is exclusive-ORed with this bit into some of the bits being shifted in the register at the inputs to the 15th, 13th, and 0th stages. The original data and the contents of the shift register (called the CRC check bits) are transmitted to the receiver. The receiver puts the received data, including the CRC check bits, through the same hardware at its end. When done, the hardware should produce a 0 in the shift register. If it doesn't, an error (CRC error) has occurred. Upon its noting a CRC error, the bisync protocol requests that the frame be resent. If the frame is good, an ACK (ASCII 0x06) is sent; but if an error is detected, a NAK (ASCII 0x15) is sent from the receiver back to the sender. If the sender gets an ACK, it can send the next frame; but if it gets a NAK, it resends the current frame.

Though developed for communication between a computer and a single RJE station, Bisync has been expanded to include *multidrop*. Several RJE stations are connected to a host computer on a half-duplex line (bus). The host is a master. It controls all transfers between it and the RJE stations. The master *polls* the stations periodically, just as we polled I/O devices after an interrupt, to see if any of them want service. In polling, the master sends a short packet to each station, so that each station can send back a short message as the master waits for the returned messages.

The Bisync protocol has some serious shortcomings. It is set up for and is therefore limited to half-duplex transmission. After each frame is sent, you have to wait for the receiver to send back an acknowledge or a negative acknowledge. This causes the computer to stutter while it waits for a message to be acknowledged. These shortcomings are improved in bit-oriented protocols. Features used for polling and multidrop connections are improved. And the information is bit-oriented to efficiently handle characters, machine code programs, or variable width data.

The first significant synchronous bit-oriented protocol was the *synchronous data link control* (*SDLC*) protocol developed by IBM. The American National Standards Institute, ANSI, developed a similar protocol, ADCCP, and the CCITT developed a third protocol, HDLC. They are all quite similar at the link-control and physical levels, which we are studying. We will take a look at the SDLC link, the oldest and simplest of the bit-oriented protocols.

The basic SDLC frame is shown in Figure 9.11b. If no data are sent, either a true bit is continually sent (idle condition) or a *flag pattern,* 0x7E (FTTTTTTF), is sent. The frame itself begins with a flag pattern and ends with a flag pattern, with no flag patterns inside. The flag pattern that ends one frame can be the same flag pattern that starts the next frame.

The frame can be guaranteed free of flag patterns by a five T's detector and F inserter. If the transmitter sees that five T's have been sent, it sends an F regardless of whether the next bit is going to be a T or an F. That way, the data FFTFFTTTTTTF is sent as FFTFFTTTTTFTTF, and the data FFTFFTTTTTFTF is sent as FFTFFTTTTTFFTF, free of a flag pattern. The receiver looks for 5 T's. If the next bit is F, it is simply discarded. If the received bit pattern were FFTFFTTTTTFTTF, the F after the five T's is discarded to give FFTFFTTTTTTTF, and if FFTFFTTTTTFFTF is received, we get FFTFFTTTTTFTF. But if the received bit pattern were FTTTTTTF the receiver would recognize the flag pattern and end the frame.

The frame consists of an 8-bit station number address, for which the frame is sent, followed by 8 control bits. Any number of information bits are sent next, from 0 to as many as can be expected to be received comparatively free of errors or as many as can fit in the buffers in the transmitter and receiver. The CRC check bits are sent next. The address, control, information, and CRC check bits are free of flag patterns as a result of the five T's detection and F insertion discussed above.

The control bits identify the frame as an *information frame, supervisory,* or *nonsequenced frame.* The information frame is the normal frame for sending a lot of data in the information field. The control field of an information frame has a 3-bit number `N`. The transmitter can send up to 8 frames, with different values of `N`, before handshaking is necessary to verify that the frames have arrived in the receiver. Like the ACK and NAK characters in Bisync, supervisory frames are used for retry after error. The receiver can send back the number *N* of a frame that has an error, requesting that it be resent, or it can send another kind of supervisory frame with *N* to indicate that all frames up to N have been received correctly. If the receiver happens to be sending other data back to the transmitter, it can send this number *N* in another field in the information frame it sends back to the transmitter of the original message to confirm receipt of all frames up to the *N*th frame, rather than sending an acknowledge supervisory frame. This feature improves efficiency, since most frames will be correctly received.

The SDLC link can be used with multidrop (bus) networks, as well as with a ring network. The ring network permits a single main, *primary station* to communicate with up to 255 other *secondary stations.* Communication is full-duplex, since the primary can send to the secondary over part of the loop, while the secondary sends other data to the primary on the remainder of the loop. The SDLC has features for the primary to poll the secondary stations and for the transmitting station to abort a frame.

The SDLC link and the other bit-oriented protocols provide significant improvements over the character-oriented Bisync protocols. Full-duplex communication, allowing up to 8 frames to be sent before they are acknowledged, permits more efficient communication. The communication is inherently transparent, because of the five T's detection feature, and can handle variable length bit data efficiently. It is an excellent protocol for moving large frames of data at a high rate of speed.

The *X.25 protocol* is a three-level protocol established by the CCITT for high-volume data transmission. The physical and link levels are set up for the HDLC protocol, a variation of the SDLC bit-oriented protocol; but synchronous character-oriented protocols can be used so that the industry can grow into the X.25 protocol without scrapping everything. This protocol, moreover, specifies the network level as well. It is oriented to packet switching. Packet switching permits frames of a message to wander through a network on different paths. This dynamic allocation of links to messages permits more efficient use of the links, increases security (since a thief would have to watch the whole network to get the entire message), and enhances reliability. It looks like the communication protocol of the future. While we do not cover it in our discussion of I/O device design, we have been using its terminology throughout this chapter as much as possible.

9.4.2 MC68HC912B32 BDLC Device

In an automobile, thousands of yards of wire connect the dashboard, sensors, microcontrollers, and actuators. A single high-speed communication line could replace most of these wires, at a significant cost savings. The Society of Automotive Engineers defined the *byte data link communications* protocol (BDLC) as SAE1850, to communicate among an automobile's tens of microcontrollers and associated systems.

Along the same lines as the SDLC, the BDLC message (Figure 9.12a) consists of a start of frame (SOF), a header containing control information, a data field, and a CRC check. The receiver may immediately send an in-frame response such as that in Figure 9.12b. The device's ports are exhibited in Figure 9.13. These are defined in C as follows:

```
volatile char BCR1@0xf8, BSVR@0xf9, BCR2@0xfa, BDR@0xfb, BARD@0xfc,
    DLCSCR@0xfd, PORTDLC@0xfe, DDRDLC@ 0xff;
```

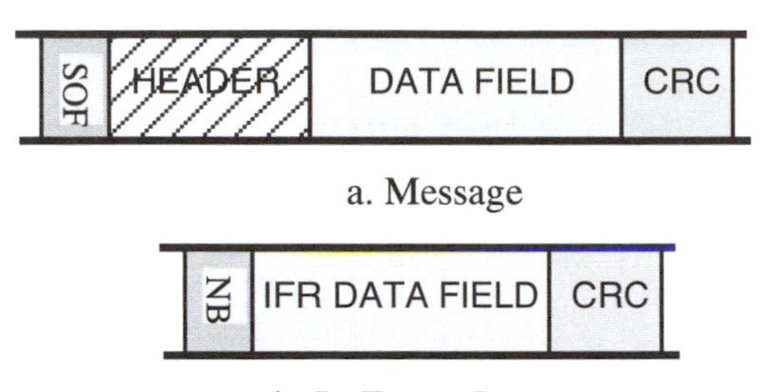

a. Message

b. In-Frame Response

Figure 9.12. BDLC Formats

The program below illustrates sending a 10-byte vector *out* and receiving this 10-byte frame into vector *in* in the same 'B32. Digital loopback mode circulates the bit stream within the 'B32. Ordinarily, the BDCL device is able to send and receive BDLC frames through an analog interface chip, the MCxxxx; without this chip, the sending and receiving of data generates `badSymbol` errors. Our program's initialization turns on the BDLC by setting bit 2 of the `DLCSCR` port; it enables digital loopback by setting bit 6 of *BCR2,* and it enables the receiver by setting bit 7, and interrupts, by setting bit 1 of *BCR1.* The frame is sent by calling a subroutine `rdrf`, which writes the first byte into the ***BDR*** port. Thereafter, interrupts are used to send the remaining bytes using subroutine `rdrf`, and receive bytes using subroutine `tdre`. Additionally, interrupts can be called to handle various error conditions. To quickly resolve all these interrupts, but not use a large number of interrupt vectors, the BDLC device provides a status port `BSVR`,which identifies the cause of the interrupt; in the handler, it is used in assembler language to index into a vector *transfer* to jump to the service subroutine.

```
unsigned char flag,iIndex,oIndex,in[10],out[10]={1,2,3,4,5,6,7,8,9,10};

void error(char c){ flag = c; do; while(1); }
void noOp(void) {}        void receiveEOF(void) {}
void receiveIFR(void error(2); BCR2 = 0x40;/* digital loopback, for testing */ }
void rdrf(void){in[iIndex++]=BDR; if(iIndex>=10){flag++; DLCSCR=BCR1=0;}}
void tdre(void){BDR=out[oIndex++];if(oIndex>=10)BCR2|=8;/* transmit EOF */ }
void arbitrationLoss(void) {error(5);} void crcError(void) {error(6);}
void badSymbol(void) {error(7);}         void wakeup(void) {error(8);}

int transfer[9]={(int)noOp,(int)receiveEOF,(int)receiveIFR,(int)rdrf,
     (int)tdre,(int)arbitrationLoss,(int)crcError,(int)badSymbol,
     (int)wakeup};

interrupt 23 void handler(void){ asm{
L1:  ldx   #transfer        ; beginning of procedure vector
     clra                   ; only D can be added to X, used indirectly
     ldab  0xf9             ; BSVR supplies operation to be done, times 4
     lsrd                   ; modify to integer length
     jsr   [D,X]            ; call the subroutine
}}
```

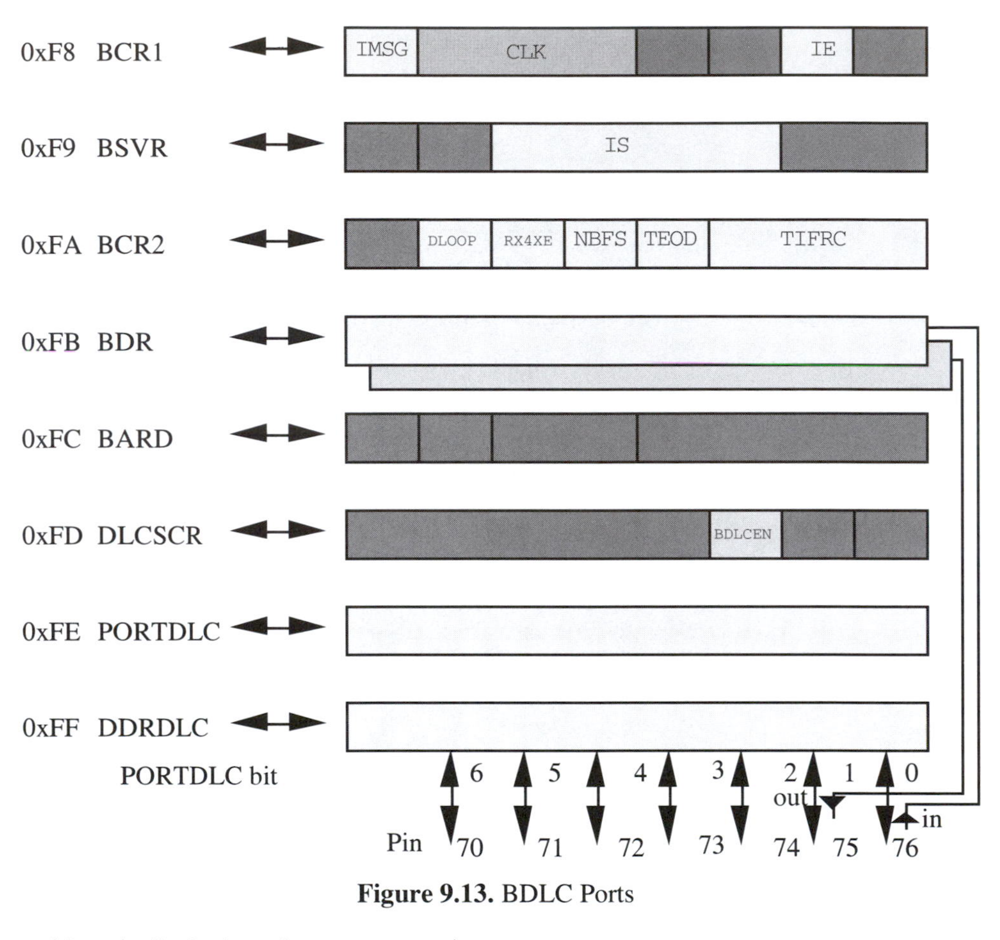

Figure 9.13. BDLC Ports

```
void main() { char dummy; asm sei
    DLCSCR = 4; /* turn on bdlc */
    BCR2 = 0x40; /* digital loopback, used only for testing */
    BCR1 = 0x72; /* enable Rcvr, set clock rate to 1 MHz, enable interrupt */
    dummy = BSVR; dummy = BDR; asm cli
    tdre(); /* send first byte */ while(flag == 0) ; /* wait for last byte */
    do ; while(1);
}
```

9.4.3 IEEE-488 Bus Standard

The need to control instruments like voltmeters and signal generators in the laboratory or factory from a computer has led to another kind of protocol, an asynchronous byte-oriented protocol. One of the earliest such protocols was the CAMAC protocol developed by French nuclear scientists for their instruments. Hewlett-Packard, a major instrument manufacturer, developed a similar standard that was adopted by the IEEE and called the IEEE-488 standard. Although Hewlett-Packard owns patents on the handshake methods of this protocol, it has made the rights available on request to most instrument manufacturers, and the IEEE-488 bus standard has been available on most sophisticated instruments, minicomputers, and microcomputers.

Communications to test equipment has some challenging problems. The communications link may be strung out in a different way each time a different experiment is run or a different test is performed. The lengths of the lines can vary. The instruments themselves do not have as much computational power as a large mainframe machine, or even a terminal, so the communications link has to do some work for them such as waiting to be sure that they have picked up the data. A number of instruments may have to be told to do something together, such as simultaneously generating and measuring signals, so they can't be told one at a time when to execute their operation. These characteristics lead to a different protocol for instrumentation buses.

The IEEE-488 bus is fully specified at the physical and link levels. A 16-pin connector, somewhat like the RS-232 connector, is prescribed by the standard, as are the functions of the 16 signals and 8 ground pins. The sixteen signal lines include a 9-bit

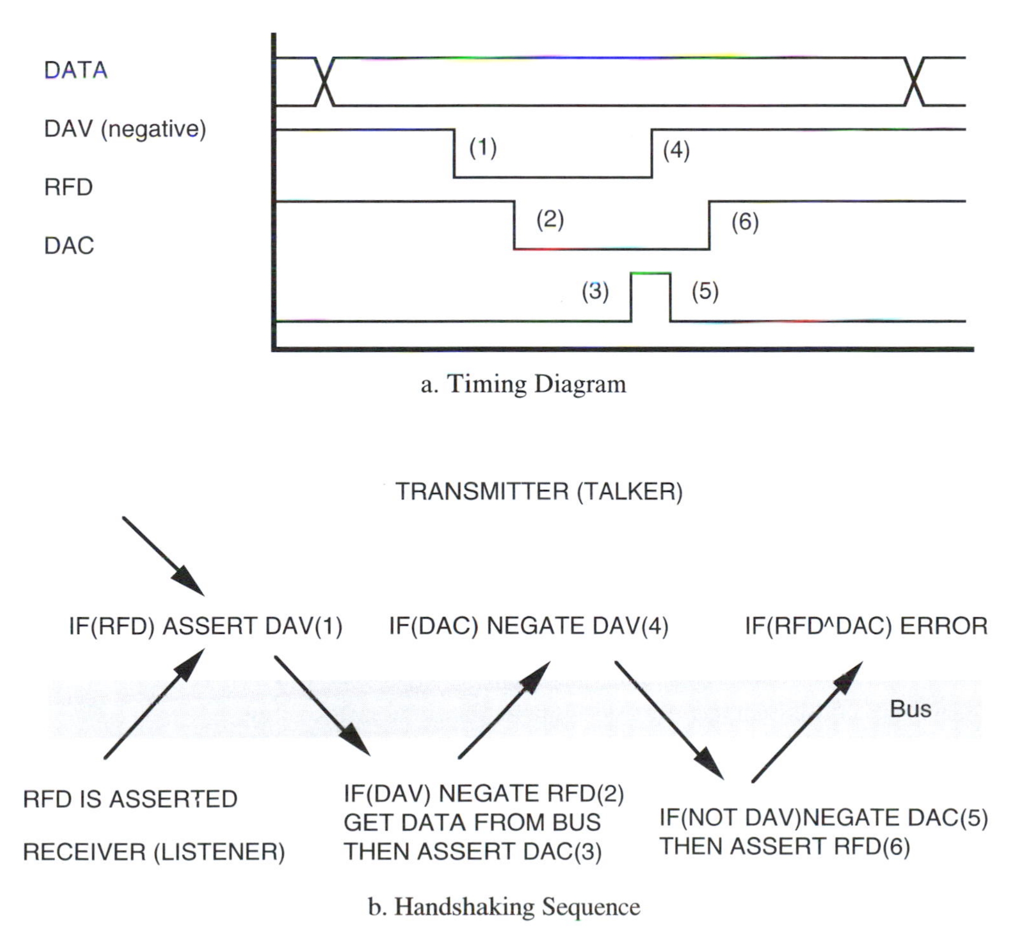

a. Timing Diagram

b. Handshaking Sequence

Figure 9.14. IEEE-488 Bus Handshaking Cycle

parallel data bus, three handshaking lines, and five control lines. The control lines include one that behaves like the system reset line in the 6812 microcomputer. Others are used to get attention and perform other bus management functions. But the heart of the bus standard is the asynchronous protocol used to transmit data on the bus.

An asynchronous bus protocol uses a kind of expandable clock signal, which can be automatically stretched when the bus is longer or shortened if the bus is shorter. During this process the "clock" is sent from the station transmitting the data to the station that receives the data on one line, then sent back to the transmitter on another line. The transmitter waits for the return signal before it begins another transmission. If the bus is lengthened, so are the delays of this clock signal. The IEEE-488 bus uses this principle a couple of times to reliably move a word on a 9-bit bus from a transmitter to a receiver. (See Figure 9.12.)

The handshake cycle is like a clock cycle. Each time a word is to be moved, the bus goes through a handshake cycle to move the word, as shown in Figure 9.12. The cycle involves negative-logic *data available* (DAV), sent by the transmitter of the data, and positive-logic *ready for data* (RFD) and positive-logic *data accepted* (DAC), sent by the receiver of the data.

If the receiver is able to take data, it has already asserted RFD high. When the transmitter wants to send a data word, it first puts the word on the bus, and then begins the handshake cycle. It checks for the RFD signal. If it is asserted at the transmitter, the transmitter asserts DAV low to indicate the data are available. This is step 1 in Figures 9.14a and 9.14b. When the receiver sees DAV asserted, it negates RFD low in step 2 because it is no longer ready for data. When the processor picks up the data from the interface, the receiver asserts DAC high to indicate data are accepted. This is step 3. When the transmitter sees DAC asserted, it negates DAV high in step 4 because it will soon stop sending data on the data bus. When the receiver sees DAV negated, it negates DAC in step 5. The data are removed sometime after the DAV has become negated. When it is ready to accept new data, it asserts RFD high in step 6 to begin a new handshake cycle.

The IEEE-488 bus (Figure 9-14) is designed for busing data to and from instruments. First, the bus is asynchronous. If the receiver is far away and the data will take a long time to get to it, the DAV signal will also take a long time, and the other handshake signals will be similarly delayed. Thus, long cables are automatically accounted for by the handshake mechanism. Second, the instrument at the receiver may be slow or just busy when the data arrive. DAC is asserted as soon as the data get into the interface, to inform the transmitter that they got there; but RFD is asserted as soon as the instrument gets the data from the interface, so the interface won't get an overrun error that a UART can get. Third, although only one station transmits a word in any handshake cycle, a number of stations can be transmitters at one time or another. Fourth, the same word can be sent to more than one receiver, and the handshaking should be able to make sure all receivers get the word. These last two problems are solved using open collector bus lines for DAV, RFD, and DAC. DAC, sent by the transmitter, is negative logic so the line is wire-OR. That way, if any transmitter wants to send data, it can short the line low to assert DAV. RFD and DAC, on the other hand, are positive logic signals, so the line is a wire-AND bus. RFD is high only if all receivers are ready for data, and DAC is high only when all receivers have accepted data.

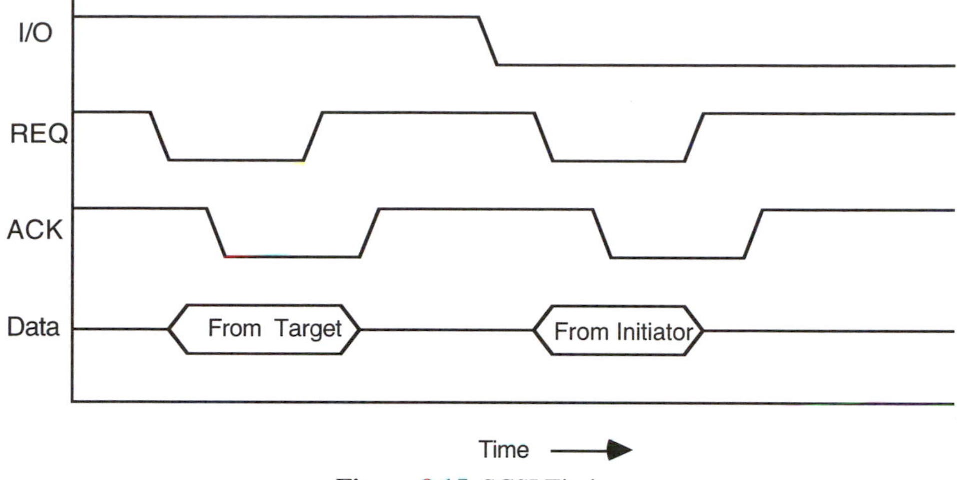

Figure 9.15. SCSI Timing

The IEEE-488 bus is well suited to remote control of instrumentation and is becoming available on many of the instruments being designed at this time. You will probably see a lot of the IEEE-488 bus in your design experiences.

9.4.4 The Small Computer System Interface (SCSI)

The microcomputer has made the intelligent I/O device economical. In a lot of systems today, a personal computer communicates with a printer that has a microcomputer to control it, or a disk that has its own microcomputer. Communications between a personal computer and the intelligent I/O device can be improved with an interface protocol specially designed for this application. The *small computer system interface (SCSI)* is designed for communications between personal computers and intelligent I/O devices. Many systems that you build may fit this specification and thus may use an SCSI interface.

The asynchronous protocol is quite similar to the IEEE-488 bus, having a 9-bit (8 data plus parity) parallel bus, a handshake protocol involving a direction signal (I/O), a request (REQ), and an acknowledge (ACK). (See Figure 9.13.) (There are also six other control signals, and the interface uses a 50-pin connector.) Up to eight bus controllers can be on an SCSI bus, and they may be *initiators* (e.g., microcontrollers) or *targets* (e.g., disk drives). A priority circuit assures that two initiators will not use the SCSI bus at the same time. After an initiator acquires the bus, a command stage is entered. The 10-byte command packet selects a target controller and is capable of specifying the reading or writing of up to 256 bytes of data on up to a 1024-GB disk. After the command packet is sent, data are transferred between the initiator and target.

Each command or data byte is transferred using the IO, REQ, and ACK signals. IO is low when the initiator wishes to write a command or data into the target, and high when it wants to read data from the target. If the initiator is sending data or a command, it begins a transfer by putting the eight-bit data or command and its parity bit on the nine-bit parallel bus; if the initiator is receiving data, it omits this step. The initiator then drops REQ low. If the target is to receive the data, it picks up the data and drops ACK low. If the target is to send the data, it puts the data on the bus and drops ACK low. When the initiator sees ACK low, if it is receiving data, it picks up the data from the data bus and raises REQ. When the target sees REQ high, it raises ACK high so the next transfer can take place. Up to 1.5 MB/s can be transferred on an SCSI bus this way, according to the original SCSI standard, but faster SCSI buses have been implemented and standardized.

9.5 Conclusions

Communications systems are among the most important I/O systems in a microcomputer. The microcomputer communicates with keyboards, displays, and typewriters, as well as with remote-control stations and other microcomputers, using the UART protocol. The microcomputer can be in a large computer system and have to communicate with other parts of the system using the SDLC protocol. It may be in a laboratory and have to communicate with instrumentation on an IEEE-488 bus. It may have to talk with other systems in a different protocol.

If you would like additional reading, we recommend the excellent *Technical Aspects of Data Communication,* by John McNamara. It embodies an exceptional amount of practical information, especially at the physical level, and also covers many of the widely used protocols. Motorola offers some fine applications notes on the SDLC protocol and its 6854 chip, such as *MC6854 ADLC, An Introduction to Data Communication,* by M. Neumann. For the IEEE-488 protocol using the 68488 chip, the applications note *Getting Aboard the 488-1975 Bus* is very informative. These applications notes are well written and take you from where this book leaves off to where you can design systems using these protocols.

For more concrete information on the 68HC11, please consult the *MC68HC11A8 HCMOS Single-Chip Microcomputer (ADI 1207)*. In particular, §5 describes the serial communication interface. As noted earlier, we have not attempted to duplicate the diagrams and discussions in that book because we assume you will refer to it while reading this book, and since we present an alternative view of the subject, you can use either or both views.

This chapter covered the main concepts of communications systems at the physical and link-control levels. You should be aware of these concepts so you can understand the problems and capabilities of specialists in this field. You should be able to handle the UART protocol – the simplest and most widely used protocol – and its variations, and you also should be able to use the SCI system in the 6812 and the ACIA chip, as well as the UART, in hardware designs. You should be able to write initialization rituals, interrupt handlers, and gadfly routines to input or output data using such hardware. Hardware and software tools like these should serve most of your design needs and prepare you for designing with the SDLC, IEEE-488, or SCSI interface protocol systems.

Do You Know These Terms?

See page 36 for instructions.

levels of abstraction
peers
end-to-end communication
network control
link control level
physical-control level
operating-system level
medium
frequency shift keying
frequency multiplexing
time multiplexing
channel
simplex
half-duplex
full-duplex
bit time period
baud rate
bit rate
synchronous
asynchronous
Manchester code
bit level
frame level
message level
protocol
handshake protocol
stack
structure
store and forward
circuit
governed
centralized
distributed
master slave system
differential line
RS-442
standard
active
passive
modem
originate modem
answer modem
data coupler
answer phone (ANS)
switch hook (SH)
ring indicator (RI)
data terminal ready (DTR)
clear to send (CTS)
break
universal asynchronous receiver transmitter (UART)
UART protocol
start bit
stop bit
parity error
framing error
double buffering
overrun error
asynchronous communications interface adapter (ACIA)
selected
hardware handshake
software handshake
remote job entry (RJE)
Bisync
transparent mode
cyclic redundancy check
synchronous data link control (SDLC)
flag pattern
information frame
supervisory
nonsequenced frame
primary station
secondary station
X.25 protocol
data available (DAV)
ready for data (RFD)
data accepted (DAC)
small computer system interface (SCSI)
initiators
targets

Problems

Problem 1 is a paragraph correction problem. See page 38 for guidelines. Guidelines for software problems are given on page 86, and for hardware problems, on page 115. Special guidelines for problems using the 6812 counter/timer modules are presented on page 390.

1.* To avoid any ambiguity, the peer-to-peer interfaces in communications systems are specified in terms of all lower level interfaces. The physical level, which is at the lowest level, is concerned with the transmission of signals such as voltage levels, with multiplexing schemes, and with the clocking of data if the clock must be sent with the data or on a separate line. The baud rate is the number of bytes of user data that can be sent per second. A channel is a data path between entities at the link-control level. It is half-duplex if every other bit is a data bit and the remainder are clock bits. Protocols are conventions used to manage the transmission and reception of data at the link-control level and to negotiate for and direct communications at the network level. A handshake protocol is one in which congratulations are extended whenever a frame is correctly sent, but the receiver is silent if the data don't appear to be correct. A store-and-forward network is one that sends frames, called packets, from node to node and stores a frame in a node before negotiating to send it to the next node closer to its destination. The bus is a particularly common half-duplex, store-and-foreward network.

2. Design a 1-bit input/output port using the SN75119 differential transceiver that is connected to a full-duplex differential line. Reading `PORTA` bit 7 will read the input data, and data written in bit 6 will pass through the transmitter for 8 cycles (1 μs) after the 'A4 writes in the output register. Use `PORTA` bit 5 to enable the transmitter.

a. Show the logic diagram of this device. Do not show pin numbers.

b. Show a self-initializing procedure `send(char d)` that outputs the least significant bit of `d`.

3. Design a differential line transmitter using two CA3140s. Both lines should be driven with OP AMP low-impedance outputs at ±5 V. Design a matching differential line receiver using a CA3140. In both designs, show pin numbers and power supply values.

4. Design an RS-232C level translator to drive an RS-232C line from a TTL level using a CA3140. Design a level translator to provide a TTL-level output from an RS-232C line using a CA3140. In both designs, show all pin numbers and power supply values.

5. A null modem is a simple module permitting one computer's RS-232C plug to connect to another computer's RS-232C plug that has the same pin connections. Suppose the computer uses only transmitted and received data, request to send, data set ready, and signal ground. Show socket connections in the null modem that can correctly interconnect the two computers so that each looks like a terminal to the other.

6. The 6812 counter/timer device 0, configured as an output compare using interrupts, and the Johnson counter of Figure 8.3, can generate modem frequencies. Device 1 as an input capture using gadfly, after appropriate filtering, can detect modem frequencies. Using device 2, time can be monitored using two global variables `int ticks, checks;`, which are incremented in `cct2()` every 5 ms. The first, `ticks`, is used to time out 17 s, and the second, `checks`, is used to check the reception time of each signal. ANS is `PORTA` bit 7 and SH is `PORTA` bit 6. A signal will be acceptable if its period is within 5% for the required time. Show self-initializing gadfly procedures that will implement the handshaking sequence of Figure 9.3:

a. `int answer()`, in the answer modem, called after answer modem's microcontroller is dialed up and data terminal is ready. The procedure returns the CTS value.

b. `int originate()` in the originate modem's microcontroller called after data terminal is ready. The procedure returns the CTS value. This procedure should call the procedure `dial()`, which was written for Chapter 8, problem 4, to dial `char numbers[7]`.

7. Use the 6812 SPI to implement a modified Pierce loop, which is an inter-microcontroller circular shift register, with, say, 8 bits of the shift register in each microcontroller. One microcontroller, the master, supplies the Pierce loop clock, the others are slaves. Frames circulate in the shift register synchronously and continuously. A simplified frame consists of a 3-bit source address (most significant bits), a 3-bit destination address (middle bits), and a 2-bit data field (least significant bits). When a frame is available to be used, the source and destination address are made TTT (111); no microcontroller sends to itself. Any microcontroller sends data to another by putting its addresses and data into such an available frame. When the frame shifts by its destination microcontroller, its data are saved there in global `char dataIn`. When a frame shifts by its source, it is rewritten as "available" again.

a. Show the logical design of a system using the SPI module to shift the data among three 'A4s, denoted microcontrollers 0 to 2. Show signal names but not pin numbers. Assume master microcontroller 0 supplies the SPI clock. The clock and slave selects are tied in common, but MISO and MOSI form a ring.

b. Write the initialization in `main()`, procedures `send(char address, char data)`, `receive()`, and interrupt handler `ccspi()` so that `main()` outputs the message 3 to destination 1; receives incoming data into local variable *i;* and sends the message 0 to destination 1 – to be used in the master microcontroller 0.

c. Write the initialization in `main()`, procedures `send(char address, char data)`, `receive()`, and interrupt handler `ccspi()` so that `main()` outputs the message 1 to destination 0, receives incoming data into local variable *i*, and sends the message 2 to destination 0 – to be used in either slave microcontroller 1 or 2.

8. Write a program to output 9600-baud UART signals, having 8 data, no parity, and 1 stop bit, using 6812 counter/timer device 0. `char` global variable `UartOut` holds output data, and `outBits` holds the output bit count. Write `main()` to initialize the counter/timer, output 0x55, and turn off the device. Output should be general so that it can be repetitively done. Write `cct0()` to interrupt each bit time, to send `UartOut`.

9. Write a program to input 9600-baud UART signals, having 8 data, no parity, and 1 stop bit, using 6812 counter/timer devices 0 and 1. `char` global variable `UartIn` holds completely shifted input data, `inBits` holds the input bit count, `shiftRegister` holds bits shifted in, and `flag` is set when data have arrived. Write `main()` to initialize the counter/timer, print an input byte in hexadecimal, and terminate the use of the device. Input should be general so that it can be repetitively done. Write `cct0()` to detect rising and falling edges that indicate the widths, and therefore the bit values, of input data, which should be shifted into `UartIn`. `cct0()` should also detect each start bit and enable device 1 when it has been found. Write `cct1()` to interrupt at the time of the middle of the stop bit to complete received data in `UartIn`, set `flag`, and disable its interrupt. Hint: call a `service()` procedure from either handler to shift input bits.

10. Show all loading of the *status* port to initialize the ACIA for the following.

a. 8 data bits, 2 stop bits, divide clock by 1, all interrupts disabled, RTS high

b. 7 data bits, even parity, 2 stop bits, divide clock by 16, only receiver interrupt enabled, RTS low

c. 8 data bits, 1 stop bit, divide clock by 16, only transmitter interrupt enabled

d. 7 data bits, even parity, 1 stop bit, divide clock by 64, all interrupts enabled

e. 7 data bits, even parity, 2 stop bits, clock divide by 1, interrupts disabled, RTS low

11. Write a simple ACIA word-oriented teletype handler, using the gadfly synchronization technique. The ACIA is at 0x200 (control) and 0x201 (data), using 'A4's CS0 to enable the ACIA pin CS2 as shown in Figure 9.7b.

a. Write the initialization routine for parts b and c. Use 7 data, odd parity, and 2 stop bits, and divide the clock by 16 to get 1200 baud. The 'A4 E clock is set to 2 MHz so that the 68B50 timing specifications are met. The 'A4 counter/timer device 0 provides the transmit and receive clocks to the ACIA.

b. Write a procedure `put(char c)` to output the character *c*. If the device is busy, wait in `put` until the word can be output.

c. Write a procedure `char get();` to return a character read from the ACIA. If no character has arrived yet, wait in `get` until it comes.

12. Write an ACIA background teletype handler to feed characters to a slow teletype using interrupt synchronization, so you can continue working with the computer as the printing is being done. A 0x100-byte `queue` contains characters yet to be printed. Part b will fill `queue` when your program is ready to print something, but the interrupt handler in part c will pull words from the queue as they are sent through the ACIA. The 'A4 E clock is set to 2 MHz so that the MC68B50 timing specifications are met. The 'A4 counter/timer device 0 clocks to the ACIA to provide 1200-baud rates.

a. Write `main()` to initialize the 'A4 and ACIA at locations 0x200 and 0x201 for 7 data, even parity, and 1 stop bit, dividing the clock by 16.

b. Write a procedure `put(char *s, char n)` that will, starting at address `s`, output `n` words by first pushing them on the queue (if the queue is not full) so they can be output by the handler in part c. If the queue is full, wait in `put` until it has room for all words. Write the code to push data into the queue in `put` without calling a subroutine to push the data.

c. Write handler `ccirq()` that will pull a word from the queue and output it, but if the queue is empty it will output a SYNC character (0x16). Write the code to pull data from the queue in the handler without calling a subroutine to pull the data.

13. Show a logic diagram of an I/O device using an IM6403 connected to an 'A4. Use 'A4 chip selects CS0 to CS3 to write data to be sent at 0x200, read data that was received at 0x280, and write control at 0x300. The OE, FE, PE, DR, and TBRE status bits can be read at location 0x380 as bits 0 to 4 respectively. Connect control and status so that the lower-numbered pins on the IM6403 are connected to lower-numbered data bits for each I/O word, and use the lower-number data bits if fewer than 8 bits are to be connected. Show signal names of all pin connections to the 'A4, and the signal names and the numbers of pins on the IM6403.

14. Show the logic diagram of a remote station that uses an IM6403 UART and a 74HC259 addressable latch so that when the number `I + 2N` is sent, latch `N` is loaded with the bit `I`. Be careful about the timing of DR and DRR signals, and with the G clock for the 74HC259. Show only signal names, and not pin numbers.

15. A stereo can be remotely controlled using an M14469. Show the logic diagrams, including pin numbers, HCMOS circuits, and component values, for the following.

a. A single volume control, whose volume is set by sending a 7-bit unsigned binary number to the M14469 with address 0x01, using the duty-cycle control technique (Figure 7.12b). Use comparator 74HC684, counter 74HC4024, and SSI chips.

b. A source and mode selector, whose source and mode are set by sending a 4-bit code to the M14469 with address 0x02. Use the select hardware in Figure 7.11.

16. Show initialization rituals in `main()` to initialize the 6812 SCI module (using an 8-MHz E clock) for the following.

a. 8 data bits, 1 stop bit, 9600 baud, all interrupts disabled, receiver and transmitter enabled

b. 9 data bits, 1 stop bit, 300 baud, only receiver interrupt enabled, receiver and transmitter enabled

c. 8 data bits, 1 stop bit, 9600 baud, all interrupts disabled, transmitter only enabled, send break

d. As in part a, but with interrupt (wake up) when the line is idle

e. As in part a, where the E clock is set to 2 MHz

17. Write an SCI device 1 background teletype handler to feed characters to a slow teletype using interrupt synchronization, so you can continue working with the computer as the printing is being done. A 0x100-byte `queue` contains characters yet to be printed. Part b will fill `queue` when your program is ready to print something, but part c's interrupt handler pulls words from the queue as they are sent through the SCI.

a. Write `main()` to initialize SCI device 1 for 8 data, no parity, and 1 stop bit, and 1200 baud (the 'A4 has an 8-MHz E clock). Only the transmitter is enabled.

b. Write a procedure `put(char *s, char n)` that will, starting at address `s`, output `n` words by first pushing them on the queue (if the queue is not full) so they can be output by the handler in part c. If the queue is full, wait in `put` until it has room for all words. Write the code to push data into the queue in `put()` without calling a subroutine to push the data.

c. Write handler `ccsci1()` that will pull a word from the queue and output it, but if the queue is empty it will output a NULL character (0). Write the code to pull data from the queue in the handler without calling a subroutine to pull the data.

18. An MC68B50 on the 'A4 is wired to the 'A4's SCI device 1. The ACIA at locations 0x200 and 0x201 and the SCI are initialized for 8 data, no parity, and 1 stop bit, at 1200 baud. The ACIA clock is generated by 'A4 counter/timer device 0 using interrupts. The 'A4 E clock is set to 2 MHz so that the 68B50 timing specifications are met. Write the `main()` procedure, that initializes all the devices, and then, using gadfly synchronization but without using any subroutines, sends 0x55 out the ACIA's transmitter, which is connected to the SCI's receiver. `main()` then moves this data from the SCI's reciever to its transmitter, which is connected to the ACIA receiver. `main()` then reads the data from the ACIA receiver into a local variable *c*. In each case, the status register is checked before data are moved into or out of the data register.

19. Implement a Newhall loop, using interrupt synchronization on SCI device 1. This loop is a ring of microcontrollers where each module's *TxD1* is connected to the next module's *RxD1,* to circulate messages. The message's first byte's most significant nibble is a module address to which the message is to be sent, and its least significant nibble is a count of the number of data bytes left in the message, which is less than 0x10. If the message address is this module's address, input data are stored into an input buffer, otherwise input data are pushed onto a 16-byte queue. Transmitter interrupts are enabled only when the queue is nonempty or output buffer data is to be sent. Upon a transmitter interrupt, if the queue is nonempty, a byte is pulled and output. Otherwise, the output buffer data is sent. If neither data are to be sent, transmitter interrupts are disabled. Write `main()` to initialize the SCI, send a 4-byte message 1, 2, 3, 4 in the 16-byte output buffer to module 12, wait until all data are moved, and disable SCI device 1. Write `ccsci1()` to buffer an input message with address 5, and move other messages through it around the loop. If no data are sent through the module, the handler will send the module's outgoing message. `push` and `pull` statements should be written into the handler code rather than implemented in separate subroutines.

20. Write gadfly-synchronized C procedures `main(), put(char *v)` and `get(char *v)` to send and receive 9-frame, 9-bit frames, with even parity rows and columns, at 4800 baud, and perform single-bit error correction using the 2-dimensional parity protocol. `main()` initializes SCI device 1, and sends and then receives these 9-frame, 9-bit frames. This protocol sends 8 bytes of data by sending even parity for each byte and, after the data are sent, by sending the exclusive-OR of the eight data bytes, in even parity. `get` corrects any single error by observing which byte and bit has a parity error.

21. Write a real-time C procedure `main()` to initialize the devices and output the stream of data bits in `char buffer[0x100]`, most significant bit of lowest-addressed word first, checking for five T's and inserting F, as in the SDLC protocol. Send the data at 100 baud, gadflying on timer device 0 to time the bits. The clock is sent on `PORTA` bit 0, and data are sent on `PORTA` bit 1 (to be determinate when the clock rises).

22. Write a gadfly routine to handshake on the IEEE-488 bus. Data can be read or written in `PORTA`, and DAV, RFD, and DAC are `PORTB` bits 2 to 0, respectively.

a. Show a C procedure `send(char i)` to initialize the ports, send *i,* and perform the handshake for a transmitter (talker).

b. Show a C procedure `receive()` to initialize the ports, perform the handshake, and return the received word (listener).

23. Write a gadfly C procedure to initialize the ports and handshake on the SCSI bus. Eight-bit data can be read or written in `PORTA` (ignore parity), and I/O, REQ, and ACK are `PORTB` bits 2 to 0, respectively. When not in use, all control lines should be high.

a. Show `initiateS(char c)` to send `c` from an initiator through its SCSI device.

b. Show `char targetR();` to return the SCSI device's data in a target.

c. Show `targetS(char c)` to send `c` from a target through its SCSI device.

d. Show `initiateR()` to return the SCSI device's data in an initiator.

The Axiom PB68HC12A4 board is fitted with female harmonica plugs and a prototyping area for a laboratory developed for this book. Experiments can be quickly connected by pushing 22-gauge wire into the harmonica plugs and prototyping areas.

10

Display and Storage Systems

The previous chapter discussed the techniques by which microcomputers can communicate with other computers. They may also have to communicate with humans, using LCD displays covered in Chapter 4 or using more complex CRT displays. We now cover CRT display technology. Also, a microcomputer may have to store data on a magnetic tape or disk. This stored data can be used by the microcomputer later, or it may be moved to another computer. Thus, on an abstract level, a magnetic storage medium can be an alternative to an electronic communications link.

This chapter covers both the CRT display and the magnetic storage device. We discuss display systems first and then storage systems. In display systems, we will use a single-chip 6812, with only an additional transistor and its resistors, to implement a primitive device. We follow this with a small but realistic bitmap display. In storage systems, we use a special chip, the Western Digital WD37C65C, to implement a very useful floppy disk controller. First using the surprisingly powerful 6812 alone lets us show the principles of these devices and allows you the opportunity to experiment with them without much expense. However, the bitmap display and special-purpose floppy disk controller are quite easy to use and to design into real systems.

In this chapter, we spend quite a bit of time on the details of video and disk formats. We also present some rather larger system designs and refer to earlier discussions for many concepts. We have somewhat less space for the important notion of top-down design than in previous chapters because the design alternatives for CRT and disk systems are a bit too unwieldy to include in this short chapter. They are important nevertheless and are covered in some of the problems at the end of the chapter.

Upon completing this final chapter, you should have gained enough information to understand the format of the black-and-white NTSC television signal and implement a single-chip or a bitmap CRT display. You should have gained enough information to understand the floppy disk and you should be able to use a floppy disk controller chip to record and play back data for a microcomputer, or use it to move data to, or form it from or to, another computer. Moreover, you will see a number of fairly complete designs similar to those you will build.

10.1 Display Systems

A microcomputer may be used in an intelligent terminal or in a personal computer. Such systems require a display. Any microcomputer requiring the display of more than the hundred or so digits an LED or LCD is able to handle can use a CRT display.

This section describes the concepts of CRT display systems. We present the format of the NTSC black-and-white signal and then show a program that enables the 6812 to display a checkerboard block on a white screen. We then illustrate a useful bitmap display and show object-oriented functions for elementary graphics applications.

10.1.1 NTSC Television Signals

A *National Television System Committee* (NTSC) signal is used in the United States and Canada for all commercial television. A computer display system consists of the *cathode-ray tube* (CRT) and its drive electronics – essentially a specialized TV set – and hardware and software able to send pulses to time the electron beam, which is a stream of bits to make the TV screen black or white at different points. Figure 10.1 diagrams the front of a TV screen. An electron beam, generated by a hot cathode and controlled by a grid, is deflected by electromagnets in the back of the CRT and made to move from left side to right and from top to bottom across the face of the CRT. More electrons produce a whiter spot. The traversal of the beam across the face is called a *raster line.* The set of raster lines that "paint" the screen from top to bottom is a field. NTSC signals use two fields, one slightly offset from the other, as shown in Figure 10.1a, to completely paint a picture *frame.*

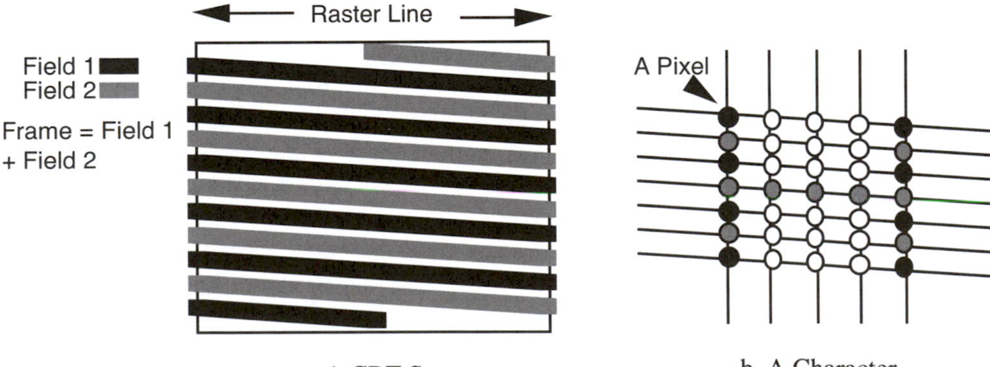

a. A CRT Screen

b. A Character

Figure 10.1. The Raster-Scan Display Used in Television

In NTSC signals, a frame takes 1/30 second and a field takes 1/60 second. The raster line takes 1/15,750 second, a field has 262 1/2 raster lines and a frame has 525 raster lines. As the beam moves from side to side and from top to bottom, the electron beam is controlled to light up the screen in a pattern. A *pixel* is the smallest controllable dot on the screen. As illustrated by Figure 10.1b, a clear circle represents a pixel having no light, and a dark circle (black for field 1 and gray for field 2) shows a lighted pixel.

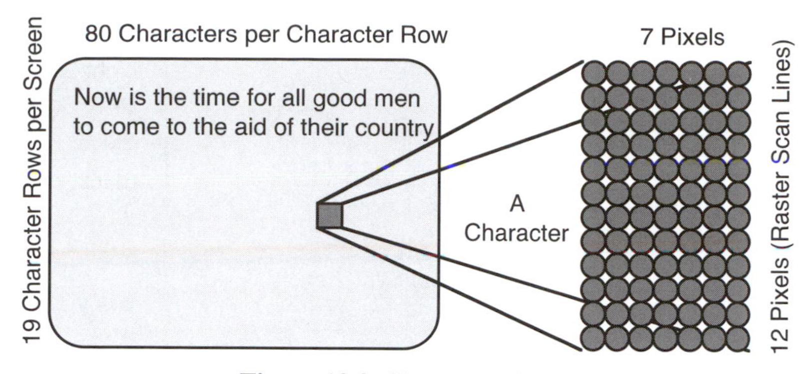

Figure 10.2. Character Display

Figure 10.1b shows how H is written in both fields of a frame. ASCII characters will be painted in a 7- by 12-pixel rectangle, 80 characters per line (Figure 10.2).

The *NTSC-composite-video signal* is an analog signal, diagrammed in Figure 10.3a. The displayed signal is an analog signal where a maximum voltage (about 1/2 V) produces a white dot, a lower voltage (3/8 V) produces gray, and an even lower voltage (1/4 V) produces a black dot. The part of the signal corresponding to the time when the electron beam is moved to the left side (*horizontal retrace*) or to the top (*vertical retrace*) occurs between the displayed parts. At these times, *horizontal sync* and *vertical sync* pulses appear as lower voltage (0 V) or "blacker-than-black" pulses. See Figures 10.3b. A CRT uses a *sync separator* circuit to extract these pulses so it can derive the horizontal and vertical sync pulses, which are used to time the beam deflections on the screen. This signal is called the composite video signal because it has the video signal and the sync signals composed onto one signal. If this signal is to be sent over the air, it is modulated onto a radio frequency (r.f.) carrier (such as channel 2). Alternatively, the separate video, horizontal, and vertical sync signals can be sent over different wires to the CRT system, so they do not need to be separated; this gives the best resolution, such as is needed in 1024-by-1024 pixel CRT displays in engineering workstations. The composite video is used in character displays that have 80 characters per line. The r.f. modulated signals are used in games and home computers intended to be connected to unmodified home TV sets, but are only capable of 51 characters per line.

The frequency of the vertical sync pulses, which corresponds to the time of a field, is generally fixed at 60 Hz, to prevent AC hum from making the screen image have bars run across it, as in inexpensive TVs. It is also about the lowest frequency at which the human eye does not detect flicker. American computer CRTs often use this vertical sync frequency. The horizontal sync frequency in computer CRTs is usually about 15,750 Hz, as specified by the NTSC standard, but may be a bit faster to permit more lines on the screen yet keep the vertical sync frequency at 60 Hz. The magnetic beam deflection on the CRT is tuned to a specific frequency, and the electronics must provide horizontal sync pulses at this frequency, or the picture will be nonlinear. The pulse widths of these horizontal and vertical pulses are specified by the electronics that drive the CRT. Thus, the CRT controller must be set up to give a specific horizontal and vertical frequency and pulse width, as specified by the CRT electronics.

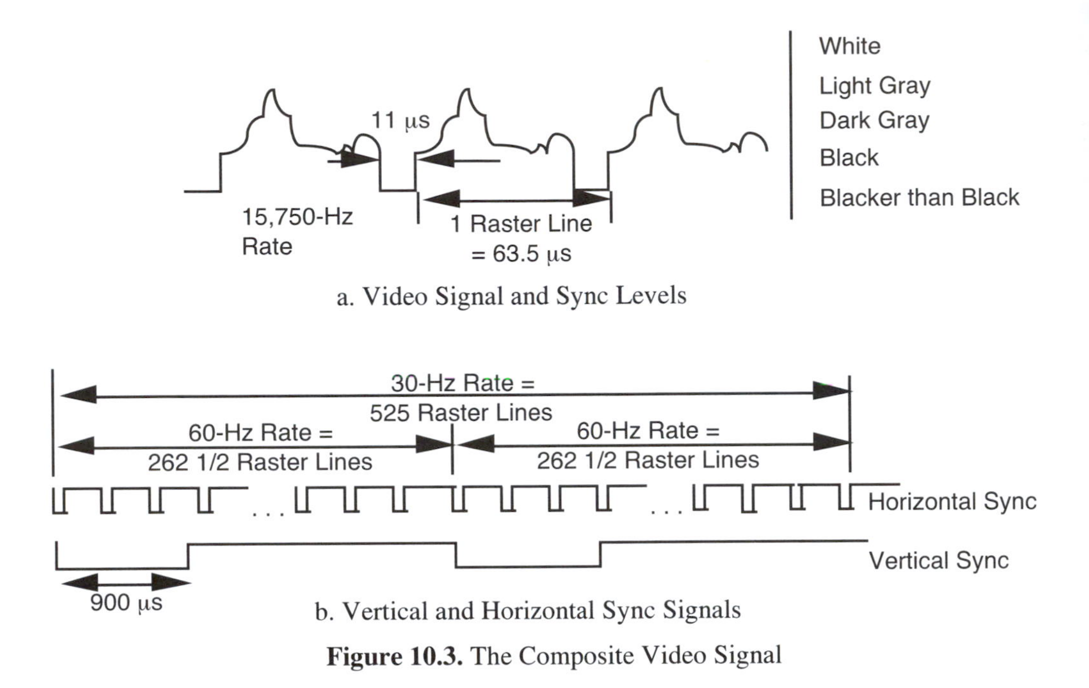

a. Video Signal and Sync Levels

b. Vertical and Horizontal Sync Signals

Figure 10.3. The Composite Video Signal

10.1.2 A 6812 SPI Display

We are fortunate that the 6812 has a built-in counter and shift register able to generate the synchronization pulses and the bit stream to implement a primitive CRT display. The 6812 output compare timers, described in Chapter 7, are capable of generating the vertical and horizontal sync pulses; and the serial peripheral interface (SPI), introduced in Chapter 4, has the capability of generating a CRT display having poor, but useful, resolution. The upcoming C procedure `main()` should produce a picture as shown in Figure 10.4, using the simple hardware diagrammed in Figure 10.5 with a single-chip 6812. It is quite useful for explaining the principles of CRT display systems, since it uses familiar 6812 peripherals. It might be useful for multicomputer systems as a diagnostic display available on each microcomputer. We have found it helpful in testing some bargain-priced CRTs when we did not have specifications on the permissible range of horizontal and vertical sync pulse widths and frequencies. This little program lets us easily test these systems to generate the specifications. We now describe how that built-in CRT generator in the 6812 can produce a CRT display.

A combined sync signal is generated that is the exclusive-OR of the vertical and horizontal sync signals. The CRT's sync separator outputs its high-frequency component to the horizontal oscillator and a low-frequency component to the vertical oscillator. By inverting the horizontal sync signal during vertical retrace, the signal's low-frequency component has a pulse during this period. The high-frequency output of the sync separator continues to synchronize the horizontal oscillator during vertical retrace, while the low-frequency component synchronizes the vertical oscillator during vertical retrace.

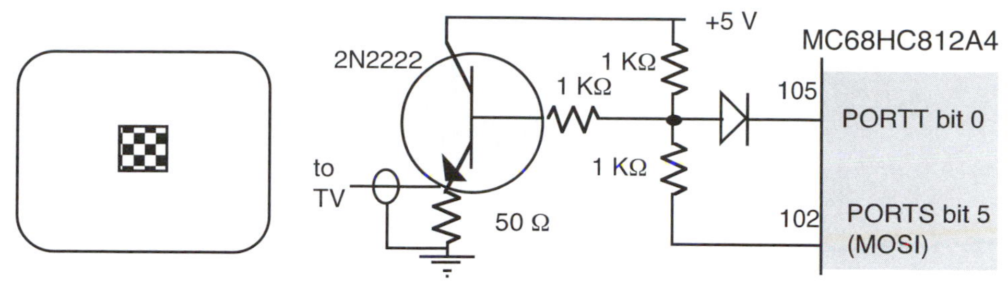

Figure 10.4. Screen Display **Figure 10.5.** Circuit Used for TV Generation

Figure 10.5 shows a simple circuit for the generation of composite video. If your CRT requires separated video and sync signals, the combined sync can be taken directly from pin 105, and the data from pin 102, and the circuit diagrammed in Figure 10.5 is not needed. The program is just two modified interrupt-based square-wave generator programs, as described in §8.2.2, with an SPI routine. An output compare 0 interrupt occurs upon each horizontal sync pulse, flipping the output signal and executing `cct0`. Since the pulse width is quite short, this handler toggles the output back to high by writing T (1) into CFORC bit 0. The vertical sync pulse is implemented by `cct1`, which uses CFORC to invert the output on bit 0 at the beginning and end of the vertical sync pulse. Using a primitive character generator an element of `char pattern[12]` is put into SPI's data port, which is shifted out as the video signal. This program produces a checkerboard box in the middle of the screen, shown in Figure 10.4.

```
#define Hw 21                    /* horizontal pulse width, µs * 2 */
#define Hp 122                   /* horizontal period, µs * 2 */
#define Vw Hp*13                 /* vertical pulse width, µs * 2 */
#define Vp Hp*262+(Hp/2)+Vw      /* vertical period, µs * 2 */
#define HPOS 10                  /* horizontal location of square */
#define VPOS 260                 /* vertical location of square */

enum{ MOSI=0x20, SCLK=0x40, SS=0x80, SPIE=0x80, SPE =0x40,  MSTR=0x10,
    CPOLA=8, CPHA=4, SSOE=2, LSBF=1, SSWAI=2, SPC0=1, SPIF=0x80];

unsigned char Vs,Hs,Fr,dLine;int lineNo; char pattern[12] =
 {0x55,0xaa,0x55,0xaa,0x55,0xaa,0x55,0xaa,0x55,0xaa,0x55,0xaa};

interrupt 8 void handler1(){ // horizontal sync pulse generator
  TCO += Hw; TFLG1=1; lineNo += 2; // interlaced display, output every other line
  if((lineNo>VPOS)&&(lineNo<=(VPOS+12))) dLine = lineNo - VPOS;
    do ; while((TFLG1&1)==0); TFLG1=1; TCO += Hp-Hw; // wait for pulse end
}

interrupt 9 void handler2(){ // vertical sync pulse generator
  if(Vs ^= 1) {TC1 += (Vp-Vw); if(Fr ^= 1) lineNo=1; else lineNo=0; }
    else TC1 += Vw;
    TFLG1 = 2; CFORC = 1;
}
```

```
void main() { char d;     asm SEI
    SP0CR1 = SPE + MSTR; SP0BR = 0;DDRS = MOSI + SCLK; /* 4 MHz clk */
    PORTS = dLine = 0;   TSCR = 0x80; TMSK2 = 2; /* 2 MHz Timer CLK */
    TFLG1 = TMSK1 = TIOS = 3; TCTL0 = 5;
    TC1 = TC0 = TCNT + 0x20; if( ! (PORTS & 1)) CFORC = 1; asm CLI
    do {
        if(dLine){ dLine = 0; // wait til next line to be displayed
        for(d = 0; d < HPOS; d++) ; // waste time to get to the square's column
        SP0DR = pattern[dLine-1] ; while(!(SP0SR & SPIF)) ; d = SP0DR;
        }
    asm WAI /* stack registers, start immediately when horizontal sync times out */
    } while(1);
}
```

The `WAI` instruction at the end of the `do while(1)` loop removes jitter from the display. `WAI` saves the registers and waits for any interrupt; the handler is entered exactly when an output compare matches the timer. Without `WAI`, the screen image would appear torn and would "dance" around because the interrupt would start on completion of the current instruction, in any of several cycles after an output compare occurred, and this would result in a mismatch between the timing of the sync signal and the data signal.

For a 7-by-12 pixel character form with 1 pixel between characters, data is shifted at 4 Mbits/s, so about 20 characters can be put on a line. The image is interlaced, so 40 lines of characters appear on the screen. While the pattern we used is a checkerboard, a different pattern can be written for each letter, and the pattern can be chosen by the ASCII representation of the letter. Thus, characters in words can be written across the screen to implement a useful display. Lines can also be drawn, to draw rectangles and other geometric figures. However, character and line shape is less than satisfactory. The problem is that the SPI's shift rate is too slow to generate the finer pixels needed to represent letters more satisfactorily. This problem is solved by using an external shift register in place of the SPI, as we will do in the next section. What is mildly surprising, though, is that the 6812, with very little external hardware, has the ability to generate CRT signals. Motorola can therefore claim that the 6812 has a built-in CRT controller.

10.1.3 A Bitmapped Display

A more realistic bitmapped display can be achieved using an external 16-bit shift register in place of the SPI. The program can write 16 bits every 8 memory cycles into its shift register so it can output a bit every 31.25 ns. This is eight times the resolution provided the system in §10.1.2. So we can get 80 characters per line. An external (8K, 16) SRAM, whose access time is less than 70 ns, holds the image of the screen. This speed is used to read the memory in one memory cycle without extending the cycle. The SRAM's size is sufficient to hold a bit image for about 2/3 of the screen. The 'A4 counter/timer system provides sync pulses. The sync signal generation technique is fundamentally the same as is used in the previous example.

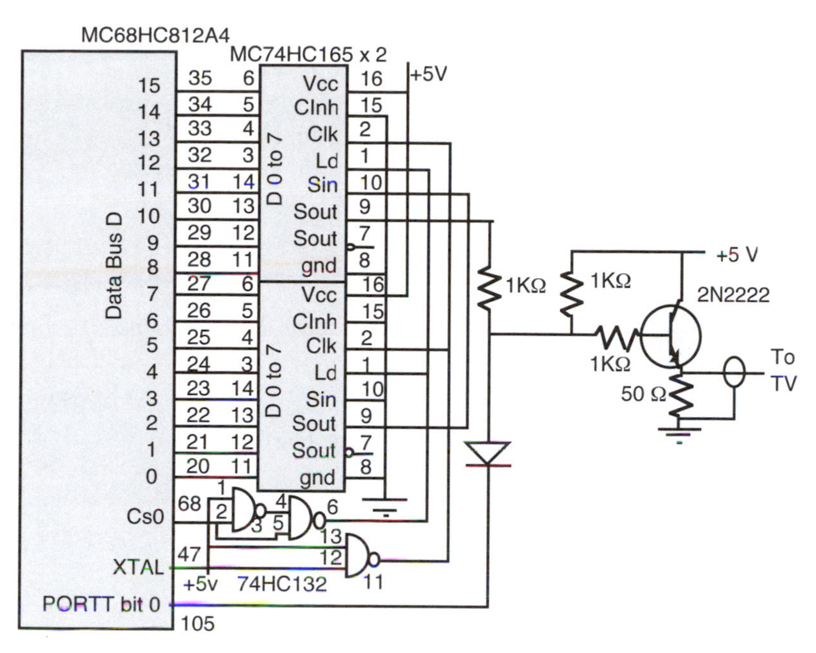

Figure 10.6. Display Hardware

Figure 10.6 shows the shift-register hardware that replaces the SPI shift register in the previous example. We used a Motorola HC12A4EVB evaluation board to implement the memory, but its chip select and RAM jumpers were altered to use CSP0 for its SRAM. Alternatively, we could use the 8K-by-16-bit memory shown in Figure 6-14.

The C program is essentially the same as that used in §10.1.2, except that an assembly-language program outputs 16-bit words from memory through the shift register to send bits to the display. Note that the expanded wide mode is set up, with access to the RW and LSTRB signals for the SRAM, chip select 0 is used with the shift register, and chip select P0 is used with the SRAM. The counter/timer is set up and its interrupts are handled in the same way as in the previous example, except that a display area of 512-by-256 pixels is provided. A white 2-pixel-wide border is then drawn around a blank screen. In the next section, an object-oriented class is illustrated that will write geometric figures and characters in the display area provided by this hardware and program. The program, beginning with its declarations, follows:

```
volatile char TIOS@0x80,CFORC@0x81,TSCR@0x86,TCTLO@0x88, TMSK1@0x8c,
    TMSK2@0x8d, TFLG1@0x8e, TFLG2@0x8f, PORTS  @0xd6, DDRS@0xd7;

volatile int TCNT@0x84,  TCO@0x90, TC1@0x92, SPOCR@0xd0,
    PEARMODE@0x0A, CSCTL@0x3C, CSSTR@0x3E;

#define Hw 21                  /* horizontal pulse width, µs * 2 */
#define Hp 122                 /* horizontal period,  µs * 2 */
#define Vw Hp*13               /* vertical pulse width, µs * 2 */
#define Vp Hp*262+(Hp/2)+Vw    /* vertical period, µs * 2 */
```

```
#define HPOS 10              /* horizontal location of square */
#define VPOS 260             /* vertical location of square */
#define TOP 128              /* top scan line of screen where display begins */
#define WIDTH   64           /* horizontal width, in bytes */
#define HWIDTH  32           /* horizontal width, in 16-bit words */
#define HEIGHT 256           /* vertical height, in scan lines */

enum {Special,Normal=0x80,Wide=0x60,RDWRE=0x400,LSTRE=0x800,CS0E=0x100,
     CSP0E=0x2000,CSP1E=0x4000,CS2E=0x400,CS3E=0x800,STR2=4,STR3=6};

unsigned char Vs,   /* vertical sync state: 0 in vertical sync, 1 outside vertical sync */
   Fr,         /* frame number: 0 or 1 */    dLine;        /* if 1, display line */
int lineNo, /* line number to be displayed */ *dPtr; /* data pointer to frame buffer */

interrupt 8 void handler1(){// horizontal sync pulse generator
    TC0 += Hw; TFLG1 = 1;
    if((lineNo > TOP) && (lineNo <= (TOP + HEIGHT))) dLine = 1;
    do ; while((TFLG1 & 1) == 0); TFLG1 = 1; TC0 += Hp - Hw;
    lineNo += 2;
}

interrupt 9 void handler2(){// vertical sync pulse generator
    if(Vs ^= 1) {
        TC1 += (Vp - Vw);
        if(Fr ^= 1) { lineNo = 1; dPtr = (int *)0x8000; }
        else { lineNo = 0; dPtr = (int *)(0x8000 + WIDTH); }
    }
    else TC1 += Vw;
    TFLG1 = 2; CFORC = 1;
}

void main() { int i; asm sei
    PEARMODE = RDWRE + LSTRE + Wide + Normal;
    CSCTL = CS0E + CSP0E; CSSTR = 0; TSCR = 0x80; TMSK2 = 2; /* init 'A4 */
    for(i=0,dPtr=(int *)0x8000;i<HWIDTH*HEIGHT;i++) dPtr[i]=0; /* clr scn */
    for(i = 0; i < HWIDTH; i++) /* draw top, bottom border */
        dPtr[i]=dPtr[i + HWIDTH]=dPtr[i + HEIGHT * HWIDTH - HWIDTH] =
        dPtr[i + HEIGHT * HWIDTH - 2 * HWIDTH] = 0xffff;
    for(i=2,dPtr=(int *)0x7fff;i<(HEIGHT-2);i++) dPtr[i*HWIDTH]=0x3c0;

    TC0 = TC1 = 100 + TCNT; TCTLO = 5; TFLG1 = TMSK1 = TIOS = 3;
    if(!(PORTS & 1)) CFORC = 1; asm cli /* if hor sync is not high, force high */
    do {
        if(dLine) {
            dLine = 0;
asm{
run:  ldx dPtr        ; get current buffer pointer
     ldy #0x200       ; destination address, shift register port
     ldd #32          ; WIDTH/2
```

```
run1: movw 2,x+,0,y; move a 16-bit word
      dbne d,run1    ; repeat for line
      leax 64,x      ; skip over next line (written in next frame)
      stx dPtr       ; replace current buffer pointer
}
        }
        asm wai /* stack registers, start immediately when horizontal sync times out */
  } while(1); /* loop forever */
}
```

10.1.4 An Object-oriented Display

The previous section's program displays the bitmapped image stored in the frame buffer, an 8K-by-16 SRAM at 0x8000. In this section, we illustrate an object-oriented class `display` that can draw geometric figures and characters in this frame buffer. Its function members should be called to draw images after the border is drawn and display area is cleared, but before interrupts are enabled.

```
unsigned char patternA[8] = {0x10,0x28,0x44,0x82,0xFE,0x82,0x82,0x82};
int points[16] = {0x8000,0x4000,0x2000,0x1000,0x800,0x400,0x200,0x100,
                  0x80,0x40,0x20,0x10,8,4,2,1};
enum{ WIDTH = 100, HWIDTH = 50, HEIGHT = 200};

class display{ int *dPtr;

   public: display(void) { dPtr = (int *)0x8000; } /* constructor */

   virtual void point(int h, int v) {
      if((v>=0) && (v < (WIDTH * 8)) && ( h >= 0) && (h < HEIGHT ))
         dPtr[( v << 5) + (h >> 4)] |= points[h & 0xf];
   }

   virtual void line(int h,int v,int dh,int dv,int n)//only for some dh, dv
        { while(n--) point( h += dh, v += dv); }

   virtual void triangle(int h, int v) {
      line(100+h, 100+v, 2, 1, 50);  line(100+h, 100+v, 1, 2, 50);
      line(200 + h, 150 + v, -1, 1, 50);
   }

   virtual void put(unsigned char *pattern,int h,int v){
      unsigned char row; int index, *p;
      if((h<0)||(h>=((WIDTH-1)*8))||(v<0)||(v>=(HEIGHT-8))) return;
      index=(int)dPtr+(h >> 3)+(v * WIDTH); // need to force char indexing
      for(row = 0, p = (int *)index; row < 8; row++, p += HWIDTH) {
           *p |= pattern[row] << (8 - (h & 7));
      }
   }
};
```

```
void main() { char i;
    display *S = new display;S->point(1,2);S->line(1,2,3,4,5);
    S->triangle(1, 2); S->put(patternA, 1, 2);
    delete S;
}
```

The class's constructor merely sets `dPtr` to the location of the frame buffer. The function member `point(int h, int v)` sets a bit to display a point at horizontal row *h* and vertical pixel column *v*, provided the point is inside the display area. The vertical coordinate, and the horizontal coordinate's high-order bits, determine which 16-bit word in the buffer is to be changed, and the horizontal coordinate's low-order bits determine which bit in that word is to be changed. This function member sets the bit chosen by the function's parameters; other variations of it can clear or complement the indicated bit.

The function member `void line(int h, int v, int dh, int dv, int n)` draws an *n*-point line from point `(h, v)`. Each time a pixel is drawn, it adds `dh` to `h` and `dv` to `v`. This simple algorithm is only suitable for lines where increment `dh` or increment `dv` is 1, and the other increment is between -2 and +2. It can draw a rectangle (see problem 8) and some triangles. The *Bresenham algorithm* is commonly used for general line drawing (see problem 9). The basic algorithm works only for the octant in which both `dh` and `dv` are positive, and `dh` is greater than `dv`. As points are drawn in consecutive columns, the algorithm keeps track of an error `e` whose sign indicates whether a point should be drawn in the same row as the last point, or the next higher-numbered row. The calculation of `e` is based on the line's differential equation, and is explained by most textbooks on computer graphics. The general algorithm determines which octant a line is in, and calls the function member that implements the basic algorithm, with operands interchanged or negated as needed, so that the basic algorithm operates in its preferred octant. A variation of this algorithm can draw ellipses and circles. The function member `triangle(int h, int v)` draws a triangle with upper left vertex at `(h, v)`. A more general triangle can be drawn using the Bresenham line drawing algorithm (see problem 10).

The function member `wchar(char *pattern, int h, int v)` writes a character whose pattern is defined by the vector `pattern` (such as `patternA` above) so its upper left pixel is at `h, v`. If the character is entirely within the display area, the pattern is ORed into the buffer a row at a time, at a 16-bit word offset determined by `v` and the high-order bits of `h`, using a shift offset determined by the low-order bits of `h`. This function member `wchar` is suited only for characters whose maximum width is 8 pixels. A slightly more general function member `wchar` can draw characters whose maximum width is 16 pixels (see problem 11). Calling `wchar` with different character patterns and offsets can write words on the screen (see problem 12). A more general function member could write null-terminated strings of characters on the screen, keeping track of the position of the last drawn character as a data member so that the calling routine need not pass this parameter to the function member.

The routines for drawing lines and characters are described here as object-oriented function members for a class `display`. While this example doesn't seem to warrant the use of object-oriented programming, a simple extension of this class will utilize object-oriented capabilities. Consider the display of multiple separate *windows*, each of which occupies a separate portion of the buffer and therefore of the screen. The class constructor can have an origin that initializes data member `dPtr`, and a horizontal and vertical range of pixels as arguments; points and characters within that range can be drawn using the offset indicated by `dPtr`. Overlapping windows are drawn from the furthest to the nearest windows, and later drawn windows overwrite the earlier drawn windows. Each window has its own horizontal and vertical axis, and when a window is moved, by modifying `dPtr`, all the line and text items are drawn relative to the new origin.

This class of graphics objects can be significantly improved and extended. Rather than drawing each window from furthest to nearest, portions of windows that will be overwritten can be *clipped.* The windows can be linked in a hierarchy so that if a parent window is moved its offspring will move. This class of graphics objects is essentially what is used in the Macintosh and Microsoft Windows operating systems. Graphics is one of the most common applications of object-oriented programming.

10.2 Storage Systems

Most microcomputers require either a communications system to move data into and out of them or a storage system to get data from or to save data in. The latter, called *secondary storage,* generally uses some form of magnetic medium. Floppy disk systems have become so cheap that they are likely to be used in many microcontrollers. This section describes techniques for data storage on floppy disks. We discuss a floppy disk format, then we will use a Western Digital WD37C65C chip, which is particularly easy to interface to the 6812, to show a floppy disk interface and an object-oriented class to read and write files in a 3 1/2" PC disk.

10.2.1 Floppy Disk Format

We now describe the 3 1/2" double-density floppy disk format. Data can be stored on the disk using either of two popular formats. Figure 10.7 shows how a bit and a byte of data can be stored on a disk, using FM (single density) and MFM (double-density) formats. The FM format is just Manchester coding, as introduced in §9.1. Figure 10.7a shows a bit cell, and Figure 10.7c shows a byte of data, in the FM format. Every 8 μs there is a clock pulse. If a 1 is sent, a pulse is put in between the clock pulses, and if a 0 is sent, no pulse is put between the clock pulses. MFM format provides half the bit cell size as FM format; it does this by using minimal spacing between pulses in the disk medium: MFM format has at most one pulse per bit cell. It is thus called double-density storage. The idea is that a 1 cell, which has a pulse in it, doesn't need a clock pulse, and a 0 cell only needs a clock pulse if the previous cell is also a 0 cell. Figure 10.7b shows a byte of data in the MFM format. High density merely doubles the density for the MFM format. For the remainder of this section, we discuss the high-density MFM format. Every 2 μs there is a data bit. If a 1 is sent, a pulse is put near the end of the bit time; if a 0 is sent after a 1, no pulse is put between the clock pulses; and if a 0 is sent after a 0, a pulse appears early in the bit time. Note that data must be read or written at the rate of 1 byte per 16 μs, which is 128 memory cycles for the 'A4 using a 16-MHz crystal.

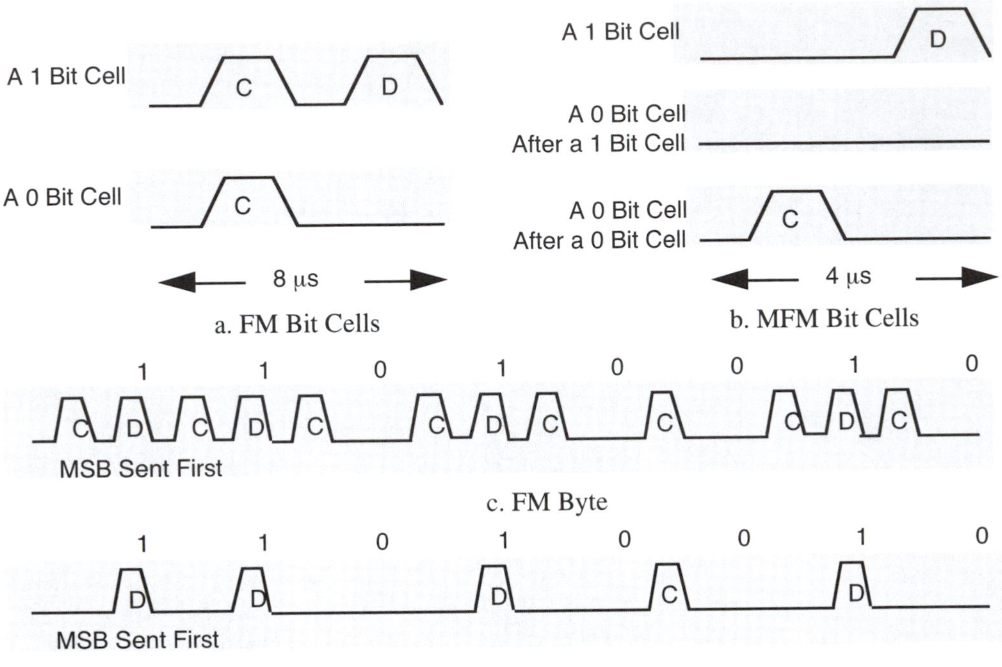

a. FM Bit Cells

b. MFM Bit Cells

c. FM Byte

d. MFM Byte

Figure 10.7. Bit and Byte Storage for FM and MFM Encoding

Data read from the disk are separated by a phase-locked loop (PLL), which synchronizes to the bit cell rather like a flywheel. Once the bit cell is locked on to, the data bits can be extracted from the input analog signal. The PLL must be designed to lock into the bit cells within 48 bit cell times.

A disk drive may have one or more disks, stacked pancake-style, and each disk may have one or two *surfaces.* Figure 10.8a shows a surface of a disk; a *track* is shown, and tracks are numbered – track 0 on the extreme outside, and track $i + 1$ next toward the center to track *i.* The track-spacing density is the number of tracks per inch and is generally 48 or 96 tracks per inch. Floppy disks have diameters of 8", 5 1/4", or 3 1/2", and these typically have 77, 35, and 80 tracks, respectively. Although disks exist that have a head on each track, generally disks have a single head per surface – used to both read and write the data on that surface – which is moved by a stepper motor to a track that is to be read or written. In a multiple-surface disk, the same tracks are accessed on each surface by a comblike mechanism holding the read-write head for each surface; the collection of tracks accessed at the same time is called a *cylinder.* We soon describe an example of a single-sided 3 1/2" disk's track format. Other disk formats are similar.

Relative to later discussions of the operation of the floppy disk controller, timing of head movements significantly affects the disk system's performance. The *step rate* is the rate at which the stepping motor can be pulsed to move the head from track `i` to track `i + 1` (step in) or to track `i - 1` (step out). There is also a *settling time,* which is the time needed to allow the mechanical parts to stop bouncing (see contact bounce in Chapter 5). Floppy disk drives have stepping rates from 2 to 30 ms and settling times of about 30 ms. If a drive has a 3-ms stepping rate and a 30-ms settling time, the time to move from track `i` to track `j` is `3 * |i - j| + 30` ms. The average time to position from track 0 to a random track is the time to move over half of the (80) tracks of the disk. There is some additional time needed to get to the data on the track, as will be discussed. Thus, on the average, about 80 ms would be used to move the head, and no data is transferred during that time.

The problem with a disk is that, to record data, a head must be energized, and the process of energizing or deenergizing a head erases the data below the head. The track is thus organized with *fill* areas where data are not stored and where the head may be energized to begin, or deenergized to end, writing, and the data between these fill areas, called (disk) *sectors,* are written in their entirety if they are written at all. A disk's indivisible storage objects thus are sectors. Figure 10.8b shows the breakdown of a typical track in terms of sectors and the breakdown of a sector in terms of its ID pattern and data. (Later, we discuss a "logical sector"; when we need to distinguish a logical sector from what we describe here, we call this a "disk sector".) There is an *index hole* on the disk (Figure 10.8a) that defines a track's beginning; it is sensed by an optical switch that provides an *index pulse* when the hole passes by the switch. The track first contains a 60-byte fill pattern. (Each fill pattern is 0x4E.) There are then 18 disk sectors on each track. The remainder of the track is filled with the fill pattern.

With respect to the timing of disk accesses, after the head moves to the right track, it may have to wait 1/2 revolution of the disk, on the average, before it finds a track it would like to read or write. Since a floppy disk rotates at 10 revolutions per second, the average wait would be 50 ms. If several sectors are to be read together, the time needed to move from one track to another can be eliminated if the data are on the same track, and the time needed to get to the right sector can be eliminated if the sectors are located one after another. We will think of sectors as if they were consecutively numbered from 0 (the *logical sector number* [LSN]), and we will position consecutively numbered sectors on the same track, so consecutively numbered sectors can be read as fast as possible. Actually, two consecutively read disk sectors should have some other sectors between them because the computer has to process the data read and determine what to do next before it is ready to read another sector. The number of disk sectors actually physically between two "consecutively numbered" logical sectors is called the *interleave factor,* and is generally about four.

We need to know which disk sector is passing under the head as the disk rotates, since sectors may be put in some different order, as just described, and we would also like to be able to verify that we are on the right track after the head has been moved. When the read head begins to read data (it may begin reading anywhere on a track), it will examine this address in an ID pattern to find out where it is.

There is a small problem identifying the beginning of an ID pattern or a data field when the head makes contact with the surface and begins to read the data on a track. To solve this, there is a special pattern whose presence is indicated by the deletion of some of the clock pulses that would have been there when data are recorded in MFM format, and there are identifying patterns called the ID address mark and data address mark. The special pattern, shown in Figure 10.9, is said to have a data pattern of 0xA1 and a missing clock pulse between bits 4 and 5. The ID address mark 0xFE is used to locate the beginning of an ID pattern on a track. The data address mark similarly identifies the beginning of data in the sector, but is 0xFB rather than 0xFE.

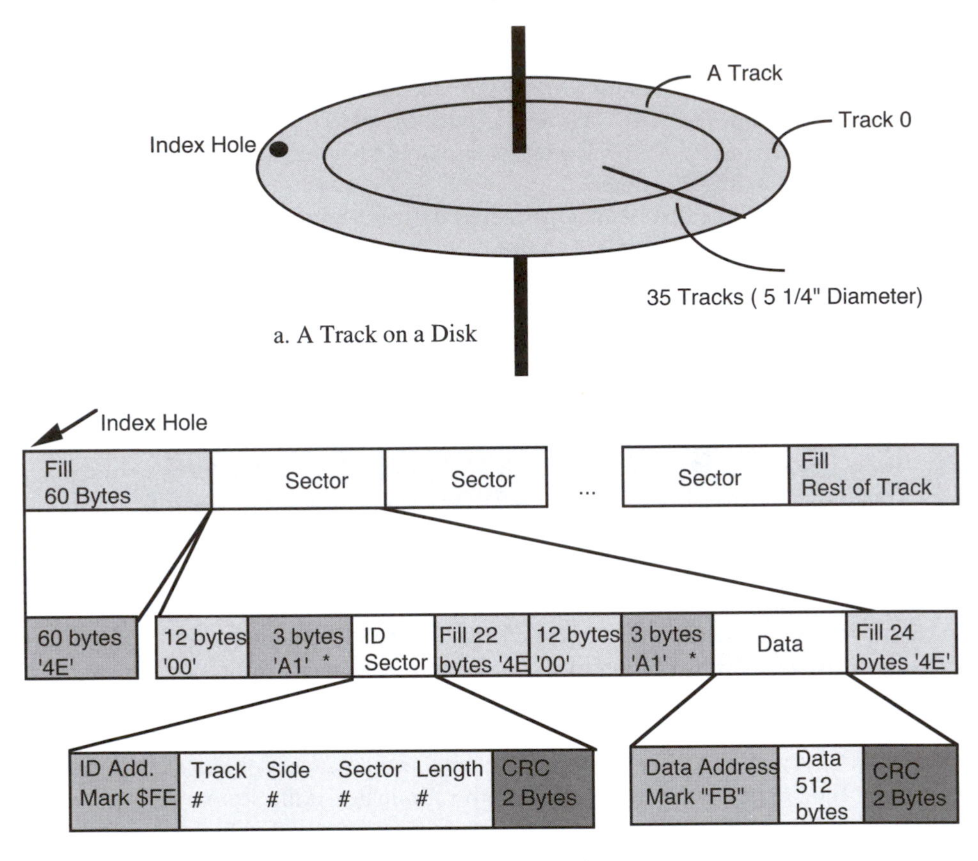

b. Sectors in a Track

Figure 10.8. Organization of Sectors and Tracks on a Disk Surface

The ID pattern consists of a 1-byte ID address mark (0xFE); a track number, side number, sector number and sector length (each is 1 byte and is coded in binary); and a 2-byte CRC check. The track number, beginning with track 0 (outermost), and the sector number, beginning with either sector 0 (zero-origin indexing) or 1 (one-origin indexing), are stored in 2 of the bytes. A simple method of mapping the logical sector number into a track and zero-origin indexing disk sector number is to divide the logical sector number by the number of sectors per track: the quotient is the track number, and the remainder is the sector number. The side number for a single-surface drive is 0, the sector length for a 256-byte sector is 1, and the sector length for a 512-byte sector is 2.

A sector is composed of a pattern of 12 0s and 3 bytes of 0xA1 (in which a clock pulse is missing), followed by an ID pattern as just described, a 22-byte fill, and another pattern of 12 0s, followed by 3 bytes of 0xA1 (in which a clock pulse is missing). The 512 bytes of data are then stored. The ID pattern and the data in a sector have some error detection information called a CRC, discussed in §9.4.1, to ensure reliable reading of the data. The track may have a total capacity of about 13K bytes, called the *unformatted capacity* of the track, but because so much of the disk is needed for fill, ID, and CRC, the *formatted capacity* of a track (the available data) may be reduced to 9216 bytes.

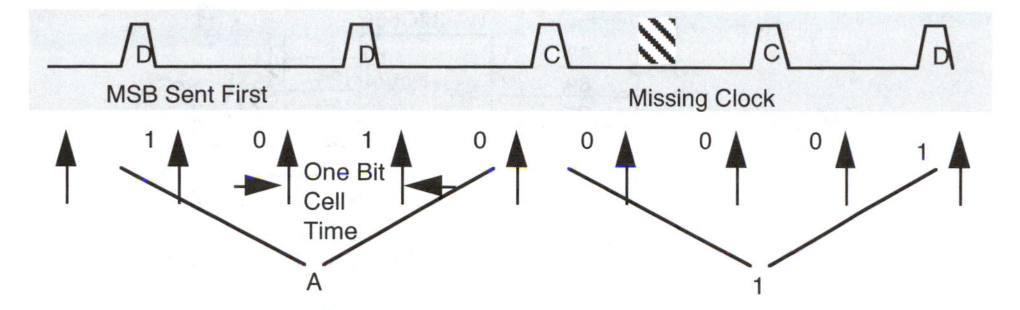

Figure 10.9. A Special Byte (Data = 0xA1, Clock Pulse Missing Between Bits 4 and 5)

The *format* of a disk is the structure just described, disregarding the content of the data field. To *format* a disk is to write this structure. Later, when data are written, only the data part of the sector, together with its data address mark and CRC check, are written. The ID pattern is not rewritten. If it is altered, data cannot be read or written because the controller will be unable to find the sector's `id` part to locate the sector.

10.2.2 The Western Digital 37C65C Floppy Disk Controller

We now examine a hardware/software system for reading and writing a double-sided high-density (HD) 3 1/2" floppy disk using the Western Digital WD37C65C chip (the '65C). We first catalog pin connections and ports in the organization of the chip. Then we list the status ports and functions that the controller executes. In the next section, we will present and describe the software used to control the chip. See Figure 10.10a.

The 34-pin cable connects the controller chip to the drive. Grounded odd-numbered pins reduce noise pickup, and all signals are in negative logic. Motor controls MOT1 and MOT2 are asserted true (grounded) to make the motor run continuously in this example, although the '65C has means to control the motor. Output Step causes the drive's stepper motor to move to another cylinder, and Dirc specifies in which direction to move the head. DS1 selects drive 1, DS2 selects drive 2, and Hs selects the head used. Wd is the write data, and We is the write enable. Input Rdd is the read data, write protect Wp is asserted if a disk tab is positioned to prevent writing in the disk, Tr00 is the track 0 signal, and Ix is the index pulse signal; inputs use a pull-up resistor.

Positive-logic `PORTE` signals communicate to the '65C control and status pins. Bit 4 resets the '65C, bit 3 senses a '65C operation completion, and bit 2 (TC) terminates counting, to stop an operation.

The '65C's organization has five ports, of which two read ports and two write ports are used herein (Figure 10.10b). The control port's two least significant bits specify disk density. The master status port has a request bit (Req), a data direction bit (Out), an execution phase bit (Exec), and five busy bits indicating the status of the control chip and up to four drives. The data port transfers data into and out of the '65C, and also sends commands into the device (as listed in Figure 10.10d). The '65C gets more detailed status from the chip by reading status bytes St0, St1, St2, and St3 (see Figure 10.10c).

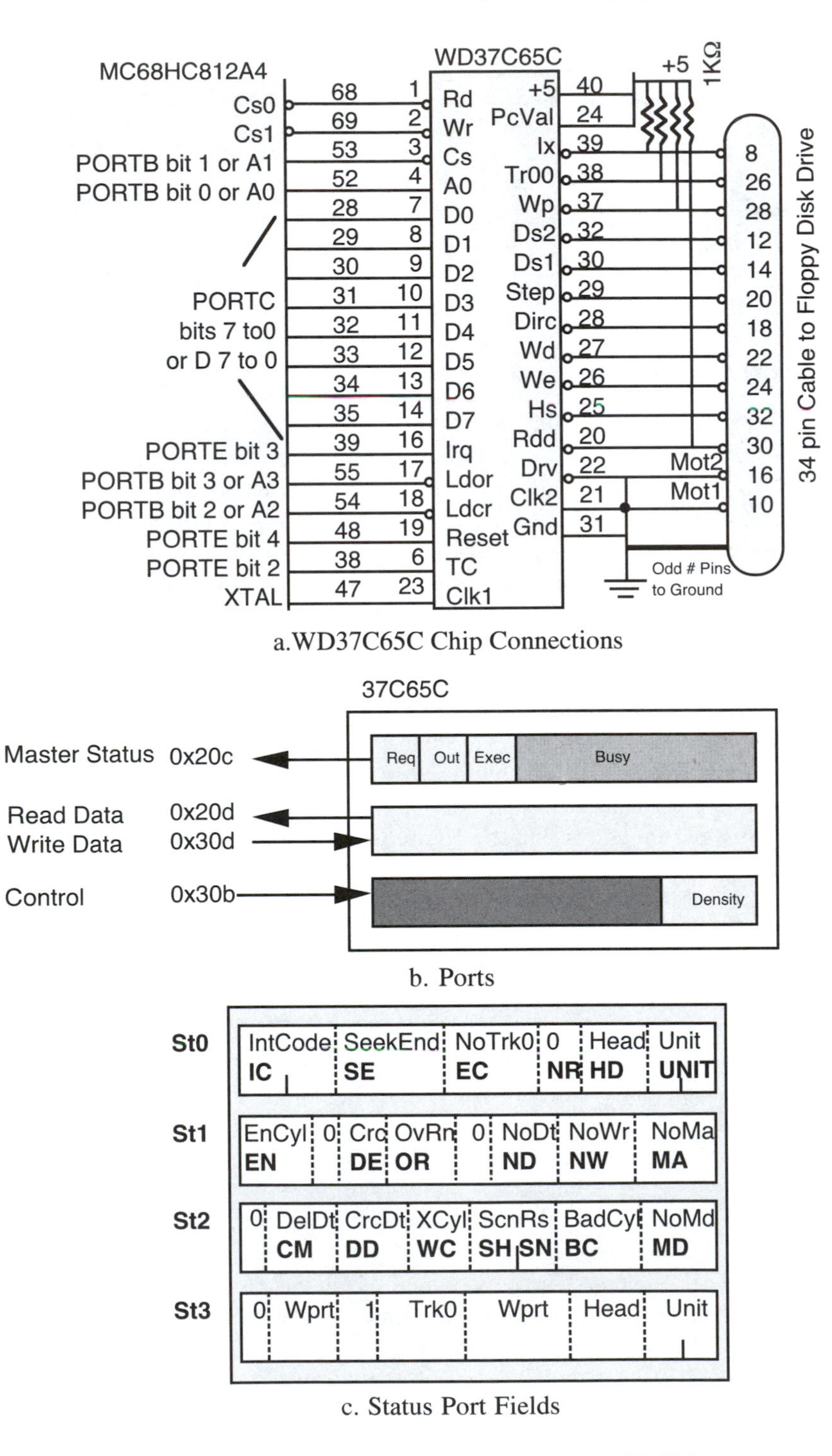

a. WD37C65C Chip Connections

b. Ports

c. Status Port Fields

Figure 10.10. The Western Digital WD37C65C

	Specify	Seek Cylinder
Command	0000 0011 StepRate Unload LoadTime NoDMA	0 0 0 0 1111 xxxxx Hs Unit[2] Cylinder
	Read Id	**Recalibrate**
Command	0 0 0 0 1010 xxxxx Hs Unit[2] Results: See Read Sector	0 0 0 0 0111 xxxxx 0 Unit[2]
	Sense Drive	**Sense Status**
Command	0 0 0 0 0 1 0 0 xxxxx Hs Unit[2]	0 0 0 0 1000
Results	St3	St0 Cylinder

	Read Sector	Write Sector	Format Track
Command	Mt Mf Sk 0 0110 xxxxx Hs Unit[2] **C** Cylinder **H** Head **R** Sector **N** Size **EOT** End **GLP** Gap **DTL** DLen	Mt Mf Sk 0 0101 xxxxx Hs Unit[2] **C** Cylinder **H** Head **R** Sector **N** Size **EOT** End **GLP** Gap **DTL** DLen	0 Mf 0 0 1101 xxxxx Hs Unit[2] Bytes/Sector Sectors/Track Gap Data
Execution	Read 2**(Size+7) Bytes	Write 2**(Size+7) Bytes	Supply (Sectors/Track)* Cylinder Head Sector Size
Results	St0 St1 St2 **C** Cylinder **H** Head **R** Sector **N** Size	St0 St1 St2 **C** Cylinder **H** Head **R** Sector **N** Size	St0 St1 St2 **C** Cylinder **H** Head **R** Sector **N** Size

d. Command Summary

Figure 10.10. Continued

To simplify the interconnection hardware, as discussed in §6.2.2, reading uses CS0 (addressed 0x200 to 0x2ff) and writing uses CS1 (0x300 to 0x3ff). The low-order address bits are negative logic enables for the master status and data registers, and control ports. So, to read master status, read 0x20c, and to read from the data port, read 0x20d; to write to the data port, write 0x30d, and to write control, write 0x30b (Figure 10.10b).

The '65C commands read and write sectors and perform auxilliary operations. (See Figure 10.10d.) The user may wish to read one or more bytes from the disk. Since it is only possible to read whole sectors, the sector or sectors that the data are at are read, and the data are extracted once the sectors are read. To read a sector, the user must *seek* the track first, then *read* the desired *sector.* The two commands, *seek cylinder* and *read sector,* are given to the floppy disk controller. Seek cylinder puts the read/write head in the disk drive over the desired cylinder, and the read sector command causes the data from a sector on the track to be read into primary memory. If the read/write head is definitely over the right cylinder, the seek command may be omitted. Also, in some floppy disk controllers having intelligence in them, the user only gives a read command regardless of where the read/write head is and an *implied seek* may be automatically generated as a result, without the user giving it, if the head is in the wrong place.

The user may wish to write one or more bytes into the disk. To write a sector, the commands to *seek cylinder* and *write sector* are given as for the read operations above. Good programs often read a sector each time right after it is written to be sure there are no errors in writing the sector. A disk can be formatted by executing the *format* command on each track. Finally, when a disk is being initialized, the position of the read/write head must be moved to track 0 (zeroed) to later establish how to move it to a desired track. This operation is called *restoring* or *recalibrating the drive.*

The '65C has additional commands. *Specify* writes the step rate, head load, and head unload times to initialize the '65C; *read id* will read the first valid `id` sector found on the track; *sense drive* will indicate the drive's status, and *sense status,* often used after an interrupt or a gadfly loop, will indicate the cause of the interrupt or termination of the loop. We will develop procedures to execute these functions and write an object-oriented class to handle these operations, as we did for other I/O devices.

10.2.3 Floppy Disk Interfacing Procedures

This section illustrates some simple software that implements the '65C commands discussed in the previous section. We begin with the innermost procedures used to write commands and to read status from the '65C. We then illustrate the initialization procedure for the controller chip, drive, and disk in the drive. Finally, we discuss procedures to read and write a sector on the disk.

This device uses a handshake protocol to issue commands and garner status information. This protocol checks the master status register, and inputs and outputs through the data port. We give these declarations for the ports, for the connections and for the map in Figure 10.10. Comments indicate the use of the variables.

```
volatile char mStatus@0x20c, rData@0x20c, control@0x30b, wData@0x30d;
```

The innermost procedure `w` writes command bytes into the '65C, and the procedure `r` reads status bytes from the '65C.

```
void w(char c){while((mStatus & 0xc0)!= 0x80) ; wData = c;} /* write data */

char r(){ while ( (mStatus & 0xc0) != 0xc0) ; return rData;} /* read data */
```

Before reading or writing, we gadfly on master status register bit 7. The '65C's microcontroller sets this bit when it is ready to move data through its data port.

The software will use error message numbers to report errors in a manner that assists the user in locating and fixing errors. These error numbers are declared as follows:

```
enum{illOp= 1,chpErr,drvError,
     dskErr,addrErr,seekErr,rdErr,wrErr,wrProtErr,eofErr};
```

The initialization of the '65C and drive hardware is shown in the procedure `init65` below. This procedure initializes the 'A4, resets the '65C, recalibrates the drive, and reads the `ID` of the disk in the drive. The 'A4 is configured to access the '65C. We need an 8-bit data bus to connect the '65C, and we need CS0 and CS1 to enable the '65C's Rd and Wr pins. We set `DDRE` to output `PORTE` control signals RESET and TC, then we pulse RESET high, then initialize RESET to L and TC to H. Compare each of the following program segment with the sequences shown in Figure 10.10d. The specify command sets up the '65C, and the recalibrate command sets up the drive hardware. The sense interrupt status command is used after recalibrate, and will later be used after each seek operation, and the sense drive status command is used to verify the success of each operation. The last program segment, which reads a sector `ID`, uses global variables introduced after this procedure is described, so it will be discussed later.

```
char init65(char drive,char step,unsigned char unload,char load){int i;
      MODE = (Narrow + Normal) << ModeFld; CSCTL = CS0E+CS1E; CSSTR=0;
      DDRE = PORTE = 0x14; PORTE = 4; /* positive logic reset of '65 */
      w(3); w(((-step) << 4) | (unload >> 4)); w(load | 1); /* specify */
      w(4); w(drive & 7); i = r(); /* sense drive status */
      if((i & 0xa0) != 0x20) return error = chpErr;

      w(7); w(drive & 3); /* recalibrate (restore) */
      while( ! (PORTE & 8)) ;  /* gadfly on drive being in execution mode */

      w(8); i = r(); C = r(); /* sense interrupt status */
      if(i != 0x20) return error = drvError;

      control=0; w(0x4a); w(0); while(!(PORTE&8)) ; check();
      if((S[0] & 0xc0) != 0) return error = dskErr;
      return 0;
}
```

Procedures that operate on sectors use global data declared below. The variables `C`, `H`, and `R` are the first three bytes in an *ID* sector field of a sector of the disk (Figure 10.8b). `C` is the cylinder number, which is the same as the track number. `H`, 0 or 1, indicates which head is used. *R* is the disk sector number to be accessed. `seek`, called by `read` and `write`, is passed a logical sector number (discussed in §10.2.1) and computes `C`, `H` and `R`. In disks formatted for the IBM PC, and used in the next section, sector numbers on a track begin with one rather than zero (one-origin indexing). Four status bytes `ST[4]` are read from the controller (Figure10.9c); `error` is nonzero when an error occurs, `verify` is 1 if we will read after writing a sector to verify it was saved. Input data is put or output data is taken from buffer *B* (`spacer` places *B* to be read in a memory dump at a 16-byte boundary, at location 0x810).

```
char C,                     // cylinder
     H,                     // head
     R,                     // sector number on track
     ST[4],                 // status
     error,                 // nonzero indicates error
     verify = 1,            // TRUE means will verify
     spacer[7], B[512];     // buffer for sectors
```

The `seek` procedure, called up at the beginning of the `get` and `put` procedures, converts the logical sector number to head, track, and disk sector numbers for the disk controller. It checks that the sector number is within range (a double-sided 80-track disk, with 18 sectors per track, has 2880 sectors), and then, dividing by the number of sectors per track, essentially obtains the disk sector number (remainder) and a combined head-cylinder number (quotient). Recall that in IBM disks, disk sector numbers begin with 1, so the remainder is incremented to get the disk sector number. The head-cylinder number is divided by two to get the disk cylinder number `C` (quotient) and the head `H` (remainder). The '65C is given a `seek` command and the status register `S[0]` is examined to determine if the seek was successful. If the drive is already on the desired track, `seek` completes quickly, but this command can take milliseconds to execute.

```
char seek(int sectorNumber) {
    if(sectorNumber>2880)return error=addrErr;/* check sector out of range */
    R = (sectorNumber%18)+1; /* get sector # (PC disks use one-origin indexing) */
    C=sectorNumber/18; /* get combined head-cylinder number, temporarily in C */
    H = C & 1; C >>= 1; /* separate into cylinder C (high bits) and head (lsb) */
    w(0xf); w(H << 2); w(C); /* give command to seek cylinder */
    while( ! (PORTE & 8)) ; /* gadfly on drive-seeking cylinder */
    w(8); ST[0] = r(); /* get interrupt status */
    if((r()!=C)||((ST[0]&0xf8)!=0x20))return error=seekErr;/* check err */
    return 0;
}
```

Procedures `get` and `put` input (read) a sector and output (write) a sector. Procedures `check` and `setup` are used in these methods to factor out some common code from them. `setup` writes the first eight parameters of a `read` or `write` command. `check` reads out the status information of the command's result phase.

```
void setup(char cmd)
    { w(cmd);w(H << 2);w(C);w(H);w(R);w(2);w(18);w(0x1b);w(0xff); }

char check() {
    if((ST[0] = r()) & 0x80) { error = illOp; return 0; }
    ST[1] = r(); ST[2] = r(); C = r(); H = r(); R = r(); N = r();
    return (ST[1] & 0x7f) || (ST[2] & 0x33);
}
```

The procedure `get` reads a sector. It calls `seek`, which computes `C`, `H`, and `R`, and executes a `seek` command. First, during a *command phase,* nine bytes are written to the data port using the `setup` procedure. The command `0x46` requests reading an MFM sector. Then the surface, which is the head and drive number, is sent. The cylinder, head, sector, and size bytes are sent exactly as they should appear in the sector's `ID` (Figure 10.8b), and the last sector, format gap size, and data length are sent to complete the command phase. The *execution phase* reads each byte of the sector's data. Asserting TC (`PORTE` bit 4) terminates this phase. Finally, the *result phase* uses the `check` procedure to read each byte of the result. The status ports `S[0]` to `S[2]` are read, then the sector's cylinder, head, sector, and size bytes are read back exactly as they appeared in the sector's `id` (Figure 10.8b), so software can verify that the right sector was read.

```
char get(int sector, register char *buffer) { int i; char *end;
    if(error||seek(sector))return 1;end=buffer+(i=512);PORTE=0;setup(0x46);
    do {
        while(!(mStatus&0x80));if(!(mStatus&0x20))break;*buffer++=rData;
    } while( --i );
    PORTE = 4;
    if( check() || (buffer != end) ) return error = rdErr; return 0;
}
```

The procedure `main()` below reads the first sector of a file that stores the floppy disk program's source code, which is logical sector 57.

```
     00 01 02 03 04 05 06 07 08 09 0A 0B 0C 0D 0E 0F
0810 23 64 65 66 69 6E 65 20 55 54 42 55 47 0D 23 69 #define UTBUG.#i
0820 6E 63 6C 75 64 65 20 3C 36 38 31 32 2E 68 3E 0D nclude <6812.h>.
0830 23 69 6E 63 6C 75 64 65 20 22 44 69 73 6B 2E 68 #include "Disk.h
```

Figure 10.11. File Dump

```
void main(){  init65(0, 3, 240, 16);  get(57, B); }
```

The *seek* procedure calculates the cylinder number as 1, the head as 1, and the sector number as 4. Then a dump (Figure 10.11) shows the data in logical sector 57.

Similarly, the *put* procedure writes a sector, and if *verify* is nonzero, the sector that was just written is read again using the *get* method to verify that it is properly stored and can likely be read later without errors. Rather than actually using *get* to verify the sector as in this example, which destroys *B*, a procedure exactly like *get* can be used, except that it writes read data into a dummy local variable. The CRC check verifies success. But reading the sector in the *verify* step slows down the writing of sectors because a sector is written and a whole disk revolution later the same sector is read. Verification is indispensable when the data being written might be lost forever if it is not written correctly. But there are times when this verify step should be omitted. When copying a whole disk, by clearing the *verify* data member, this verify step is omitted to speed up writing of sectors. The destination disk is first fully written, without verifying each sector, and then each sector on the disk is read just to verify it.

```
char put(int sector, register char *buffer){ int i;
    if(error || seek(sector)) return 1; i = 512;
    PORTE = 0; setup(0x45);
    do {
        while(!(mStatus&0x80));if(!(mStatus&0x20))break; wData=*buffer++;
    } while( --i );
    PORTE = 4;
    if(i||check()){if(ST[1]&2)return error=wrProtErr;return error=wrErr;}
    if(verify) if(get(sector, B)) return error = wrErr;
    return 0;
}
```

We deferred a discussion of *init65*'s *read id* command, which reads any valid *ID* sector. It uses the same status reporting mechanism that *get* and *put* to check the disk in the drive for readability.

From these examples, the reader should see that movement of data to or from the '65C always uses handshaking, such as gadflying on some master status bits. The *read sector* (also *read id, write sector,* and *format*) commands go through three phases: command, in which control values are sent to the chip; execution in which data are read or written; and result in which status is read from the chip. Other operations (*sense drive, sense status*) have only a command and a result phase, and the remainder (*specify, recalibrate,* and *seek*) have only a command phase.

Hex Address	Length Bytes	Meaning/Contents
0	3	0xe9xxxx or 0xebxx90
3	8	OEM name
b	2	# Bytes/sector
d	1	# Sectors/cluster
e	2	# Sectors in boot record
10	1	# FATs
11	2	# Root directory entries
13	2	# Logical sectors
15	1	Medium descriptor = 0xf0
16	2	# Sectors/FAT
18	2	# Sectors/track
1a	2	# Heads
1c	2	# Hidden sectors
1e	...	Boot program

a. The Boot Sector (logical sector 0)

Logical Sectors	Contents
0	Boot sector
1 - 9	FAT
10 - 18	Duplicate FAT
19 - 32	Root directory
33 +	Data files

b. Logical Sectors

Figure 10.12. PC Disk Organization

The example above can be significantly improved. For instance, it used gadfly synchronization; by connecting '65C's interrupt request to a `PORTJ` bit rather than `PORTE` bit 3, a key wakeup interrupt at the end of a seek could permit other programs to run while a long seek is in progress. See problem 27 at the end of the chapter.

10.2.4 Personal Computer Disk Data Organization

Each operating system organizes its disks differently; herein we briefly describe the organization of the popular IBM PC 3 1/2" HD disk, which we formatted on a Macintosh for a PC. Important data on logical sector 0, called the *boot sector,* are listed in Figure 10.12a. Key logical sectors are arranged as in Figure 10.12b.

The boot sector, which is logical sector 0, or disk sector 1 in track 0 using head 0, is shown in Figure 10.12a. Besides storing a boot program that loads and starts an operating system, this sector stores parameters that can be used to unambiguously read the disk. Figure 10.13 shows a dump of the boot sector of the disk we are using. You can examine it to locate Figure 10.12a's parameters discussed following Figure 10.13.

```
     00 01 02 03 04 05 06 07 08 09 0A 0B 0C 0D 0E 0F
0810 EB 34 90 50 43 58 20 32 2E 30 20 00 02 01 01 00 .4.PCX 2.0 .....
0820 02 E0 00 40 0B F0 09 00 12 00 02 00 00 00 00 00 ...@...........
```

Figure 10.13. Dump of a Boot Sector

A *cluster* is a fixed number of sectors (one in this case) that are allocated or deallocated as a unit. The two bytes at location 0xb indicate the number of bytes per sector, and the byte at location 0xd indicates the number of sectors per cluster. The two bytes at location 0xe indicate the number of sectors, starting at sector 0, that store these parameters and the boot program used to install an operating system; in this case it is one. The number of FATs is two, so that a duplicate is available if one is corrupted when there is a disk crash. Any 32-byte entry in a *directory,* the *root directory* being in logical sectors 19 to 32, describes one file listed in the directory, as illustrated in Figure 10.14a. (Other directories are subdirectories.) The number of root directory entries, each of which is 32 bytes (Figure 10.14a), is always a multiple of 16 to use up an integral number of sectors, and here is 224. The number of logical sectors on the disk is stored in two bytes at 0x13. Figure 10.14b shows how the *file allocation table* (FAT), stored in logical sectors 1 to 9 and duplicated in logical sectors 10 to 18 to aid in file recovery after a disk crash, is associated with sectors storing data in the file. The number of sectors in each FAT is in the two bytes at location 0x16, and the number of sectors in each track is in the two bytes at location 0x18. The number of heads, or surfaces, is in the two bytes at location 0xa. The number of hidden sectors is in the two bytes at location 0x1c. The boot program is stored after location 0x1e.

Incidentally, each two-byte value, and all multibyte values stored on the disk, are in the Intel format: least significant byte first. For instance, the number of bytes per sector may be 0x12 at location 0xb and 0 at 0xc. The value is not 0x1200, but rather 0x0012.

Boot sector parameters unambiguously determine the root directory's beginning logical sector. If the boot-record length at location 0xb is 1, the number of FATs stored in 0x10 is 2, and the number of sectors per FAT stored in 0x16 is 9, then the root directory begins at 1 + (2 * 9) = 19. Each 32-byte directory entry, shown in Figure 10.14a, begins with an 8-character name and a 3-character extension, each in ASCII, and each entry's location 0x1a stores a two-byte "data area logical sector number" (DALSN).

Hex Address	Length Bytes	Meaning/Contents
0	8	Name
8	3	Extension
b	1	Attribute
c	10	Reserved
16	2	Time of modification
18	2	Data of modification
1a	2	Start cluster
1c	4	File length

a. Directory Entry

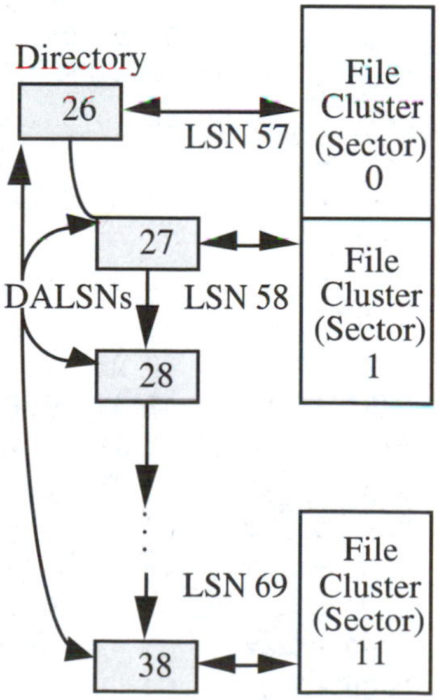

b. FAT Mapping

Figure 10.14. PC File Organization

After the boot sector is loaded into buffer *B*, key locations needed in locating the file are computed by the following program segment.

```
void getFat(){ int fat, dir, entries;
    fat = B[0xe] + (B[0xf] << 8); /* get boot record size, this is LSN of FAT */
    dir = fat+(B[0x10]*(B[0x16]+(B[0x17] << 8))); /* is base of directory */
    entries = B[0x11] + (B[0x12] << 8); /* is # directory entries */
    base=dir+(entries>>4)-2; /* is base of data sectors ("2" is discussed later)*/
}
```

To read the first byte of a file having a given name, compare the desired name with each directory entry name until there is a match. Our disk's root directory dump is shown in Figure 10.15. The file name DISK.C is in the seventh entry. This file's beginning DALSN is in the two bytes in this entry's location 0x1a, and its value is 0x1a or 26. Its length is 0x00001770 bytes.

After the first directory sector is loaded into buffer `B`, it can be searched to get the starting `firstDALSN` and `length` using the following program segment. This program segment will only search a root directory with fewer than 32 files. Problem 24 expands this program segment to a 14-sector root directory.

```
void Search(char fileName[]){ int i, j, k, length;
    for(i=0;i<512;i+= 32){ /* should search 14 sectors; here we search 1 sector */
    for(j = k = 0; j < 11; j++)
        if((B[i + j] & 0xff) != (fileName[j] & 0xff)) { k = 1; break; }
    }
    if(k == 0) { firstDALSN = B[i + 0x1a] | (B[i + 0x1b] << 8); }
    length=B[i+0x1c]|(B[i+0x1d]<<8) | (B[i+0x1e]<<16)|(B[i+0x1f] << 24);
}
```

```
     00 01 02 03 04 05 06 07 08 09 0A 0B 0C 0D 0E 0F
0810 46 49 4E 44 45 52 20 20 44 41 54 22 00 00 00 00 FINDER  DAT"....
0820 00 00 00 00 00 00 3C 77 15 23 02 00 B8 02 00 00 ......<w.#......
0830 44 45 53 4B 54 4F 50 20 20 20 20 22 00 00 00 00 DESKTOP    "....
0840 00 00 00 00 00 00 41 77 15 23 00 00 00 00 00 00 ......Aw.#......
0850 53 61 6D 70 6C 65 20 20 20 20 20 28 00 00 00 00 Sample     (....
0860 00 00 00 00 00 00 D4 76 15 23 00 00 00 00 00 00 .......v.#......
0870 52 45 53 4F 55 52 43 45 46 52 4B 12 00 00 00 00 RESOURCEFRK.....
0880 00 00 00 00 00 00 D5 76 15 23 03 00 00 00 00 00 .......v.#......
0890 46 49 4C 45 49 44 20 20 44 41 54 22 00 00 00 00 FILEID  DAT"....
08A0 00 00 00 00 00 00 9B A4 26 23 05 00 80 00 00 00 ........&#......
08B0 E5 44 45 53 4B 54 4F 50 46 4F 4C 10 00 00 00 00 .DESKTOPFOL.....
08C0 00 00 00 00 00 00 97 A4 26 23 06 00 00 00 00 00 ........&#......
08D0 44 49 53 4B 20 20 20 20 43 20 20 20 00 00 00 00 DISK    C   ....
08E0 00 00 00 00 00 00 54 2F 26 23 1A 00 70 17 00 00 ......T/&#..p...
08F0 E5 52 41 53 48 20 20 20 20 20 20 10 00 00 00 00 .RASH       .....
0900 00 00 00 00 00 00 99 A4 26 23 08 00 00 00 00 00 ........&#......
```

Figure 10.15. Dump of a Directory

Since the data area follows the root directory (Figure 10.12b), its logical sector number is essentially the logical sector number of the beginning of the directory, computed previously, added to the length of the directory in sectors. The 2-byte value in boot sector location 11 is the number of directory entries; this divided by 16 gives the number of sectors in the root directory. In our example the data-area base is sector 31. If our directory search located a file whose DALSN is 26, then the beginning of the file is in logical sector 31 + 26 = 57. The first 512 bytes of the file are stored therein.

Each consecutive logical sector is associated with a consecutive 12-bit DALSN in the FAT and its duplicate (Figure 10.16). Each sector's corresponding FAT entry gives the DALSN of the file's next consecutive sector, or gives a value greater than 0xff7 when the file doesn't have a next sector. This is effectively a linked list used to find the remainder of the file. The total number of valid bytes in the file is the four-byte number at the end of the directory entry for the file; it must be less than the file's number of DALSN entries times 512. DALSNs herein are 12-bit binary numbers in Intel format; for every three consecutive bytes in the FAT, such as 0x12 0x34 0x56, there are two DALSNs, 0x412 and 0x356. Peculiarly, each FAT's first three bytes are 0xf0 0xff 0xff, and the first DALSN is numbered 2. This initial 3-byte pattern has an advantage of ensuring that the sector being examined is actually the beginning of the FAT. Correspondingly, because the first two FAT entries do not correspond to logical sectors in the data area, the DALSNs are converted to LSNs by adding the base address, which is the address of the beginning of the data area less two sectors. This is accommodated in the calculation of *base* in the program segment earlier in this section. We suspect that small PC disks had a boot and a directory sector that were included in the "data area," and this FAT "signature" pattern was kept, as larger disks had larger boot files and root directories, to retain this convenient distinctive pattern for verification purposes.

The file DISK.C is stored beginning at DALSN 0x01a = 26. The next segment of the file is determined by the "left half" of the 13th triple-byte entry in dump bytes 0x837 to 0x839. The value of this left half is 0x01b. The next segment of the file is determined by the "right half" of the 13th triple-byte entry in dump bytes 0x837 to 0x839. The value of this right half is 0x01c. Our file's DALSNs are 0x01a, 0x01b, 0x01c, 0x01d, 0x01e, 0x01f, 0x020, 0x021, 0x022, 0x023, 0x024, and 0x025. These can be dumped to examine the contents of the file.

After a file's `firstDALSN` is determined by the directory search and the first FAT sector is loaded into buffer B, LSNs for the file can be stored in `int list[64]`. This program segment is able to build the list of LSNs if all the disk's DALSNs are held in only the first sector. Problem 26 expands this program segment to a 9-sector FAT. These program segments are built into the object-oriented class in the next section.

```
     00 01 02 03 04 05 06 07 08 09 0A 0B 0C 0D 0E 0F
0810 F0 FF FF 07 F0 FF 0B F0 FF 00 F0 FF 00 00 00 00 ................
0820 C0 00 0D E0 00 0F 00 01 11 20 01 13 40 01 15 60 ......... ..@..
0830 01 17 80 01 FF 0F 00 1B C0 01 1D E0 01 1F 00 02 ................
0840 21 20 02 23 40 02 25 F0 FF FF 0F 00 00 00 00 00 ! .#@.%.........
```

Figure 10.16. Dump of an Initial FAT Sector

```
void FindSector(){ int i, j, curDALSN, list[10];
    for(i=0,curDALSN=firstDALSN; (curDALSN < 0xff0) && (i < 64); i++){
        list[i] = curDALSN + base;  j= (curDALSN >> 1) * 3;
        if(curDALSN & 1)
            curDALSN=(((B[j+1]>>4) & 0xf) | (B[j+2]<<4)) & 0xfff;
        else curDALSN = (B[j] | ((B[j + 1] << 8) & 0xf00)) & 0xfff;
    }
}
```

10.2.5 Object-oriented Disk I/O

Object-oriented programming encapsulates data members for a file so that multiple files can be easily read from or rewritten in the same procedure. We consider files that are only read from, or only rewritten, to simplify the functions. The class `file` below is made a derived class of `Port<char>` so that bytes can be read or written by the `put` and `get` function members and overloaded assignment, cast, <<, and >> operators. The constructor executes the program segments of the previous section, leaving the list of LSNs for the file in the data member `list[64]`. Procedures `init65`, `get`, and `put` from §10.2.3 are also called in this class. Subsequent member functions use the *list* prepared by the constructor to read a sector into a data member buffer ***B*** in preparation for reading each 512 bytes, or write a sector into ***B*** into sectors after each 512 bytes are written into it. The destructor writes out ***B*** into the last sector if it is partially rewritten.

```
enum{ rd = 1, wr};

class Drive { char C, H, R, N, ST[4]; public : int errors;

    Drive(char drive,char step,unsigned char unload,char load){int i;
        PEARMODE = Narrow + Normal; CSCTL = CS0E + CS1E; CSSTR = 0;
        DDRE = PORTE = 0x14; PORTE = 4; /* positive logic reset of '65 */

        w(3);w(((-step)<<4) | (unload>>4)); w(load|1); /* specify */
        w(4); w(drive & 7); i = r(); /* sense drive status */
        if((i & 0xa0) != 0x20) { errors = chpErr; return; } errors = 0;

        w(7); w(drive & 3); /* recalibrate (restore) */
        while( ! (PORTE & 8)) ;  /* gadfly on drive being in execution mode */

        w(8); i = r(); C = r(); /* sense interrupt status */
        if(i != 0x20) { errors = drvErr; return; }

        control=0;w(0x4a);w(0);while(!(PORTE & 8));check();/* read id */
        if((ST[0] & 0xc0) != 0) errors = dskErr;
        return;
    }

    void w(char c){ while ((mStatus&0xc0)!=0x80);wData=c;} /* write data */

    char r(){while((mStatus&0xc0)!=0xc0) ; return rData;} /* read data */
```

```
void setup(char cmd)
    {w(cmd);w(H << 2);w(C);w(H);w(R);w(2);w(18);w(0x1b);w(0xff);}

char check() {
    if((ST[0] = r()) & 0x80) { errors = illOp; return 0; }
    ST[1] = r(); ST[2] = r(); C = r(); H = r(); R = r(); N = r();
    return (ST[1] & 0x7f) || (ST[2] & 0x33);
}

char seek(int sectorNumber) {
    if(sectorNumber>2880)return errors=addrErr; /* sect out of rng ? */
    R=(sectorNumber%18)+1; /* get sector # (one-origin indexing) */
    C=sectorNumber/18;  /* get cylinder-head number, temporarily in C */
    H=C & 1; C >>= 1; /* separate into cylinder C (high bits) and head (lsb)  */
    w(0xf); w(H << 2); w(C); /* give command to seek cylinder */
    while( ! (PORTE & 8)) ; /* gadfly on drive seeking cylinder */
    w(8); ST[0] = r(); /* get interrupt status */
    if((r()!=C)||((ST[0]&0xf8)!=0x20))return errors=seekErr;
    return 0;
}

char get(int sector, register char *buffer) { int i; char *end;
    if(errors||seek(sector))return 1;end=buffer+(i=512);PORTE=0;
    setup(0x46);
    do {
        while( ! (mStatus & 0x80)) ;
        if( ! (mStatus & 0x20)) break;
        *buffer++ = rData;
    } while( --i );
    PORTE = 4;
    if(check()||(buffer != end)) return errors = rdErr; return 0;
}

char put(int sector, register char *buffer, char verify){ int i;
    if(errors || seek(sector)) return 1; i = 512;
    PORTE = 0; setup(0x45);
    do {
        while(!(mStatus&0x80));
        if(!(mStatus&0x20))break;
        wData=*buffer++;
    } while( --i );
    PORTE = 4;
    if( i || check() )
        {if(ST[1]&2)return errors=wrProtErr;return errors=wrErr;}
    if(verify&&get(sector,buffer))return errors=wrErr;
    return 0;
}
};
```

```
class File : public Port<char>{ char B[512], mode; Drive &D;
    unsigned int curDALSN,firstDALSN,base,fat,dir,entries,list[64];
    long length, position; public : int errors; char verify;

    File(char *fileName,Drive &D,char mode):Port(0x200) { int i, j, k;
        errors = position = 0; verify = 1; this->D = D;

        // locate fat and read directory
        if(D.get(0, B)) return; /* read boot sector */
        fat=B[0xe]+(B[0xf] << 8); /* get boot record size; it is LSN of FAT */
        dir=fat+(B[0x10]*(B[0x16]+(B[0x17]<<8))); /* is directory base */
        entries = B[0x11] + (B[0x12] << 8); /* is # directory entries */
        base=dir+(entries>>4)-2; /* data sector base ("2" is discussed later)*/

        // search directory for file name
        if(D.get(dir, B)) return; /* read directory */
        for(i = 0; i < 512; i += 32){ /* search one sector of directory */
            for(j = k = 0; j < 11; j++)
                if((B[i+j]&0xff)!=(fileName[j]&0xff)) {k=1; break; }
            if(k==0) {firstDALSN=B[i+0x1a]|(B[i+0x1b]<<8); break; }
        }
        length = B[i + 0x1c] | (B[i + 0x1d] << 8);

        // get list of sectors
        if(D.get(fat, B)) return; /* read FAT */
        for(i=0,curDALSN=firstDALSN;(curDALSN<0xff0)&&(i < 64); i++){
            list[i] = curDALSN + base;  j= (curDALSN >> 1) * 3;
            if(curDALSN & 1)
                curDALSN=(((B[j + 1]>>4)&0xf)|(B[j+ 2]<<4)) & 0xfff;
            else curDALSN=(B[j]|((B[j + 1]<<8) & 0xf00))&0xfff;
        }
    }

    virtual char get(void){ // input
        if(!(mode&&rd)||(position>(length-1))) {errors=1;  return; }
        if((position & 0x1ff) == 0) D.get(list[position >> 9], B);
        return B[((position++) & 0x1ff)];
    };

    virtual void put(char data) {// output
        if(!(mode&&wr)||(position>=length)){errors=1; return; }
        B[((position++) & 0x1ff)] = data;
        if((position&0x1ff)==0)D.put(list[(position>>9)-1],B,verify);
    };

    void seek(long position){ // seek a location in the file
```

```
        if(position >= length) {errors = 1;  return; }
        if((mode == wr) && ((position & 0x1ff) == 0))
            D.put(list[(position >> 9) - 1], B, verify);
        this->position = position; D.get(list[position >> 9], B);
    }

    File &operator = (File &f); // copy file
        { do put(f.get());while(!(errors|=f.errors));return *this; }

    ~File(){ if(position&0x1ff)D.put(list[position>>9],B,1);}// destructor
};

Drive d1(0,3,240,16); // delare the object; call the constructor to initialize the disk

void main() {
    File f1("F1",d1,wr),f2("F2",d1,rd);// delare objects  to open files
    f1 = f2;  // copy files
    f1.~File; f2.~File; // close the files
}
```

This class can be used to read or write data from several files, as the following program illustrates. This program copies file F1 to file F2.

```
file f1("F1"), f2("F2"); // delare the objects; call the constructor to open the files
void main(){do f1=f2;while(!f1.error);~f1();~f2();} // copy til end of file err
```

The object-oriented disk access class `file` makes the disk appear like any other I/O device described in this book. Using device independence, a disk can be substituted for another I/O device at run time. Using I/O independence, a disk can be substituted for another I/O device at compile time. Objects encapsulate all the data and functions needed to access a disk, so that two or more files can be accessed in the same program. These objects have many of the advantages of operating-system device drivers, but with much lower overhead. These capabilities illustrate the advantages of object-oriented I/O.

10.3 Conclusions

This chapter introduced two common interfaces: CRT display and secondary storage. These rather complete case studies give a reasonably full example of common interface designs. They also embody the techniques you have studied in earlier chapters. Besides presenting these important interfaces, this chapter serves to complete the book by showing how the techniques in the other chapters will find extensive application in almost any interface design.

For further reading on floppy disks, we strongly recommend the data sheets for the '65C from Western Digital. Harold Stone's "Microcomputer Interfacing" has additional general information on the analog aspects of storage devices. These can be consulted for further examples and inspiration.

This text has been fun for us. Microcomputers like the 6812 are such powerful tools that it challenges the mind to dream up ways to use them well. We sincerely hope you have enjoyed reading about and experimenting with the 6812 microcomputer.

Do You Know These Terms?

See page 36 for instructions.

National Television System Committee (NTSC)
raster line
frame
pixel
NTSC composite video signal
horizontal retrace
vertical retrace
horizontal sync
vertical sync
sync separator
Bresenham algorithm
window
clipped
secondary storage
surface
track
cylinder
step rate
settling time
fill
sector
index hole
index pulse
logical sector number (LSN)
interleave factor
unformatted capacity
formatted capacity
format
seek
read sector
implied seek
write sector
restoring
recalibrating the drive
specify
read id
sense drive
sense status
command phase
execution phase
result phase
verify
boot sector
cluster
directory
root directory
file allocation table (FAT)

Problems

Problems 1 and 13 are paragraph correction problems. See page 38 for guidelines. Guidelines for software problems are given on page 86, and for hardware problems, on page 115.

1.* A TV screen is a series of fields; and in the NTCS format, a field takes 1/30 second. There are about 500 raster lines in a field, with each line scanning from top to bottom of the screen, and each raster line takes about 60 μs. Sync pulses are incorporated into the composite video signal as gray-level signals, and these are used to synchronize the horizontal and vertical oscillators that cause the electron beam to scan the screen. CRT controllers use either character or graphics display modes at any time. The former can use an independent mode, where the CRT gets characters from the primary memory of the processor using DMA; or the shared mode, where the processor writes into a separate display memory only during the horizontal retrace periods.

2. Rewrite `char pattern[12]` in §10.1.2 to display

a. a solid black 8-by-12 square

b. an 8-by-12 black-outlined white square; the outline is 1 pixel wide

c. a horizontal line two pixels high across the top of the 8-by-12 black square

d. a vertical line two pixels wide on the left of the 8-by-12 black square

e. a letter A on the top eight lines of the 8-by-12 black square, with the right column blank, with four blank lines on the bottom of the 8-by-12 black square

3. Rewrite `#define`s in §10.1.2 to display the square (giving approximate values ±10%)

a. at the top left corner of the screen

b. at the top right corner of the screen

c. at the bottom left corner of the screen

d. at the bottom right corner of the screen

4. Rewrite the program in §10.1.2 that outputs the same TV picture as in Figure 10.4, using gadfly synchronization rather than interrupt synchronization, to implement horizontal- and vertical- sync pulses. Use the same counter and SPI modules.

5. Write a C procedure `border()` that will draw a border that is four pixels wide around §10.1.3's display, rather than two pixels wide, and fill the display with gray rather than black. Use §10.1.3's constants, such as `HEIGHT`, `HWIDTH`, and `WIDTH`.

6. When displaying on a low-bandwidth CRT, so that horizontal and vertical lines have equal brightness, the video signal should be "chopped" by ANDing it with the shift clock because horizontal lines have more low-frequency signal than vertical lines. The 74HC132's remaining NAND gate can "chop" the video signal, requiring two hardware changes and changes in the accompanying software to display the same white border on black background as in §10.1.3. Write a paragraph accurately describing these changes.

7. The program in §10.1.3 displays 256 lines of 512 pixels per line. By displaying the same line in both fields, for instance, so that location 0x8000 appears on the top left of the first scan line of the first field and again on the top left of the first scan line of the second field, our (8K, 16) SRAM can display a 496-by-512 screen image. Show the program needed to display this 496-by-512 screen, wherein each memory location is displayed twice, in the same relative location of each field.

8. Write a function member `rectangle(int h,int v,int w,int ht)` for §10.1.4's class `screen` to draw a rectangle whose top left corner is at row `v`, column `h`; and whose width is `w` and height is `h`, using the function member `line` given in §10.1.4.

9. The function member `line(int h,int v,int dh,int dv,int n)` for §10.1.4's class `screen` can only draw lines where either `dh` or `dv` is 1, and the other, `dh` or `dv`, is between -2 and +2. Use the Bresenham algorithm.

a. Write a function member `line1(int h1,int v1,int h2,int v2,int dh, int dv)` where `dh = h2 - h1, dv = v2 - v1, dh > dv` and `dh > 0`, to draw a continuous line from row `v1`, column `h1` to row `v2`, column `h2`.

b. Write function member `lineto(int h1,int v1,int h2,int v2)` to draw a continuous line from row `v1` column `h1` to row `v2` column `h2`. Write a modification of part a's `line1`, with a fifth argument `reverse`, so that if `reverse` is 0, `line1` calls `point` with unsubstituted `h` and `v`; if 1, `line1` calls `point` exchanging `h` and `v`; if 2, `line1` calls `point` negating `h`; and if 3, `line1` calls `point` exchanging `h` and the negative of`h`.

10. Write a function member `triangle(int h1,int v1,int h2,int v2,int h3,int v3)` using problem 9's `lineto`, that will draw a triangle with a vertex in row `v1` column `h1`, in row `v2` column `h3`, and in row `v3` column `h3`.

11. Write function member `wchar(int *a,int h,int v)` for §10.1.4's class `screen` to draw a character whose pattern is pointed to by `a`, with top left corner at row `v`, column `h`, and with width up to 16 pixels wide and 16 pixels tall. Use only `int` variables and pointers in your solution. Also write a vector to draw the letter 'A'.

12. Write a program *main()* to write MISSISSIPPI in the middle of the screen. Show vectors `patternM`, `patternI`, `patternS`, and `patternP`, analogous to §10.1.4's vector `patternA`, to draw the letters M, I, S, and P in a 7-pixel-wide and 8-pixel-high, font. Write `main()` to bless a pointer `SCN` to an object of the class `screen`, and write "MISSISSIPPI" in the middle line and around the middle column of the display area.

13.* A surface of a typical floppy disk is divided into concentric rings called sectors, and each sector is divided into segments. A sector may be read or written as a whole, but individual bytes in it may not be read or written. The format of a sector has only some 0s, a 0xA1 flag pattern, a data-address mark, the data, and a CRC check; counters are used to keep track of the track and sector. To read (write) a sector, it is necessary to first give a command to seek the track, then give a command to read (write) the sector.

14. Trace the pattern for the following bytes: 0x80, 0x55, 0xcc, 0xca, 0x1 (assume the bit previous to this byte is a 0).

a. Use FM encoding. b. Use MFM encoding.

15. Show timing diagrams of the middle special byte 0xA1 in Figure 10.9 and relevant parts of the beginning and end of the previous and following special bytes as they shift through a byte-size window. Show that the shifting byte appears to match the required pattern exactly once, which defines the byte boundary.

16. Determine how many total bits are in a sector, and how long does it take to read it.

17. The operations register OR is written when LDOR, attached to address line A3, is asserted and WR is asserted.

a. At which locations is this operations register and only this register written into?

b. OR bit 4, output in negative logic on pin 33, runs drive 0's motor. Similarly, OR bit 5 on pin 34 runs drive 1's motor. Show drive-cable connections so that asserting OR bit 4 runs drive 0's motor, and asserting OR bit 5 runs drive 1's motor.

18. For the logic diagram shown in Figure 10.10, determine which addresses can be used to uniquely select the ports. Identify all addresses in which the following can be done.

a. Read the master status register. b. Read the data register.
c. Write the data register. d. Write the control register.

19. Deleted data has a different delete data mark $FD in place of Figure 10.8's data mark $FB, and a sector with this mark is skipped when reading data. Write this mark by the command `write deleted data` (command 0xc) if the sector has defective media. Such a deleted sector can still be read using the command `read deleted data` (command 0x9). Otherwise, commands to write and read deleted data are the same as the commands `write data` (command 0x7) and `read data` (command code 0x6). Show pictoral descriptions that can be added to Figure 10.10d to describe these two commands.

20. Write a multithread scheduled procedure `seek` and a handler `cckj` for the '65C IRQ pin attached to `PORTJ` bit 0. When a seek cylinder operation is begun (see the centronics printer, §5.3.3), if the '65C asserts IRQ within 12 µs, `seek` exits, but otherwise *seek* puts the thread to sleep. When this operation is complete, a key wakeup interrupt executes handler `cckj` to awaken the sleeping thread. Assume thread 1 is used.

21. Write procedures that try five times to read or write a sector, until the sector is read or written correctly. When writing, if `verify` is 1, verify the written sector without destroying `B`. After the second attempt, and if the head is not on cylinder 0, move the head to the next lower cylinder, then read or write the sector. After the fourth attempt, and if the head is not on cylinder 80, similarly move it out a cylinder and back in. "Jiggling" the head this way facilitates reading or writing a misaligned cylinder.

a. Write a `get` procedure to read a sector. b. Write a `put` procedure to write a sector.

22. Write a procedure `format(int c, char h)` to format cylinder `c` on side `h` of drive 0. Before this procedure is executed, the drive should have been initialized using `init65`, but at the beginning of this procedure, a seek cylinder command is given. Note that errors will occur, during initialization and seeking, which should be ignored. To format a track for a HD disk, the '65C is given the command 0x4d, a byte containing the head (bit 2) and drive number (bits 1 and 0), the number `N` of bytes per sector, the number of sectors per track, a gap width, and the byte used to fill the data portion of each sector. For an HD disk, `N` is 2, there are 18 sectors per track, the gap width is 0x54, and the data portion of each sector is filled with 0x46. The execution phase waits for the index pulse, then formats an entire track, and then asserts IRQ. While the '65C formats the track, the 'A4 writes each sector's bytes C, H, R, and N into the '65C's data port as in the `write sector` command. That is, for the 18 sectors, write 72 bytes. Your procedure should write the track with an interleave factor of four. The status phase returns the status bytes ST[0], ST[1], and ST[2], which should be checked for errors, and four bytes (C, H, R, and N) that are not used in this case. Use §10.2.3's variables.

23. In Figure 10.15, determine the DALSN and length of the following files.

a. FINDER.DAT b. DESKTOP c. RESOURCE.FRK

24. Write a program segment to search a 14-sector root directory, analogous to the program segment following Figure 10.15.

25. From Figure 10.16, determine the next (hex) DALSN when the current DALSN is

a. 2. b. 3. c. 4. d. 0xb. e. 0xc. f. 0xd.

26. Write a program segment to construct the file's DALSN `list` from a 9-sector FAT, analogous to the program segment following Figure 10.16. Note that DALSNs of consecutive sectors in a file are not necessarily consecutive, and may even be nonmonotonic (they may skip around). Take care to handle the special case where a three-byte sequence containing two DALSNs overlaps a sector boundary.

27. Write a function member `seek(long a)` for class `file` to position the read or write in `position a` so the next byte read by `char get()` or written by `put(char)` is the *a*th byte of the file. If a sector needs to be output to save bytes written before `seek` is executed, do so, and then read in the sector in which byte `a` appears.

28. The class `file` can be modified to permit either reading of data in the file, writing of data in the file, or reading and writing of data in the same file (called updating the file); however, we must be careful about putting back sectors that may have been partially over-written when we read the data, and about putting a sector into the buffer before writing a byte into it, in case we will read bytes from this sector later.

a. Write a function member `char get()` for class `file` that will output the next byte of the file (at location `position`), but if this requires reading in another sector, the sector previously stored in the buffer is written back.

b. Write a function member `put(char c)` for class `file` that will write `c` into the next byte of the file (at location `position`), but if this requires writing into another sector, that sector is read into the buffer.

c. Explain why an object of class `file` should be declared or blessed with "permissions" read-only, write-only, or update to make the file both readable and writeable at the same time. In particular, comment on how long reading or writing can take, in the worst case, for each example. How can our class `file` be modified to permit this capability to be declared in the constructor and used in the function members?

The Adapt812 is connected to an M68HC12B32EVB board which is configured in POD mode, which in turn connects to a PC. We used this configuration to download and debug using HIWAVE, using the ASCIIMON target interface.

Appendix
Using the HIWARE CD-ROM

This appendix helps you use the accompanying CD-ROM to simulate your programs, and to download and debug them on EVB Boards and other target microcontrollers.

A-1 Loading HIWARE Software

Open the CD ROM, check "installation", and choose the Motorola HC12 target. If you have 60 Megabytes of disk space, load all parts of the tool chain.

A-2 Running the Simulator

You can use the software on the CD-ROM to simulate your programs on a PC running Windows 95 or later, or Windows NT 5.0 or later, without using any extra hardware. Using Acrobat Reader 3.0 or later, run the \hiware\docu\hc12\demo12.pdf file. This file provides a tutorial guide on how to load and run the compiler, linker, and simulator. Following this guide, compile, link, and simulate the program Fibo.c.

A very simple way to experiment with other programs is to modify the file Fibo.c. Using any text editor, such as NOTEPAD, rewrite the Fib.c file with a program that you wish to study. Compile, link, and simulate the modified program Fib.c. You can rewrite Fibo.c each time you wish to study a new program. You can use more sophisticated techniques, but this simple technique can get you started with minimal effort.

A-3 Running Examples from This Book

Note that the folder EXAMPLES on the CD-ROM has files in it such as Ei2.txt. These files contain examples from this text book, which you can copy-and-paste into Fibo.c, so that you can run these examples on the Hiwave simulator. The file Ei2.txt contains all the examples in Chapter 2 of this textbook, and the file Ei4.txt contains all the examples in Chapter 4 of this textbook, and so on. Copy this folder into your hard disk; most conveniently, put it into your HIWARE folder.

A-4 Downloading to a 'B32 Board

You can use the HIWARE software to download and debug Fibo.c. on the Motorola M68HC12B32EVB board (abbreviated the 'B32 board) as your target. Begin by simulating Fibo.c. on the Hiwave simulator, as described above. After you are comfortable with the simulator's operation, follow the procedures described in the \hiware\docu\hc12\manual\MWb2.pdf file. You should always apply the 5V power after all connections are made, and you should never change a connector while power is applied to the 'B32 board.

A-5 POD-Mode BDM Interface

You can run HIWARE on a PC running Windows 95 or later, or Windows NT 5.0 or later, using the Motorola M68HC12B32EVB board in its POD mode, to connect a different target, such as an Technological Arts Adapt-812 board, or an Axiom PB68HC12A4 board, (called the target) to run experiments.

This technique utilizes the state-of-the-art background debug module BDM in your target, providing a debugger that runs in the M68HC12B32EVB board (called the POD) that is isolated from the target. If the target is not fully functional, the POD still functions and can help you debug the target. This technique also provides additional functionality to Hiwave, such as the ability to profile and analyze coverage. However, since more things can go wrong with a PC, a POD, and a target, than with just a PC, or a PC and a target M68HC12B32EVB, we recommend using this technique after you have had experience with the two simpler techniques described above.

Begin by runnin Fibo.c. on the Hiwave simulator, and then running Fibo.c. on the 'B32 board, as described above. After you are comfortable with the simulator's and 'B32 board's operation, reconnect the W3 to its 0 position and W4 to its 1 position to configure the board for POD mode, and reset the POD. Select the Asciimon target. Load Fibo.abs. You should be able to dupliciate what you did in the simulator and 'B32 board, running it on the Adapt-812 or PB68HC12A4 board.

You can use the Motorola SPI module, a more powerful BDI debugger, in place of the POD. Other similar BDI interface modules, but not all, are also compatible with Hiwave. Other target microcontrollers can be run using the POD or similar board.

A-5 Techniques for HIWARE Tools

We have had some experiences with HIWARE tools, which might help you use them more efficiently. We add a note here on our suggestions, to help you with this powerful software.

A problem with the current version is that when you change project files, the compiler/linker/hiwave debugger may read or write the wrong files, or fail to find the files it needs. We found that by shutting down all HIWARE programs, and starting them up again, the problem goes away. But you do not have to restart the computer. If you have verified that the paths to the files are correct, but you are unable to access them through the compiler/linker/hiwave debugger, then try restarting all HIWARE programs "from scratch". The same remedy is suggested when the Hiwave simulator or debugger fails to execute single-step commands, or breakpoints, correctly.

When dealing with different environments such as your own PC running Windows 95, workstations running Windows NT, and a PC running Windows 98 in the laboratory, keep separate complete project folders for each environment, and copy the source code from one to another folder. That way, you will spend less time readjusting the paths to your programs and HIWARE's applications when you switch platforms.

We hope that the CD-ROM supplied through HIWARE makes your reading of this book much more profitable and enjoyable. We have found it to be most helpful in debugging our examples and problem solutions.

INDEX

LIMITED WARRANTY AND DISCLAIMER OF LIABILITY

ACADEMIC PRESS ("AP") AND ANYONE ELSE WHO HAS BEEN INVOLVED IN THE CREATION OR PRODUCTION OF THE ACCOMPANYING CODE ("THE PRODUCT") CANNOT AND DO NOT WARRANT THE PERFORMANCE OR RESULTS THAT MAY BE OBTAINED BY USING THE PRODUCT. THE PRODUCT IS SOLD "AS IS" WITHOUT WARRANTY OF ANY KIND (EXCEPT AS HEREAFTER DESCRIBED), EITHER EXPRESSED OR IMPLIED, INCLUDING, BUT NOT LIMITED TO, ANY WARRANTY OF PERFORMANCE OR ANY IMPLIED WARRANTY OF MERCHANTABILITY OR FITNESS FOR ANY PARTICULAR PURPOSE. AP WARRANTS ONLY THAT THE OPTICAL DISK(S) ON WHICH THE CODE IS RECORDED IS FREE FROM DEFECTS IN MATERIAL AND FAULTY WORKMANSHIP UNDER THE NORMAL USE AND SERVICE FOR A PERIOD OF NINETY (90) DAYS FROM THE DATE THE PRODUCT IS DELIVERED. THE PURCHASER'S SOLE AND EXCLUSIVE REMEDY IN THE EVENT OF A DEFECT IS EXPRESSLY LIMITED TO EITHER REPLACEMENT OF THE DISK(S) OR REFUND OF THE PURCHASE PRICE, AT AP'S SOLE DISCRETION.

IN NO EVENT, WHETHER AS A RESULT OF BREACH OF CONTRACT, WARRANTY OR TORT (INCLUDING NEGLIGENCE), WILL AP OR ANYONE WHO HAS BEEN INVOLVED IN THE CREATION OR PRODUCTION OF THE PRODUCT BE LIABLE TO PURCHASER FOR ANY DAMAGES, INCLUDING ANY LOST PROFITS, LOST SAVINGS OR OTHER INCIDENTAL OR CONSEQUENTIAL DAMAGES ARISING OUT OF THE USE OR INABILITY TO USE THE PRODUCT OR ANY MODIFICATIONS THEREOF, OR DUE TO THE CONTENTS OF THE CODE, EVEN IF AP HAS BEEN ADVISED OF THE POSSIBILITY OF SUCH DAMAGES, OR FOR ANY CLAIM BY ANY OTHER PARTY.

ANY REQUEST FOR REPLACEMENT OF A DEFECTIVE DISK MUST BE POSTAGE PREPAID AND MUST BE ACCOMPANIED BY THE ORIGINAL DEFECTIVE DISK, YOUR MAILING ADDRESS AND TELEPHONE NUMBER, AND PROOF OF DATE OF PURCHASE AND PURCHASE PRICE. SEND SUCH REQUESTS, STATING THE NATURE OF THE PROBLEM, TO ACADEMIC PRESS CUSTOMER SERVICE, 6277 SEA HARBOR DRIVE, ORLANDO, FL 32887, 1-800-321-5068. AP SHALL HAVE NO OBLIGATION TO REFUND THE PURCHASE PRICE OR TO REPLACE A DISK BASED ON CLAIMS OF DEFECTS IN THE NATURE OR OPERATION OF THE PRODUCT.

SOME STATES DO NOT ALLOW LIMITATION ON HOW LONG AN IMPLIED WARRANTY LASTS, NOR EXCLUSIONS OR LIMITATIONS OF INCIDENTAL OR CONSEQUENTIAL DAMAGE, SO THE ABOVE LIMITATIONS AND EXCLUSIONS MAY NOT APPLY TO YOU. THIS WARRANTY GIVES YOU SPECIFIC LEGAL RIGHTS, AND YOU MAY ALSO HAVE OTHER RIGHTS WHICH VARY FROM JURISDICTION TO JURISDICTION.

THE RE-EXPORT OF UNITED STATES ORIGIN SOFTWARE IS SUBJECT TO THE UNITED STATES LAWS UNDER THE EXPORT ADMINISTRATION ACT OF 1969 AS AMENDED. ANY FURTHER SALE OF THE PRODUCT SHALL BE IN COMPLIANCE WITH THE UNITED STATES DEPARTMENT OF COMMERCE ADMINISTRATION REGULATIONS. COMPLIANCE WITH SUCH REGULATIONS IS YOUR RESPONSIBILITY AND NOT THE RESPONSIBILITY OF AP.